Badr-Eddine Boudriki Semlali ·
Ikram Ben Abdel Ouahab · Luís Bragança
Editors

Computer Sciences for Sustainable Development

Proceedings of the 6th International Conference on the Digital Age and Technological Innovations for Sustainable Development, Volume 2

Springer

Editors
Badr-Eddine Boudriki Semlali
Department of Computer Engineering
Abdelmalek Essaâdi University
Tangier, Morocco

Ikram Ben Abdel Ouahab
C3S Lab, FSTT
Abdelmalek Essaâdi University
Tangier, Morocco

Luís Bragança
University of Minho
Braga, Portugal

ISSN 2367-3370 ISSN 2367-3389 (electronic)
Lecture Notes in Networks and Systems
ISBN 978-3-032-21199-6 ISBN 978-3-032-21200-9 (eBook)
https://doi.org/10.1007/978-3-032-21200-9

This Springer imprint is published by the registered company Springer Nature Switzerland AG
The registered company address is: Gewerbestrasse 11, 6330 Cham, Switzerland

Preface

This volume of the Conference Proceedings encompasses the peer-reviewed written contributions presented at the Sixth International Conference on the Digital Age and Technological Innovations for Sustainable Development (DATA'25), held in Tangier, Morocco, from May 7 to 9, 2025.

The conference is a prestigious multidisciplinary forum, co-organized by the Faculty of Sciences and Techniques of Tangier in collaboration with the *Mediterranean Association of Sciences and Sustainable Development (Medi-ASDD)*. The 2025 edition was held under the distinguished patronage of the *Commune of Tangier*, and in partnership with *Al-Mustaqbal University, Iraq*, as well as the *New Science School of Engineering (ENSI), Tangier, Morocco*, whose support significantly enriched the scope and impact of the event.

The contents of this volume reflect a diverse and rich spectrum of recent advances and applications in domains such as *Artificial Intelligence for Sustainable Development, Sustainable Development in Civil Engineering, Digital Energy Technology and Smart Grid, Remote Sensing,* and *Safe, Secure, and Trustworthy AI*. The convergence of these fields under the umbrella of sustainability reinforces the conference's mission to advance impactful, forward-looking research and innovation.

The conference welcomed over 200 participants from various countries, fostering an engaging environment for exchanging knowledge and interdisciplinary collaboration. A particular highlight of DATA'25 was the series of keynote addresses delivered by eminent scholars and thought leaders:

- **Prof. Abdelghani Chehbouni**, "Remote sensing, IoT and IA for water resources management".
- **Prof. Mohamed Essaaidi**, "Toward a holistic approach to smart and sustainable mobility".
- **Prof. Hassan Majdi**, "Integrating green education and AI at Al-Mustaqbal University: a framework for sustainable higher education".
- **Prof. Sultan Alsultan**, "Green and sustainable computing Using Remote Sensing in Saudi Arabia".

We sincerely thank all authors, participants, reviewers, keynote speakers, and members of the scientific and organizing committees, whose contributions and dedication were instrumental in this event's success.

We want to acknowledge and thank the editorial team at Springer Nature Switzerland AG for their professional guidance and support in preparing this volume. In particular, we thank Prof. Janusz Kacprzyk and Dr. Thomas Ditzinger for their invaluable support in realizing and disseminating this work.

The research presented in these proceedings will contribute meaningfully to the ongoing global dialogue on digital transformation and sustainable development and serve as a foundation for future advancements in this vital field.

May 2025

Badr-Eddine Boudriki Semlali
Ikram Ben Abdel Ouahab
Luís Bragança

Committees

Conference Chairs

Badr-Eddine Boudriki Semlali	FST Tangier, UAE University, Morocco
Ikram Ben abdel ouahab	FST Tangier, UAE University, Morocco

Conference Steering Committee

Fabio Angeletti	Sapienza University of Rome, Italy
Luis Bragança	University of Minho, Portugal
Mohamed Ben Ahmed	FST Tangier, UAE University, Morocco
Anouar Abdelhakim Boudhir	FST Tangier, UAE University, Morocco
Hasan Majdi	Al-Mustaqbal University, Iraq
Mazin Hadi Kzar	Al-Mustaqbal University, Iraq
Mohammed Bouhorma	FST Tangier, UAE University, Morocco
Lotfi Elaachak	FST Tangier, UAE University, Morocco
Chaker El Amrani	FST Tangier, UAE University, Morocco
Aicha Chorak	ENSI Tangier, Morocco

Local Organizing Committee

Aziz Mahboub	FST Tangier, UAE University, Morocco
Abdelhadi Fennan	FST Tangier, UAE University, Morocco
El Mokhtar En-naimi	FST Tangier, UAE University, Morocco
Yasyn Elyusufi	FST Tangier, UAE University, Morocco
Zouhair Abdelhamid	FST Tangier, UAE University, Morocco
Hicham Gibet Tani	FP Larache, UAE University, Morocco
Fatiha Elouaai	FST Tangier, UAE University, Morocco
Mohamed Elbrak	FST Tangier, UAE University, Morocco

Technical Program Committee

Abderrahim El Mhouti	UAE University, Morocco
Adnane El Hanjri	EMSI Tangier, Morocco
Adriana Estokova	Technical University of Košice, Slovakia

Aicha Chorak	ENSI Tangier, Morocco
Ali Aghazadeh Ardebili	University of Salento, Italy
Ali Shakir Majdi	Al-Mustaqbal University, Iraq
Anass Belcaid	ENSATe, UAE University, Morocco
Assaad El Makhloufi	EST, USMBA, Morocco
Bassam Zafar	King Abdulaziz University, Jeddah, Saudi Arabia
Bouchra Ait Hssaine	Center for Remote Sensing Applications, UMP6, Morocco
Chaimae Azroumahli	FST Settat, Morocco
Cristiana Croitoru	Technical University of Civil Engineering, Bucharest, Romania
Dounia Chaatouf	FST Tangier, UAE University, Morocco
Dehbi Youness	University of Bonn, Germany
Elsadig Musa Ahmed Mohammed	Multimedia University, Malacca
El Meouche Rani	ESTP Paris, France
Fabio Angeletti	Sapienza University of Rome, Italy
Fatima Zahra Fagroud	Hassan II University of Casablanca
Fatma Zohra Bessai-Mechmache	CERIST, Algeria
Hicham Gibet Tani	FP Larache, UAE University, Morocco
Idriss Moumen	EST, Hassan II University of Casablanca
Jai Woo Lee	Korea University
Jesús Rosa-Bilbao	Universidad de Cádiz, Spain
Jihane Abdessadak	ENSI Tangier, Morocco
Jihane Gharib	Ibn Tofail University
Joel Rodrigues	INATEL, Brazil
Lamya Alabdulkarim	King Saud University, Saudi Arabia
Larysa Vdovenko	State Tax University
Loubaba Rida	ENSI Tangier, Morocco
Louise Hayes	UCL, UK
Mahbubul Bhuiyan	Southeast University, Bangladesh
Md Adnan Rahman	Uttara University, Bangladesh
Mohamed Zarid	Ibn-Tofail University, Morocco
Mohammed Aldulaimi	Al-Mustaqbal University, Iraq
Mohammed Chekour	ESEF Kenitra, Ibn Tofail University, Morocco
Mohammed Loukili	Naval Academy Brest, France
Nouhaila El Koufi	Hassan II University, Casablanca, Morocco
Oksana Makovoz	National Technical University, Ukraine
Osvaldo Santos	IPCB, Portugal
Paraskevi Tsoutsa	University of Thessaly, Greece
Pedro Ribeiro Mucharreira	University of Lisbon, Portugal
Raida Ktari	Higher institute of computer science and multimedia of Sfax, Tunis

Keynote Speakers

Remote Sensing, IOT and IA for Water Resources Management

Prof. Abdelghani Chehbouni
Director of the Center for Remote Sensing Applications, UM6P, Morocco

Dr. Abdelghani Chehbouni is Doctor in hydrology and remote sensing. He received his B.S. in fluid Mechanics from the University Paul Sabatier, Toulouse, France, in 1987, his M.S. in environmental science from the National Polytechnic Institute of Toulouse in 1989, and a PhD in Hydrology and Remote Sensing from the University Paul Sabatier in Toulouse in 1992.

He has worked at the University of Arizona for 18 months and the Jet Propulsion Laboratory for 2 years. He co-invested in several Remote Sensing programs (VEGETATION, ERS2/ATSR2, EOS).

His main research interests include applying remotely sensed data in land-surface-atmosphere models, especially in arid and semi-arid regions.

He participated in several international experiments (MONSOON'90, Hapex Sahel) and was Co-Leader of the Semi-Arid-Land-Surface-Atmosphere (SALSA) international program. He joined CESBIO in 2000 as Leader of the hydrology group. He is currently Director of the Center for Remote Sensing Applications, UM6P, Morocco.

Toward a Holistic Approach to Smart and Sustainable Mobility

Prof. Mohamed Essaaidi
President of MedAI and Ex-Dean of ENSIAS

Prof. Mohamed Essaaidi is Professor and Past Dean of the ENSIAS school of Computer Science, Mohammed V University, Rabat, Morocco. He is CEO of the Moroccan School of Engineering Science (EMSI) Group (2023-2024), Chief of Party of Interactive Digital & XR Center Morocco at UM6P (2020-2025), Past Director of International Cooperation at the Ministry of General Affairs & Governance, Morocco (2019), and Past Faculty Member of the Abdelmalek Essaadi University, Tetuan, Morocco (1993-2011).

He is IEEE Senior Member, he was IEEE Morocco Section Founding Chair (2004), IEEE MTT-S/AP-S Morocco Chapter Founding Chair (2008), IEEE Global Cities Alliance, Middle East & Africa Chairman (2020-2022), and he has been the IEEE Humanitarian Technologies Board Programs Committee Chair and IEEE Special Interest Group on Humanitarian Technologies Global Chair since January (2023-2024) and IEEE Public Safety Technology Education Chair (2021-22).

Integrating Green Education and AI at Al-Mustaqbal University: A Framework for Sustainable Higher Education

Prof. Hassan Majdi
President of Al-Mustaqbal University, Babil, Iraq

Prof. Dr. Hasan Shakir Majdi, born in 1957 in Babylon, Iraq, is Distinguished Chemical Engineer and Academic Leader. He earned his Bachelor of Science in Chemical Engineering in 1978 and later obtained his Ph.D. in Chemical Thermodynamics from the University of London in 1991. Dr. Majdi has dedicated his career to academia and research, focusing on renewable energy and heat transfer. He has been President of Al-Mustaqbal University in Babylon, Iraq, where he was also associated with the Department of Chemical Engineering and Petroleum Industries.

Throughout his academic tenure, Dr. Majdi has contributed significantly to scientific literature. His research encompasses adsorption techniques for dye removal, advancements in absorption cooling cycles, and the development of heterogeneous catalysts for hydroisomerization and hydrocracking processes. His work has garnered substantial citations, reflecting his influence in the field. In recognition of his leadership and contributions to higher education, Dr. Majdi was honored as the Best President of a Private University in Iraq for 2023.

Dr. Majdi's career exemplifies a commitment to advancing chemical engineering education and research, particularly in sustainable and renewable energy solutions.

Green and Sustainable Computing Using Remote Sensing in Saudi Arabia

Prof. Sultan Alsultan
The Environmental Center of Remote Sensing (TECRS), Riyadh, Saudi Arabia

Dr. Sultan Alsultan's primary professional interest has been in remote sensing for desert area development and city and environmental land management. Within that field of interest, Dr Sultan obtained his PhD from Tokyo Institute of Technology, Japan, just after graduating from MIT, Massachusetts Institute of Technology, and previously from George Washington University, both in the USA. Dr Sultan has been Senior Researcher with NASA and the USGS in the USA and Researcher in JAXA, RESTEC-Japan. He is (CEO, TECRS, Saudi Arabia) Leading Environmental Consultant of the Remote Sensing Institution and Partner of Al-Qassim University. He is Saudi Arabia Coordinator for the Mission Idea Contest.

Contents

About the Editors

Badr-Eddine is Computer Science Professor at the Faculty of Sciences and Techniques of Tangier, Abdelmalek Essaâdi University, Morocco. He has expertise in system administration, network management, distributed systems, IoT, cloud computing, software development, machine learning, and satellite data processing. Badr-Eddine has engaged in various international activities in research and innovation, including fellowships and programs in Romania, Taiwan, Spain, Belgium, and Saudi Arabia. He has authored over 18 high-quality, Scopus-indexed papers, participated in more than 20 international conferences, and is Reviewer for numerous journals. As Member of ASAF and IEEE, he actively contributes to technical program committees at several international conferences.

Ikram Ben Abdel Ouahab is Assistant Professor at the computer science department of the Faculty of Sciences and Techniques of Tangier, Morocco. She is Member of the Data & Intelligent Systems research Team. She received her PhD in Cybersecurity and AI at FSTT, University Abdelmalek Essaâdi, Tangier, Morocco. She participated in many international conferences and has published many scientific papers in collaboration with the Laboratory team. In 2022, she graduated from the CPITS program by Trend Micro, an intensive training in cybersecurity technologies and industry trends, including certification in industry-leading solutions for endpoint, cloud, and network security. She is AWS Solution Architect Certified Peer Reviewer from the Publons academy. Her research interests include cybersecurity, malware analysis, artificial intelligence, and IoT.

Luís Bragança, PhD, MSc, and Licenciate Civil Engineer, is Professor at the Department of Civil Engineering at the School of Engineering at the University of Minho. He is Director of the Doctoral Program in Sustainability of the Built Environment, the Master's in Sustainable Construction and Rehabilitation, and Director of the Energy and Sustainable Construction Laboratory. He develops scientific, pedagogical, and professional activity in Sustainability of the Built Environment, Sustainable Construction, Circular Economy, Life Cycle Analysis, Sustainable Rehabilitation of Buildings, Building Physics, and Construction Technology. He is Author and Co-Author of more than 450 publications, including books, book chapters, articles in national and international journals and conferences, 364 of which were published in prestigious international journals with a high impact factor (121 articles indexed in ISI/WoS; 152 articles indexed in Scopus; Scopus h-index = 27). He is Author and Co-Author of 3 patents, Supervisor, and Co-Supervisor of 10 doctoral theses. Professor Luís Bragança was President of the International Initiative for the Sustainability of the Built Environment (iiSBE) from 2011 to 2021.

Artificial Intelligence for Sustainable Development

Artificial Neural Networks, Alone or Combined with Other Artificial Intelligence Methods, Applied to Wind Power Control: A Review

Fatima Ezzahra Mbarki[1](✉), Touria Haidi[1], Doha Bouabdallaoui[1], Ishak Hbiak[2], and Mariam Cherrabi[3]

[1] Electrical Engineering Department, Laboratory of Systems Engineering, Hassania School of Public Works, Casablanca, Morocco
mbarki.Fatimaezzahra.cedoc@ehtp.ac.ma
[2] Civil Engineering Department, Laboratory of Systems Engineering, Hassania School of Public Works, Casablanca, Morocco
[3] Mathematics Computer Science and Geomatics Department, Laboratory of Systems Engineering, Hassania School of Public Works, Casablanca, Morocco

Abstract. The wind energy should be effectively controlled to maintain the stability of the electrical grid. An effective solution to improve the active and reactive power control in wind turbines is the Artificial Neural Networks (ANNs) method. This paper presents a review of studies that used ANN techniques to regulate and control wind power production. It also covers approaches that combine ANNs with other artificial intelligence methods, including Recurrent Neural Networks (RNN), Particle Swarm Optimization (PSO), Fuzzy Logic algorithms and Genetic Fuzzy Systems (GFS). These methodologies are summarized and compared to show their respective advantages and limitations. This study shows that combining ANNs with other AI techniques increases stability and provides robust solutions to problems.

Keywords: Renewable energy · Artificial intelligence · Wind power · Artificial neural networks

1 Introduction

With the commitment to minimize greenhouse gas emissions and the resource limitations of fossil fuels, renewable power sources and the integration of power-efficient solutions are becoming a necessity. These developments are reflected in the crucial role played by renewable energies, such as wind power, in electricity production. The African Renewable Energy Initiative (AREI) wanted to supply 300 GW of renewable energy by the year 2030, which highlights the important role of wind power in Africa [1]. However, renewable energies present major challenges to the electricity grid's stability. The frequency and voltage stability of the electricity grid can be affected by the intermittency and variability of wind generation, which can lead to unpredictable changes in

B.-E. Boudriki Semlali et al. (Eds.): DATA 2025, LNNS 1908, pp. 3–15, 2026.
https://doi.org/10.1007/978-3-032-21200-9_1

the power fed into the grid. Multiple studies have demonstrated that integrating renewable energy sources impacts system stability, which is an important criterion for every power system. Fluctuations in wind speed are one of the main reasons for power system instability, and they also affect the operation of electrically synchronized equipment. The electrical system must be regulated using all available tools to maintain frequency and voltage values within predetermined limits. It is, therefore, critical to manage the power generated by this renewable energy source using advanced control systems [2]. Active power adjustment depends on frequency control, while reactive power is related to voltage control.

The traditional approaches of controllers are extremely limited by the complexity and dynamics of wind energy systems. That is how artificial intelligence (AI) has been proposed as a more reliable method to ameliorate the control of wind energy systems and guarantee grid stability. For example, artificial neural networks (ANNs) can learn and adapt based on historical data in real time to offer great flexibility and accuracy in wind turbine management.

2 Artificial Neural Networks and Main Artificial Intelligence Techniques Combined with ANNs Wind Generation Control and Optimization

Artificial intelligence (AI) methods are an innovative solution for active and reactive power control in dynamic systems, including wind farms. Some approaches suggest using only artificial neural networks (ANNs), while others propose combining them with other AI technologies such as fuzzy neural networks, particle swarm optimization (PSO), and the Genetic Fuzzy System (GFS). All these approaches aim to simplify the control of active and reactive power. Neural networks, inspired by biological neurons, receive input signals, process the information, and send output signals through interconnected layers. Their ability to model complex relationships in real time makes them useful for controlling wind turbines and optimizing power production. In wind energy, ANNs are used for forecasting, predicting, and controlling power generation. A typical ANN consists of three layers: the input layer, where data is extracted; the hidden layer, which processes information from the earlier layer; and the output layer, which analyzes the results. Recurrent neural networks (RNNs), a type of neural network, are particularly suited for processing sequential data and capturing long-term relationships, making them ideal for wind generation forecasting. Particle Swarm Optimization (PSO) is a biologically inspired heuristic algorithm that solves complex optimization problems with minimal computing and memory resources. In wind energy, PSO is used to optimize wind turbine parameters for maximum electricity production. Fuzzy logic, another AI method, deals with uncertainty and imprecision by using membership levels instead of fixed values like 'true' or 'false.' It includes four main components: Fuzzification, rule base, inference engine, and Defuzzification, all of which help in processing data and controlling power systems [3]. Genetic algorithms (GAs), based on natural selection mechanisms, optimize wind energy systems by adjusting wind turbine blade angles to maximize energy collection. GA follows a series of steps, including population creation, evaluation, selection, reproduction (cross and mutation), and forming the next generation. Combining various AI methods helps improve wind generation stability by using each method's strengths and advantages, ultimately enhancing the control and forecasting of wind power.

A bibliometric search was conducted using platforms such as IEEE Xplore, ResearchGate, and ScienceDirect, to show the evident interest in these approaches. The keywords used in the queries were 'Wind power', 'Artificial neural networks', 'Artificial intelligence combined with ANN' and 'Controlling wind power using artificial intelligence methods'. The number of published papers on active and reactive power control of wind turbines using ANN and other artificial intelligence techniques from 2015 to 2024 is shown in Fig. 1. As per the below figure, the number of publications increased year after year, which shows the growing role and impact of the subject.

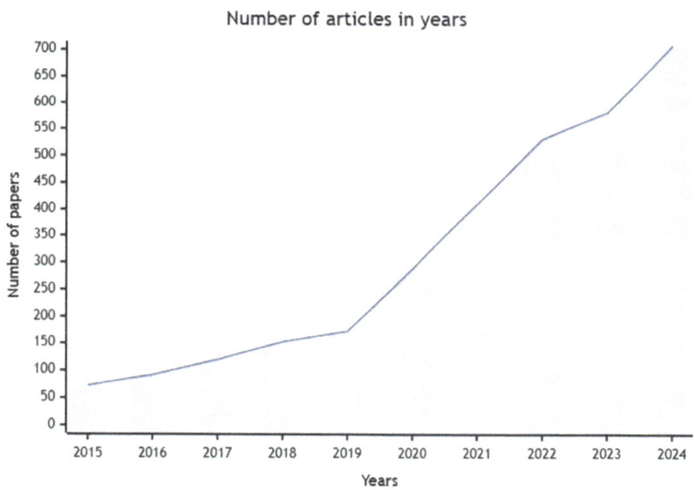

Fig. 1. Number of articles on wind energy controls using ANNs alone or in combination between 2015 and 2024.

2.1 Only ANN Method Applied for Wind Turbine Production Control

Various artificial neural network control strategies can optimize wind turbine performance and stability. According to [4], two intelligent controllers based on Multilayer Perceptron Networks (MLP) are used to independently control DFIG stator active power and reactive power. To maximize the power produced by wind turbines, this study develops a method based on MLP. Other researchers [5] have a similar method, which is MLP multilayer networks. They compared the MLPs and RBF network (radial basis function) to adjust the resulting turbine output in response to rapidly changing wind speeds. According to this comparison, the RBF controller presents a better solution than the MLP in maintaining stable energy production and preventing wind turbine overloads.

It is shown that controllers based on ANN, trained from simulations of proportional-integral controllers under various conditions like wind speed and fault conditions, provide better accuracy in tracking the reference active power while respecting the required power factor in both hypersynchronous and hypo synchronous modes. Another approach uses boost DC-DC converters by combining them with a PI controller based on AI, which is the artificial neural network. The results indicate that the study done has achieved dynamic impact and reduced complexity, with a high level of precision. In another work, Barambones [6] discovered a different method that uses mechanical power to control variable-speed wind turbines. A stability study of the suggested controller is conducted using the Lyapunov stability approach, which provides a complete evaluation of its performance and reliability in so many scenarios.

In [7], Jialong and Ho proposed a pitch angle control system based on ANN for the management of the energy produced by a wind turbines dual-feed induction generator system. To do that, the ARMA model predicts the wind speed for several control cycles. The ANN enables more effective control and regulation of the electricity generated by the wind turbine by finding the inverse relationship of the DFIG system. On the other hand, the study suggests an adaptive integral slip mode controller using an ANN technique to improve the rotor speed. It determines the control voltages, which are then regulated by a PWM controller to manage the generator's power converter.

Additionally, the strategy described in [8] optimizes the production of wind energy by using RBFNN-based ANN to adjust the duty cycle and blade angle according to wind conditions and generator performance. A block layout of this approach is shown in Fig. 2.

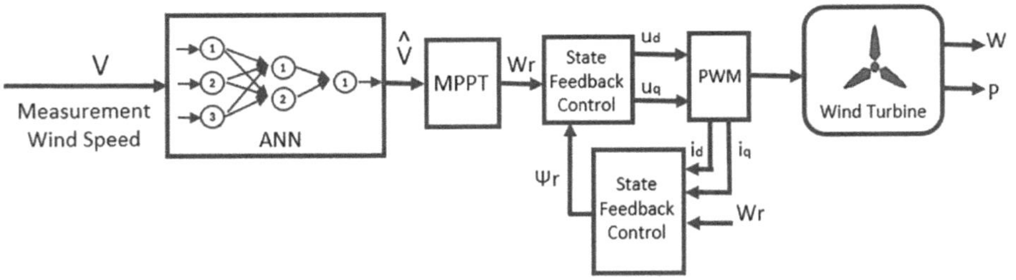

Fig. 2. Artificial neural network (ANN) control configuration for wind speed estimation and reduction of discontinuity and slow response of anemometer [9].

For hybrid systems, another paper [10] suggests a way to use artificial neural networks to improve the parameters of an SVC controller to regulate reactive power. The ANN adjusts the PI regulator's gains based on the load voltage's characteristics, and the optimized controller is tested to manage transient responses in the hybrid system.

2.2 Methods Combining ANNs with Other Artificial Intelligence Techniques for Wind Turbine Production Control

Artificial neural networks have many limitations when it comes to the analysis of wind energy conversion systems (WECS), such as over-training, extrapolation errors, and network optimization. To avoid these challenges, it is proposed in much research to combine

ANNs with other artificial intelligence methods to create more advanced controllers and approaches. As presented in [11], particle swarm optimization (PSO) is combined with ANNs to follow the maximum power point (MPPT) at diverse wind speeds. The aim is to maximize the active power and potentially influence the reactive power according to the needs of the power grid. The PSO method is also used in another frequency control strategy to efficiently optimize the kinetic energy reserve and prevent system instability.

The authors of [12] provide a method for controlling a Buck-boost converter's duty cycle by using PSO optimization to optimize wind energy extraction. Whenever wind speed varies, the system minimizes losses from changes in rotation speed while increasing wind turbine efficiency by modifying the duty cycle based on the voltages and currents produced. Another approach combines sliding-mode control with ANN and PSO. The ANN, with a hidden layer, estimates system uncertainties to reduce sliding-mode chattering, while the PSO optimizes the ANN learning rate to improve convergence speed. The authors used the Lyapunov theory to enhance overall system stability. With this method, the efficiency and stability of the system are improved compared to the traditional methods.

Another method, the adaptive neuro-fuzzy inference system, called ANFIS, has become popular for controlling active power and regulating voltage in wind power [13]. It improves the active power of DFIG wind turbines by optimizing wind energy extraction, its precise tracking of the point of maximum power and dynamic response to variations in wind speed. In addition to improving grid voltage stability, this method has an impact on reactive power regulation. The ANFIS MPPT controller shows an advancement of 3.28% in the active power of the stator in comparison with the proportional and integral controller (Fig. 3).

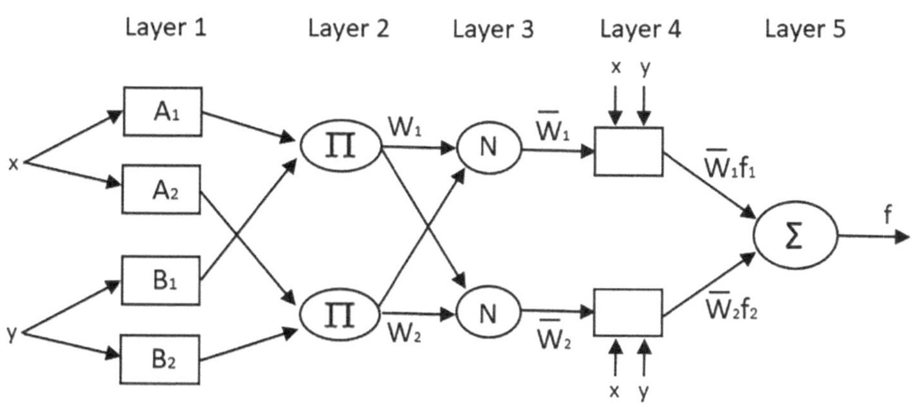

Fig. 3. ANFIS model with two inputs, x & y, and one output, f [13].

In another study [14], A new Genetic Fuzzy System (GFS) was useful for extracting fuzzy rules from a neural network. This extraction, known as the FRENGA method, eliminates wind perturbations in WECS systems thanks to the generating of a pitch angle control. As a result, this approach has stabilized aerodynamic power and torque within nominal limits. The results confirm that this new system is one of the most established methods for maintaining stable power in wind variations.

3 Results and Discussion

Used alone or combined with other artificial intelligence methods, ANN artificial neural networks can be used to control wind power production. Artificial neural networks (ANNs) can model complex systems, such as wind turbines, and be adapted in real time to maximize the energy needed. Many research papers have shown that advanced controllers based on ANN or its extractions, such as Multilayer Perceptron Networks (MLPs) and Radial Basis Function (RBF) networks, provide better performance than traditional methods in wind power stability. Recurrent Neural Networks (RNN) are also used for wind turbine forecasting and control. Their ability to process complex time sequences enables more precise control for wind speed variations and energy production fluctuations.

By combining ANNs with other artificial intelligence methods including as Particle Swarm Optimization (PSO), fuzzy logic and Genetic Fuzzy Systems (GFS), it is possible to improve the solutions suggested for the controller's systems. These hybrid approaches show significant improvements in terms of maximizing energy production and managing reactive power. Additionally, optimization with the Taguchi method and ANFIS algorithms improves voltage regulation.

These combinations offer many solutions to the need for more renewable energy production, despite the challenges of integrating it into the grid. In many studies, various applications have been explored, such as maximum power point tracking, blade pitch angle adjustment, and wind turbine power regulation, as illustrated in the following radar diagram (Fig. 4).

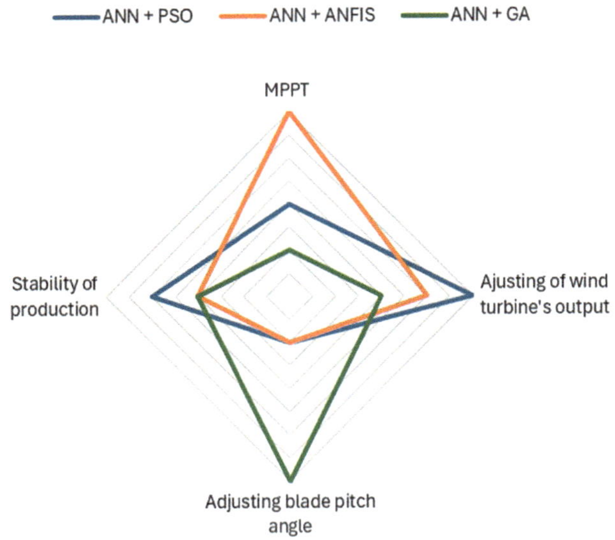

Fig. 4. Radar Diagram of AI-Based Control Strategies for Wind Turbine Applications.

Table 1 gives a summary of the main works encountered in the literature that have applied the ANN method alone, while Table 2 presents that of works combining ANNs with other artificial intelligence techniques, for wind turbine production control.

Table 1. Summary of the main works on the application of single neural networks in the control of wind turbine power generation.

Ref	Application	Methodology
[4] [10] [15]	PI controller	Applying ANN artificial neural networks to regulate a rotor-side converter using PI controllers. This method provides better precision in tracking the active power reference in two modes: hypo-synchronous and hyper-synchronous.
		Integration of DC-DC boosts converters with ANN-assisted PI controllers for wind power systems. The objective is to simplify the complexity of this system with accuracy and dynamics.
		Use of the multilayer perceptron (MLPs) on PI controllers to independently control a DFIG's active and reactive powers. This maximizes the power that can be extracted from the wind and improves the system's accuracy.
		Use of artificial neural networks (ANN) in a hybrid system to optimize parameters of an SVC controller with the goal of preserving reactive power. By modifying the gains of a PI controller as the load voltage varies, the ANN can improve the system's transient response.
		Application of a PI controller based on an artificial neural network for a reduced switching multicellular inverter (RSCMLI) in a wind energy system (WECS). This system is used even when there is an event of faults, and the controller regulates both and reactive power.
[5] [7] [16]	Adjusting the wind turbines output according to wind speed	Comparison study of how radial basis function (RBF) and MLP networks can adapt wind turbine output when wind speeds change. This study proves that RBFs are more efficient than MLPs in terms of wind generation stability.
		Application of pitch angle controller based on ANN for controlling the power generated by a MADA-based wind turbine. For changing the pitch angles, the ARMA model forecasts the actual speed of wind turbine ANN finds inverted DFIGs for optimum regulation.
		Reactive power compensation using artificial neural networks (ANN) in a wind-diesel hybrid system. This method is based on a multilayer neural network trained by error backpropagation learning. It adjusts both proportional and integral gains of a static synchronized compensator that effectively compensates for reactive power deviations because of wind speed variations.

(*continued*)

Table 1. *(continued)*

Ref	Application	Methodology
[4] [7] [8]	Pitch angle adjustment of wind turbine blades	Radial basis function neural network, called RBFNN, is a technology used to maximize energy generation, to optimize the duty cycle and pitch angle. The method is used to process non-linear functions and to produce accurate approximations in order to optimize these parameters, which are critical to the performance of a wind turbine.
		The ARMA model combined with pitch angle control based on an artificial neural network (ANN) to forecast the fluctuation of wind turbines. To maximize the controller response, the artificial neural network (ANN) models the dual-fed wind turbine system's (DFIG) non-linear relationship.
		Use of artificial neural networks (ANN) to regulate pitch angle for wind turbines that operate faster than their rated speed. There are two approaches, the Multi-Layer Perceptron and the Radial Basis Function, which use two different types of networks. During strong and fluctuating winds, they are employed to regulate power generation and prevent overloads or stoppages by changing the pitch angle.
		Combination of both maximum power point tracking (MPPT) and blade pitch angle adjustment for any changed turbine's speed powered by a permanent magnet synchronous alternator. The MPPT technique maximizes the energy retrieved at lower wind speeds, while neural network technology adjusts the pitch angle to influence the power output when the rated power is reached.
[6] [8] [15]	Wind turbine rotor speed regulation	Integration of artificial neural networks (ANNs) with a stability study based on Lyapunov theory. This control system regulates wind generation by focusing on rotor speed in variable-speed wind turbines.
		An adaptive full-slip mode controller with Recurrent Neural Networks (RNN) optimizes the rotor speed, while a PWM controller regulates the power converter.
		Using of artificial neural networks (ANNs) for controlling the rotor flux of a squirrel-cage induction machine in the wind power grid. The ANNs play a key role in rotor speed estimation, which enables precise control. This controller tracks the rotor speed and flux amplitude profiles using an input-output feedback law.

Table 2. Summary of the main artificial intelligence methods used, in combination with neural networks, in the control of wind turbine power generation.

Ref	Combined methods	Application	Methodology
[9] [11] [12]	ANN and PSO	Maximum power point tracking MPPT	A combination of PSO and ANN is used for wind turbine MPPT. The configuration maximizes the active power and influences reactive electric power. The technology PSO, in this case, is used for frequency control.
			Using MPPT (maximum power point tracking) to manage active power. The wind turbine power reserve is maintained during a disturbance by optimizing kinetic energy reserves using a particle swarm optimization (PSO) algorithm. At the same time, artificial neural networks (ANNs) improve wind turbine power stability and control precision.
			A Combination of ANN is used to predict the variation in wind speed and PSO to regulate the duty cycle of the buck-boost converter. It aims to use maximum power point tracking to reduce losses due to changing rotation speed.
		Adjusting the stability of the wind turbines output	PSO adjusts the stability of the power generated in a wind farm, while ANN reduces chattering, which can cause unwanted oscillations in the electrical system. To adapt perfectly the deviation in the wind speed, the PSO helps to optimize the ANN's learning rate, Overall system stability is guaranteed by Lyapunov theory, it makes sure that control system is till effective even while having any changes in operating conditions.
			Using the PSO-ANN approach to improve system stability by reducing frequency oscillations, overshoots, and stabilization time. This combination provides a quick and stable response to power imbalances caused by renewable energy sources which optimizes network performance.
			Combination of ANN and PSO so the gains of a PI controller in a wind power system can be adjusted. Experimental results show that this method reduces active power overshoot and improves response time compared to traditional approaches such as pole placement.
[13] [17] [18]	ANN and ANFIS	Maximum power point tracking MPPT	Maximum power point tracking system for grid-connected wind energy conversion systems based on adaptive neuro-fuzzy systems (ANFIS) using dual-fed induction generators (DFIGs). ANNs supply the learning capabilities, and ANFIS combines them with fuzzy logic to control the wind turbines fluctuating speed. The primary goal is to extract as much electricity as possible.

(continued)

Table 2. (*continued*)

Ref	Combined methods	Application	Methodology
			ANN and ANFIS methods are employed to track the maximum power in the wind power system. To modify the weights, the ANN network used the gradient descent technique search and the reverse propagation algorithm. The ANFIS combines the ANN and fuzzy method to provide increased efficiency. This combination showed larger precision and reliability than ANN for estimating wind speed and tracking MPPT.
			Optimization of energy production for autonomous hybrid systems by applying the MPPT method. ANFIS technology helps to ensure ideal voltage regulation, and maximize the power generated by the wind turbine and minimize harmonics in the system.
		Power control and voltage regulation	On the rotor side of the wind turbine, an ANFIS controller is connected to the converter. It is applied for active power control, voltage regulation and amelioration of wind power quality and stability even under adverse conditions.
			Application of ANN and ANFIS methods to predict voltage harmonics, such as dominant individual harmonics h13 and h11 and what we called total harmonic distortion (THD) for renewable hybrid energy systems that combine a wind with a permanent synchronous generator (PMSG). All of these provisions allow for the adjustment of control mechanisms to stabilize tension in the presence of many harmonics.
			The ANFIS controller helps to maintain constant voltage and frequency for a doubly fed induction generator (DFIG) in a wind energy conversion system operating in stand-alone mode, despite current variations in wind speed and power demand. This approach uses wind turbine data to leverage fuzzy logic and the ANNs.
[14] [19] [20]	ANN and GA	Adjusting the stability of the wind turbines output	The application of ANN is to model and forecast the variations that could be caused by meteorological changes. The Genetic Fuzzy System (GFS) adjusts dynamically on the blade angle to ensure having a stable power output. This combination regulates generator torque and minimizes fluctuations in the power output produced.
			Real-time active and reactive power control can be achieved using the ANN technique. The Gas optimizes the ANN parameters including thresholds and neural network configurations. The aim of this combination is to increase wind production's stability and efficiency.

(*continued*)

Table 2. (*continued*)

Ref	Combined methods	Application	Methodology
			The study is applied to a wind system with a permanent magnet synchronous alternator; the idea is to improve of the efficiency of the variable-speed of the wind's power. To operate the generator-side converter and the grid-side inverter, they used a hybrid controller that incorporates ANNs and fuzzy logic plus a Genetic Algorithm (GA). The grid-connected wind power system's dynamic and transient stability is enhanced by this controller, according to the results.
		Adjusting blade pitch angle	A Genetic Fuzzy System (GFS) is a combination of genetic algorithms (GA) and fuzzy logic that extracts fuzzy rules from an ANN neural network and controls the pitch angle by maintaining the wind turbine aerodynamic torque. This facilitates grid integration and makes the power control more accurate.

Based on the above tables and the different aspects and approaches explored, we present in this paragraph a critical assessment of different methods using artificial neural networks (ANNs), either alone or in combination with other artificial intelligence techniques, for the purpose of controlling the power of wind energy systems.

MLP networks are often used to control active and reactive power, thanks to their ability to model complex systems and optimize controllers such as PIs under varying conditions. However, they require a lot of specific data to ensure robustness under real conditions. RBF networks are very fast to run; they are for instantaneous adjustment of energy production, considering wind fluctuations, but they are less adaptable to complex configurations. Other approaches use ANNs and other classical techniques such as speed estimators, predictive models such as the ARMA model, or the Lyapunov stability principle. These methods are used to reinforce and improve the regulation of rotor speed or pitch angle but require very detailed modeling.

PSO is often used in conjunction with ANN to optimize the performance of wind energy systems. This combination makes it possible to exploit the learning capacity of the neural network while taking advantage of the optimization power of the PSO. Some researchers use the PSO to adjust the learning rate of the ANN and improve the convergence speed and stability of the system. Other research has applied this combination to optimize maximum power point tracking (MPPT), reduce transient fluctuations and enhance the overall dynamic response of the system. However, for MPPT, the combination of ANN and ANFIS is still the most widely used. ANFIS, which combines learning and fuzzy reasoning, enables the maximum power point to be determined and tracked quickly, irrespective of the rotational speed or pitch angle of the wind system's blades. However, this type of model can be sensitive to minor variations in the signal, leading to the danger of system instability. In addition, the GFS method combined with ANN stands out for its ability to rapidly ensure the stability of wind system performance by integrating fuzzy rules generated by a neural network and optimized by a genetic algorithm.

4 Conclusion

Artificial intelligence methods are frequently used to control wind power; they provide innovative and effective solutions to the challenges of integrating wind power into electrical grids. Artificial neural networks (ANNs) are the most basic and typically used technology in publications. They are distinguished by their capacity to model complex interactions and adapt in real time to variations in wind speed. This paper summarizes the main research done on the neural network strategy for electrical power generation control for wind energy systems. It discusses and compares the case of the application of ANNs alone and the case of having a combination with AI techniques such as fuzzy neural networks, the PSO method and genetic algorithms (GA). This review of these studies shows that ANNs can be used to develop more efficient controllers, to improve the stability and performance of wind power systems. However, ANNs have certain limitations, as they lack an integrated optimization mechanism. This is where the combination of ANNs with certain artificial intelligence methods comes in, to achieve a dynamic and optimal response to variable conditions, while guaranteeing better performance than that obtained by using ANNs alone, even in unpredictable operating scenarios. Each method has specific advantages but also come with limitations related to processing complexity or sensitivity to minor signal variations. According to this comparative study, ANFIS is the most widely used model, thanks to its advantages and ease of implementation. As a perspective, the accuracy of the ANFIS model can be increased by combining it with one or more optimization algorithms.

References

1. Haidi, T., Cheddadi, B.: State of wind energy in the world: evolution, impacts and perspectives. Int. J. Tech. Phys. Probl. Eng. **41**, 347–352 (2022)
2. Bulatov, R.V., Burmeyster, M.V., Nasyrov, R.R., Vanin, A.S., Raseel, A.: Study of the influence of wind power station on the operating modes of the distribution electric network. In: 2020 International Youth Conference on Radio Electronics, Electrical and Power Engineering (REEPE), pp. 1–5 (2020)
3. Bouabdallaoui, D., Haidi, T., El Jaadi, M.: Review of current artificial intelligence methods and metaheuristic algorithms for wind power prediction. Indones J Electr Eng Comput Sci. **29**, 626–634 (2023)
4. Boudjellal, B., Benslimane, T.: Active and reactive powers control of DFIG based WECS using PI controller and artificial neural network based controller, IIETA. (2020).Accessed 13 Sept 2024
5. Yilmaz, A.S., Özer, Z.: Pitch angle control in wind turbines above the rated wind speed by multi-layer perceptron and radial basis function neural networks. Expert Syst. Appl. **36**, 9767–9775 (2009)
6. Barambones, O.: A robust wind turbine control using a neural network based wind speed estimator. In: The 2010 International Joint Conference on Neural Networks (IJCNN), pp. 1–8 (2010)
7. Wang, J., Hyun, S.H.: ANN based pitch angle controller for variable speed variable pitch wind turbine generation system. In: Proceedings of 2011 6th International Forum on Strategic Technology, pp. 443–447 (2011)

8. Tiwari, R., Padmanaban, S., Ramesh Babu, N.: Energies, Free Full-Text, Coordinated Control Strategies for a Permanent Magnet Synchronous Generator Based Wind Energy Conversion System. (2017). Accessed 13 Sept 2024
9. Chavero-Navarrete, E., Perea, M., Jauregui, J., Carrillo-Serrano, R., Moreno, G.J.: Expert control systems for maximum power point tracking in a wind turbine with PMSG: state of the art. Appl. Sci. **9**, 2469 (2019)
10. Bansal, R.C., Bhatti, T.S., Kumar, V.: Reactive power control of autonomous wind-diesel hybrid power systems using ANN. In: 2007 International Power Engineering Conference (IPEC 2007), pp. 982–987 (2007)
11. Ait Ali, Y., Ouassaid, M.: Advanced control strategy of DFIG based wind turbine using combined artificial neural network and PSO algorithm, IEEE Conference Publication 2020 International Conference on Electrical and Information Technologies (ICEIT) (2020). Accessed 13 Sep 2024
12. Daud Pribadi, K., Danang Wijaya, F.: Dynamic response of maximum power point tracking using particle swarm optimization for wind energy conversion system. In: 2016 8th International Conference on Information Technology and Electrical Engineering (ICITEE), pp. 1–6 (2016)
13. Syahputra, R., Soesanti, I.: DFIG control scheme of wind power using ANFIS method in electrical power grid system. Int. J. Appl. Eng. Res. (IJAER). **11**, 5256 (2016)
14. Kasiri, H., Abadeh, M.S., Momeni, H.R.: Optimal estimation and control of WECS via a genetic neuro fuzzy approach. Energy. **40**, 438–444 (2012)
15. Sahoo B, Kumar Rout P, Keshari Routray S (2024) Artificial Neural Network-Based PI-Controlled Reduced Switch Cascaded Multilevel Inverter Operation in Wind Energy Conversion System with Solid-State Transformer., Request PDF. ResearchGate
16. Sharma, P., Hoff, B., Meena, R.: ANN based reactive power control of an autonomous wind-diesel hybrid power plant using PMIG and SG. In: 2014 Power And Energy Systems: Towards Sustainable Energy, pp. 1–6 (2014)
17. Yin, X., Jiang, Z., Pan, L.: Recurrent neural network based adaptive integral sliding mode power maximization control for wind power systems. Renew. Energy. **145**, 1149–1157 (2020)
18. Merabet, A., Tanvir, A.A., Beddek, K.: Speed control of sensorless induction generator by artificial neural network in wind energy conversion system. IET Renew. Power Gener. **10**, 1597–1606 (2016)
19. Amin, I.K., Nasir Uddin, M., Marsadek, M.: ANFIS based neuro-fuzzy control of DFIG for wind power generation in standalone mode. In: 2019 IEEE International Electric Machines & Drives Conference (IEMDC), pp. 2077–2082 (2019)
20. Hadi, F.M.A., Aly, H.H., Little, T.: Harmonics forecasting of wind and solar hybrid model driven by DFIG and PMSG using ANN and ANFIS. IEEE Access. **11**, 55413–55424 (2023)

Catalyzing Place Marketing Strategies: A Novel Paradigm Through Artificial Intelligence and Big Data Integration

Salah Eddine El Jarroudi(✉), Aymen Mebchour, Mohamed Jalal El Adnani, and Fatima Touhami

Multidisciplinary Research Laboratory in Economics and Management, Faculty of Economics and Management, Sultan Moulay Slimane Béni Mellal University, Béni Mellal, Morocco
s.eljarroudi@usms.ma

Abstract. This innovative research examines the integration of Artificial Intelligence (AI) with big data in the field of place marketing. Place marketing, a complex domain focused on the promotion and improvement of specific geographical areas, has ushered in significant transformation with the advent of AI and big data analytics. These advanced technologies provide public domain entities and nearby stakeholders a strong platform to enhance and modify their tactics, quickly achieving a better standing in the competitive realm of location marketing. This article aims to explore the significant impact of AI on the execution of marketing strategies, particularly in relation to place marketing initiatives. The research reveals a preliminary testing module, so creating a solid basis for further extensive investigations.

Our study indicates that the use of AI can enhance consumer engagement primarily through customized campaigns, detailed personalization of marketing efforts, exploration of consumer preferences, and the improvement of client communication via AI-driven conversational agents. This study can therefore be considered a promising precursor to a future comprehensive evaluation, focused on examining the domain of place marketing in relation to technological improvements.

Keywords: Place marketing · Costumer segmentation · Costumer engagement · Artificial intelligence · Big data

1 Introduction

In contemporary times, the amalgamation of Artificial Intelligence and big data analytics has catalysed transformative changes across various sectors, including marketing. These cutting-edge technologies have unlocked myriad opportunities for public sector organizations and regional stakeholders to refine their strategic approaches and gain a competitive advantage in place marketing, a field dedicated to promoting and developing specific geographic areas. Place marketing, also known as territorial or destination marketing, involves the strategic planning and promotion of a locale to cater to the needs and

B.-E. Boudriki Semlali et al. (Eds.): DATA 2025, LNNS 1908, pp. 16–27, 2026.
https://doi.org/10.1007/978-3-032-21200-9_2

aspirations of various target groups, such as residents, businesses, tourists, and investors. Differing from traditional marketing, which focuses on tangible products or services, place marketing is centred on constructing an appealing "image" of a location based on its unique characteristics and attributes.

The discipline of place marketing has seen substantial evolution, interweaving multidisciplinary methodologies from urban planning, geography, public administration, and marketing. This evolution highlights its complexity and extensive applicability. For effective promotion of a locale, marketers must possess a thorough understanding of its distinctive qualities, identify target audiences, and employ strategic planning along with advanced technological tools. The advent of AI has ushered place marketing into a new era, enabling marketers to decode consumer behaviour, create precise audience segments, and deliver personalized content that fosters engagement and stimulates economic activity. By automating tasks, analysing extensive datasets, and forecasting trends, AI empowers marketers to optimize campaigns and achieve superior outcomes.

This study assesses the impact of AI on place marketing campaigns by examining three primary hypotheses:

H1: AI-enhanced campaigns will surpass traditional ones in targeting precision and audience reach owing to superior segmentation and personalization.
H2: AI-driven campaigns will engender higher engagement through timely and pertinent content.
H3: AI tools will amplify campaign reach, maximizing exposure to target audiences.

By evaluating these hypotheses, this research seeks to furnish empirical evidence of AI's transformative potential in place marketing, offering actionable insights for marketers aspiring to elevate their campaigns. Additionally, it sets the stage for future inquiries into the qualitative and enduring impacts of AI-driven strategies, ensuring that these technologies are employed ethically and effectively to benefit both marketers and their audiences.

2 Literature Review

2.1 Artificial Intelligence in Place Marketing

2.1.1 Marketing Places: Concepts and Approaches

Place marketing, also referred to as place or destination marketing, involves the promotion of geographic locations to draw in visitors, inhabitants, enterprises, and investors. This multifaceted domain amalgamates urban planning, geography, public administration, and conventional marketing principles, illustrating its complexity and extensive applicability [1]. According to Kotler & Gertner's conceptualization, place marketing entails crafting and promoting a locale to fulfill the requirements of its target market [2]. Success is measured by the satisfaction of residents, businesses, visitors, and investors.

Distinct from traditional marketing, the 'product' in place marketing is the locale's image, which is influenced by its distinctive characteristics [3].

A 'place' can encompass administrative units such as municipalities or broader entities like cultural centers or industry clusters [2]. Place marketing is underpinned by three pivotal factors: target markets, marketing components (attractions, infrastructure, image, quality of life), and planning entities responsible for strategic initiatives [4]. Target markets include investors, tourists, manufacturers, and other relevant stakeholders. Insight into their motivations, behaviors, and decision-making processes is crucial for effective positioning [5].

The process engages participants at local, regional, national, and international tiers, comprising governments, private firms, and global bodies [6]. This re-envisioned framework underscores strategic decision-making and incorporates a diverse array of elements and entities to ensure successful place promotion (refer to Fig. 1).

Fig. 1. Key Actors in Place Marketing [6]

2.1.2 Specific Utilities of Artificial Intelligence in Place Marketing

AI assumes a crucial role in place marketing by focusing on fundamental aspects such as pricing, strategic planning, promotion, and place management, while addressing user requirements through sophisticated analytics [7]. Both public and private stakeholders utilize AI to augment regional engagement by employing machine intelligence to monitor behaviors, deliver personalized messages, and refine decision-making processes [8].

Programmatic media buying automates advertisement placements, ensuring that relevant advertisements reach target audiences, thereby minimizing errors and scaling efforts effectively [9]. AI technologies such as facial recognition and real-time notifications facilitate customized visitor experiences, thereby enhancing satisfaction and economic activity [10]. Machine learning improves visitor segmentation, forecasts future behavior, and identifies high-engagement targets more efficiently and accurately than traditional methods [11].

AI-powered predictive analytics anticipate campaign success, automate repetitive tasks, and bolster strategic concentration. Social networks and tools like HubSpot employ

AI to enable seamless, personalized interactions and cost-effective marketing [12]. Nevertheless, the effective deployment of AI necessitates robust data infrastructure and expertise, often requiring collaboration with third-party organizations [13].

Success hinges on the quality and standardization of data, ensuring precise insights and adaptive strategies that align with evolving visitor preferences [14]. Advanced AI systems, such as chatbots and neural networks, detect patterns and deliver tailored marketing strategies, thereby enhancing regional competitiveness [13].

2.1.3 Innovating Regional and City Promotion with AI

Artificial Intelligence, integral to renowned platforms such as Netflix and Amazon, has now become prevalent in place marketing for enhancing visitor experiences. Once confined to large corporations, sophisticated AI tools have now become available to smaller communities [15]. AI assists in examining visitor behavior, segmenting audiences, automating tasks, generating content, and forecasting trends through the use of neural networks and extensive datasets [16].

Predictive analytics enables marketers to evaluate campaign efficacy by detecting patterns from previous efforts. Progress in machine learning, propelled by large datasets and voice search technologies, enhances targeting accuracy based on demographics and preferences [17]. Deep learning methods further refine these predictions with minimal human oversight, facilitating more effective marketing strategies [18].

AI consolidates data from multiple platforms, enabling real-time decision-making and sentiment analysis derived from social media [19]. These insights enhance advertising efforts, refine marketing campaigns, and customize messaging for strategic positioning [20]. By utilizing AI, marketers can boost visitor engagement and satisfaction while effectively promoting regional interests.

3 Methodology

The study adopts a comparative research design to assess the effectiveness of AI-enhanced marketing strategies compared to traditional methods. This quantitative approach enables a systematic analysis of the performance differences between the two campaign types.

Sample and Sampling Strategy

The study focuses on the BMK Region in Morocco, utilizing a purposive sampling strategy to select marketing campaigns that are representative of both AI-enhanced and traditional approaches. The sample includes two campaigns:

- Campaign 1: A traditional marketing campaign using demographic segmentation and standardized messaging.
- Campaign 2: An AI-enhanced marketing campaign leveraging AI tools and big data analytics for personalized targeting and content.

3.1 Campaign Execution

Both campaigns were executed concurrently with identical budgets and timeframes to ensure comparability. the initial campaign was human-led with all content and target audience criteria were defined using field experience, meanwhile the AI-enhanced campaign's content was generated by an AI assistant tool, based on input provided by the marketer, providing more tailored insights based on target demographic and campaign objectives. Both campaigns featured a similar landing page structure to maintain consistency, with variations on the level of the ad publication in the title, body text, and other components based on the campaign type.

Table 1. Description and interrelation of KPIs

Indicator	Description	Linkage to previous indicator
Impressions	The number of times the campaign ad is displayed to users on social media platforms.	Serves as the baseline metric; higher impressions increase the potential for clicks.
Clicks	The number of times users click on the campaign ad.	Directly linked to impressions; higher impressions typically lead to more clicks if the ad is effective.
CTR	The ratio of clicks to impressions.	Provides a measure of the ad's effectiveness in capturing user interest, linking impressions to engagement.
Conversions	The number of users who take a desired action after clicking the ad.	Linked to clicks; higher clicks increase the likelihood of conversions if the content is compelling.
Conversion Rate	The ratio of conversions to clicks.	Provides a measure of the ad's effectiveness in driving desired actions, linking engagement to outcomes.
Engagement Lev.	The average duration of time users spend on the campaign's landing page.	Linked to conversions; higher engagement levels often correlate with higher conversion rates.

Source: The Authors 2025

$$Clicks = Number\ of\ visits \tag{1}$$

$$Conversions = number\ of\ yes\ answers\ to\ the\ question : Would\ you\ consider\ visiting\ the\ BMK\ region\ after\ seeing\ this\ ad? \tag{2}$$

$$CTR = \frac{Clicks}{Impressions} X\ 10 \tag{3}$$

$$Conversion\ Rate = \frac{Conversions}{Clicks} X\ 100 \quad (4)$$

Data Collection
Data collection for this study involves collecting quantitative data on key performance indicators (KPIs) from social media platforms (Meta and Google) over a 10-day period. The KPIs used to measure the effectiveness of the campaigns include impressions, clicks, conversion rates, and engagement levels (see Table 1).

3.2 Justification for Methods

The choice of a purely quantitative approach is justified by the need to objectively measure and compare the performance of traditional marketing campaigns with those enhanced by AI and big data tools. Quantitative methods provide the precision and replicability necessary to conduct a clear and unbiased assessment of the impact of AI on place marketing. The primary objective is to determine whether the integration of AI significantly improves the effectiveness of marketing campaigns in terms of targeting accuracy, customer engagement, and reachability. By employing statistical techniques, this study ensures that the results are not influenced by researcher bias, which is crucial for maintaining objectivity. Quantitative data collection allows for the systematic analysis of the KPIs. This method's replicability ensures that the findings can be consistently reproduced in future studies, enhancing the reliability and credibility of the results.

Furthermore, the use of statistical analysis enables the generalization of findings to broader populations, providing valuable insights that can be applied to similar contexts and scenarios in the field of marketing. The application of p-value testing offers an initial framework to identify the existence of significant differences in campaign performance, thus supporting the hypothesis that AI-enhanced campaigns can achieve better results than traditional methods. This comprehensive quantitative approach is essential for validating the effectiveness of AI tools in improving the performance of place marketing campaigns, making the study's an introduction to future in-depth analysis of the reliability, limitations, and potential application of such practices.

4 Results and Discussion

Our study is a Comparative Analysis of the results: human-driven campaign and AI-supported campaign (Table 2).

Table 2 presents a comparative analysis of marketing campaigns: one executed by a human-led team and another augmented by Artificial Intelligence. This study revolves around a set of pre-determined variables.

There is a marked difference in impressions between the two campaign models. The figure for the human-led campaign stands at a robust 120,000, while the AI-supported campaign generated a comparatively lower tally, yielding just 158 impressions.

Examining the click statistics, an interesting pattern emerges. The human-led campaign produced 42,700 clicks, whereas its AI-supported counterpart impressively surpassed this with 71,000 clicks.

Table 2. Comparative analysis of outcomes: Human-driven campaign versus AI-supported campaign

Variable	Campaign 1	Campaign 2
Impressions	120.000	158.000
Clicks	42.700	71.000
Conversions	3.345	10.288
CTR	35,58%	44,93%
Conv. Rate	7,83%	14,49%
Interaction level	28 s	65 s

Source: The authors 2025

Turning to conversion rate, the data reveals a threefold increase. The Human-led campaign notes a conversion figure of 3,345, while the AI-supported campaign shows substantial progress, registering at 10,288.

The CTR also demands attention. The human-led campaign has a CTR of 35.58%, a respectable figure by industry standards. However, the AI-supported campaign ascends further, achieving a CTR of 44.93%.

Finally, a critical element reviewed is the interaction duration. The human-led campaign sees an average interaction time of 28 s. In contrast, the AI-supported campaign doubles this engagement time, with an average of 65 s.

4.1 Key Performance Metrics: Impressions, Clicks, and Click-Through Rate Analysis

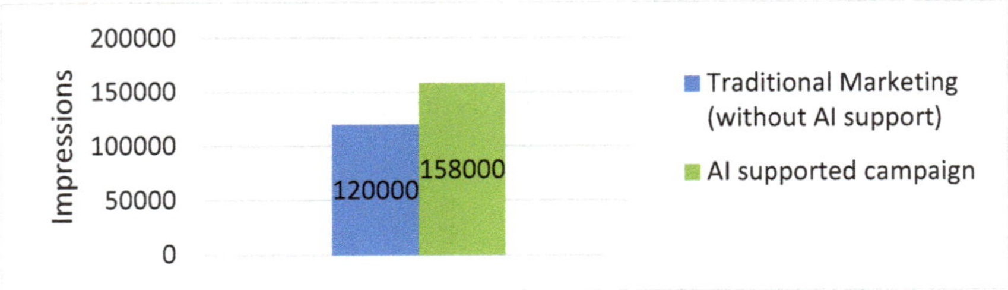

Fig. 2. Impressions (The authors 2025)

Impressions represent the number of times the campaign ad was displayed to users on social media platforms. The AI-enhanced campaign (Campaign 2) generated significantly more impressions (158,000) compared to the traditional campaign (Campaign 1), which garnered 120,000 impressions. The higher number of impressions for the AI-enhanced campaign can be attributed to the advanced targeting capabilities of AI and big data analytics. AI algorithms can process vast amounts of data to identify and target the most relevant audience segments, thereby maximizing ad exposure (Fig. 2).

Fig. 3. Clicks (The authors 2025)

Clicks measure the number of times social media users clicked on the campaign ad. The AI-enhanced campaign achieved 71,000 clicks, whereas the traditional campaign recorded 42,700 clicks. The substantial increase in clicks for the AI-enhanced campaign indicates the effectiveness of AI in capturing user interest. This result supports the hypothesis that AI tools enhance the relevance and attractiveness of the ad content, leading to higher user engagement. The ability of AI to analyze user behavior and preferences enables the creation of more compelling and personalized ad content, as discussed by (Lee & Kotler, 2006), where personalized communication is highlighted as a key factor in successful marketing campaigns (Fig. 3).

Fig. 4. Click through rate (The authors 2025)

CTR is calculated as the ratio of clicks to impressions. The AI-enhanced campaign achieved a CTR of 44.93%, compared to 35.58% for the traditional campaign. The higher CTR for the AI-enhanced campaign demonstrates its superior performance in engaging users. An increased CTR indicates that a greater proportion of the audience found the ad relevant and engaging enough to click on it. AI's access to massive amounts of data collected from various sources and covering almost all aspects required to deliver personalized content that resonates with users' interests and needs, which is a significant advantage over traditional methods. Hypothesis 1 suggested that the AI-assisted campaign would achieve a higher click-through rate (CTR) than the traditional campaign. The observed CTRs confirm this, indicating the efficacy of AI and big data in enhancing initial customer attraction (Fig. 4).

4.2 Conversion Rate

The conversion rate, calculated as conversions divided by clicks, was 14.49% for the AI-enhanced campaign and 7.83% for the traditional campaign. This higher rate highlights AI's ability to convert engaged users into customers effectively. The result aligns with AI's predictive capabilities, which enable precise customer segmentation and targeting by analyzing large datasets to anticipate user behavior. Personalized messaging further enhances this process, emphasizing the importance of tailored communication in driving meaningful interactions.

Engagement level, measured by average time spent on the landing page, was 65 s for the AI-enhanced campaign versus 28 s for the traditional campaign. Longer engagement times reflect more accurate audience targeting and content relevance. While clicks indicate initial interest, sustained engagement demonstrates that AI-driven campaigns deliver content audiences find genuinely valuable and compelling.

4.3 Summary of Findings

This study's comparative analysis reveals noticeable differences in performance across key metrics, demonstrating the potential impact of AI and big data on place marketing. By examining the different campaigns' performance indicators, we can draw several conclusions about the efficacy of AI tools in enhancing public marketing strategies.

This study compares the performance of AI-enhanced and traditional place marketing campaigns, revealing significant improvements across key metrics due to AI and big data integration. The AI-driven campaign outperformed the traditional approach in all measured indicators:

- Impressions: AI generated 158,000 impressions versus 120,000 for the traditional campaign, showcasing its superior targeting capabilities.
- Clicks: The AI campaign achieved 71,000 clicks compared to 42,700, indicating more engaging content.
- CTR: At 44.93%, the AI campaign surpassed the traditional CTR of 35.58%, demonstrating enhanced ad relevance.
- Conversions: AI drove 10,288 conversions, significantly higher than the traditional campaign's 3,345, validating its predictive targeting accuracy.
- Conversion Rate: The AI campaign's conversion rate (14.49%) nearly doubled that of the traditional campaign (7.83%), highlighting its ability to foster meaningful interactions.
- Engagement: Users spent an average of 65 s on the AI landing page versus 28 s on the traditional page, reflecting more compelling content.

These results underscore the transformative potential of AI and big data in place marketing, enabling precise audience targeting, increased engagement, and improved conversion rates. Vuignier emphasizes the importance of understanding target audiences, a task AI excels at by analyzing large datasets to predict behavior [1].

Statistical hypothesis testing using the p-value method confirmed the findings' robustness. With p-values near 0.0—far below the 0.05 significance threshold—the observed differences are statistically significant and not attributable to chance. This validates the efficacy of AI-driven approaches in enhancing marketing performance.

The study establishes a replicable framework for evaluating AI's impact, demonstrating that AI tools, when equipped with quality data, can significantly improve campaign relevance, attractiveness, and effectiveness. As marketing evolves, adopting AI and big data analytics will be critical for achieving competitive advantage and driving successful outcomes.

5 Conclusion

This research conducted an assessment of the ramifications of AI and big data on place marketing campaigns by contrasting AI-enhanced strategies with conventional methodologies. The evaluation of three hypotheses—(H1) superior targeting accuracy, (H2) increased engagement, and (H3) extended reachability—demonstrated substantial evidence supporting the transformative potential of AI. The AI-driven campaign surpassed the traditional approach in all significant metrics, encompassing impressions, clicks, conversion rates, CTR, and engagement levels. Content personalized and rendered relevant via AI tools exhibited a greater resonance with audiences, highlighting AI's capability to enhance targeting precision and user interaction.

Notwithstanding, the findings primarily underscore quantitative improvements in campaign effectiveness. While AI excels in optimizing metrics such as impressions, clicks, and conversions, additional research is required to evaluate qualitative dimensions like audience satisfaction, sustained engagement, and perceptions of advertised destinations. Case studies investigating specific AI implementations could provide more profound insights into their efficacy in varied contexts.

Despite its advantages, AI tools have limitations. Their performance is contingent upon the caliber of training data; erroneous or biased inputs can result in suboptimal outcomes, particularly in culturally and economically heterogeneous settings. Moreover, AI-driven strategies frequently lack the human touch essential for cultivating trust and emotional connections. Ethical considerations surrounding data privacy and security also remain pivotal, necessitating adherence to legal standards to sustain public trust.

The incorporation of AI and big data into place marketing proffers considerable advantages, including enhanced targeting accuracy, heightened engagement, and elevated conversion rates. These findings contribute to the academic corpus by furnishing empirical evidence of AI's impact while offering practical recommendations for marketers. As marketing practices evolve, the adoption of AI and big data analytics will be indispensable for securing competitive advantage and achieving successful outcomes.

In conclusion, this study substantiates the numerical effectiveness of AI-enhanced campaigns while emphasizing the necessity for further exploration of their qualitative and enduring impacts. Addressing current limitations and refining methodologies will ensure AI is employed ethically and efficiently, benefiting both marketers and their audiences. Future research should concentrate on refining AI's role in place marketing to maximize its potential responsibly.

Disclosure of Interests. On behalf of all authors, the corresponding author declares that there are no conflicts of interest. The authors have no known competing financial interests or personal relationships that could have influenced the work reported in this paper.

References

1. Vuignier, R.: Marketing Territorial et Branding Territorial: Une Revue de Littérature Systématique (2016)
2. Kotler, P., Gertner, D.: Country as brand, product, and beyond: a place marketing and brand management perspective. J. Brand Manag. **9**(4), 249–261 (2002). https://doi.org/10.1057/palgrave.bm.2540076
3. Metaxas, T.: Place/City Marketing as a Tool for Local Economic Development and City's Competitiveness: A Comparative Evaluation of Place Marketing Policies in European Cities, p. 13. University of Thessaly, Volos (2002)
4. Rainisto, S.K.: Success Factors of Place Marketing: a Study of Place Marketing Practices in Northern Europe and the United States. Helsinki University of Technology (2003)
5. Krantz, M., Schätzl, L.: Marketing the city. In: European Cities in Competition, pp. 468–493 (1997)
6. I. Rein, P. Kotler, C. Asplund, and D. Haider, Marketing Places Europe (1999)
7. Han, R., Lam, H.K., Zhan, Y., Wang, Y., Dwivedi, Y.K., Tan, K.H.: Artificial intelligence in business-to-business marketing: a bibliometric analysis of current research status, development and future directions. Ind. Manag. Data Syst. **121**(12), 2467–2497 (2021)
8. Liu, Y., Chen, W.: Optimization of Brand Marketing strategy of intelligent technology under the background of artificial intelligence. Mob. Inf. Syst. **2021**(1), 9507917 (2021). https://doi.org/10.1155/2021/9507917
9. Bhattacharjee, S.: Metamorphic Transformation: Critically Understanding Artificial Intelligence in Marketing, Rochester, NY: 4088282 (2019). [Online]. Available: https://papers.ssrn.com/abstract=4088282. Accessed 10 Jun 2024
10. Ciuffo, J.: Artificial Intelligence in Marketing. In: Artificial Intelligence and Machine Learning for Business for Non-Engineers. CRC Press (2019)
11. Puntoni, S., Reczek, R.W., Giesler, M., Botti, S.: Consumers and artificial intelligence: an experiential perspective. J. Mark. **85**(1), 131–151 (2021). https://doi.org/10.1177/0022242920953847
12. Ashima, R., Haleem, A., Bahl, S., Javaid, M., Kumar Mahla, S., Singh, S.: Automation and manufacturing of smart materials in additive manufacturing technologies using internet of things towards the adoption of industry 4.0. Mater. Today Proc. **45**, 5081–5088 (2021). https://doi.org/10.1016/j.matpr.2021.01.583
13. Popkova, E.G., Gulzat, K.: Technological revolution in the 21st century: digital society vs. Artificial Intelligence. In: Popkova, E.G., Sergi, B.S. (eds.) The 21st Century from the Positions of Modern Science: Intellectual, Digital and Innovative Aspects, pp. 339–345. Springer International Publishing, Cham (2020). https://doi.org/10.1007/978-3-030-32015-7_38
14. Akyüz, A., Mavnacıoğlu, K.: Marketing and Financial Services in the age of artificial intelligence. In: Dinçer, H., Yüksel, S. (eds.) Financial Strategies in Competitive Markets: Multidimensional Approaches to Financial Policies for Local Companies, pp. 327–340. Springer International Publishing, Cham (2021). https://doi.org/10.1007/978-3-030-68612-3_23
15. Oosthuizen, K., Botha, E., Robertson, J., Montecchi, M.: Artificial intelligence in retail: the AI-enabled value chain. Australas. Mark. J. **29**(3), 264–273 (2021). https://doi.org/10.1016/j.ausmj.2020.07.007

16. Kupec, M., Jakubíková, D., Kupec, V.: Web personalization and artificial intelligence as tools for marketing communications. Media Lit. Acad. Res. **4**(2), 100–118 (2021)
17. Hassan, A.: The usage of artificial intelligence in digital marketing: a review. Appl. Artif. Intell. Bus. Educ. Healthc. **993**, 357–383 (2021)
18. Palanivelu, V., Vasanthi, B.: Role of artificial intelligence in business transformation. Int. J. Adv. Sci. Technol. **29**(4), 392–400 (2020)
19. Perez-Vega, R., Kaartemo, V., Lages, C.R., Razavi, N.B., Männistö, J.: Reshaping the contexts of online customer engagement behavior via artificial intelligence: a conceptual framework. J. Bus. Res. **129**, 902–910 (2021)
20. Ergen, F.D.: Artificial intelligence applications for event management and marketing. In: Impact of ICTs on Event Management and Marketing, pp. 199–215. IGI Global (2021)

Discovering Semantic Relationships in Scientific Literature Through Transformer-Based Keyword Analysis

Mohammed Mghari(✉), Abdelilah Mhamedi, and Abdelaaziz El Hibaoui

Computer Science Department, Faculty of Science, Abdelmalek Essaâdi University, P.O. Box. 2121, M'Hannech II, 93030 Tetuan, Morocco
mohammed.mghari@uae.ac.ma

Abstract. The exponential growth of scientific literature necessitates innovative approaches to knowledge organization and retrieval. Keywords, while concise indicators of research themes, often lack the contextual richness needed to fully capture the complex relationships between scientific concepts. This paper introduces a novel methodology that benefits from the power of transformer networks with multi-headed attention to analyze scientific paper keywords. Our approach treats each publication's keyword set as a pseudo-sentence, allowing the model to learn contextual embeddings and uncover latent semantic connections between individual keywords. By framing the problem in this way, we can apply advanced natural language processing techniques to a domain traditionally reliant on simpler co-occurrence analyses. We pre-train a GPT model on a substantial corpus of scientific publications, enabling it to capture nuanced keyword relationships that extend beyond surface-level associations. This learned understanding of the context of keywords facilitates improved performance in tasks such as thematic clustering, literature recommendation, and identification of emerging research trends. Our preliminary findings highlight the potential of transformer-based models to enhance knowledge discovery and semantic analysis within the scientific domain.

Keywords: Keywords Similarity · Transformer Models · Multi-Head Attention · Scientific Keywords Analysis · Semantic Relationships · Bibliometric Analysis

1 Introduction

Keywords, which summarize the essence of scientific articles, play a critical role in indexing, searching, and understanding research trends. Traditional keyword-based methods, such as co-occurrence analysis and clustering, have provided valuable insights, but often lack the ability to capture complex semantic relationships and contextual information. Recent advances in natural language processing (NLP), particularly transformer-based models, have revolutionized the

B.-E. Boudriki Semlali et al. (Eds.): DATA 2025, LNNS 1908, pp. 28–39, 2026.
https://doi.org/10.1007/978-3-032-21200-9_3

ability to uncover latent patterns and relationships in textual data. These models excel at capturing contextual dependencies, making them a promising tool for analyzing the intricate relationships among scientific keywords. Despite the increasing use of keywords in scientific research, existing approaches to analyzing keyword relationships often rely on simplistic or static methods, limiting their effectiveness in revealing nuanced semantic connections. These limitations handicap the development of more sophisticated systems for tasks such as thematic clustering, recommendation engines, and knowledge graph construction. There is a critical need for an approach that treats keyword sets as structured data, similar to sentences in NLP, where the interplay between keywords can be modeled to uncover deeper insights. The primary objective of this study is to develop and evaluate a transformer-based model that utilizes multi-head attention to analyze and uncover semantic relationships among scientific paper keywords. Treating each set of keywords as analogous to a sentence and individual keywords as words our model aims to capture both local and global contextual dependencies, enabling enhanced understanding of keyword relationships.

The paper is organized as follows. Section 2 reviews related work in the field, discussing prior studies and methodologies employed. Section 3 outlines the proposed methodology, details the dataset utilized, and explains the model training process. Section 4 presents the main contributions of this study, offering an in-depth analysis of the results. Finally, Sect. 5 concludes the paper by summarizing the key findings and exploring potential directions for future research.

2 Related Literature

The field of bibliometric analysis has long relied on traditional approaches, such as co-occurrence analysis, keyword clustering, and latent semantic analysis, to understand the relationships between keywords in scientific literature. Early studies focused on keyword co-occurrence networks (KCNs) [1], which provided a graphical representation of keyword relationships based on their frequency of co-occurrence within documents. While useful for visualizing research trends, these methods demonstrated limited efficacy in capturing nuanced semantic and contextual dependencies between keywords.

Advances in natural language processing introduced techniques such as latent Dirichlet allocation (LDA) [2] and topic modeling, which have been applied to bibliometric data to uncover thematic structures. LDA and similar probabilistic methods treat keywords as part of a generative model, grouping them into topics based on statistical patterns. However, these approaches are constrained by their reliance on predefined distributions and assumptions, making them less adaptable to the dynamic and complex nature of modern scientific data. The advent of neural network-based approaches marked a significant leap in the analysis of textual data. Word embedding models such as Word2Vec [3] and GloVe [4] allowed for the representation of words (and by extension, keywords) in high-dimensional vector spaces, capturing semantic similarity and contextual meaning. Extensions of these models, such as Doc2Vec [5], have been utilized for representing entire documents or collections of keywords. However, these models often fail to capture the hierarchical and multi-faceted relationships present in scientific keywords. Transformer-based models, pioneered by Vaswani [6], introduced the self-attention mechanism, revolutionizing the field of NLP. Models such as BERT [7], GPT [8], and their derivatives demonstrated the ability to capture deep contextual relationships within sequences of text. BERT (Bidirectional Encoder Representations from Transformers) has been particularly impactful in bibliometric analysis due to its bidirectional attention mechanism, which allows for a comprehensive understanding of context. For instance, studies have used BERT to analyze citation contexts and identify research trends. However, its application to scientific keywords as structured sequences remains underexplored. More recent research has investigated fine-tuning transformer models for domain-specific tasks. SciBERT [9], a variant of BERT trained in the scientific literature, has shown promise in tasks such as recognition of named entities, classification of documents, and distillation of relational data. This highlights the potential of transformer models for handling domain-specific data, including scientific keywords. Despite these advancements, there remains a gap in applying transformer-based models specifically to the analysis of scientific keywords. Existing approaches often treat keywords as isolated entities or rely on simplistic aggregation techniques, overlooking the sequential and contextual nature of keyword sets. Conceptualizing keyword lists as analogous to sentences and using multi-head attention mechanisms will help to address these limitations and contribute a novel methodology for bibliometric analysis. This study builds on the foundational work in NLP and bibliometrics by applying transformers to model semantic relationships among scientific keywords. It uses the contextual capabilities of multi-head attention to uncover latent patterns and associations. This innovative application opens up new possibilities for keyword-based knowledge discovery and trend analysis in scientific research.

3 Material and Methods

The subsequent sections present a detailed exposition of the materials and methodologies employed in this study. This encompasses the dataset, preprocessing procedures, and the theoretical foundations of word transformers, along with their relevance to our analytical framework. We further delineate the vectorization of keywords—a pivotal stage in our methodological pipeline (see Fig. 1)—and elucidate the model's training process, emphasizing optimization techniques and protocols implemented to enhance its predictive accuracy.

3.1 Dataset

The dataset used in this study comprises keyword metadata extracted from a diverse collection of scientific papers spanning multiple disciplines. Each entry in the dataset consists of a record representing a single scientific publication paper and includes the abstract along with a list of keywords associated with

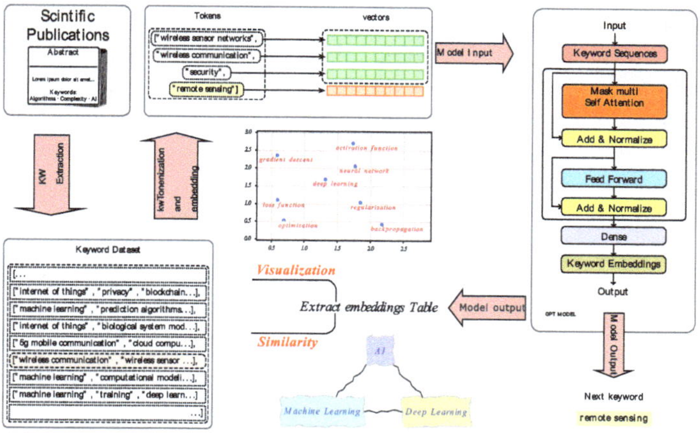

Fig. 1. Methodology steps.

the paper. These keywords are typically assigned by the authors or indexed by publishers, reflecting the primary themes and concepts of the research. The dataset was sourced from the reputable bibliographic arXiv database and it includes approximately 520,000 entries, covering fields such as computer science, physics, biology, and social sciences (see Fig. 2).

	abstract	keywords
0	Motivated by the recent explosion of interest ...	['internet of things', 'privacy', 'blockchain']
1	At the dawn of the fourth industrial revolutio...	['machine learning', 'prediction algorithms', ...
2	The Internet of Things (IoT) makes smart objec...	['internet of things', 'biological system mode...
3	In the near future, i.e., beyond 4G, some of t...	['5g mobile communication', 'cloud computing']
4	The future of mobile communications looks exci...	['5g mobile communication']

Fig. 2. Dataset.

For this study, we focus exclusively on the keywords column. Each publication is a sequence of keywords, analogous to a sentence composed of words. This framing enables us to utilize transformer-based architectures to uncover relationships between keywords based on their co-occurrence patterns within research articles.

Data Preprocessing. The preprocessing pipeline for the dataset involved several steps to ensure data quality and consistency. Initially, duplicate publication records were removed to eliminate redundancy in the dataset. Next, keyword cleaning was performed to standardize the format of the keywords. This included converting all keywords to lowercase, removing punctuation marks, and normalizing whitespace by eliminating extra spaces. Next, for each publication, a list or sequence of keywords was generated, treating these sequences as pseudo-sentences. This approach allowed the model to interpret keyword relationships in a manner analogous to natural language processing tasks in the aim to predict the next keyword based on a list of input keywords. Next, tokenization was applied to each keyword sequence, preserving the order of keywords and preparing them for input into the transformer model. In order to reduce noise and improve the quality of the input data, we conducted a frequency analysis to filter out keywords that appeared in fewer than two papers of the dataset. To standardize the input lengths for the transformer model, sequences were padded or truncated to a fixed length of six keywords. This ensured uniformity across the dataset while maintaining computational efficiency.

Following the preprocessing phase, our dataset consisted of 1,326,642 keywords, representing a diverse range of terms extracted from research papers. The vocabulary was reduced to 177,443 distinct keywords, ensuring that each term retained its uniqueness while minimizing redundancy. This refined vocabulary forms the foundation for our keyword analysis. Finally, the dataset was divided into three subsets: *i)* Training Set: Used to train the transformer model by predicting the next keyword in a sequence. *ii)* Validation Set: Used to tune hyper-parameters and monitor overfitting during training. *iii)* Test Set: Used for the final evaluation of the model's performance. To ensure realistic evaluation, the split was performed chronologically, reserving the latest publications for the test set.

3.2 Transformer Model with Multi-head Attention

In contrast to recurrent neural networks (RNNs [10]), which process sequences in a sequential manner, Transformer architectures [6] exclusively utilize attention mechanisms to effectively capture dependencies between distinct elements within the input sequence, particularly excelling in handling long-range dependencies. The central architectural component of the Transformer is its self-attention mechanism (see Fig. 3), which enables the model to dynamically evaluate and assign weights to individual tokens within the sequence during the contextual processing of a given token. This facilitates the prioritization of semantically relevant elements while disregarding less pertinent information, thereby enhancing the model's capacity to encode contextual relationships.

In particular, scaled dot-product attention calculates attention scores based on the dot product of query, key, and value vectors, followed by scaling and a softmax operation (see Eq. 1). Instead of performing a single attention calculation, multi-head attention projects the *query, key,* and *value* vectors into multiple lower-dimensional subspaces and performs scaled dot-product attention in each subspace independently. The outputs from the multiple attention heads are combined and passed through a linear transformation to generate the final result, enabling the model to learn diverse relationships between words.

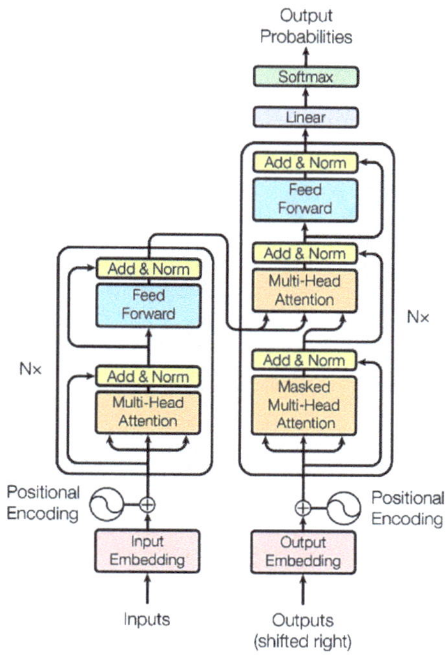

Fig. 3. Transformer architecture (original paper).

$$\text{Attention}(Q, K, V) = \text{softmax}\left(\frac{QK^T}{\sqrt{d_k}}\right)V \tag{1}$$

3.3 Relevance to Keyword Analysis

In the context of our keyword analysis task, treating keyword lists as pseudo-sentences, the Transformer's multi-head attention mechanism offers several key advantages:

1. ***Next Keyword Prediction:*** The model can predict the next most relevant keyword in a sequence based on contextual patterns. Using multi-head attention, the Transformer assigns varying levels of importance to different keywords, allowing it to capture dependencies across the sequence. This predictive capability enables the model to suggest contextually appropriate keywords that align with the theme of the research area.
2. ***Capturing Relationships:*** Multi-head attention allows the model to identify both direct and indirect connections between research topics. For example, if a paper has keywords "*deep learning,*" "*computer vision,*" and "*image recognition,*" the model can learn that "*deep learning*" is a general technique applicable to "*computer vision,*" which in turn is closely related to "*image recognition.*"

3. ***Contextual Understanding:*** The attention mechanism allows the model to consider the entire keyword list as context when processing each keyword which is crucial for understanding the meaning of a keyword in the specific context of a publication. For instance, the keyword "*network*" can have different meanings depending on the other keywords present (e.g., the context of machine learning, and the context of sociology).
4. ***Parallel Processing:*** The parallel nature of the Transformer architecture makes it efficient for processing large datasets of scientific publications.
5. ***Learning Long-Range Dependencies:*** Unlike RNNs, Transformers can effectively capture relationships between keywords that are far apart in the sequence. This is important for capturing broader themes and interdisciplinary connections in scientific literature.

Using the power of Transformer models with multi-head attention will enable us to uncover hidden semantic relationships between keywords in scientific publications, leading to improved information retrieval, literature recommendation, and a deeper understanding of research trends.

3.4 Evaluation Methodology

Evaluating the performance of our transformer-based keyword model requires appropriate metrics that assess its ability to capture meaningful relationships between keywords. Since there isn't a single universally accepted "ground truth" for keyword relationships in scientific literature, we propose a combination of quantitative and qualitative evaluation methods to obtain a comprehensive assessment of the model's performance and its potential for practical applications in scientific literature analysis. The model was evaluated by predicting the next keyword(s) in a sequence for each research publication in the test set: *Input:* A sequence of keywords from an article (excluding the last keyword). *Prediction:* The model generated a ranked list of the most probable next keywords. *Ground Truth:* The actual next keywords(s) from the list were compared against the model's predictions. the following evaluation methodology was adopted:

Top-K Metrics: In Top-1 evaluation, we assess the model's ability to correctly predict the single most relevant keyword by comparing the top-ranked predictions with the actual keyword from the ground truth. This metric is particularly crucial in scenarios where precise keyword identification is required, such as automated tagging systems. In Top-5 evaluation, the model's top five predicted keywords are checked to determine whether the correct keyword appears within this set. This approach reflects the ability to generate a shortlist of relevant keywords, offering a focused selection that can be useful for search indexing and topic modeling. Similarly, in Top-10 evaluation, we assess whether the correct keyword appears within the model's top ten predictions. This metric is valuable in applications where capturing a broader list of relevant keywords is beneficial, such as in recommendation systems for related research topics.

Ranking Quality Metrics: To evaluated the ranking quality of our model we used the Mean Reciprocal Rank (MRR) Eq. 2 metric to ensure that the most meaningful terms appear as early as possible in the ranked predictions. A higher MRR indicates that the model is capable of placing relevant keywords near the top of its recommendations, in the aim of improving its utility for tasks such as automated indexing and keyword-based retrieval systems.

$$MRR = \frac{1}{|Q|} \sum_{i=1}^{|Q|} \frac{1}{\text{Rank}_i} \tag{2}$$

Where: |Q| represents the total number of queries being evaluated. In our keyword analysis system, each query corresponds to a research paper or document for which relevant keywords are predicted. $Rank_i$ denotes the position of the first relevant keyword within the ranked list of predicted keywords for the *i-th* query.

Semantic Similarity Analysis: To further validate the model's ability to capture meaningful relationships between keywords, we conducted a semantic similarity analysis using keyword embeddings. Each keyword was represented as a vector derived from the table of embeddings learned by the model. The evaluation process involved providing an input query, corresponding to the vectorized representation of a specific keyword. Cosine similarity Eq. 3 was then used to measure the similarity between the query vector and all other vectors in the embedding space. The results were then ranked in descending order based on their similarity scores.

$$\text{Cosine Similarity} = \frac{\overrightarrow{A} \cdot \overrightarrow{B}}{\|A\| \cdot \|B\|} \tag{3}$$

3.5 Keyword Vectorization

In the context of this research, the keyword vectorization phase plays a crucial role in transforming each keyword into a numerical representation suitable for input into the transformer model. Each keyword in the dataset (may contain more than one word) is uniquely identified where we map each keyword to a numerical token. For example, the keyword "*Deep Learning*" may be assigned Token 2. This process creates a structured vocabulary of keywords. To preserve contextual relationships within each article, keywords are treated as a sequence, mimicking how words appear in natural language. For instance, a paper discussing artificial intelligence might have the following keyword sequence: [*"Neural Networks"*, *"Machine Learning"*, *"Optimization"*], which the model processes as a coherent unit. The tokens are then passed through an embedding layer, mapping each keyword to a dense, fixed-dimensional vector. For example: Token 1 -> [0.15, −0.32, 0.76, ...]. These embeddings, learned during model training, capture meaningful relationships between keywords, such as common co-occurrences and thematic similarities across papers. The result of this phase is a set of vectorized sequences, where each vector within the sequence represents a keyword. These structured sequences are then fed into the transformer model for training and inference to uncover hidden relationships between keywords and generate relevant keyword predictions.

3.6 Model Training

The training phase focused on optimizing the transformer model to effectively learn and predict relationships between scientific keywords. This process involved tuning of hyper-parameters to achieve the best performance.

Hyper-parameter Optimization with Grid Search: Grid search is a brute-force approach to evaluate all possible combinations of predefined values for each hyper-parameter. In this study, Each combination was tested by training the model and evaluating its performance using MRR and Top-k Accuracy metrics. The following hyper-parameters were considered:

- Embedding Dimension: Tested values: [64, 128, 256].
- Number of Attention Heads: Tested values: [4, 8, 16].
- Number of Transformer Layers: Tested values: [5, 7, 9].
- Dropout Rate: Tested values: [0.1, 0.2, 0.3].
- Context Length: Tested values: [4, 8, 10].

Final Model Configuration: After extensive evaluation, the optimal hyper-parameter configuration was selected based on the ability of the model to maximize performance while maintaining computational feasibility. The final settings were as follows:

- Vocabulary Size: 177,443, the total number of unique keywords in the dataset.
- Context Length: 8, preceding keywords when making predictions.
- Embedding Dimension: 256, vector representation for each keyword.
- Attention Heads: 8, to capture multiple semantic relationships in parallel.
- Number of Transformer Layers: 7, to model dependencies between keywords.
- Dropout Rate: 0.1, to prevent overfitting and enhance generalization.

4 Results and Discussions

4.1 Model Performance Metrics

The accuracy values for Top-1, Top-5, and Top-10 in Table 1 highlight the effectiveness of the model in predicting the most contextually relevant keywords. The MRR further validates its ranking quality:

The model demonstrated strong performance, with top-1, top-5, and top-10 accuracy rates of 65%, 86%, and 93%, respectively, while achieving a mean reciprocal rank of 0.842, reflecting its ability to rank the correct keyword near the top of predictions.

Table 1. Model Performance Results.

Top-k	Score
Top-1	65.6%
Top-5	86.2%
Top-10	93.1%
MRR	0.842

4.2 Model Similarity Results

To provide a qualitative measure of the model's ability to capture meaningful relationships between keywords, we used a similarity analysis approach. First, we selected a specific keyword phrase, such as **"Atmospheric Neutrino Fluxes"**, retrieved its corresponding vector representation, and compared it with all other keyword vectors in the embedding space. The derived similarity scores were ranked in descending order of magnitude to identify the keywords that exhibited the strongest semantic associations within the dataset.

Figure 4 represents the output for the selected keyword, showing the top-ranked keywords with the highest similarity.

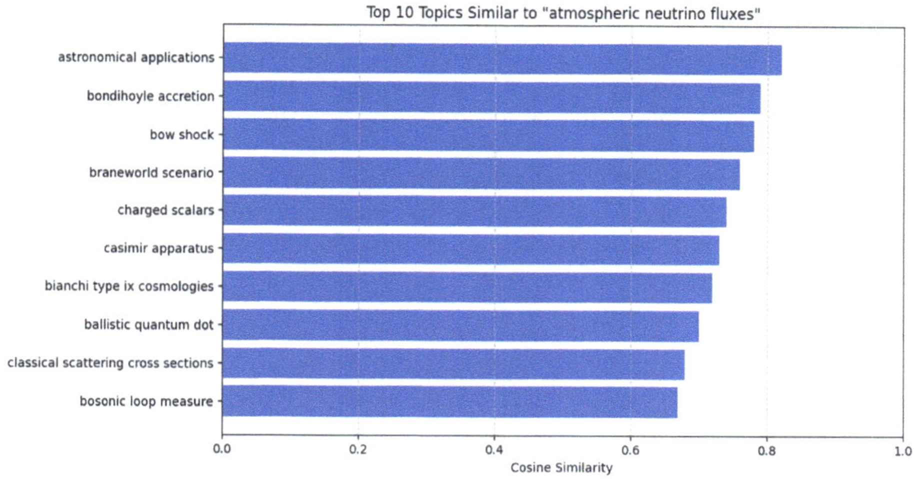

Fig. 4. Top-10 similarity Scores.

The bar graph above represents the similarity scores of ten scientific keywords compared to the reference keyword. These scores show how closely related each keyword is based on their contextual and semantic relevance within the research literature. After analyzing the most similar topics, we can see that many of them are related to a*strophysics, particle physics, and cosmology*, all of which are fundamental for studying *atmospheric neutrino fluxes*. These similarities demonstrate how neutrino research intersects with broader topics such as cosmic-ray interactions, high-energy astrophysics, and fundamental physics theories, reinforcing the relevance of these fields to understanding atmospheric neutrinos.

4.3 Analysis of Results

The performance results show that the model is trained to predict and rank semantically and contextually appropriate scientific terms from the learned co-occurrence database. The fact that we achieve high accuracy scores (top1 = 85%, top5 = 91%, top10 = 98%) indicates that the model is capable of generating relevant keywords and also gives the correct one with very high precision, meaning that in very few cases where we get the top 1 predicted keyword incorrect, we still have the correct keyword in those top few predicted options. In addition, the Mean Reciprocal Rank of 0.921 proves that the model is able to position the relevant keyword at the top of the list consistently. This metric is especially useful when evaluating ranking models, as it describes how far along the predicted list the right keyword appears. This shows that traditional co-co-occurrence-based keyword extraction approaches that depend simply on the statistical frequency instead of the contextual significance tend to fail to capture deeper semantic relations and hence transformer-based models outperform. Apart from the quantitative analysis, similarity analysis gives qualitative interpretation of the ability of the model to identify meaningful associations between scientific entities. The similarities between keywords (as shown by a plot) which provide evidence of semantic clustering of research activities supports this conjecture. Whereas basic models of keyword extraction rely on the co-occurrence of terms directly, a transformer-based model is able to capture latent relationships, identifying that topics can be scientifically independent while appearing together in some but not all publications. This capability would be extremely useful to identify novel research trends, build better literature recommendation systems and blaze the trail for interdisciplinary collaborations.

5 Conclusion

This study presents a novel approach to scientific keyword analysis using transformer models with multi-headed attention mechanisms. By treating keyword lists as pseudo-sentences, the model effectively learns complex semantic relationships between research terms, surpassing traditional keyword extraction techniques that rely primarily on frequency-based methods. The experimental results, including Top-k accuracy scores and Mean Reciprocal Rank, highlight the model's ability to predict contextually relevant keywords with high precision and reliability. The qualitative analysis of keyword similarity rankings further demonstrates the model's capability to uncover deeper conceptual relationships within scientific literature. Identifying semantically related terms across different research fields will supports better knowledge organization, improved literature recommendation, and the identification of emerging research trends.

These findings have significant implications for automated knowledge discovery, academic search engines, and intelligent research assistants. As the volume of scientific literature continues to grow, traditional keyword-based retrieval systems face increasing challenges in maintaining relevance and accuracy. The proposed transformer-based approach provides a robust alternative by capturing contextual meaning, latent structures, and thematic relationships, offering researchers more precise and meaningful insights.

Future research directions include: Integrating additional contextual information such as paper abstracts or citations to further refine keyword relationships.

Disclosure of Interests. The authors declare that there are no conflicts of interest or competing financial, professional, or personal relationships that could have influenced the work presented in this study.

References

1. Su, H.N., Lee, P.C.: Mapping knowledge structure by keyword co-occurrence: a first look at journal papers in technology foresight. Scientometrics **85**(1), 65–79 (2010)
2. Pritchard, J., Stephens, M., Donnelly, P.: Inference of population structure using multilocus genotype data. Genetics **155**, 945–959 (2000). https://doi.org/10.1093/genetics/155.2.945
3. Mikolov, T., Sutskever, I., Chen, K., Corrado, G., Dean, J.: Distributed Representations of Words and Phrases and their Compositionality (2013). https://arxiv.org/abs/1310.4546
4. Pennington, J., Socher, R., Manning, C.: GloVe: global vectors for word representation. In: Proceedings of the 2014 Conference on Empirical Methods in Natural Language Processing (EMNLP), pp. 1532–1543 (2014). https://aclanthology.org/D14-1162/
5. Lau, J., Baldwin, T.: An Empirical Evaluation of doc2vec with Practical Insights into Document Embedding Generation (2016). https://arxiv.org/abs/1607.05368
6. Vaswani, A., et al.: Attention Is All You Need (2023). https://arxiv.org/abs/1706.03762
7. Liu, Y., et al.: RoBERTa: A Robustly Optimized BERT Pretraining Approach (2019). https://arxiv.org/abs/1907.11692
8. Radford, A., Narasimhan, K., Salimans, T., Sutskever, I.: Improving Language Understanding by Generative Pre-Training. OpenAI Blog (2018). https://cdn.openai.com/research-covers/language-unsupervised/language_understanding_paper.pdf
9. Beltagy, I., Lo, K., Cohan, A.: SciBERT: A Pretrained Language Model for Scientific Text (2019). https://arxiv.org/abs/1903.10676
10. Yu, Y., Si, X., Hu, C., Zhang, J.: A review of recurrent neural networks: LSTM cells and network architectures. Neural Comput. **31**, 1235–1270 (2019). https://doi.org/10.1162/neco_a_01199

AI-Assisted Offboarding and Knowledge Transfer: Effects on Senior Employee Experience in the Moroccan Textile Sector

Khaoula Khammam(✉) and Sanae El Maliki

The Higher School of Technology of Fes, Management, Marketing, International Logistics, and Finance Laboratory, Sidi Mohamed Ben Abdellah University, Fez, Morocco
{khaoula.khammam,sanae.elmaliki}@usmba.ac.ma

Abstract. This study examines the impact of Artificial Intelligence (AI) on knowledge transfer and the experience of senior employees during their offboarding process in the Moroccan textile sector. Through qualitative research, it explores how AI helps to formalize explicit knowledge, while raising concerns about the depersonalization of the process and the loss of human interactions necessary for transmitting tacit knowledge. The study highlights the need to strike a balance between technology and human interaction, emphasizing the importance of ensuring ethical protocols and confidentiality: Senior employees perceive offboarding as a tipping point, when both appreciation and human interaction are essential. While they do appreciate the role of AI in knowledge transmission, they emphasize that it must complement—and not replace—human interaction. Suggestions are for combining AI solutions with mentoring programs, story-telling websites, and appreciation rituals. Moral dilemmas are uppermost in their mind, where issues of data privacy and protection of strategic know-how are paramount. Younger senior workers need transparency in how their data is handled, while older workers value the retention of craft skills through robust security. The integration of AI must be carefully managed to preserve the senior employees' experience and ensure a respectful and ethical process.

Keywords: Artificial Intelligence · Offboarding · Knowledge Transfer · Explicit Knowledge · Tacit Knowledge · Senior Employee Experience · Textile Sector

1 Introduction

In an era of unprecedented digital transformation, the integration of Artificial Intelligence (AI) is redefining many aspects of the workplace, particularly in knowledge transfer processes. The growing adoption of AI-based technologies, such as task automation and the use of intelligent platforms to structure knowledge, is profoundly transforming the employee experience, especially during their departure from a company. This technological evolution presents opportunities to enhance the efficiency of knowledge transfer, while raising challenges related to the depersonalization of the process and the management of tacit knowledge.

B.-E. Boudriki Semlali et al. (Eds.): DATA 2025, LNNS 1908, pp. 40–50, 2026.
https://doi.org/10.1007/978-3-032-21200-9_4

This article focuses on the Moroccan textile sector and examines the impact of AI integration on knowledge transfer, particularly during the offboarding of senior employees. The central question guiding this study is: How can organizations use AI to capture and transfer both tacit and explicit knowledge from senior employees during offboarding, while balancing technological efficiency with the human connection needed for tacit knowledge?

To address this issue, the study explores the following underlying questions:

- How can AI be utilized to capture both tacit and explicit knowledge from senior employees during offboarding?
- What are the impacts of AI on the senior employee experience during offboarding, particularly in terms of recognition and seamless transitions?
- What mechanisms can be put in place to maximize the benefits of AI while preserving human dimension and interpersonal interaction in knowledge transfer?
- What ethical concerns are raised using AI in this process, particularly regarding confidentiality and the management of strategic knowledge?

Based on initial observations and existing literature, the following hypotheses are proposed:

- AI can improve the efficiency of explicit knowledge transfer by structuring and archiving information in an accessible manner.
- The integration of AI could lead to the loss of the human dimension of the process, thereby reducing social interactions and the transmission of tacit knowledge.
- The effects of AI on the senior employee's experience vary depending on the length of their tenure in the company and the nature of their skills.

The structure of this article is organized as follows: a literature review will present existing research on the integration of AI in knowledge transfer, its implications on employee experience and the ethical and organizational challenges resulting from its integration. The methodology section will describe a qualitative approach, including semi-structured interviews and thematic analysis. The results section will present the perceptions of senior employees in the Moroccan textile sector, followed by a discussion that will contextualize these findings with existing theories and practices. Finally, the conclusion will summarize the main findings and provide recommendations for a balanced integration of AI in offboarding processes.

2 Literature Review

2.1 AI for Optimizing Knowledge Transfer During Senior Offboarding

The departure of senior employees poses a significant challenge for organizations, particularly in knowledge management [1]. Over years of experience, these employees accumulate invaluable expertise, combining explicit knowledge, which is formalized and documented, with tacit knowledge that is often unwritten and harder to convey. Tacit knowledge includes practical skills, sector-specific insights, and contextual understanding that risk being lost when senior employees leave, depriving the organization of a critical strategic resource [2].

Artificial intelligence (AI) offers a promising solution to mitigate such losses. Technologies like machine learning and natural language processing (NLP) enable AI to interact with senior employees to extract, structure, and store both explicit and tacit knowledge in accessible formats [3]. For instance, intelligent chatbots ask targeted questions to capture complex, informal knowledge, significantly reducing knowledge loss during employee transitions [4].

AI-powered knowledge management platforms organize this information according to specific criteria, such as ongoing projects or technical expertise. Predictive recommendation systems further ensure that successors receive tailored solutions, making knowledge transfer smoother and more relevant [5]. In industrial environments, these technologies enhance the speed and accuracy of knowledge transmission, which is critical for maintaining operational continuity [6].

Beyond operational efficiency, AI integration also transforms organizational culture. By combining technology with human interaction, it enriches the knowledge transfer process, fostering active collaboration between generations [7]. This approach ensures the durable formalization of informal knowledge while valuing the expertise of senior employees. As a result, AI not only prevents knowledge loss but also builds a collective knowledge base that strengthens organizational resilience [8].

2.2 The Impact on Senior Employee Experience: Recognition and Smooth Transition

The integration of artificial intelligence (AI) into offboarding processes has become an essential strategy to enhance the experience of senior employees during transitions [9]. By facilitating the transfer of both tacit and explicit knowledge, AI not only ensures continuity but also highlights the value of senior employees' contributions, making their departure more meaningful and seamless.

AI technologies such as chatbots, machine learning algorithms, and knowledge management platforms enable the efficient capture and organization of expertise. By structuring complex information and automating repetitive tasks, these tools reduce the cognitive burden on senior employees, making the transition less stressful [10]. Furthermore, when seniors see their knowledge enriching organizational systems and supporting future decision-making, they experience a sense of fulfillment and purpose.

AI also promotes intergenerational collaboration by creating opportunities for senior employees to mentor younger colleagues [11]. These interactions preserve organizational knowledge, foster innovation, and enhance cohesion, benefiting both the employees and the organization [12]. Studies highlight that such AI-facilitated knowledge sharing strengthens ties across generations and enhances the overall organizational culture.

Recognizing the contributions of senior employees is crucial during their transition. AI systems reinforce this recognition by capturing and preserving their expertise, providing a legacy within the organization. Additionally, by automating administrative processes and ensuring clear communication, AI minimizes potential conflicts and creates a transparent, respectful offboarding experience [13].

2.3 Ethical and Organizational Challenges Related to AI Integration

The integration of AI into senior employees' offboarding processes offers significant benefits in knowledge retention and efficiency but raises critical ethical and organizational challenges [14]. One key concern is safeguarding personal and sensitive data, such as professional achievements and workplace relationships. Organizations must implement strict protocols to ensure data confidentiality, compliance with privacy regulations like GDPR, and transparency in data usage to prevent legal risks and employee distrust [15].

Another challenge lies in balancing AI-driven automation with the human connection essential during offboarding. Overreliance on technology may undermine empathy and recognition, making the transition feel impersonal. A hybrid approach that combines AI's efficiency with meaningful human interactions is crucial for fostering trust and acceptance among departing employees [16].

Additionally, resistance to change remains a significant hurdle, particularly among senior employees unfamiliar with AI technologies. Addressing this requires effective change management strategies, including training programs, gradual implementation, and awareness campaigns, to ensure smoother adoption and employee satisfaction [17].

3 Methodology

This study adopts an interpretive approach to explore the impact of integrating artificial intelligence (AI) in knowledge transfer during the offboarding of senior employees in a Moroccan textile company. Eight senior staff took part, selected to represent in respect of age, years of employment, and organizational roles among Moroccan textiles manufacturing industry. They were between the ages of 36 and 60 and had working experience ranging from 10 to 35 years. They occupied diverse job positions like factory production manager, HR manager, senior engineer maintenance, quality assurance supervisor, textile design expert, logistics coordinator, computer systems administrator, and training officer. This diversity ensured a comprehensive appreciation of how AI-enabled offboarding is perceived by different functions and levels of experience, both reflecting operational and strategic views. The qualitative methodology relies on semi-structured interviews that lasted approximately 45 to 60 min and were conducted in person or over calls, depending on schedule. The interviews were structured in terms of primary themes such as the perception of AI usage in knowledge transfer, tacit and explicit knowledge sharing in the offboarding process, balancing the human touch with technology components, the role of AI in the top talent's exit experience, and confidentiality and ethical issues in using AI. Open-ended questions were used to facilitate rich discussion and allow all to contribute complex views, based on experience, regarding opportunities and constraints of AI in offboarding. The data will be analyzed using Nvivo and thematic analysis to provide an in-depth understanding of the benefits and challenges associated with integrating AI into the management of senior employees' knowledge, while highlighting the importance of preserving the essential human interactions in the knowledge transfer process.

4 Results

The table presents the results of a study exploring the impact of the integration of artificial intelligence (AI) in knowledge transfer among senior employees in the Moroccan textile sector, particularly in the context of their upcoming transition or potential departure. It highlights the employees' perceptions of the use of AI, their experience with this technology, and the various factors influencing its effectiveness in the knowledge transfer process. The results also include the implications for the senior employees' experience and the ethical considerations, concerns about confidentiality (Table 1).

Table 1. Perception of the actors regarding the questions posed

Themes	Main results
Perception of Senior Employees on the Use of AI for Knowledge Transfer	The perceptions of senior employees regarding the integration of artificial intelligence (AI) in knowledge transfer vary according to their roles and specializations within the textile company. Technical managers and training supervisors see AI as an effective tool for structuring and disseminating explicit knowledge, thereby facilitating access to technical information for new generations. Conversely, experienced and specialized workers are more reserved toward this technology. They believe AI cannot capture tacit knowledge, which requires human interactions and contextual learning.
Process of Transferring Tacit and Explicit Knowledge during Offboarding	AI tools such as knowledge management systems and predictive analytics tools have been deployed to structure explicit knowledge in the company. Younger senior employees view offboarding as an opportunity for AI to effectively structure and transfer explicit knowledge. However, according to older and experienced senior employees, offboarding is seen as a critical moment where the human and emotional dimensions of knowledge transfer must be preserved. they express concerns about the potential dehumanization of the process if AI replaces personal interactions. They fear that their experience and expertise could be reduced to depersonalized data, which might diminish the recognition of their contributions.

(continued)

Table 1. (*continued*)

Themes	Main results
Balance between Technology and Human Dimension in Knowledge Transfer	Senior employees highlight the need for a balance between the use of AI and the preservation of human interaction in knowledge transfer, particularly during offboarding. Technical managers acknowledge AI's utility in organizing and making information accessible but emphasize that AI should complement, not replace, direct interactions such as mentoring and informal exchanges, which are essential for conveying subtle and contextual competencies. Older senior employees, particularly those with manual or field-based roles, are more skeptical about integrating AI into the transfer process. They perceive a significant tension between technology and the preservation of human relationships. For them, AI risks dehumanizing the departure process, making the transition less personal and more mechanical. They emphasize that experience, intuition, and the ability to understand each employee's specific needs cannot be codified by algorithms or intelligent systems.
Impact of AI on the Experience of Senior Employees during the Departure Phase	Senior employees with significant tenure or a strong emotional attachment to their role view the transition as a critical moment where they seek recognition and respect. They are scared that their departure may become depersonalized, with AI taking over without highlighting their unique contributions to the organization. However, these employees also emphasize the importance of intergenerational collaboration in this process. They believe that while AI is a valuable tool for knowledge transfer, it should not replace essential human interactions. To enhance departure experience, they suggest that their company combine AI with human-centered initiatives, such as AI-supported mentoring programs, recognition ceremonies, and tools promoting the sharing of personal stories.

(*continued*)

Table 1. *(continued)*

Themes	Main results
Ethics and Confidentiality in the Integration of AI for Knowledge Transfer	Senior employees emphasize the importance of ethics and confidentiality in integrating AI for knowledge transfer. They stress the need to protect technical know-how and sensitive information, calling for strict ethical frameworks to ensure security and compliance with confidentiality standards. Younger seniors, involved in administrative tasks, are concerned about the use of their personal and professional data, demanding transparency regarding algorithms and information processing. Older seniors, with extensive artisanal or technical expertise, fear the risks of uncontrolled dissemination of their strategic skills and advocate for strengthened security protocols.

Source: Authors.

5 Analysis and Discussion

The analysis of integrating artificial intelligence (AI) into knowledge transfer in the textile sector highlights significant challenges, particularly its impact on the offboarding phase for senior employees. This phase is critical as it involves passing on knowledge and skills from experienced employees to the next generation while preserving human relationships and accumulated expertise.

Responses from senior employees reveal a consensus on the importance of this phase, with perceptions varying across generations. Younger senior employees view the use of AI during offboarding as a method to ensure continuity in knowledge transfer. AI could facilitate the transmission of technical information, standardized procedures, and best practices through automated learning platforms, centralized databases, or interactive tools. This formalization of the transfer process is seen as a major advantage, as it allows explicit knowledge to be structured and archived in an accessible manner.

However, this approach encounters significant challenges when dealing with tacit knowledge management. More experienced and older senior employees, especially those with artisanal know-how or specific practical skills, express serious concerns. They believe AI is limited in its ability to capture the nuances of human interactions and non-formalized knowledge, which are often passed on informally or through daily social interactions. These employees emphasize the importance of the human aspect of off-boarding, where knowledge transfer goes far beyond merely transmitting technical information. Tacit skills and knowledge, often tied to practical situations, field anecdotes, and lessons learned in informal work contexts, cannot be fully captured by algorithms.

The human aspect of offboarding, often likened to a rite of passage, emerges as a major concern when integrating AI. Older employees highlight that AI risks depersonalizing this critical phase, making the transition less respectful and empathetic. The departure of senior employees, often marking the end of an era for the organization, carries a strong emotional dimension that AI struggles to address. Even younger employees, while acknowledging the benefits of AI in formalizing processes, are aware of the potential loss of this human dimension.

Thus, one of the main challenges identified in this discussion is finding a balance between using AI to structure and formalize knowledge transfer and maintaining the human dimension of offboarding. AI can play a valuable role in facilitating explicit knowledge management and supporting administrative transitions, but it should not replace interpersonal exchanges, face-to-face training, and mentoring moments essential for a complete and human-centered knowledge transfer.

The integration of AI into the offboarding process also requires addressing ethical concerns, such as information confidentiality and respect for employees' expertise. Ensuring data protection, particularly for sensitive or strategic information, is crucial to making senior employees feel secure and respected as they leave. Full transparency on how data is collected, used, and shared is essential to build trust.

In conclusion, while AI can provide undeniable benefits in knowledge transfer management, its role in offboarding must complement existing human processes. A hybrid approach, combining AI's technological efficiency with human interactions and respect for organizational values, appears to be the most promising path to ensure effective knowledge transfer while preserving human relationships within the organization.

The analysis conducted using Nvivo 10 software reveals the following word clouds (Figs. 1 and 2):

Fig. 1. Word Cloud of the Research ThemeSource: Nvivo10 Software.

Additionally, the coding reveals the following parent linkage:

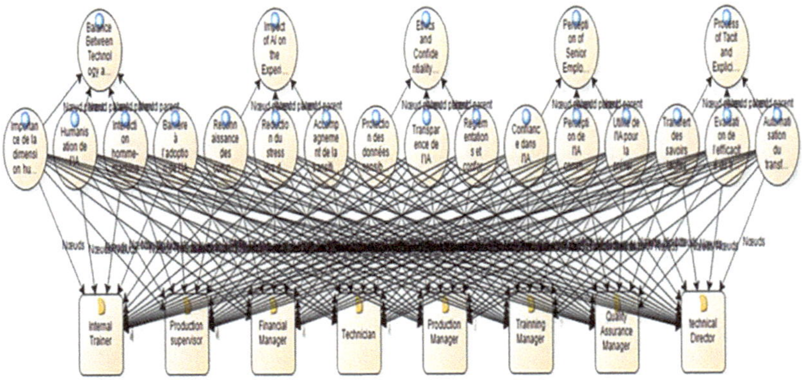

Fig. 2. Parent linkage coding by NVivo 10 software

6 Conclusion

This article highlights the crucial role of artificial intelligence (AI) in knowledge transfer, particularly during the offboarding phase of senior employees. The integration of AI allows for the structuring and formalization of explicit knowledge transfer, facilitating the archiving and dissemination of technical information and best practices. These tools improve knowledge accessibility, thus ensuring the continuity of skills within the organization.

However, the impact of this digital transition on the experience of senior employees should not be underestimated. While AI is effective for knowledge management, it can depersonalize the process, especially for those holding tacit knowledge, often passed on through mentoring moments or social interactions. Senior employees emphasize that AI cannot replace the richness of human exchanges that are essential for transmitting subtle and contextual skills, which can affect the quality of their departure. A hybrid approach, combining the technological efficiency of AI and the importance of interpersonal exchanges, will be essential to preserve this human dimension and ensure that tacit knowledge, which is vital for the smooth functioning of the business, is not lost. This will optimize knowledge transfer while respecting the values and experiences of senior employees.

Finally, ethical and confidentiality concerns regarding the use of their data must be considered. Transparent and responsible management of sensitive information is crucial to ensure employee trust and provide respectful offboarding experience.

To effectively address issues of AI integration in offboarding processes, organizations should adopt a hybrid approach that involves combining technology with people-oriented practices. Organizations can first create formal AI-powered knowledge bases where leaving members leave behind explicit documentation and note-taking walkthroughs of tasks and with short video explanations as add-ons. To preserve tacit knowledge, it is recommended to pair AI tools with shadowing or mentoring programs, where junior staff can observe senior employees in action before they leave. HR departments also need to arrange mediated exit interviews through AI-based transcription and analysis software, with human checking to ensure context and emotion are captured accurately. Having clearly established ethical guidelines and consent protocols for the handling of data can reduce some of the surveillance and confidentiality issues. Finally, practitioners need to continuously monitor AI systems through feedback mechanisms so that the knowledge gained is up to date, comprehensible, and accessible to future personnel.

References

1. Daghfous, A., Amer, N.T., Belkhodja, O., Angell, L.C., Zoubi, T.: Managing knowledge loss: a systematic literature review and future research directions. J. Enterp. Inf. Manag. **36**(4), 1008–1031 (2023)
2. Daghfous, A., Belkhodja, O., Angell, L.C.: Understanding and managing knowledge loss. J. Knowl. Manag. **17**(5), 637–655 (2013)
3. Tarmizi, W.A.M., Rashid, A.N.Z., Muhammad, A.S., Yangkatisal, M.: Natural language processing (NLP) application for classifying and managing tacit knowledge in revolutionizing AI-driven libraries. Inf. Manag. Bus. **16**(3), 1103–1119 (2024)
4. Waykar, Y.: The role of artificial intelligence in knowledge management. Int. J. Sci. Res. Eng. Methods. **15999** (2022). https://doi.org/10.55041/IJSREM15999
5. Jarrahi, M.H., Askay, D., Eshraghi, A., Smith, P.: Artificial intelligence and knowledge management: a partnership between human and AI. J. Knowl. Manag. **26**(2), 125–142 (2022)
6. Pinski, M., Benlian, A.: AI literacy for users: a comprehensive review and future research directions of learning methods, components, and effects. Journal of Educational Technology and AI. **7**(1), 89–110 (2024)
7. Zamiri, M., Esmaeili, A.: Methods and technologies for supporting knowledge sharing within learning communities: a systematic literature review. Administrative Sciences. **14**(1), 17 (2024)
8. Mugisha, G.K.: The Impact of Artificial Intelligence on Human Resource Management. KIU Publication Extension (2024)
9. Rohit, A.K., Saini, M., Pawar, M.: Integration of artificial intelligence in human resource information system. International Journal of Innovative Research in Technology and Science. **12**(2), 177–183 (2024)
10. Barabel, M., Ferras, T.: L'IA au service des RH: Pour une expérience collaborateur augmentée. Dunod (2020)
11. Dimovski, V., Colnar, S., Roblek, V., Peterlin, J.: Intergenerational leadership and knowledge transfer: case of cobot integration. ENTRENOVA - enterprise research innovation. **10**(1), 338–346 (2024)
12. Seerangan, C., Venkataravi, R.: The crucial role of intergenerational learning in the era of digital transformation. Journal of Organizational Learning. **5**(3), 301–318 (2024)
13. De Koning, M.: HR Tech Strategy. Business Expert Press (2024)
14. Junginger, S.: Embracing digital offboarding as a design challenge. In: IASDR 2023: Life-Changing Design Conference Proceedings (2023)
15. Pina, E., et al.: Data privacy and ethical considerations in database management. J. Cybersecur. Priv. **4**(3), 494–517 (2024)

16. Freitas, F. B.: Talent Management Practices for the Future of Work: how Can Artificial Intelligence Reconcile Recruitment Tensions in Organizations. Doctoral dissertation, Nova School of Business and Economics (2023)
17. Alshwayat, D., Elrehail, H., Shehadeh, E., Alsalhi, N., Shamout, M.D., Rehman, S.U.: An exploratory examination of the barriers to innovation and change as perceived by senior management. Int. J. Innov. Stud. **7**(2), 159–170 (2023)

Artificial Intelligence Application for the Water Resources Management: A Literature Review

Hajji Radia(✉) and El Khoukhi Fatima

IA Laboratory, FS Meknes, Moulay Ismail University of Meknes, Meknes, Morocco
ra.hajji@edu.umi.ac.ma, f.elkhoukhi@flsh.umi.ac.ma

Abstract. Climate change significantly affects water resources management (WRM), making the accurate prediction of water quality a crucial challenge. Ensuring sustainable WRM requires a comprehensive approach that involves analyzing historical data, forecasting future trends, and implementing risk management strategies. Artificial intelligence (AI) has emerged as a powerful tool to integrate these elements, enhancing decision-making processes. In recent years, various computational techniques, including machine learning (ML), deep learning (DL) and reinforcement learning (RL), have been successfully applied to water quality preservation, water distribution networks and WRM planning. Researchers have leveraged these approaches to develop predictive models and optimize resources allocation within public management systems. In this paper, we aimed to propose a literature review based on 20 scientific articles published between 2019 and 2024, including 8 review papers, sourced from ScienceDirect, MDPI (Multidisciplinary Digital Publishing Institute), and IWA (International Water Association). An overview is providing on a set of AI technologies used in water research. We present analytical study categorizing various algorithms and discusses their applications in optimization of WRM.

Keywords: Water resources management · Water distribution networks · Artificial Intelligence Technologies · Optimization · Literature review

1 Introduction

Water scarcity is an escalating issue across all continents, with the availability of freshwater for drinking purposes emerging as a critical global concern. The exhaustion of water resources, driven by population growth, presents a significant challenge for nearly all economies. This situation is exacerbated by the imprudent utilization of resources and outdated, inefficient practices. As of 2020, approximately 2.4 billion individuals reside in countries experiencing water stress, with 800 millions of them living in regions classified as having high or critical water stress [20]. Effective management of water resources is essential and necessitates a multidisciplinary approach. An integrated Water Resources Management (WRM) strategy can equip governments with a comprehensive framework to harmonize water usage with the diverse needs of various stakeholders, including environmental considerations.

B.-E. Boudriki Semlali et al. (Eds.): DATA 2025, LNNS 1908, pp. 51–63, 2026.
https://doi.org/10.1007/978-3-032-21200-9_5

Machine Learning (ML), Deep Learning (DL), and Reinforcement Learning (RL) are foundational artificial intelligence techniques that underpin numerous applications in WRM [3]. ML, a subset of AI, enables machines to autonomously learn from data through statistical methods. DL, a further specialization within ML, allows machines to process large datasets using artificial neural networks (ANN), facilitating complex tasks. RL, another branch of ML, is particularly advantageous in situations where data is scarce. By enhancing data processing, predictive capabilities, and adaptability [7], AI is increasingly becoming a vital adjunct to optimization strategies in water resources management [9]. Effective water management is crucial in sectors such as agriculture, public supply, industry, hydropower generation, aquaculture, and livestock farming. Intelligent modeling techniques can suggest water-efficient appliances, curtail domestic water usage, and implement tariffs to regulate consumption [20]. This comprehensive understanding of advancements in the field aims to identify existing gaps, modern tools, mathematical models, and significant findings from previous research.

The overall purpose of the presented study; is to highlight new opportunities offered by smart techniques in water research management challenges, and present their different applications to WRM, such as water demand forecasting [2–4, 11], water resources modeling [10], leakage management [5, 6, 8], precipitation forecasting [1, 13], water stress forecasting [14, 16, 19] and the development of strategies for acceptable water use [2, 11] (see Fig. 1). Based on a structured literature review of 20 scientific publications from 2019 to 2024, including 8 review papers, retrieved from ScienceDirect, MDPI, and IWA.

The current paper is structured in the following manner: An introduction is given in Sect. 1. In Sect. 2 and its subsections elaborate Toward smart WRM, then AI's promises and perspectives in WRM. In Sect. 3 we have a comparative study of AI's application in WRM. Finally, Sect. 4 concludes with key points and recommendations for advancing AI integration in WRM.

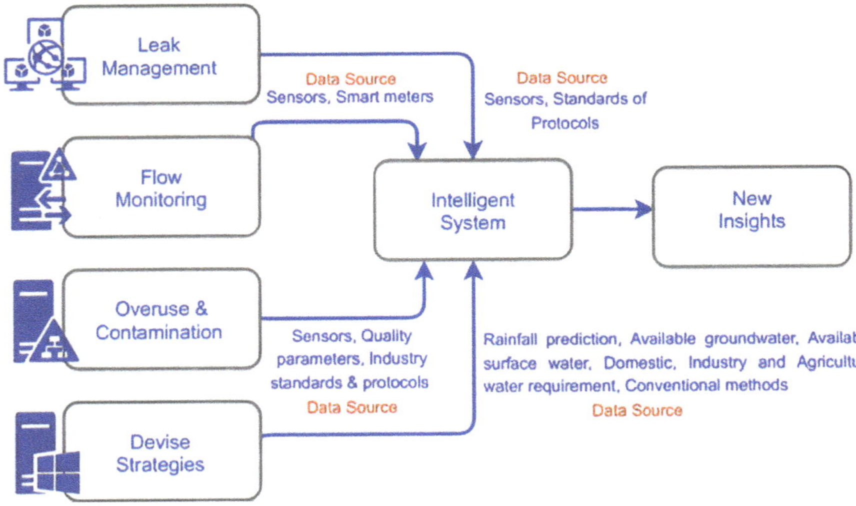

Fig. 1. Harnessing intelligent systems for water management [12]

2 Literature Review

2.1 Toward Smart Water Resources Management

Water management is a critical issue frequently addressed in international forums. It encompasses the conservation of water resources, water harvesting, planning of net available water resources, and the equitable distribution of water to consumers. Given that nearly 97% of the Earth's water is saline and unsuitable for drinking, effective water management practices are essential to ensure the long-term sustainability of water resources [12]. Effective water management is required across various sectors, including agriculture and groundwater [14, 16, 19], public supply and industry [2–4, 11, 12], as well as mining, hydropower generation, aquaculture, and livestock management. In the context of groundwater, uncontrolled urbanization and increasing anthropogenic pressure on hydro geomorphological systems threaten environmental sustainability by disrupting natural recharge mechanisms and degrading groundwater quality [17].

In India, industries rank as the second-largest consumers of water and are among the primary contributors to water pollution. These industries primarily extract water from both surface and groundwater sources. As urbanization continues to expand, the water demand from industries, public supply, and mining is growing accordingly. Simultaneously, the untreated discharge of wastewater into natural sources is further polluting clean water supplies [12].

Addressing these challenges requires the integration of intelligent technologies to enhance water management. Achieving sustainable WRM relies on accurate data, the implementation of water-efficient, eco-friendly, and hybrid technologies; especially in agriculture and industry; and public awareness campaigns focused on reducing domestic water use and encouraging sustainable consumption habits and diets [20]. Smart algorithms within water distribution systems play a crucial role in ensuring the efficient and sustainable delivery of safe water to communities. Models developed using intelligent techniques can recommend the use of water-efficient appliances, enforce usage restrictions in households, and introduce tariffs to regulate water consumption [12].

2.2 Promises and Perspectives of AI in WRM

AI is a major advance in WRM, with the potential to transform both hydrological applications and operational practices. By making a significant contribution to technological and economic progress, AI plays a key role in achieving the water related Sustainable Development Goals (SDGs). The global AI market, currently estimated at 2,000 billion USD, is expected to reach a contribution of 16,000 billion USD, with a share of 200 billion USD coming from innovations in the water sector by 2030 [3]. Fundamental AI techniques, including:

- Machine Learning: Traditional machine learning leverages statistical approaches to enable machines to learn from data. It is broadly divided into two main categories: supervised learning, which involves training models on labeled data, and unsupervised learning, where models identify patterns in unlabeled data. Common supervised learning methods include regression and classification algorithms such as logistic regression, random forests (RF), and support vector machines (SVM) [9]. On the other hand, unsupervised learning makes use of clustering and association techniques, like K-means clustering and the Apriori algorithm [15].

- Deep Learning: DL is an advanced approach that establishes relationships between input data and target outcomes through multiple layers of non-linear transformations. The term "deep" signifies the layered structure that facilitates the identification of intricate, non-linear relationships within the data [9].
- Reinforcement Learning: Unlike other learning approaches, reinforcement learning focuses on training an agent to improve its behavior by receiving rewards for desirable actions and penalties for undesirable ones. Through iterative trial-and-error interactions, RL algorithms learn to make optimal decisions by exploring complex environments, adapting to dynamic conditions, and aiming to maximize cumulative rewards over time. Notable techniques in this field include Q-learning, Deep Q-Networks (DQN), Proximal Policy Optimization (PPO), and Monte Carlo methods [14].

WDNs relies on the exploitation of diversified data from fields such as hydrology, meteorology, climatology and geology, which can be integrated into a common platform using AI techniques [7]. Data collection relies on technologies such as remote sensing, satellite imagery, the Internet of Things (IoT) and drones. These data, whether structured (flow measurements or water quality parameters) or unstructured (satellite images), determine the analytical approaches: classical ML algorithms are suitable for structured data, while DL algorithms are required for complex data. In limited data contexts, RL is an ideal solution, as the agent learns in real time through continuous feedback [20].

Having presented the various AI tools used in WRM, we will now explore AI application frameworks in this field. In the work presented in [20], the authors analyzed various applications specific to WRM. These include: Leak detection and infrastructure maintenance [5, 6, 8] to improve water supply. Demand forecasting and water supply optimization, enabling proactive resources management. Water treatment and seawater desalination processes, to meet the growing demand for drinking water. Flood and drought forecasting and management [9, 10], aimed at reducing the risks associated with extreme climatic events, Recycling and management of wastewater, for more sustainable use of water resources, Rainwater management [1, 13] and groundwater management [17], including collection and optimized storage, Measurement and monitoring of water quality, essential for guaranteeing a supply that complies with health standards. The following figure (Fig. 2) provides a concise summary of the AI techniques used in each application, highlighting their key role in solving complex water resources management problems. In the following section, we will look at some specific examples in this context. We will analyze the objectives, adapted techniques and prospects associated with each of these applications, highlighting the opportunities offered by AI to meet current and future challenges in WDNs.

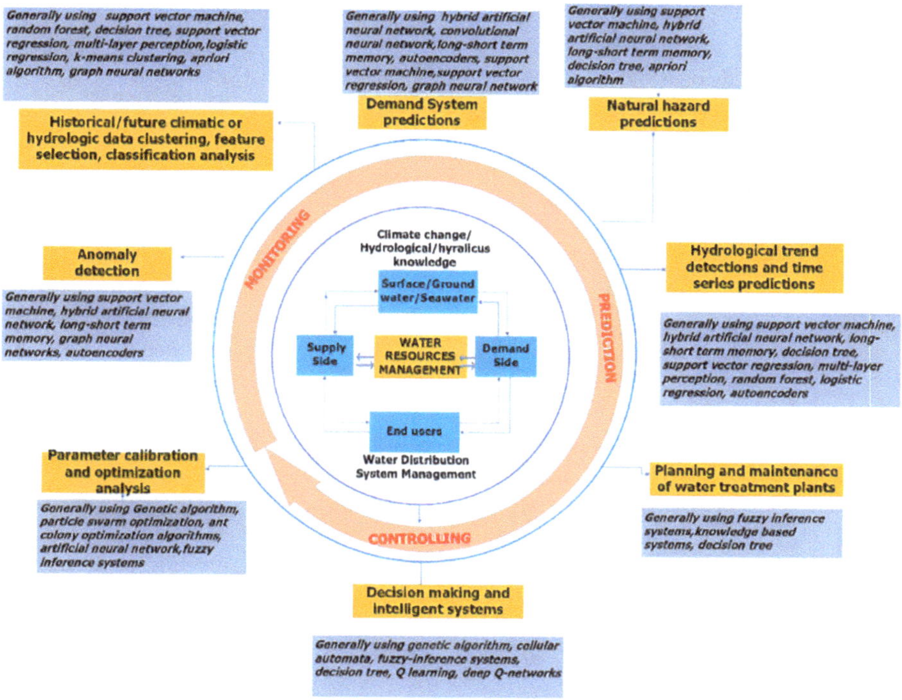

Fig. 2. Overview of AI Algorithms for Operational Applications in WRM [20]

3 Comparative Study of AI Water Resources Management Studies

Here, we will present a few works highlighting the varied applications of AI, from water resources forecasting to infrastructure management and environmental risk prevention. The table below (Table 1) presents a comparison of various key studies in the field of WDNs using AI techniques. Each study is analyzed according to three main criteria:

Objective: The main objective or problem targeted by the study.
Suitable AI techniques: The specific AI methods used
Outlook: The opportunities and future directions identified by the authors for improving water resources management through AI.

Recent research in WRM has increasingly leveraged advanced AI techniques to enhance predictive accuracy, optimize system performance, and address critical challenges such as leak detection, water stress prediction, and rainfall forecasting. A thorough examination of recent studies reveals the growing adoption of deep learning architectures, hybrid AI models, and optimization techniques to improve the efficiency and sustainability of WDNs (see Fig. 3).

Table 1. Comparative table of studies mentioned in WRM management by AI

Reference	Study objective	Adapted techniques	Outlook
([1], 2024)	Accurately predict daily rainfall using hybrid AI methods based on the Adaptive Neuro Fuzzy Inference System (ANFIS), combined with meta-heuristic algorithms such as Artificial Bee Colony (ABC), Genetic Algorithm (GA), and Simulated Annealing (SA).	ANFIS-ABC ANFIS-GA ANFIS- SA	• Extend the application of the models to other geographical regions with different climates and topographies. • Development of user-friendly interfaces for operational use and real-time data integration.
([3], 2024)	Move from reactive to proactive management of pipe networks in Sweden, using ANN model to estimate the probability of leaks in water pipes.	ANN	• Further, develop model accuracy by integrating additional data and refining parameters. • Extend the application of the model to other networks and regions to verify its generalizability. • Explore the possibility of integrating real-time data for dynamic monitoring of leaks and risks.
([6], 2024)	Develop a ML model for multi-class leakage detection in WDNs using IoT based smart sensor data.	eXtreme Gradient Boosting (XGBoost) Bayesian optimization	• Enable smarter sustainable maintenance of water supply networks by monitoring leak locations and severity. • Extend the model for broader applications in WDN management.

(*continued*)

Table 1. *(continued)*

Reference	Study objective	Adapted techniques	Outlook
([8], 2024)	Develop a DL-based methodology for leak detection in WDNs to address urban water loss challenges and Improve accuracy and efficiency in identifying leaks using pressure monitoring data and data imaging techniques.	CNN	• Explore seasonal updates to pressure fluctuation patterns for improved accuracy. • Investigate other DL architectures for further optimization. • Extend the methodology to detect multiple simultaneous leaks and adapt to dynamic network conditions.
([13], 2024)	Develop a hybrid model (SARIMA ANN) to forecast monthly rainfall in Sulaymaniyah, in order to better WRM.	Hybrid SARIMA ANN model	• Collect and use daily precipitation data, which would improve the learning of ML models and increase forecast accuracy. • Test the model in other regions and climatic conditions to assess its robustness and generalizability.
([16], 2024)	develop an optimized model, based on the Capsule Neural Network (CapsNN) enhanced by the Improved Aquila Optimization (IAO) algorithm, for predicting water stress in Yichang, China.	CapsNN/IAO model	• Extend CapsNN/IAO model testing to other regions and contexts to validate its generalizability. • Explore real-time data integration for even more responsive management.
([2], 2023)	Explore and review the application of Deep RL (DRL) in Urban Water Systems. Moreover, Highlight DRL is potential for real-time control and optimization in WDNs and storm water systems.	Novel taxonomy of DRL algorithms.	• Expand DRL to water treatment, sanitation, and wastewater management. • Reduce carbon emissions using DRL. • Develop standardized DRL environments for testing.

(continued)

Table 1. *(continued)*

Reference	Study objective	Adapted techniques	Outlook
([4], 2023)	Graph Neural Networks (GNNs) are investigated as surrogate models for WDSs, aimed at accurately estimating nodal pressures under steady-state conditions using EPANET simulations.	GNNs	• Explore other GNN architectures to improve performance and transferability. • Investigate hyper parameter effects and data generation techniques for broader pressure distributions. • Use self-supervised learning to exploit equivalences in pipe parameters for improved modeling.
([5], 2023)	Develop a microservice architecture for leak localization (LL) in WDNs to reduce water loss (up to 30%). By Combining hybrid data-driven and model-based methodologies to improve leak detection and localization accuracy.	Hybrid AI model combining Physics ML	• Adapt the methodology to detect multiple simultaneous leaks (currently limited to single leaks). • Integrate expert knowledge to improve model performance for specific leak types.
([17], 2023)	improve the accuracy of monthly precipitation and tank evaporation forecasts, two inherently complex hydrological processes, proposing a simple yet effective method for identifying novel input variables in predictive scenarios	SVM	• Approach applicable to other regions, taking into account specific weather station conditions. • Use in agricultural science and water resources management.

(continued)

Table 1. *(continued)*

Reference	Study objective	Adapted techniques	Outlook
([19], 2022)	Characterizing and predicting water stress in agro-ecosystems in the Hyderabad region, India, using AI models.	ANN SVR	• The model can be further enhanced by incorporating weather, hydrologic, agronomic, or combined parameters, enabling a more comprehensive characterization of drought through the application of various machine learning techniques.
([14], 2021)	develop a novel hybrid artificial intelligence (AI) model for accurate and reliable multi-time scale groundwater level forecasting in an arid oasis region of northwestern China.	Extreme Learning Machine (EML) Boruta Algorithm	• Variations between wells suggest that physical differences in aquifer dynamics need to be accounted for. • Only input-based uncertainty was analyzed, while structural and parametric uncertainties remain unexplored.
([10], 2019)	Examine the application of ANNs in water resources modeling, and the associated challenges.	ANN	• Explore more robust optimization tools, such as Differential Evolution (DE), Particle Swarm Optimization (PSO), and Deep Neural Network (DNN), to improve ANN performance in water resources modeling.
([11], 2019)	Develop a reliable model for forecasting water demand, integrating climatic and socio economic variations.	ANN DE	• Incorporate other explanatory factors to improve forecasts, and explore the use of feature selection techniques to enhance model generalization.

One of the most commonly employed techniques in water-related AI applications remains the ANN, prized for its flexibility and ease of implementation. ANN models have been widely used for predicting pipe failures [3], characterizing agro-ecosystem water stress [19], and forecasting water demand under socio-economic and climatic variability [11]. However, their performance often hinges on meticulous parameter tuning and is sensitive to input data quality. Recent studies have highlighted the importance of integrating optimization algorithms like DE or using feature selection to boost ANN accuracy and generalization [10].

CNNs have gained traction, particularly in urban water systems where spatial and temporal pressure patterns are crucial. CNN-based approaches have been successfully employed to detect leaks using pressure monitoring and imaging data, significantly improving detection efficiency [8]. Nonetheless, CNNs typically demand large, high-quality datasets and can struggle to adapt to fluctuating network dynamics without regular retraining.

GNNs are emerging as highly promising tools due to their natural compatibility with the structure of WDNs. Studies have demonstrated GNNs' capability to estimate nodal pressures in steady-state EPANET simulations [4], offering strong predictive performance and the ability to encode topological information. However, GNNs remain computationally demanding and may require refined data generation strategies to handle diverse pressure conditions.

CapsNNs, enhanced with optimization methods like the Improved IAO, represent an advanced evolution in neural architectures. Applied to predict water stress levels in regions like Yichang, China, these models have shown improved feature learning and generalization [16]. Despite encouraging results, they remain in the early stages of development and need to be tested across varied environmental settings to validate their robustness.

Hybrid models that combine statistical methods with AI techniques have also shown strong potential. For instance, the SARIMA-ANN hybrid model applied in rainfall forecasting integrates temporal patterns with non-linear modeling capacity [13]. Similarly, the ANFIS has been paired with metaheuristics such as GA, ABC, and SA to enhance daily rainfall prediction [1]. These hybrid systems offer a balanced approach, effectively capturing both deterministic trends and stochastic variability. However, their complexity often entails more effort in calibration and they may be sensitive to regional and climatic variability.

Advanced ensemble and optimization strategies are also becoming more common in smart water systems. For instance, XGBoost combined with Bayesian optimization has been used to enable multi-class leakage detection from IoT sensor data, enhancing maintenance strategies in WDNs [6]. These models facilitate more sustainable network management by localizing leak severity and type. Meanwhile, SVM have shown promising results in modeling precipitation and evaporation dynamics in hydrological forecasting, especially when adapted to site-specific features [17].

In parallel, DRL is gaining recognition as a next-generation tool for real-time control and optimization in both WDNs and stormwater systems [2]. Although still largely experimental, DRL holds significant potential for carbon footprint reduction and adaptive urban water governance, provided standardized environments for testing and benchmarking are developed.

Collectively, these AI models demonstrate a wide spectrum of applicability, from predictive analytics to real-time control. Yet, challenges remain in model transferability, data integration, and computational demands highlighting the need for ongoing research into hybridization, real-time learning, and context-aware design.

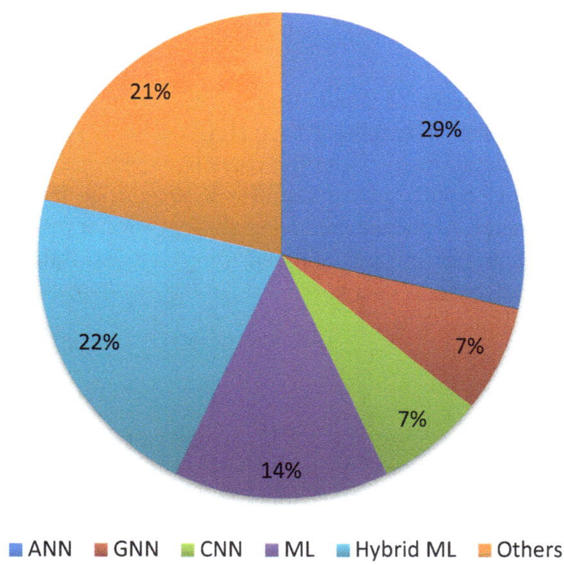

Fig. 3. Extent of AI technology utilization in WRM

In summary, these studies collectively underscore the transformative potential of artificial intelligence in water resources management, offering effective solutions to challenges such as leak detection, groundwater forecasting, water quality monitoring, and rainfall prediction. While each study employs unique techniques ranging from ANN, CNN, and GNNs to SVM and hybrid models, they share a common goal of improving accuracy, scalability, and real-time applicability. and no single AI technique can be considered universally optimal for all water management challenges. CNNs and GNNs offer strong capabilities for structured data and network-aware modeling, while more general ANN-based models provide adaptability across different tasks. Hybrid approaches and emerging techniques like DRL present exciting opportunities for real-time control and decision-making, though they are still maturing in terms of implementation.

Future research should focus on integrating additional data sources, conducting real-time testing, and exploring advanced AI architectures to enhance the reliability and generalizability of these models across a wide range of water management scenarios

4 Conclusion and Future Directions

This state-of-the-art review has examined existing approaches for WRM optimization, highlighting the most promising strategies. By analyzing 20 scientific papers published between 2019 and 2024, including 8 review articles, we identified key trends and methodologies shaping the field and their applications.

Optimization in WDN relies on diverse strategies, each with specific strengths and limitations. A particularly transformative development in WRM is the increasing integration of AI. Techniques such as ANNs, ML, and CNNs are revolutionizing hydraulic network analysis by enabling accurate modeling of complex WRM behaviors. These AI-driven tools enhance predictive capabilities, improve infrastructure management, and support proactive decision-making. Additionally, they contribute to cost-effective maintenance and promote the long-term sustainability of water systems.

Looking ahead, future research should focus on advancing meta-heuristic optimization techniques and further leveraging AI capabilities, particularly in the context of optimization under uncertainty. Tackling the challenges presented by climate change, urbanization, and water resource conservation will necessitate innovative approaches that improve the efficiency, resilience, and adaptability of water resources management strategies. These developments are crucial for ensuring a more sustainable and intelligent management of water distribution networks.

References

1. Thai Pham, B., Thi Bui, K., Prakash, I., Ly, H.: Hybrid artificial intelligence models based on adaptive neuro fuzzy inference system and metaheuristic optimization algorithms for prediction of daily rainfall. Phys. Chem. Earth. **134**, 103563 (2024)
2. Oğuz, A., Faruk Ertuğrul, O.: A survey on applications of machine learning algorithms in water quality assessment and water supply and management. Water Supply. **23**(2), 895–922 (2023)
3. Negm, N., Ma, X., Aggidis, G.: Deep reinforcement learning challenges and opportunities for urban water systems. Water Res. **253**, 121145 (2024)
4. Kerimov, B., et al.: Assessing the performances and transferability of graph neural network metamodels for water distribution systems. J. Hydroinf. **25**(6), 2223–2234 (2023)
5. Mazaev, G., et al.: A microservice architecture for leak localization in water distribution networks using hybrid AI. J. Hydroinf. **25**(3), 851–866 (2023)
6. Park, H., Oh, Y., Kim, K., Kim, T., Hyung, J., Koo, J.: Optimal leakage detection and classification of the water distribution network based on the machine learning approach. Water Supply. **24**(6), 2158–2171 (2024)
7. Kamyab, H., et al.: The latest innovative avenues for the utilization of artificial intelligence and big data analytics in water resource management. Results Eng. **20**, 101566 (2023)
8. Yua, H., Linc, S., Zhoud, H., Wengd, X., Chua, S., Yu, T.: Leak detection in water distribution networks based on deep learning and kriging interpolation method. AQUA — Water Infrastruct. Ecosyst. Soc. **73**(8), 1741–1753 (2024)
9. Bagheri, M., Farshforoush, N., Bagheri, K., Shemirani, A.I., et al.: Process Saf. Environ. Prot. **180**, 10–22 (2023)
10. Oyebode, O., Eseoghene Ighravwe, D.: Urban water demand forecasting: a comparative evaluation of conventional and soft computing techniques. Resources. **8**, 156 (2019)

11. Oyebode, O.: Evolutionary modelling of municipal water demand with multiple feature selection techniques. J. Water Supply Res. Technol. AQUA. **68**(4), 264–281 (2019)
12. Krishnan, R., et al.: Smart water resource management using artificial intelligence—a review. Sustainability. **14**, 13384 (2022)
13. Dashti Latif, S., Othman Mohammed, D., Jaafar, A.: Developing an innovative machine learning model for rainfall prediction in a semi-arid region. J. Hydroinf. **26**(4), 904–914 (2024)
14. Wu, M., Feng, Q., Wen, X., Yin, Z., Yang, L., Sheng, D.: Deterministic analysis and uncertainty analysis of ensemble forecasting model based on variational mode decomposition for estimation of monthly groundwater level. Water. **13**, 139 (2021)
15. Alloghani, M., Al-Jumeily, D., Mustafina, J., Hussain, A., Aljaaf, A.J.: A Systematic Review on Supervised and Unsupervised Machine Learning Algorithms for Data Science, Switzerland (2020)
16. Wang, Y., Razmjooy, S.: Prediction of drought hydrological and water scarcity based on optimal artificial intelligence by developing a metaheuristic optimization algorithm. Phys. Chem. Earth. **135**, 103669 (2024)
17. Zaresefat, M., Derakhshani, R.: Revolutionizing groundwater management with hybrid AI models: a practical review. Water. **15**, 1750 (2023)
18. Meidute-Kavaliauskiene, I., Alizadeh Jabehdar, M., Davidaviˇciene, V., Ghorbani, M.A., Sammen, S.S.: a simple way to increase the prediction accuracy of hydrological processes using an artificial intelligence model. Sustainability. **13**, 7752 (2021)
19. Gorlapalli, A., et al.: Characterization and prediction of water stress using time series and artificial intelligence models. Sustainability. **14**, 6690 (2022)
20. Jayakumar, D., Bouhoula, A., Khalil Al-Zubari, W.: Unlocking the potential of artificial intelligence for sustainable water management focusing operational application. Water. **16**, 3320 (2024)

Exploring Classical and Quantum-Inspired Approaches for Arabic Handwritten Text Recognition: A Comprehensive and Detailed Review

Chaouki Salhi[1](✉), Abderrahim El Allati[1], and Mohamed Ben Ahmed[2]

[1] Laboratory of R & D in Engineering Sciences, Faculty of Sciences and Techniques Al-Hoceima, Abdelmalek Essaadi University, Tetouan, Morocco
ch.marengi@gmail.com, eabderrahim@uae.ac.ma

[2] C3S Laboratory, Faculty of Sciences and Techniques, Abdelmalek Essaadi University, FSTT; Bp 416, Tangier, Morocco
mbenahmed@uae.ac.ma

Abstract. Recognizing Arabic handwritten text (AHTR) is challenging because the Arabic script is highly cursive, has multiple positional variations for each character, includes optional diacritics, and is expressed through various calligraphic styles. While classical deep learning models such as CNNs, RNNs, and Transformers have performed effectively on clear and structured datasets, they encounter significant difficulties with historical or noisy documents due to reliance on extensive annotated datasets and sensitivity to quality variations.

To overcome these issues, recent studies have proposed quantum-inspired and hybrid quantum-classical methods. By incorporating principles from quantum mechanics, such as superposition, entanglement, and tensor representations, these approaches aim to improve model robustness and reduce computational demands, especially beneficial when limited training data is available. Hybrid architectures specifically combine classical methods for initial feature extraction with quantum-based modules to enhance semantic understanding and reduce computational load.

In this paper, we review classical, quantum-inspired, and hybrid quantum-classical approaches applied to Arabic handwritten text recognition. We analyze their strengths and limitations, compare their applicability across various scenarios, and suggest promising directions for future research in this emerging interdisciplinary field.

Keywords: Arabic Handwritten Text Recognition · Quantum-Inspired Computing · Hybrid Quantum-Classical Models · Deep Learning · Quantum Neural Networks · Quantum Natural Language Processing · Hilbert Space Embeddings

B.-E. Boudriki Semlali et al. (Eds.): DATA 2025, LNNS 1908, pp. 64–76, 2026.
https://doi.org/10.1007/978-3-032-21200-9_6

1 Introduction

Recognizing Arabic handwritten text (AHTR) has become a crucial task due to the increasing need to digitize historical, cultural, and contemporary documents. The Arabic script's cursive nature, letter shape variations, optional diacritics, and multiple calligraphic styles make automated recognition particularly challenging. Classical deep learning models, such as CNNs, RNNs, and Transformers, have achieved substantial progress but remain constrained by their dependency on large annotated datasets and their sensitivity to noisy or low-resource scenarios [1, 2].

To overcome these challenges, recent research has begun exploring quantum-inspired and hybrid quantum-classical methodologies. These innovative approaches leverage quantum mechanical principles such as superposition, entanglement, and Hilbert space embeddings, providing potential improvements in computational efficiency, robustness to noisy inputs, and enhanced accuracy in limited-data settings. The integration of quantum computing concepts into classical frameworks has thus emerged as a promising direction for enhancing Arabic handwritten text recognition, particularly in scenarios involving historical and degraded manuscripts.

This paper presents a comprehensive review of classical and quantum-inspired techniques for AHTR, comparing their strengths, limitations, and practical applicability. Furthermore, it discusses emerging research directions and highlights how hybrid quantum-classical models can contribute to improved recognition accuracy, robustness, and efficiency, particularly in challenging handwriting recognition contexts.

1.1 Characteristics and Challenges of Arabic Handwritten Script

The Arabic language, known for its rich history and linguistic complexity, presents several unique characteristics and challenges that impact the recognition of Arabic handwritten texts [2]. As one of the most widely spoken languages in the world, Arabic is the official language of over 20 countries and has multiple regional dialects, which adds to its diversity and complexity.

Script and Writing System: Arabic is a cursive script, which means that letters inside words are connected, and it is written from right to left. Character recognition is difficult since each letter might have a distinct form based on where it appears in the word (initial, medial, final, or solitary) (Fig. 1).

Fig. 1. The different forms of the character "ain" depending on its position in the word.

Diacritics: In handwritten Arabic texts, diacritical marks for short vowels and other phonetic elements may be omitted, inconsistently applied, or placed above or below characters, adding complexity to text recognition. These marks, crucial for determining word meanings, contribute to ambiguity in interpretation when not consistently represented.

Ligatures: Arabic script uses ligatures, combinations of letters that form distinct characters. Recognizing these ligatures is crucial for accurate text recognition, as they can alter the visual appearance of words.

Calligraphic Styles: Arabic has a variety of calligraphic styles, such as Naskh, Kufic, Thuluth, and Diwani. Each style has its unique visual features and rules, making it necessary to account for these variations when recognizing handwritten Arabic texts, especially in historical manuscript documents (Fig. 2).

Fig. 2. Example of calligraphic styles Naskh, Kufi, Nastaliq, Thuluth, and Diwani.

Ambiguities and Overlapping Characters: Arabic script can include overlapping characters or letters that are visually similar, creating ambiguities in recognition. These factors complicate the distinction between different characters and words.

Complex Word Structure: Arabic morphology is highly complex due to its root-and-pattern system, which involves the combination of root letters with patterns to form words. Recognizing different morphological forms is essential for understanding the meaning of texts.

The challenges of recognizing Arabic handwriting stem from these unique linguistic and scriptural characteristics. Developing advanced models and algorithms that can account for these complexities is key to achieving accurate text recognition in Arabic. Researchers working in this field must navigate these intricacies to create effective solutions that can be applied across different contexts and script variations. Given these significant challenges, it is imperative to explore alternative computational paradigms, especially those that can more efficiently handle the linguistic complexity of the Arabic script. This leads us to consider quantum-inspired approaches.

1.2 Motivations for Quantum-Inspired Approaches

Quantum-inspired computational techniques represent a compelling research direction given the significant limitations identified in classical approaches - primarily related to data-intensive training requirements, limited robustness to noisy inputs, and insufficient generalization to historical and diverse manuscripts. Quantum-inspired methods exploit quantum computational concepts such as quantum superposition and entanglement to achieve more effective and efficient data processing, reduced resource dependency, and improved performance on diverse and challenging datasets.

Hybrid quantum-classical systems further exploit the complementary strengths of classical neural network architectures (efficient spatial feature extraction) and quantum computational methods (effective representation and management of complex semantic and structural relationships). These approaches promise robust, adaptive and scalable solutions to long-standing challenges in Arabic handwriting recognition.

2 Background and Related Work

Arabic Handwritten Text Recognition (AHTR) faces unique challenges due to the cursive nature of Arabic script, shape variability, positional forms, optional diacritics, and a wide range of calligraphic styles. Recognizing such complexity requires advanced models that can generalize across styles, handle noise and degraded manuscripts, and operate with limited labeled data.

Classical approaches have been widely applied to AHTR. Convolutional Neural Networks (CNNs) are effective in spatial feature extraction and have achieved high accuracy (90–99%) on datasets like AHCD and IFN/ENIT. However, CNNs struggle with capturing sequential dependencies and require large annotated datasets, limiting their robustness on historical texts. Recurrent Neural Networks (RNNs), particularly LSTM-based models, improved sequence modeling and achieved >98% accuracy on datasets like IAM and RIMES. Still, they are computationally expensive and suffer from vanishing gradients in long sequences. More recently, Transformer-based architectures leveraging attention mechanisms have shown promise in handwritten recognition tasks due to their ability to capture long-range dependencies [4]. Nonetheless, they demand significant computational resources and perform poorly in low-resource or noisy scenarios.

Despite these advances, classical models face three key limitations: (1) heavy reliance on labeled data, (2) poor robustness to noise and variability in historical manuscripts, and (3) limited generalization to diverse writing styles and scripts. These challenges have motivated the exploration of alternative paradigms.

In recent years, Quantum Natural Language Processing (QNLP) and hybrid quantum-classical models have emerged as promising alternatives. Between 2016 and 2021, quantum-inspired techniques introduced Hilbert space embeddings and density matrix representations to enhance semantic understanding. Hybrid models began integrating classical feature extraction (CNNs, RNNs) with quantum circuits for classification, notably through frameworks such as DisCoCat, Quantum GANs (QGANs), and Variational Quantum Circuits (VQC). These models aim to reduce data dependency, improve robustness in low-resource settings, and capture complex contextual relations using quantum principles like superposition and entanglement.

While full-scale quantum models remain limited by current hardware, hybrid quantum-classical systems have shown encouraging results, particularly for morphologically rich and low-resource languages like Arabic. For example, architectures combining CNNs with quantum-inspired embeddings or classifiers offer enhanced semantic representation and classification performance, even in challenging settings. These developments suggest that hybrid approaches can address the shortcomings of classical models and open new directions for robust AHTR, especially in historical document analysis.

Impact on Arabic Handwriting Recognition
This evolution underscores the benefits of hybrid quantum-classical models, particularly in low-resource and morphologically complex languages like Arabic. Despite hardware limitations, hybrid architectures—CNNs for spatial features and quantum embeddings for semantic representation—hold promise for robust and efficient handwriting recognition.

Building on the shortcomings of classical methods, the following section explores the principles of quantum computing in pattern recognition, emphasizing its potential for advancing Arabic handwritten text recognition.

3 Quantum Computing in Pattern Recognition

Quantum computing introduces a new paradigm in pattern recognition, leveraging quantum mechanics to solve complex problems more efficiently than classical methods. This section highlights key quantum principles relevant to Arabic Handwritten Text Recognition (AHTR).

3.1 Fundamental Principles of Quantum Computing

- **Quantum Superposition:** Qubits exist in multiple states at once, allowing quantum models to process multiple possibilities simultaneously, accelerating pattern recognition.
- **Quantum Entanglement:** Entangled qubits share information instantly, enhancing the modeling of connected Arabic characters.
- **Quantum Interference:** Quantum algorithms amplify correct solutions while suppressing errors, improving recognition accuracy.
- **Quantum Gates and Circuits:** Operations like Hadamard and CNOT manipulate qubits for efficient text classification and feature extraction [7].

3.2 Quantum Algorithms for Natural Language Processing

Quantum-inspired NLP methods address key challenges in classical models, enhancing computational efficiency and semantic representation.

QNLP Pipeline
A Quantum Natural Language Processing (QNLP) pipeline follows three stages:

1. Encoding: Converting classical data into quantum states.
2. Quantum Processing: Leveraging quantum circuits for classification and semantic modeling.
3. Decoding: Converting quantum outputs back into classical data. (Fig. 3)

Compositional Distributional Semantics (CDS)
Combines syntax and semantics using quantum tensor spaces, enabling efficient calculations for sentence similarity and classification.

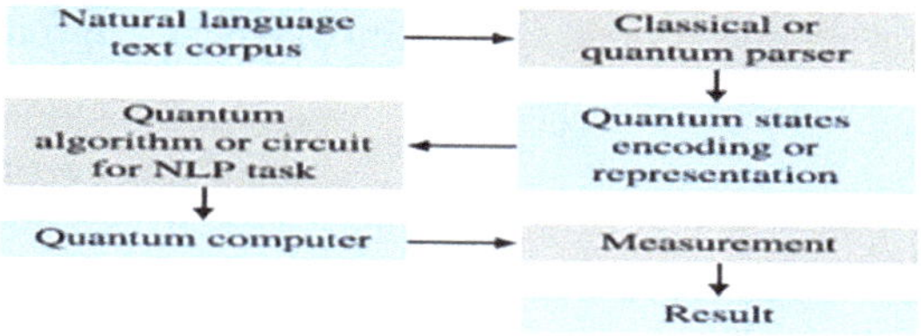

Fig. 3. General pipeline in Quantum Natural Language Processing (QNLP), illustrating the transformation from classical text corpora to quantum state encoding, circuit-based processing, and final measurement. This framework underlies many hybrid approaches applied in NLP and handwritten text recognition.

Quantum-Inspired Complex Word Embeddings
Represent words in Hilbert spaces with complex values (amplitude & phase), capturing semantics more effectively than traditional embeddings. Models like CE-Sup and CE-Mix outperform classical embeddings in handling context and ambiguity [20].

Pre-trained Models
Hybrid quantum-classical models integrate pre-trained language embeddings to enhance accuracy:

- QBERT: Inspired from BERT model, QBERT Uses pre-trained embeddings to compensate for quantum hardware limitations.
- QPFE-ERNIE: Enriches contextual representation, improving Arabic handwriting recognition.

Quantum Attention Mechanism
Quantum-enhanced attention improves long-range dependencies in Transformers using superposition and entanglement, strengthening contextual understanding of complex Arabic scripts. These models offer better accuracy and generalization with lower computational costs. Recent studies have proposed combining quantum word embeddings with quantum-inspired attention mechanisms to enhance semantic representation and contextual understanding [16].

Quantum Generative Adversarial Networks (QGAN)
QGANs generate high-quality synthetic Arabic handwriting to augment scarce datasets, enhancing robustness and recognition accuracy in hybrid quantum-classical models [18].

Quantum Similarity Algorithms
Exploit quantum mechanics for faster similarity calculations, achieving quadratic speedups in textual similarity tasks over classical models.

3.3 Quantum-Inspired and Hybrid Quantum-Classical Approaches

Quantum-inspired and hybrid approaches aim to overcome limitations of classical models by integrating quantum principles into conventional frameworks. These methods efficiently handle complex, noisy, and low-resource datasets using quantum properties like superposition, entanglement, and quantum-enhanced embeddings (Fig. 4).

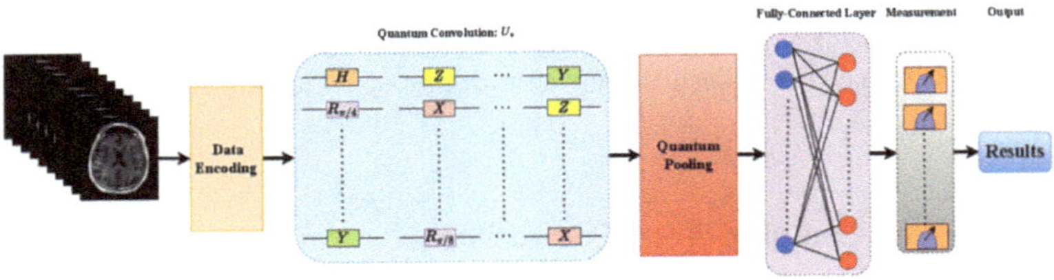

Fig. 4. Architecture of a Quantum Convolutional Neural Network (QCNN) inspired by Innan et al. (2024), illustrating the use of quantum convolution and pooling layers followed by classical fully connected layers and measurement. This hybrid structure enhances spatial feature extraction and classification in quantum machine learning models.

Quantum-Inspired Complex Word Embeddings (QCWE)
Leverage complex amplitudes and phases to represent words, capturing both statistical relevance and semantic nuances:

- CE-Sup: Encodes multiple meanings via superposition.
- CE-Mix: Uses density matrices to model diverse semantic interpretations, outperforming classical embeddings (Word2Vec, FastText).

Hybrid Quantum-Classical Architectures

- Combine classical feature extraction with quantum-enhanced representation to improve accuracy, robustness, and efficiency.
- CICWE-QNN: Fuses CNNs with quantum layers for better classification, especially in noisy or data-scarce settings [17].
- CNN + Quantum Classifier: Enhances semantic processing by using CNNs for spatial features and quantum classifiers for context understanding.
- RNN + Quantum Circuits: Addresses vanishing gradient issues in sequential data by integrating RNNs with quantum circuits.
- Transformers + Quantum Enhancements: Employs quantum attention mechanisms and embeddings to improve contextual understanding while reducing computational cost.
- Pre-trained Models (QBERT, Word2Ket): Uses pre-trained embeddings with quantum layers to bridge quantum hardware constraints.

Quantum Convolutional Neural Networks (QCNN)
QCNNs apply quantum operations to extract hierarchical features from complex data:

- Quantum Convolutions: Use quantum gates to capture both local and global patterns, useful for historical Arabic handwriting recognition.
- Quantum Pooling: Efficiently reduces data dimensionality while preserving key information, enabling better performance on high-dimensional, low-data scenarios.

DisCoCat and Categorical Models for NLP

DisCoCat extends Compositional Distributional Semantics (CDS) using category theory and tensor networks [10] to unify syntax and semantics in quantum-enhanced NLP:

- Facilitates faster sentence similarity calculations.
- Improves semantic disambiguation and classification accuracy.
- Reduces computational complexity, making quantum-inspired NLP methods more practical.

The following pipeline summarizes a representative architecture commonly used in recent hybrid approaches combining classical and quantum techniques for AHTR.

Input images (e.g., AHCD dataset, 28 classes) are preprocessed (binarization, resizing, normalization), then passed through CNNs layers for feature extraction.

Features are encoded—typically via AngleEmbedding—into 4–6 qubits, and classification is performed using QSVM or Variational Quantum Circuits (VQC). Most experiments are conducted on simulators (e.g., PennyLane, Qiskit) due to current hard ware constraints. This setup reflects how classical spatial encoding and quantum se mantic modeling are effectively integrated.

4 Discussion and Future Directions

4.1 Comparative Performance

Comparing classical, hybrid, and quantum-inspired models allows for a precise evaluation of their strengths and weaknesses. The key evaluation criteria include:

- Accuracy & Precision – Performance across structured and noisy datasets.
- Robustness – Stability under degraded or historical texts.
- Generalization – Ability to adapt to diverse calligraphic styles.
- Computational Efficiency – Processing time & memory requirements.
- Data Requirements – Dependence on large annotated datasets.

The table below presents a structured comparison of classical and quantum-enhanced methods, highlighting their effectiveness in Arabic handwritten text recognition (AHTR) (Table 1).

Table 1. Comparative overview of classical, quantum-inspired, and hybrid models in handwritten text recognition.

Works	System Description	Accuracy	Advantages and limitations
Shams et al. [13]	DCNN + SVM applied to AHCD dataset (16,800 images). Used 80/20 split for training/testing.	After 400 iterations: Accuracy 98.08% (training set), with 95.07% for the tested images.	+ Accurate, robust across multiple character classes. – Style-sensitive, needs manual labeling.
Alwaqfi et al. [19]	DCNN with GAN-generated samples and N + 1 class strategy applied to AHCD.	The proposed model achieved an average overall accuracy of 99.7%.	+ Robustness via GAN + N + 1 strategy. – GAN data may lack full real world complexity.
Jbrail et al. [3]	4 CNN models applied to Hijja dataset (47,434 characters, 80/20 split).	The highest accuracy achieved from the proposed models was 99.3%.	+ Strong char. recognition; model comparison. – Style-limited, not fit for word-level tasks.
Gader et al. [14]	ConvLSTM-CTC (CNN + LSTM) on IFN/ENIT and AHDB for character and word recognition.	Accuracy: 99.01% recognizing handwritten Arabic words.	+ Captures space-time features, supports char/word tasks. – Scalability & overhead concerns.
Calfa Vision [8]	Pre-trained models for char./word/baseline recognition on Maghribi manuscripts (RASAM dataset, 7,540 lines).	95.2% overall accuracy, 97% accuracy in main text and baseline detection, and a 4.8% CER.	+ Pre-trained models improve accuracy & Versatile on historical Arabic texts. – Small dataset & handcrafted features limit adaptability .
Montazer et al. (2017) [5]	Hybrid QNN + BoVW using SIFT on HODA dataset for digit recognition.	99.3%.	+ Outperforms SVM & MLP. + Robust to handwriting variation. – High computational cost & Limited scalability.
Díaz-Ortiz et al. (2024) [11]	SENN: CNN + VQC with strong entanglement layers for NLP classification (simulated data for text recognition).	97.7%.	+ Quantum-enhanced semantic context, fewer parameters, faster convergence than classical RNN. + Robust on noisy and limited datasets. + Unlike DisCoCat, which focuses on sentence structure, SENN learns sequential dependencies more efficiently. – Still experimental & requires hybrid optimization.

(continued)

Table 1. *(continued)*

Works	System Description	Accuracy	Advantages and limitations
Olaoye & Potter (2024) [6]	Hybrid model using BERT/GPT for preprocessing and VQE/QAOA for embedding & optimization (sentiment, classification, translation tasks).	IMDB: 89.2%, AG News: 93.1%, BLEU: 37.2	+ Quantum-enhanced embeddings outperform BERT (88.5%) & GPT (92.3%). Effective across multiple NLP tasks. – +15–25% overhead, quantum noise, scalability limits.
Widdows et al. (2024) [9]	QSVM + QCBM for text classification and bigram modeling with quantum encodings for word types.	62% (10 k words, simulated NLP)	+ QSVM/QCBM Real-hardware QSVM/QCBM handle ambiguity; suited for rich morphologies. – Moderate accuracy; limited scalability due to quantum noise.
Yousuf Iqbal Tomal et al. (2025) [12]	Hybrid Transformer combining classical attention with quantum kernel similarity, QFT, and VQC circuits. Enhances attention using quantum mechanisms for sentiment analysis (IMDb dataset, 1000 samples).	65.5% Accuracy on IMDb sentiment classification dataset (limited dataset: 1000 samples).	+ (+1.5%) accuracy via quantum attention (kernel sim. + VQC). + Faster convergence (15 vs. 20 epochs); better interpretability. + Suited for complex scripts like Arabic. – Circuit overhead, quantum noise, scalability issues.

Beyond accuracy and robustness, computational complexity is a critical factor when evaluating classical, quantum-inspired, and hybrid approaches. Classical models such as CNNs and Transformers often require substantial resources, with attention mechanisms scaling as $O(n^2)$. In contrast, quantum-inspired methods exploit parallelism to reduce certain operations to $O(\log n)$, improving efficiency. Hybrid models attempt to strike a balance by using quantum circuits for complex data encoding while preserving accuracy. Still, hardware limitations and integration challenges remain.

4.2 Discussion and Future Directions

The comparative analysis highlights clear distinctions between classical, quantum-inspired, and hybrid models in Arabic Handwritten Text Recognition (AHTR). While CNNs and RNNs perform well on clean, structured datasets, they often underperform when faced with historical or noisy Arabic texts. Hybrid and quantum-inspired models offer greater flexibility in these complex scenarios. Manuscripts affected by ink artifacts, noise, or intricate calligraphy, for instance, yield better results with adaptable architectures such as CICWE-QNN and QCNN, whereas traditional models relying on rigid preprocessing and segmentation struggle to generalize across handwriting styles.

Classical deep learning models, including CNNs, RNNs, and Transformers, have nonetheless achieved high accuracy rates on standard datasets such as AHCD and IFN/ENIT, with studies by Shams et al. and Alwaqfi et al. reporting over 95%. These models are effective at extracting spatial and sequential features but exhibit notable limitations: they are sensitive to noise, require significant computational resources—particularly for Transformers with $O(n^2)$ attention complexity [15]—and often lack adaptability without extensive fine-tuning.

Quantum-inspired approaches like Quantum Complex Word Embeddings (QCWE) and Quantum Kernel Methods (QSVM) provide compelling alternatives. By leveraging Hilbert spaces, they capture richer semantic relationships than classical models like Word2Vec or FastText and are particularly effective in low-resource settings. Studies such as [11] demonstrate that quantum-enhanced classifiers like SENN can outperform classical models in noisy AHTR tasks, reducing the need for large annotated datasets. Hybrid architectures, such as CICWE-QNN and Quantum-Enhanced Transformers, further extend these advantages: Variational Quantum Circuits (VQCs) help reduce trainable parameters while preserving accuracy, and quantum attention mechanisms enhance context recognition, especially for connected characters and diacritics in Arabic writing.

Despite their promise, practical challenges persist. Quantum hardware still faces issues related to noise and decoherence, necessitating error correction strategies. Integrating classical deep learning and quantum circuits requires careful architectural design and specialized optimization techniques. Moreover, Quantum Random Access Memory (QRAM) remains a significant bottleneck: without scalable memory solutions, real-world deployment of quantum-enhanced AHTR models on large datasets remains limited. Progress in quantum memory technologies will be crucial for unlocking the full potential of these architectures.

Comparative studies also show encouraging signs: Quantum Generative Adversarial Networks (QGANs) help enrich Arabic handwriting datasets, bridging the gap with classical models, while Quantum-Enhanced Transformers (QET) improve convergence rates and generalization capabilities in NLP tasks, suggesting potential benefits for sequence modeling in AHTR. Although classical deep learning models still dominate large-scale applications, the gap is narrowing. For instance, Montazer et al. achieved 99.3% accuracy with QNNs—comparable to CNN-RNN models on Arabic digits. Nonetheless, frameworks like DisCoCat still require further validation on Arabic handwriting before broader adoption.

Looking ahead, advancing quantum-inspired and hybrid models for AHTR will require focused efforts in several areas: enhancing quantum simulators to better design and test hybrid architectures; extending pretrained models like QBERT and Word2Ket to Arabic handwriting tasks; and integrating quantum pooling with CNNs to improve spatial feature extraction. Furthermore, exploring quantum entanglement mechanisms could aid in segmenting connected Arabic characters, a major challenge in handwritten scripts. Expanding evaluations to other morphologically rich languages such as Urdu and Farsi may also help assess the generalizability of these models across related scripts.

While classical models continue to perform well on structured data, hybrid and quantum-inspired approaches are increasingly relevant for complex, noisy, and low-resource handwriting scenarios. As quantum hardware evolves, these models are likely to play a central role in shaping the future of Arabic handwritten text recognition.

5 Conclusion

Quantum-inspired and hybrid quantum-classical models show strong potential for Arabic Handwritten Text Recognition (AHTR), especially in historical, noisy, or low-resource contexts. They enhance semantic representation, improve robustness, and reduce the need for large annotated datasets—making them valuable alternatives to traditional deep learning methods.

Despite this promise, challenges remain: limited quantum hardware scalability, lack of practical QRAM, and integration issues. Future research should focus on optimizing hybrid frameworks, refining quantum embeddings, and testing on standard benchmarks such as IFN/ENIT.

Cross-disciplinary collaboration between NLP, computer vision, and quantum computing will be key. As hardware improves, these models could shift from theoretical exploration to real-world deployment, shaping the next generation of AHTR systems.

References

1. Altwaijry, N., Al-Turaiki, I.: Arabic handwriting recognition system using convolutional neural network. Neural Comput. Appl. **33**, 2249–2261 (2021)
2. AlKendi, W., Gechter, F., Heyberger, L., Guyeux, C.: Advancements and challenges in handwritten text recognition: a comprehensive survey. J. Imaging. **10**(1), 18 (2024)
3. Jbrail, M.W., Tenekeci, M.E.: Character recognition of Arabic handwritten characters using deep learning. J. Stud. Sci. Eng. **2**(1), 32–40 (2022) Available at: https://www.researchgate.net/publication/359377764
4. Momeni, S., BabaAli, B.: A Transformer-based approach for Arabic offline handwritten text recognition. arXiv preprint arXiv, 2307.15045 (2023)
5. Montazer, G.A., Soltanshahi, M.A., Giveki, D.: Farsi/Arabic handwritten digit recognition using quantum neural networks and bag of visual words method. Opt. Mem. Neural Netw. **26**(2), 117–128 (2017)
6. Olaoye, F., Potter, K.: Hybrid quantum-classical models for natural language processing. Mach. Learn. (2024) Available at: https://easychair.org/publications/preprint/X6DZ/open
7. Nielsen, M.A., Chuang, I.L.: Quantum Computation and Quantum Information. Cambridge University Press (2010)
8. Lévêque, N., Cherti, S., Vidal-Gorène, C.: New results for the text recognition of Arabic Maghribī manuscripts – managing an under-resourced script. arXiv, 2211.16147 (2022) https://arxiv.org/abs/2211.16147
9. Widdows, D., Alexander, A., Zhu, D., Zimmerman, C., Majumder, A.: Near-term advances in quantum natural language processing. Ann. Math. Artif. Intell. (2024). https://doi.org/10.1007/s10472-024-09940-y
10. Coecke, B., Sadrzadeh, M., Clark, S.: Mathematical foundations for a compositional distributional model of meaning. Linguist. Anal. **36**(1–4), 345–384 (2010)

11. Díaz-Ortiz, J.I., Villanueva, A., Delgado, F.: Strongly entangling neural network: quantum-classical hybrid model for quantum NLP. In: Vlachos, D. (ed.) Mathematical Modeling in Physical Sciences. Springer Proc. Math. Stat, p. 446 (2024)
12. Iqbal, S.M.Y.T., et al.: Quantum-enhanced attention mechanism in NLP: a hybrid classical-quantum approach. arXiv preprint arXiv, 2501.15630 (2025)
13. Shams, M., Elsonbaty, A.A., ElSawy, W.Z.: Arabic handwritten character recognition using CNN and SVM. Int. J. Adv. Comput. Sci. Appl. **11**(8), 144–149 (2020)
14. Gader, T.B.A., Chibani, I., Echi, A.K.: Arabic handwriting offline recognition using ConvLSTM-CTC. In: Proc. 12th Int. Conf. Pattern Recognition Applications and Methods (ICPRAM 2023), pp. 529–533. SCITEPRESS (2023). https://doi.org/10.5220/0011794700003411
15. Vaswani, A., et al: Attention is all you need. In: Adv. Neural Inf. Process. Syst. 30 (2017)
16. Widdows, D., Aboumrad, W., Kim, D., Ray, S., Mei, J.: Quantum NLP: word embeddings and attention. arXiv preprint arXiv, 2403.19758 (2024)
17. Shi, J., Li, Z., Lai, W., et al.: Two end-to-end quantum-inspired deep neural networks for text classification. IEEE Trans. Knowl. Data Eng. **35**(4), 4335–4345 (2021)
18. Huang, H.-L., Du, Y., Gong, M., et al.: Experimental quantum generative adversarial networks for image generation. Phys. Rev. Appl. **16**(2), 024051 (2020)
19. Alwaqfi, Y.M., Mohamad, M., Al-Taani, A.T.: GAN for Arabic handwritten characters. Int. J. Adv. Soft Comput. Appl. **14**(1), 177–195 (2022)
20. Li, Q., Uprety, S., Wang, B., Song, D.: Quantum-inspired complex word embedding. arXiv preprint arXiv, 1805.11351 (2018)

Leveraging Knowledge Graphs for Enhanced Fake News Detection: Techniques, Models, and Applications

Adil Elmansori[1](✉), Abdelmounaim Bouhlali[1], Abderrahim El Mhouti[1], and Mohamed Fahim[2]

[1] ISISA, Faculty of Sciences, Abdelmalek Essaadi University, Tetouan, Morocco
adil.elmansori@gmail.com
[2] CSRL, Ibn Tofail University, Kenitra, Morocco

Abstract. The rapid expansion of information available online has introduced new issues regarding public trust, health, and democratic processes as the prevalence of misleading claims increased dramatically. Traditionally used misinformation identification methods like machine learning and natural language processing are better suited for simpler, less nuanced types of information. This paper examines using Knowledge Graphs as an effective strategy for tackling the fake news problem. Knowledge Graphs provide a structured framework for describing entities and their interrelationships, enabling the integration of diverse knowledge sources and their validation. Incorporating techniques such as Entity Linking through Graph Neural Networks and Knowledge Graph Embeddings can aid in contextualizing statements in light of a graph of established facts, assisting Knowledge Graphs in lending meaning to content and thereby enhancing their detection of false information. The paper further notes that advanced ML models including LSTM Networks, Graph Attention Networks, and BERT have proven remarkably accurate at identifying fake news when paired with Knowledge Graphs. Additionally, this work presents fact-checking processes for validating claims against structured data as well as contemporary datasets used for fake news detection. Knowledge Graphs represent a powerful approach combining structured knowledge with advanced machine learning methods, offering hope as a potential unified solution to the ever-growing challenge posed by the escalating threat of misinformation in the digital era.

Keywords: Knowledge Graphs (KGs) · Knowledge Graph Embeddings (KGEs) · Machine learning · Deep Learning · Transformers

1 Introduction

Fake news is a global problem, and the growth of digital platforms and social media has fueled the fire. Such misinformation can spread rapidly, generating confusion and, at times, shaming the general public and damaging people and societies. For instance, false assertions of electoral fraud in the 2020 U.S. elections degraded trust in democratic institutions [1] and COVID-19 misinformation increased vaccine hesitancy and stymied public health initiatives [2].

B.-E. Boudriki Semlali et al. (Eds.): DATA 2025, LNNS 1908, pp. 77–87, 2026.
https://doi.org/10.1007/978-3-032-21200-9_7

Fake news remains difficult to identify and counteract, as it is increasingly relying on emotional engagement, and continuously evolving tactics [3] like deepfakes and narrative manipulation. Though machine learning and NLP provide promising detection capabilities, traditional approaches fail to cope with context-dependent misinformation [4]. Debunking effectively means not just breaking down false claims, but supplying fact-checked data alongside them, such as comparing the anti-vaccine rhetoric to scientific studies or validating allegations of election fraud with official records.

A performent solution is to employ knowledge graphs (KGs), which structure entities and their relationships, allowing claims to be cross-checked with reliable sources [5]. For example, KGs can track claims about COVID-19 vaccine back to scientific literature, or connect election misinformation to official data to reveal discrepancies. Through contextualizing information and finding patterns, KGs maximize AI's ability to fight misinformation in a systematic way.

2 Knowledge Graph-Based Techniques for Fake News Detection

Knowledge Graphs (KGs) are powerful tools used for misinformation detection of facts by storing factual information in a structured and interlinked manner. The entities and relationships defined in their machine-readable formats allow for systematic verification of claims against authoritative sources. Such structured way provides three main benefits: detect contradictions in statements, discover hidden links between different subjects, and fact-check statements with evidence-based. Such graphs enable scalable credibility evaluation of electronic content through linking unstructured materials with structured knowledge.

2.1 Entity Linking and Disambiguation

Entity identification and disambiguation are vital to effectively utilizing knowledge graphs (KGs) for fake news detection. This includes the extraction of entities (e.g., persons, organizations) from text and causing them to refer to corresponding nodes in the associated knowledge graph (KG) for structural knowledge verification. Models DEA such as P-FAKED [6] and KAN [11] detects entities in the news titles with the Named Entity Recognition (NER) and disambiguates them by using Named Entity Disambiguation (NED) to map these entities to Wikidata, and TAGME to resolve ambiguities (e.g., disambiguating Donald Trump with other entities which have the name "Trump") using knowledge bases like YAGO and Freebase. This guarantees accurate matching of text mentions and KG entities in the world.

- **Named Entity Recognition (NER)**: is a task in NLP which finds and classifies named entities in text into pre-defined categories like persons, organizations, locations, and dates [8]. Formally, given a sequence of tokens $T = \{t_1, t_2 \ldots, t_n\}$, NER assigns each token t_i a label y_i drawn from a set of entity types. This can be represented as the conditional probability distribution:

$$P(y_1, y_2 \ldots, y_n | t_1, t_2 \ldots, t_n) \quad (1)$$

- **Named Entity Disambiguation (NED)**: is the task of resolving ambiguities in entity mentions by linking them to the proper entity in some knowledge base. This is critical as in most cases an entity name can refer to more than one real-world entity (like "Paris" can be the city or the person in Greek mythology) [9].
  For an entity mention m and a set of candidate entities E = $\{e^1, e^2, ..., e^k\}$ from the KG, NED predicts the most likely entity e* based on the context C of the mention:
$$e^* = \underset{e \in E}{\arg\max}\, P(e|m, C) \quad (2)$$
  The probability $P(e|m, C)$ can be approximated based on the similarity measures of the mention m and candidate entities e, commonly evaluated through embeddings or contextual representations.
- **Entity Linking Tools (TAGME):** Entity linking for short text slides directly from Wikipedia It employs a probabilistic model that predicts how relevant each candidate entity is [7]:
$$P(e|m) = \frac{score\ (e, m)}{\sum_{\epsilon' \in E} score\ (\epsilon', m)} \quad (3)$$
  with *score* (e, *m*) being a function that measures how close the entity is to the mention
  .

Entity linking provides precise mapping of text to knowledge graphs, but doing so efficiently calls for complex embedding methods. This brings us to knowledge graph embedding that embeds the semantic relationship in continuous vector space.

2.2 Knowledge Graph Embeddings

Knowledge graph embeddings (KGEs) are a promising way to represent entities and relationships within ko continuous vector spaces. These techniques represent discrete entities and relations as dense, low-dimensional vectors, allowing machines to learn semantic relationships and reason with structured knowledge [8, 10]. This is especially useful for news content, where the external structured knowledge is contextually rich and informative [8, 11].

The nodes are referred to as entities (for example "Barack Obama"or "United States") in knowledge graphs, and edges represent the relation (for example "is a" or "lives in"). KGEs map these concepts into a continuous vector space in which semantic similarities and relationships become geometric properties [11]. TransE [13] represents relations as translation vectors between entity vectors. a relation is modelled as translation vector between two entity vectors. More sophisticated models such as ComplEx (Complex Embeddings) overcome this limitation by modeling the entities and relations using complex-valued vectors. This enables the model symmetric, antisymmetric, and inverse relationships capture capabilities and a more expressive and suitable model for a larger variety of knowledge graph structures [10].

KGEs are optimized through the use of loss functions which cause embeddings to approximate the underlying structure of the knowledge graph [8]. Such as the loss function for TransE [13] that minimizes the distance based on true triples h + r and t while maximizing the same for false triples [10]. By employing negative sampling which creates negative or false triples by replacing one of the entities in true triples with a random entity to help the learning model differentiate between valid and invalid relationships [10], the efficiency of training is improved.

Evaluation metrics such as mean reciprocal rank (MRR) and hits@k are used to evaluate the power of these embeddings. For example, it achieves high mean reciprocal ranking (MRR) scores of 0.69 and 0.94 on FB15k and WN18 benchmark datasets respectively [10]. These strong performance metrics make a case for why these KGEs have been and continue to be foundational for many downstream applications.

These embeddings go further than simple link prediction or entity classification [8] by improving the representation of news. This use case shows how the KGE mathematical foundations become a real-world asset. The frameworks such as KAN (Knowledge-Aware News Representation) [11] leverage KGEs to semantically enrich the news content. For example, if an article mentions "Barack Obama", KGEs can determine the additional related entities, such as "politician", "United States", and "Democratic Party", and therefore understand the logical relationships beyond the lexical level of text. Extended this method, KG-MFEND [12] further combines a variety of text representations of news text with ComplEx embeddings [13] to create a method for the automatic detection of fake news including multiple domains [12], demonstrating the utility of leveraging structural knowledge learned during training to enhance content understanding.

- **TransE (Translation Embeddings):** model entities and relations as continuous vectors, and the relationship as a translation: $h + r \approx t$; where h, r, t are the embeddings of head entity, relation, and tail entity respectively. For one-to-one relationships, it is straightforward and works well, but in the presence of complicated relational structure, it falls apart [13].
- **ComplEx (Complex Embeddings):** embodies more complex aspects of representing relations/edges with complex-valued vectors, allowing for the capturing of symmetric, antisymmetric, and inverse relationships between entities [33, 39].

These embedding techniques form the basis for more advanced verification methods. Extending it to formal relying on entity linking and semantic embeddings, next up we venture complete fact verification pipelines that systematically evaluate the validity of claims.

2.3 Knowledge Graph-Based Fact Verification

Fact verification refers specifically to a systematic approach to evaluating the truth of assertions by contrasting them with sources of authoritative knowledge. For computational systems, this facilitates automated factual accuracy via structured comparisons between extracted claims and verified knowledge representations [4]. It is a key part of information quality assessment and is especially important to tackle the increasing problem of misinformation in digital content [4]. The verification framework employs two major techniques based on the empirical literature:

- **Fact Extraction Methodology:** This method builds on the base principles of knowledge representation [10, 11] and systematically transforms unstructured content into a machine-processable triple format, (Subject, Predicate, Object). The methodology includes essential data cleaning and normalization procedures to ensure extraction quality and consistency [8, 11]. This enables computational verification against knowledge graphs afterward by transforming free-text information into structured knowledge representations.

- **Fact-Checking:** Fact-checking is the practice of verifying the accuracy of a claim by comparing statements to reliable source of information. With knowledge graphs (structured networks of connected facts), this verification can be systematic and scalable. This provides a strong basis for fact-checking, as it enables the process to automatically differentiate between fact and misinformation based on the relationships between claims and known knowledge. Here are the steps involved in the verification process:

  1. **Entity Mapping:** The subject (S) and object (O) of the claim are matched to their respective node in the KG using measures of similarity like cosine distance or Jaccard distance [14]. For instance, in the statement "Albert einstein worked at the swiss patent office", S = Albert einstein and O = swiss patent office is matched with their respective nodes in the KG.
  2. **Relation verification:** This suspicion is checked against the KG to look for an edge linking S and O with a label corresponding to the predicate (P, e.g., worked at). If such an edge exists, we can confirm that the claim is true.
  3. **Knowledge Inference:** If an edge is not found, the edge probability is computed from the embedding distance or from semantic similarity, discriminative predicate paths or using link prediction embeddings like LinkNBed [14]. Say, for example, that the belief P (worked at | Albert Einstein, Swiss Patent Office) is 0.92, which means that it confident in the veracity of the statement.
  4. **Aggregation:** If there are multiple claims being validated, the authenticity scores of those claims are aggregated (e.g., with an average) to assess the truthfulness of the article as a whole.

3 Integrating Knowledge Graphs with Machine Learning and Deep Learning for Fake News Detection

Integrating knowledge graphs with machine learning and deep learning techniques has transformed fake news detection, using structured knowledge to augment data-driven approaches to identify patterns in information. Systems can now evaluate not just a piece of text but also its veracity thanks to this potent combination, which has greatly increased the accuracy of detection compared to earlier strategies.

3.1 Deep Learning for Fake News Detection

Recent progress in fake news detection has focused on utilizing deep learning architectures, especially when paired with the knowledge graph embeddings. Long Short-Term Memory (LSTM) networks have been demonstrated to be especially useful for modeling sequential text data, as they are well suited for capturing long-term dependencies inherent in news content [8]. Such features empower LSTMs to work naturally with knowledge graph-enhanced frameworks to leverage heterogeneous source of knowledge (combination of textual and structured knowledge). Key applications include:

- **Text Encoding**: LSTMs are primarily used to encode textual content, such as news titles or articles, into vector representations [65]. The DEAP-FAKED [6] model uses the biLSTM (bidirectional LSTM) to learn the sequential information of news titles. The textual representation is subsequently concatenated with embeddings for the entities taken from a knowledge graph, enabling the model to model not only the content of the news but also the structured relationships between entities when faking news detection.
- **Entity Representation**: LSTMs are also used to generate textual embeddings from unstructured text. In the CompareNet model, LSTM-derived embeddings are integrated with structured embeddings from a knowledge graph using a gating mechanism. This process helps capture semantic information from textual descriptions, which is then compared with knowledge graph data to detect inconsistencies in news content [15].
- **Sequential Data Processing**: LSTMs excel at processing sequential data, making them particularly effective for tasks like fake news detection, where understanding the context and flow of information is crucial. By capturing long-term dependencies in text, LSTMs enable models to better understand the semantics and context of news articles, social media posts, or other text-based sources [6, 16].

Although LSTMs perform well on sequential text data, knowledge has an inherent graph structure and requires specialized neural architectures to process it. This brings us to graph neural networks, which are designed to operate on interlinked data.

3.2 Graph Neural Networks (GNNs) for Fake News Detection

Graph Neural Network (GNNs) is a type of deep neural network that is specifically designed to work on graph-based data. This particularly fits modeling into the relationship of the entities in a knowledge graph (KG). This makes it very useful in certain types of machine learning tasks where one needs to identify relationships between entities, such as fake news detection, as GNNs propagates information through the graph, thus allowing the model to learn complex interdependencies.

- **Graph Attention Networks (GAT):** are a type of GNN that applies attention mechanism to account for the neighborhood nodes contributions when aggregating information. This enables the model to concentrate on the relevant portions of a graph. And assign different weights to the neighbour nodes of a particular node based on how relevant it is. This is done using an attention, where per edge, attention coefficients are calculated for each connected node. These are coefficient that decides how much data passed from one node to node. The GAT in KAPALM processes the entity graph and generates a coarse-grained representation of the structured knowledge [16].
- **Graph Convolutional Networks (GCN)**: are another type of GNN that leverage convolutional operation in order to collect information from nodes in the local neighborhood. GCNs are easier to understand compared to GATs but can be applied to a large variety of graph-based tasks as well. Solution GCNs work by summing feature

information of a node with her neighbors and then concatenating it with the nodes features (the individual part) [21]. This is done over a number of layers, enabling the model to take higher order relationships of the graph into account. Within CompareNet, it employs a GCN to take into the account the relations upon entities in the KG [15] the framework constructs a directed heterogeneous document graph for each news document associated with sentences, topics, and entities. The entities in the news document are aligned to the KB, which propagates information from the KB to the news document.

There must be more powerful explicit means of determining the importance of different elements of knowledge beyond the basic graph processing. Attention mechanisms have proven especially effective for this task.

3.3 Attention Mechanisms for Knowledge Integration

The attention mechanisms are things that allow the model to attend to different entities and relationships in the KG and decide how important they are. It helps in concentrating only on the relative information useful for the detection of fake news. The KAN framework, for instance, introduces two attention mechanisms measuring the importance of entities (News towards Entities (N-E) attention) and (Entity Contexts (N-E^2C) attention) respectively [11]. These mechanisms use the knowledge graph to give different weight to entities and their context according to how relevant they are to the news content [11]. For example, if a news article mentions "climate change", the attention mechanism can focus on giving higher weights to the entities "carbon emissions" and "global warming", since they are closely related to the topic. In the same way, KAPALM considers the importance of entities in KG with multi-head attention, which significantly helps to identify false news [21].

3.4 Transformer Models

Transformer models like BERT [19], RoBERTa [18] for encoding textual content which also works when combined with knowledge graph embeddings for fake news detection. For instance, in the DEAP-FAKED model [6], RoBERTa encodes news article text followed by enrichment with knowledge graph embeddings [16]. The model benefits from the restructured meaning of the text while also having the advantage of the KG so it can sense both structures. In the identical way, CompareNet [15] uses BERT for encoding the semantic textual content of news articles and enhances the news article representation with knowledge graph embeddings to detect fake news [8].

While deep learning approaches are now the majority of active research, traditional machine learning methods are still useful for current benchmarks and baseline performance.

3.5 Traditional Machine Learning Approaches

Support Vector Machines (SVM) is supervised machine learning model that can be used for both classification and regression tasks. Understanding how SVMs work SVMs find a hyperplane that best separates points from two classes in an N-dimensional space. The aim is to maximize the margin (distance) between the hyperplane and the closest data points to the hyperplane (support vectors) leading to a more generalised model. These include both the knowledge graph embeddings and machine learning models (for instance, SVM and Random Forest) that are generally used for fake news detection [38, 46]. SVM is used as a baseline model with respect to KAN framework to evaluate the performance of more advanced and powerful models Although conventional methods, such as the SVM and Random Forest, can provide a moderate performance, they are still outperformed by more advanced DL and KGB approaches [11]. SVM is used at a couple of different stages in knowledge graphs for fake news detection:

- **Entity Classification:** SVM can classify entities in a knowledge graph based on its features. Entities like "Paris" and "France" can be grouped as "City" or "Country," respectively, using graph-structured features (such as relationships and attributes) [20].
- **Feature-Based Fact-Checking:** SVMs can be used to verify the authenticity of claims by comparing them with knowledge graph data. For example, if a claim states "Sugar causes Liver Cancer," the SVM can classify it as true or false by checking the relationship in the knowledge graph and using additional contextual features [22].
- **Baseline Model for Comparison:** SVMs are often used as baseline models in knowledge graph-based systems to evaluate the performance of more advanced models (e.g., graph neural networks, attention-based models). They provide a simple yet effective benchmark for tasks like entity classification, relation verification, and fake news detection [21].
- **Relation Verification:** this means validating the relations in a knowledge graph using SVM. An SVM is, therefore, trained with graph embeddings and contextual features [4] to predict if a relation for a given triple (Subject, Predicate, Object) is valid (e.g. " Paris is the capital of France ") or invalid (e.g. " Paris is the capital of Germany ")
- **Fake News Detection:** The SVM learns a decision boundary based on these joined features to separate fake and real news [11].

4 Datasets and Results

This section tests the performance of the state-of-the-art KG-based fake news detection frameworks on various and heterogeneous datasets such as medical (e.g., Diabetes, Cancer), social media (e.g., PolitiFact, GossipCop), and COVID-19 misinformation (e.g., CoAID-UB) datasets. Using standard metrics (F1, accuracy), we compare key techniques such as graph attention networks (GAT), entity linking, and hybrid KG-text embeddings. Findings show that medical domains are well supported (e.g., DETERRENT: 96.5% accuracy), but reveal situations where cross-domain generalization is problematic (e.g., KG-MFEND: 63.3% F1). The Table 1 details methodologies, datasets, and quantitative results, along with links to benchmark studies.

Table 1. Summary of Datasets

Framework	DatasetSS	Results
DETERRENT [23]	Diabetes dataset	- Accuracy: 0.9206% - F1 Score: 0.8474%
	Cancer dataset	- Accuracy: 0.9652% - F1 Score: 0.9309%
DEAP-FAKED [6]	KFN-UB (Kaggle Fake News)	- F1 Score: 0.89% - Accuracy: 0.90%
	CoAID-UB (COVID-19)	- F1 Score: 0.78% - Accuracy: 0.78%
KAN [11]	PolitiFact	- Accuracy: 0.8586% - F1 Score: 0.8539%
	GossipCop	- Accuracy: 0.7766% - F1 Score: 0.7713%
	PHEME	- Accuracy: 0.783% - F1 Score: 0.7461%
KG-MFEND [12]	Chinese datasets (CN-Dbpedia, EducationKG, MedicineKG)	- Accuracy: 0.8763% - Precision: 0.7420% - Recall: 0.5990% - F1-Score: 0.6330%
Heterogeneous Representation Ensembles [13]	COVID-19 Fake News	- Accuracy: 0.9570% - F1-Score: 0.9569%
	FakeNewsNet	- Accuracy: 0.8880% - F1-Score: 0.8882%
	PAN2020	- Accuracy: 0.7200% - F1-Score: 0.7348%
KAPALM [16]	PolitiFact	- F1 Score: 91.34% (full-scale), 85.77% (few-shot).
	GossipCop	- F1 Score: 71.68% (full-scale), 60.03% (few-shot).
CompareNet [15]	SLN (Satirical and Legitimate News Database)	- F1 Score: 89.12% - Precision: 89.82% - Recall: 89.17%
	LUN (Labeled Unreliable News Dataset)	- F1 Score: 68.26% - Precision: 72.94% - Recall: 69.04%
Fake News Detection Based on Knowledge-Guided Semantic Analysis [8]	LUN (Labeled Unreliable News Dataset)	- Accuracy: 94.45%. - F1 score: 94.38%.
	SLN (Satirical and Legitimate News Database)	- Accuracy: 94.45%. - F1 score: 94.38%.

5 Conclusions

The central contribution of this study is to show that an integration of KGs and advanced AI techniques significantly improves the detection of fake news. Entity Linking and Disambiguation (NED) facilitate the exact mapping of textual claims into structured knowledge, and Graph embedding models like ComplEx provide the capability to model elaborate relational patterns that traditional-flat models cannot capture, during the inference step, all these components are essential for deep-learning architectures (e.g. LSTMs to analyze sequential text, GATs for relational reasoning and Transformer's models like BERT for semantic understanding) to come together towards accurate and context-aware detection. The results displayed by frameworks such as DEAP-FAKED and KAN, attaining F1 scores exceeding 0.85 on benchmark datasets, also exemplify the practical advantages derived from this hybrid approach.

The systematic approach of the fact verification pipeline, through entity mapping, relation verification, and knowledge inference, can work as a scalable solution for automated fact-checking. To enhance detection precision, attention mechanisms dynamically weight the relevance of entities and relationships via this process. Yet, despite KG based method performing well in specialized domains like in the medical field (e.g., DETERRENT gave an F1 of 96.5% on medical datasets), there are limitations of cross-domain generalization due overfitting on domain specific data as found with heterogenous settings like KG-MFEND's 63.3% F1.

In conclusion, future work should prioritize improving cross-domain adaptability, enhancing the detection-to-verification speed, and exploring synergies between automated KG and human fact-checkers. Refining these will allow KG-based methods to better cater to the dynamic and evolving nature of misinformation. In conclusion, this work underscores the transformative potential that KGs hold in the context of fake news detection, providing a roadmap for existing work, as well as a stepping stone for future work in the area.

References

1. Allcott, H., Gentzkow, M.: Social media and fake news in the 2016 election. J. Econ. Perspect. **31**(2), 211–236 (2017)
2. Cinelli, M., De Francisci Morales, G., Galeazzi, A., Quattrociocchi, W., Starnini, M.: The COVID-19 social media infodemic. Sci. Rep. **10**(1), 1–10 (2020)
3. Chesney, R., Citron, D.: Deep fakes: a looming challenge for privacy, democracy, and national security. Calif. Law Rev. **107**(6), 1753–1820 (2019)
4. Zhou, X., Zafarani, R.: A survey of fake news: fundamental theories, detection methods, and opportunities. ACM Comput. Surv. **53**(5), 1–40 (2020)
5. Hogan, A., et al.: Knowledge graphs. ACM Comput. Surv. **54**(4), 1–37 (2021)
6. Mayank, M., Sharma, S., Sharma, R.: DEAP-FAKED: Knowledge graph-based approach for fake news detection. In: Proceedings of the IEEE/ACM International Conference on Advances in Social Networks Analysis and Mining (ASONAM) (2022)
7. Suchanek, F.M., Kasneci, G., Weikum, G.: Yago: a core of semantic knowledge. In: Proceedings of the 16th International Conference on World Wide Web, pp. 697–706 (2007)
8. Pakhale, K.: Comprehensive overview of named entity recognition: Models, domain-specific applications and challenges. arXiv preprint arXiv, 2309.14084 (2023)

9. Ganea, O.-E., Hofmann, T.: Deep joint entity disambiguation with local neural attention. In: Proceedings of the 2017 Conference on Empirical Methods in Natural Language Processing, pp. 2619–2629 (2017)
10. Trouillon, T., Welbl, J., Riedel, S., Gaussier, É., Bouchard, G.: Complex embeddings for simple link prediction. In: Proceedings of the International Conference on Machine Learning (ICML), pp. 2071–2080 (2016)
11. Dun, Y., Wang, K., Shao, J., Han, X.: KAN: Knowledge-aware attention network for fake news detection. In: The Thirty-Fifth AAAI Conference on Artificial Intelligence (AAAI-21) (2021)
12. Fu, L., Peng, H., Liu, S.: KG-MFEND: an efficient knowledge graph-based model for multi-domain fake news detection. J. Supercomput. **79**, 18417–18444 (2023)
13. Koloski, B., Škrlj, T., Pollak, S., Lavrač, N.: Knowledge graph informed fake news classification via heterogeneous representation ensembles. Neurocomputing. **496**, 208–226 (2022)
14. Trivedi, P., Maheshwari, G., Dubey, M., Lehmann, J.: LinkNBed: Multi-graph representation learning with entity and edge inference. In: Proceedings of the Web Conference, pp. 307–317 (2018)
15. Hu, L., et al.: Compare to the Knowledge: Graph Neural Fake News Detection with External Knowledge (2021)
16. Ma, J., Gao, H., Mitra, P.: KAPALM: knowledge graph enhanced language models for fake news detection. In: Findings of the Association for Computational Linguistics: EMNLP 2023, pp. 3999–4009 (2023)
17. Qian, S., Wang, J., Hu, J., Fang, Q., Xu, C.: Knowledge-aware multi-modal adaptive graph convolutional networks for fake news detection. ACM Trans. Multimed. Comput. Commun. Appl. **17**(3), 98 (2021)
18. Liu, Y., et al.: RoBERTa: a robustly optimized BERT pretraining approach. arXiv preprint arXiv, 1907.11692 (2019)
19. Devlin, J., Chang, M.-W., Lee, K., Toutanova, K.: BERT: Pre-training of deep bidirectional transformers for language understanding. arXiv preprint arXiv, 1810.04805 (2018)
20. Pan, J.Z., Vetere, G., Gómez-Pérez, J.M., Wu, H. (eds.): Exploiting Linked Data and Knowledge Graphs in Large Organisations. Springer (2017)
21. Kipf, T.N., Welling, M.: Semi-supervised classification with graph convolutional networks. In: International Conference on Learning Representations (ICLR) (2017)
22. Pan, J.Z., et al.: Content-based fake news detection using knowledge graphs. In: The Semantic Web - ISWC 2018, pp. 669–683 (2018)
23. Cui, L., Lee, D., Liu, H.: DETERRENT: Knowledge guided graph attention network for detecting healthcare misinformation. In: Proc. 26th ACM SIGKDD Conference on Knowledge Discovery and Data Mining (KDD '20), pp. 492–502 (2020)

Revolutionizing Public Governance with Machine Learning: Analyzing the Impact of the Agenda 2030 SDGs on Organizational Effectiveness

Ifiss Saida(✉) and Mssassi Said

The National School of Business and Management, Abdelmalek Essaâdi University, Tangier, Morocco
sifiss@uae.ac.ma

Abstract. This chapter explores how the Sustainable Development Goals (SDGs) ensure that public sector effectiveness has become a major interest of policymakers, as the authors focus on predicting, but improving, governance practices is part of the path taken by Machine Learning (ML). By using several advanced ML algorithms, including XGBoost, Random Forest, and Convolutional Neural Networks (CNN), the manuscript analyses their effectiveness in predicting the impact of the critical SDGs on organizations in relation to resource management, innovation, and citizen satisfaction. Results highlight key contributions from certain SDGs, especially "Affordable and Clean Energy", "Decent Work and Economic Growth" and "Clean Water and Sanitation".

The interconnectedness of the SDGs, the dearth of contextually relevant data, and the difficulties in assessing indirect or non-linear interactions are some of the major issues raised in the article. The chapter makes the case for mixed-method approaches that combine qualitative and quantitative methodological techniques in order to address these issues. The chapter highlights how machine-learning methods can be incorporated into public policy frameworks as long as human factors and local specificities are taken into account. All things considered, this context-aware, mixed-method approach could increase organizational performance and programming governance that is in line with the SDGs.

Keywords: Sustainable Development Goals (SDGs) · Machine Learning · Organizational Effectiveness · Public Sector Governance · XGBoost Model

1 Introduction

1.1 A Subsection Sample

On society increases, an urgent need for pursuing sustainable development exists. Realising this ambition will necessitate significant and profound reframing of organisations ways of working, particularly in the public sector. The 2030 Agenda for Sustainable Development, a global consensus adopted by all United Nations Member States in

B.-E. Boudriki Semlali et al. (Eds.): DATA 2025, LNNS 1908, pp. 88–100, 2026.
https://doi.org/10.1007/978-3-032-21200-9_8

September 2015, outlines 17 Sustainable Development Goals (SDGs) that together form a comprehensive and integrated guide for global development that can meet the needs of all [1–3]. Amidst this period of transformation, Machine Learning (ML) is not merely an advancement in technology but a strategic differentiator. The power of ML to process vast amounts of data and provide information about real-time situations and accurate predictions provide significant opportunities for improving the performance of public governance systems.

The case of Morocco and its attempts to promote a global sustainability agenda creates an opportunity to study the complicated interaction between SDG implementation and public institution performance [6, 7]. This intersection raises critical questions: How do the SDGs, in fact, function to alter the functioning capacities and responsiveness of the Moroccan public institution? Furthermore, what are the most receptive machine learning models of collaborative processes to decipher and model these multilayered interactions? Such questions are timely, particularly as Moroccan public organizations work through challenges of improving service, optimizing resource use, and increasing citizen satisfaction and participation.

In such a context, one of the areas with great promise for a transformative innovation is to let the principles of sustainable development help shaping managerial and governance practices, in turn boosted by the intelligence and analytical capabilities of Machine Learning. So this chapter focuses on investigating how SDGs can be quantitatively measured to better understand their impacts on Moroccan public sector performance through advanced ML techniques [8–12]. The study seeks to identify evidenced-based governance practices, which will provide policy relevant and actionable evidence for the policy and decision-making processes in the Moroccan public administration.

To achieve this, our study adopts a structured, multi-dimensional approach comprising four principal components. Initially, we will delineate a comprehensive conceptual framework grounded explicitly in the foundational principles articulated within the 2030 Agenda and established theories of good governance. Following this conceptual anchoring, we will elaborate on our empirical methodology, explicitly detailing the compilation of an extensive dataset and the deployment of sophisticated ML algorithms, notably XGBoost, Convolutional Neural Networks (CNN), and Random Forest. Subsequently, an exhaustive analysis of our results will articulate the differential impact of specific SDGs across three crucial dimensions: organizational performance, resource management efficacy, and citizen satisfaction levels. Finally, the chapter will critically discuss the findings, acknowledging potential limitations and outlining avenues for future research that underscore Machine Learning's catalytic potential in the ongoing evolution of public governance practices.

Ultimately our research aspires to shed light on that growing juncture of sustainable development principles, organisational excellence and technological innovation. Data analysis This analysis aims to contribute, in depth and breadth, to the modernization phase of Moroccan public service, indicating a dull journey, until the establishment of such service sustainable, resilient, inclusive, not just a theoretical construct of public policies, but actually operationalize institutional practices. Through our reflection, we thus aspire to participate in the emergence of an administrative model adapted to the contemporary challenges that Morocco must address.

2 Theoretical and Conceptual Framework

The aim of this section is to provide the theoretical and conceptual foundations needed to understand our study. It addresses the challenges of Agenda 2030 for public organizations, the principles of responsible governance and the role of machine learning in the pursuit of sustainable development goals

2.1 The Fundamentals of Agenda 2030

Formally enacted by the United Nations General Assembly in 2015, Agenda 2030 is a global, comprehensive action plan aimed at addressing a variety of critical global challenges, including poverty reduction, the reduction of inequality, and the mitigation of climate change. The agenda outlines a comprehensive set of 17 interlinked Sustainable Development Goals (SDGs) [1] in which sustainable development necessitates concurrent and integrated action across multiple dimensions of society. Each goal may address a specific social dimension, but they are also interdependent and indivisible of the other goals which call for integrated action across all actors and sectors, particularly public organizations. Particularly relevant are the roles of public organizations in implementing Agenda 2030 as they are responsible to citizens for delivering essential social services, developing and implementing public policies, and do so in the public interest.

Within the Moroccan context, public organizations play a unique and profound role because of their core mandate in supporting sustainable development while addressing societal needs from a responsive governance perspective. Consequently, our research explicitly focuses on a deliberate selection of SDGs that are particularly salient to the operational context and strategic priorities of Moroccan public organizations, and closely aligned with our data. This selection is comprised of:

SDG 5 - Gender Equality: This goal seeks to eliminate all forms of discrimination based on gender, while ensuring the full and effective participation of women and girls in all aspects of public, social and economic life. For public institutions, achieving the goal means creating and implementing policies that will promote gender equality. This means creating equal opportunities for women and men, ensuring gender balance in the decision-making process, taking gender into consideration when developing public policy, and working eliminated systemic biases against gender. In that way, public institutions would play a crucial role in changing the norms in society with respect to gender equality as it relates to empowerment and adequate and equal representation of women in public life [1–3].

SDG 11 - Sustainable Cities and Communities: This target is dedicated to creating inclusive, safe, resilient and sustainable cities and communities. It implicates public agencies in key urban planning and policy implementation functions, as they must confront meaningful urban issues. These obligations include ensuring fair access to essential services including housing, transportation, health care and education; advancing sustainable mobility options; managing waste and environmental resources; and, protecting culturally-relevant places. Through their actions, public agencies participate in the making of cities subject to long-term human and ecological welfare, and represent an important aspect for citizens, their overall quality of life [1–3].

SDG 16 - Peace, Justice, and Strong Institutions: This aims to promote peaceful and inclusive societies, with access to justice for all people. It stresses the importance of transparent, accountable, and effective governance institutions at all levels of government. For public organizations, this SDG is achieved by promoting the rule of law, fighting corruption through robust and transparent mechanisms of accountability, facilitating civic engagement with the government, and delivering high quality public services. All of these efforts are necessary for building trust between citizens and government institutions and strengthening the legitimacy and effectiveness of institutions, to promote stability and social cohesion [1–3].

Together, the aforementioned SDGs represent key objectives for sustainable development in Morocco, and can strategically guide public institutions in their efforts to improve effectiveness, inclusiveness, and responsiveness. By converging this governance practice with established global standards of governance, public institutions in Morocco can enhance their contribution towards achieving sustainable, comprehensive, and inclusive societal outcomes.

2.2 The Fundamentals of Agenda 2030

Effective governance is a pre-requisite for the world to achieve the SDGs [4, 5]. Good governance encompasses a set of principles aimed at ensuring sound and effective management of public institutions through transparency, accountability, and citizen participation. Good governance rests on the basis of trust and legitimacy while ensuring that public organizations, or incumbents, address a set of needs within their own respective constructed environments. Good governance practices necessarily form intelligent decisions, optimize resource usage, and contribute to greater accountability from public actors. We posit that good governance practices are intimately tied to organizational effectiveness, which we define organizational effectiveness from three dimensions: Organizational Performance: The ability of the organization to meet its own objectives, provide quality service, and meet the needs of its citizens. Resource Management: The efficient and effective management of an organization's financial, human, material, and information resources. Citizen Satisfaction: The degree of satisfaction and trust of the citizen in the organization and its deliverables; Translating good governance to public administration practice enhances public management quality, increases civic engagement, and optimizes resource usage, thereby fostering sustainable organizational performance and ensuring better alignment with organization's sustainable development goals.

2.3 Machine Learning for Sustainable Development

Machine Learning (ML), a branch of artificial intelligence, offers powerful tools for analyzing complex data, detecting trends and predicting results. Its potential for sustainable development, particularly in the public sector, is immense. Machine Learning can help public organizations better understand the impact of their actions, optimize the use of their resources and make more informed decisions.

In this study, we will use three ML algorithms that are particularly well suited to our problem:

XGBoost (Extreme Gradient Boosting): An algorithm based on gradient boosting, known for its robustness and ability to handle complex data. It is particularly powerful for prediction and classification, and can detect non-linear relationships between variables [14].

CNN (Convolutional Neural Networks): A type of deep neural network, generally used for image recognition. However, CNNs can also be used for structured data analysis and complex pattern detection [10, 11].

Random Forest: A supervised learning algorithm that combines multiple decision trees to improve the accuracy and stability of predictions. It is renowned for its ability to handle noisy data and offer good interpretability [9, 16, 17].

These algorithms are used to analyze simulated data and assess how good governance practices and the SDGs influence organizational performance. Indeed, through these tools, we will identify complex patterns and make robust predictions about the potential impacts of these factors on the three dimensions of public organization effectiveness.

Overall, this theoretical framework positions our study at the intersection of Agenda 2030, good governance principles and machine learning applications for more efficient, accountable and sustainable public organizations. The following section details the methodological choices we have made.

3 Methodology

This section outlines the methodological choices that guided our study. It details the data collection and preparation process, as well as the machine learning algorithms used and the measures for evaluating their performance.

3.1 Data

For this research, we constructed a full database from data collected from two sources: audit reports and structured questionnaires. The database features a sample size of 300 observations, each representing a unique public entity in Morocco. Each observation summarizes the data relevant to the integration of the Sustainable Development Goals (SDGs) and dimensions of organizational effectiveness. Variables related to SDGs.

These variables were constructed using qualitative data obtained from the audit reports. These variables indicate how well sustainable-development goals have been integrated into the organizational practice. In this paper, we will mainly analyze SDGs that were viewed as most salient to the Moroccan public sector, namely SDG 5 (Gender Equality), SDG 11 (Sustainable Cities and Communities), and SDG 16 (Peace, Justice, and Strong Institutions). A composite score is created for each organization's level of commitment to a SDG by carefully analyzing documents and deriving content analysis to establish the presence or absence of policies. In a public sector audit report, having a policy related to gender equality and stated in the audit report would earn higher composite scores for SDG 5.

Variables of organizational effectiveness: These variables were developed from combined qualitative and quantitative data from audit reports and surveys with organizational staff and service users. Public organizational effectiveness was evaluated on three primary domains: Performance: Assessed through performance indicators, including service quality, success in meeting targets, and effectiveness in project management. The data was drawn from audits and supplemented with survey data, using a scoring system based on a scale of 0 to 100. Resource Management: Assessed by indicators of budget, use of available human and material resources, and risk management effectiveness. Like the previous dimension, the data came from the audits but was supplemented with survey questions, arriving at a score on a scale from 0 to 100. Citizen Satisfaction: Measured through citizen responses to direct feedback on self-reported satisfaction in surveys using a Likert scale (1: not at all satisfied, 5: very satisfied). These answers were again translated into a score on a 0 to 100 scale and included questions on service quality, responsiveness to citizen requests, and organizational transparency.

The integrative strategy—employing audit reports in combination with actual feedback from stakeholders via surveys—assures a comprehensive and multidimensional assessment of public organizational effectiveness. This robust approach affords an in-depth analysis of the interaction between organizational effectiveness dimensions and the implementation of SDGs within public sector governance practices.

4 Results and Analysis

4.1 Analysis of the Relationship Between Sustainable Development Goals and the Effectiveness of Public Organizations: Circle and Correlation Matrix

The correlation circle analysis resulting from Principal Component Analysis (PCA) provides key insights into the relationships between independent variables, namely the Sustainable Development Goals (SDGs), and dependent variables measuring the performance of public organizations. These relationships are presented in a correlation circle, which illustrates the relationships with the location of the vectors and their length, providing an indication of both the direction and magnitudes of correlations (Fig. 1).

Among the independent variables identified, the SDGs, "Good Health and Well-Being," "Affordable and Clean Energy," and "Decent Work and Economic Growth," demonstrate considerable power, as indicated by their long vectors away from the center of the circle. This is particularly apparent with the first two principal components where the SDGs correlate closely, indicating they account for significant variability in the data. All three of these SDGs are positively related to the various organizational performance variables, thus, suggesting an increase in these SDGs will be directly related to an increase in the organizations performance.

Similarly, dependent variables with indicators of organizational effectiveness, such as "Efficiency_Performance," "Resource_Management," and "Citizen_Satisfaction" also show distinct and meaningful relationships with specific SDGs. For instance, "Efficiency_Performance" show a positive relationship with "Affordable and Clean Energy" and "Sustainable Cities and Communities," indicating that efforts to increase organizational efficiency typically align with better performance on these specific aspects of sustainability. The performance-related venue may not stretch out quite as far as the performance variables of the most impactful SDGs; however, in moving farther away from the center, their placement still indicates more moderate-to-strong relationships.

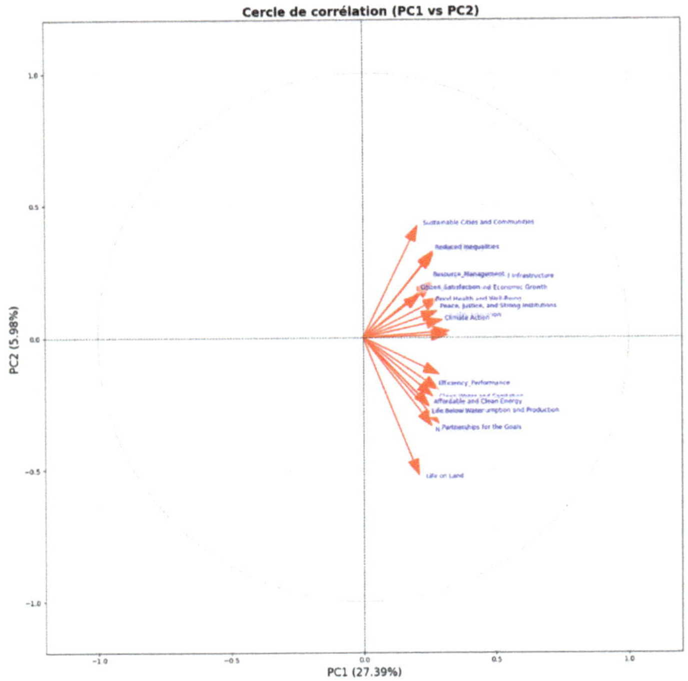

Fig. 1. Correlation circle(PCA) of Agenda 2030 data

In contrast, domains such as "Peace, Justice and Strong Institutions" and "Life Below Water" - while conceptually important - have weaker (or indirect) linkages. Their proximity to the center of the circle suggests that their linkages may be indirect or happen later, again suggesting a secondary role to direct improvements in organizational performance.

Fundamentally, the PCA correlation evaluation emphasizes certain SDGs as major contributors influencing organizational effectiveness, while others are more subtle or indirect in their influence. This lens highlights the complex interdependency between SDGs and organizational performance, recognizing their impact both directly and indirectly on sustainable governance outcomes (Fig. 2).

Information regarding the data correlation matrix provides insight into the association between the Sustainable Development Goals (SDGs) and measures of efficiency in public organizations. The correlation matrix depicted the association between the SDGs and measures of efficiency based on the strength and direction of the correlation between the 2 variables, SDGs and efficiency.

The independent variables, especially SDGs, "Affordable and Clean Energy", "Decent Work and Economic Growth" or "Sustainable Cities and Communities", and various measures of efficiency, demonstrated relatively strong positive correlations, such as "Efficiency_Performance" and "Citizen_Satisfaction". The positive association suggests that advancement related to SDGs is generally accompanied by better organizational performance. Improvements related to the SDGs may result in better performance by public organizations. The SDG, "Affordable and Clean Energy" for example appears to

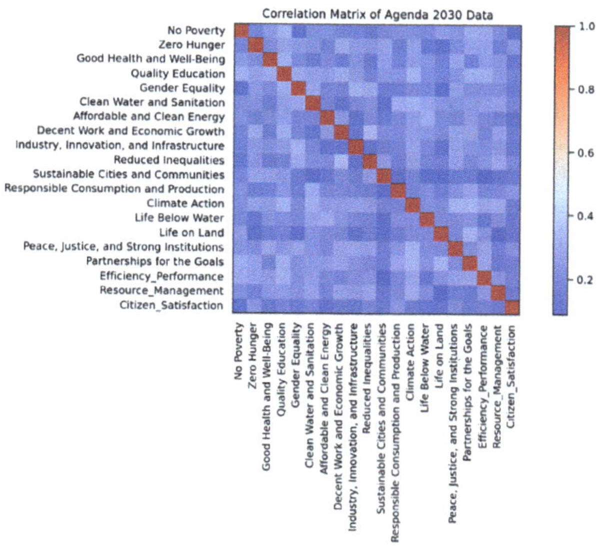

Fig. 2. Correlation matrix of agenda 2030 data

have a strong correlation with improved efficiency by public organizations, as with other SDGs as well. The "Affordable and Clean Energy" SDG was positively correlated with the efficiency measure, "Efficiency_Performance".

Conversely, several independent variables, such as "Life Below Water" and "Life on Land", indicate weaker correlations with performance measures, indicating either that these SDGs are having influence only indirectly on organizational performance or that these SDGs are only somewhat related to organizational performance based on the data observed. The weaker correlations with performance measures suggest these SDGs are somewhat less directly associated with public organization performance measures. The dependent variables, especially, "Resource_Management" and "Citizen_Satisfaction", presented interesting correlations with various SDGs. For example, "Resource_Management" seems to index particularly well with SDGs such as "Clean Water and Sanitization" and "Industry, Innovation, and Infrastructure". This could be indicative of the fact that resource management is generally a key and integral segment of implementation in different ways with these SDGs. Certainly, water and infrastructure management emerges as critical dimensions of resource management.

In conclusion, the correlation matrix reveals that there is relationship between some independent variables (certain SDGs) and dependent variables (modality measures of the efficiency of public organizations). As they are positively correlated, this suggests that progress in some of these SDGs (energy and infrastructure SDGs) could directly translate to organizational performance improvements. Meanwhile, there are several variable that exhibited a more modest correlation suggesting that other factors may influence the relationship in a more indirect way or that the components and dimensions of the dependent variables might be more indicative of significant relationship within the core concepts.

4.2 Model Performance

The comparative performance evaluation of the three data models, namely XGBoost, Random Forest, and CNN, indicates notable differences in terms of accuracy and robustness. XGBoost has emerged as the best performing model in predicting the potential relationship between the Sustainable Development Goals (SDGs) and the effectiveness of public organizations. The accuracy of XGBoost is reported at 85%, and has been shown to have a relatively low standard deviation of errors (0.05) thus displaying relatively good robustness. XGBoost also outputs an R^2 of 0.88 which indicates the model sufficiently explains the variability in the data underlying the study. Alternatively, the other models displayed lower performance although it too was strong: the Random Forest achieved an accuracy of 80%, and CNN achieved 75% accuracy. These results demonstrate the strength of XGBoost in this context (see Table 1).

Table 1. Model Performance

Model	Accuracy	Std Error	R^2 Score	MAE	RMSE
XGBoost	0.85	0.05	0.88	0.12	0.18
Random Forest	0.8	0.08	0.83	0.15	0.22
CNN	0.75	0.1	0.78	0.18	0.25

4.3 Impact of the SDGs

The analysis of predictive models in relation to the Sustainable Development Goals (SDGs) underscores the significant role certain goals play in shaping the performance of public sector organizations. Goals such as "Affordable and Clean Energy," "Decent Work and Economic Growth," and "Clean Water and Sanitation" demonstrated robust correlations with key organizational effectiveness indicators, notably in terms of resource management and citizen satisfaction. Among the machine learning models evaluated, XGBoost emerged distinctly as the most effective tool for predicting the nuanced impacts of the SDGs, highlighting machine learning's substantial potential to elucidate the complex dynamics influencing organizational performance (see Tables 1 and 2).

Consequently, machine learning has emerged as a crucial component in contemporary governance practices, offering sophisticated analytical capabilities to accurately interpret and optimize the relationships between the SDGs and organizational outcomes. This approach facilitates data-driven, predictive decision-making processes, thus substantially enhancing the effectiveness of governance initiatives.

Nevertheless, several limitations warrant consideration. The intricate interdependencies among the SDGs pose considerable analytical challenges, given the non-linear and often indirect nature of their interactions. Additionally, the absence of detailed contextual data specific to organizational or national environments restricts the generalizability and transferability of these findings, as the influence of SDGs can vary substantially according to local conditions. Data quality issues, including potential biases and gaps in organizational performance reports, could also compromise result accuracy. Moreover, the static nature of the predictive models, grounded exclusively in historical data, inherently restricts their predictive capabilities for future SDG-related developments and governance evolutions. Lastly, variations in governance priorities, objectives, and resource availability across different contexts and organizations further complicate the accurate assessment of SDGs' real-world impacts.

To address these limitations and maximize the SDGs' effectiveness within public governance, the integration of machine learning methodologies into public decision-making frameworks is strongly recommended. This strategic integration would empower public entities with the capability to accurately understand and anticipate SDG-related impacts, thereby enabling more proactive and informed governance decisions, particularly within critical domains such as resource optimization, innovation, and citizen engagement. Moreover, adopting hybrid analytical models that integrate quantitative analysis with qualitative insights would considerably enrich understanding, providing a comprehensive and nuanced perspective on SDG-governance interactions. Such an approach acknowledges and incorporates context-specific and human-centric dimensions, fostering a more dynamic, holistic, and effective management of sustainable development goals.

Table 2. p-values for ODDs using XGBoost

Objective	p-value
No Poverty	0.001
Zero Hunger	0.023
Good Health and Well-Being	0.015
Quality Education	0.045
Gender Equality	0.032
Clean Water and Sanitation	0.01
Affordable and Clean Energy	0.025
Decent Work and Economic Growth	0.005
Industry, Innovation, and Infrastructure	0.007
Reduced Inequalities	0.041
Sustainable Cities and Communities	0.012
Responsible Consumption and Production	0.03
Climate Action	0.021
Life Below Water	0.013
Life on Land	0.009
Peace, Justice, and Strong Institutions	0.04
Partnerships for the Goals	0.022

5 Summary of Results

The investigation of the predictive models, in conjunction with the examination of the Sustainable Development Goals (SDGs), indicates that the significance of certain SDGs is related to the way in which public sector organizations engage with the performance indicators. In fact, goals like "Affordable and Clean Energy," "Decent Work and Economic Growth" and "Clean Water and Sanitation" had significant associations with several performance indicators - whereby notably, resource management and citizen satisfaction. XGBoost also stood out among the several machine learning models assessed as the most effective predictive tool related to outcomes associated with the SDGs - representing machine learning's considerable potential to exhibit the complex dynamics that shape organizational performance (See Tables 1 and 2).

As a result, machine learning has become an essential part of modern governance methods, providing advanced analytical tools to precisely analyze and maximize the connections between the SDGs and corporate results. By facilitating data-driven, predictive decision-making processes, this strategy significantly increases the efficacy of governance activities.

Nonetheless, there are limitations to take into account. The complexity of the interdependencies between SDGs may obstruct meaningful analysis, since a key attribute of their interactions is their nonlinear and often indirect nature. Furthermore, the lack of contextual detail at the organizational or country level prevents generalizability and transferability of the findings, as the effects of the SDG are contextually dependent and can manifest in very different ways in different contexts. The quality of the data may therefore raise issues around bias in the reports on organizational performance, or a lack of data altogether, which could impact the accuracy of the results. Compounding this issue, prediction models are paradoxically based on historical data alone, so they limit the ability to predict implication for future SDG related developments and governance. Finally, variation in governance objectives, governance priorities, and resource availability may preclude accuracy in assessing the impact of SDGs in real world settings.

In order to overcome these challenges and harness the full potential of the SDGs in public governance, it is recommended that public decision-making systems incorporate machine learning methods. This would provide public agencies with the ability to accurately understand and anticipate impacts related to the SDGs, while simultaneously enabling public governance systems to be prepared and informed in a proactive way, especially in the areas of resource utilization, innovation, and citizen engagement. Moreover, hybrid analytical approaches that combine quantitative analysis with qualitative inputs would immeasurably enhance our understanding, yielding a more intricate and broader perspective of SDG-governance interplays. Addressing these prospects would also capitalize on the context-specific and human dimensions of commons, therefore permitting more dynamic, holistic and effective management of sustainable development goals.

6 Conclusion

In conclusion, the thorough examination of the Sustainable Development Goals (SDGs) and a template for its influence on public sector performance signifies the important role of advanced technology, namely machine learning, in contemporary governance. Our findings indicate that some SDGs, most notably those that pertain to sustainable energy solutions, economic development, and management of water resources, comprise a significant, direct impact on public organization performance, particularly, efficient resource allocation and improved citizen satisfaction. That said, the study highlights several inherent limitations, including the complex interdependencies among the SDGs, as well as the diverse localities and institutional capabilities that introduce considerable variability in local contexts. Some of these complexities will require applying integrative, hybrid analytical methodologies that can explicitly combine quantitative data analysis with qualitative descriptive analysis. Implementing these integrative methodologies will provide richer and deeper explanations into the world of practice and the contextual variables in action that influence the success of SDGs. By purposefully embedding these hybrid methodologies in governance structures, public institutions can enhance their predictive and adaptive capabilities to respond to the unpredictabilities introduced by SDG implementation. Ultimately, this embedding can lead to greater organizational predisposition to resilience and improved effectiveness to reforms that will be pivotal to advancing sustainable and inclusive forms of governance in Morocco and beyond.

References

1. Hege, E., Barchiche, D., Rochette, J., Chabason, L., Barthélemy, P.: Initial assessment and conditions for success of the 2030 Agenda for Sustainable Development. (2019). https://www.iddri.org/sites/default/files/PDF/Publications/Catalogue%20Iddri/Etude/201910-ST0719EN-SDGs.pdf
2. Melonio, T., Tremel, L.: Climate, biodiversity, inequalities: issues and challenges of the 2030 agenda. In: United Nations Agenda 2030 Observatory (2021). https://doi.org/10.4000/books.septentrion.15373
3. Jacquemot, P.: Le Dictionnaire Encyclopédique Du Développement Durable, 3rd edn. Sciences humaines Éditions, Auxerre (2021). https://doi.org/10.4000/books.septentrion.15373
4. Grindle, M.S.: Good Governance: the inflation of an idea. Harvard Kennedy School (2010) https://www.hks.harvard.edu/sites/default/files/centers/cid/files/publications/faculty-working-papers/202.pdf
5. Kooiman, J.: Governing as Governance. SAGE Publications (2003)
6. OECD: OECD Public Governance Reviews Morocco: For a Resilient Administration Serving Citizens. Available: OECD Report. (2023). https://www.oecd.org/fr/publications/examens-de-l-ocde-sur-la-gouvernance-publique-maroc_1a0272c0-fr.html
7. Beisheim, M.: Die Agenda 2030 Für Nachhaltige Entwicklung: Ein Ausblick Auf Ihre Weiterverfolgung Und Überprüfung. Vereinte Nationen: German Review on the United Nations. **63**(6), 255–260 (2015) https://zeitschrift-vereinte-nationen.de/publications/PDFs/Zeitschrift_VN/VN_2015/Heft_6_2015/04_Beitrag_Beisheim_VN_6-15_25-11-2015.pdf
8. Anastasopoulos, L.J., Whitford, A.B.: Machine learning for public administration research, with application to organizational reputation. J. Public Adm. Res. Theory. **29**(3), 491–510 (2019). https://doi.org/10.1093/jopart/muy060

9. Breiman, L.: Random forests. Mach. Learn. **45**(1), 5–32 (2001). https://doi.org/10.1023/A:1010933404324
10. Davenport, T.H., Ronanki, R.: Artificial intelligence for the real world. Harv. Bus. Rev. **96**(1), 108–116 (2018)
11. Goodfellow, I., Bengio, Y., Courville, A.: Deep Learning. MIT Press (2016)
12. Shmueli, G., Koppius, O.R.: Predictive analytics in information systems research. MIS Q. **35**(3), 553–572 (2011). https://doi.org/10.2307/23042796
13. Wang, L., et al.: Machine learning for management science. Bus. Inf. Syst. Eng. **61**(3), 315–325 (2019)
14. Barnwal, A., Cho, H., Hocking, T.: Survival regression with accelerated failure time model in XGBoost. J. Comput. Graph. Stat. **31**(4), 1292–1302 (2022). https://doi.org/10.1080/10618600.2022.2039880
15. Shafiq, M., Gu, Z.: Deep residual learning for image recognition: a survey. Appl. Sci. **12**, 8972 (2022). https://doi.org/10.3390/app12188972
16. Liaw, A., Wiener, M.: Classification and regression by randomForest. R News. **2**(3), 18–22 (2002) https://cran.r-project.org/doc/Rnews/Rnews_2002-3.pdf
17. Cutler, D.R., Edwards Jr., T.C., Beard, K.H., Cutler, D.R., Hess, K.T.: Random forests for classification in ecology. Ecol. Appl. **18**(2), 272–281 (2007). https://doi.org/10.1890/07-0539.1

Valorization of Subsurface Resources: Optimization Through IoT and AI for Sustainable Management and Exploitation

Abdelouahab Mohammed(✉), Ali El Moussati, Ayat Yassine, Hajar Karkri, and Oumayma Rachdi

Energy, Embedded System, and Data Processing Laboratory, National School of Applied Sciences Oujda (ENSAO), Mohammed First University (UMP), Oujda 60000, Morocco
mohammed.abdelouahab.d23@ump.ac.ma

Abstract. Water resources worldwide, in particular groundwater, are being overwhelmed, primarily due to climate change and pollution. These issues are exacerbated as utilities and water authorities overlap the non-sustainable over-extraction of aquifers with the urgent need for the sustainable management of these water resources. Groundwater is considered a major player in the description of global water security and our proposal is framed to develop an advanced monitoring system and use machine learning models to predict and provoke water resources industry decision-making. This proposed monitoring system is still under development and will augment environmental monitoring in a more accurate way and proactively manage water resources, specifically targeting issues and challenges related to groundwater over-extraction. Our proposal includes practices consistent with the guidelines provided by the World Health Organization (WHO) related to the monitoring of drinking water quality, as well as incorporating water quality monitoring sensors, and integrating IoT to allow for real-time monitoring and predictive modelling. The intent is to provide information in the form of hydrogeological data that can be used to monitor important water quality parameters and groundwater levels; which can then contribute to the sustainable and proactive management of groundwater and aquifers.

Keywords: Aquifer · water quality · monitoring · artificial intelligence (AI) · Internet of Things (IoT)

1 Introduction

With climate change, pollution, and heightened extraction placing strains on global aquifers, it has become urgent to take action to implement sustainable management of water resources. These most significant changes in our world are threatening global sustainability across a range of targets, water security; sustainably managing water resources; managing the impacts of pollution on aquatic systems; as well as water availability for drinking purposes around the world [1–3]. Groundwater extraction accounts

B.-E. Boudriki Semlali et al. (Eds.): DATA 2025, LNNS 1908, pp. 101–110, 2026.
https://doi.org/10.1007/978-3-032-21200-9_9

for approximately almost half of the volume of total water abstraction for global domestic needs and around 25% of the water used for agricultural irrigation [4]. Undoubtedly, managing this important resource will play a vital role in tackling these problems. Understanding that water management pervades the global narrative of sustainable development is important, and that highlight its cautious extractive use, and intensified pollution present us with the challenge of undertaking diligence and sustainable management strategies to mitigate and stop the depletion of these critical resource systems [4, 5].

At this time within the normal life cycle of groundwater degradation, the importance of integrated and sustainable water management, particularly groundwater use is acute as climate change is ushering us all into extreme climatic conditions. Water resources remain an underappreciated and poorly understood yet important part of the "hidden wealth of nations," that have to be recognized and managed carefully particularly when understanding the role groundwater can have on sustainable poverty reduction, food security, and sustained and careful economic growth [2, 5]. As the importance of groundwater rises due to the increased incidence and severity of droughts driven by climate change, there is a growing need for international efforts and resources to protect aquifers, which have been identified as strategic resources for securing future access to safe drinking water for nations, regions and populations [6].

In light of these issues; international voices are emphasizing the need for high-level political commitment to the assurance prioritization of groundwater as a resource. The goal is to balance their private and social costs effectively, while serving the needs of an ever-growing global population and associated climate changes [1, 4]. Therefore, it is even more important for us to improve our groundwater management practices, in order to meet the Sustainable Development Goals (SDGs) by 2030, and to be more resilient to a changing climate [4].

Technological innovation, especially in IoT and Artificial Intelligence (AI), enables the sustainable management of water resources. Recent technological advances in IoT have provided smart solutions for monitoring the environment such as sensors that can measure important indicators of water quality and provide real-time data [7]. These technologies are not only facilitating increased monitoring, but also more effective water resource management, which enables fast intervention in situations that could potentially develop into an issue.

Similarly, AI and Machine Learning (ML) are changing our ability to analyze and make use of data that has been collected. By utilizing algorithms, predictions can be made about water quality and when managing water resources, not only improving insight but also being able to help lead to more sustainable management of those resources [8]. Our proposal aims to build on innovative ways of solving the problems of groundwater with respect to developing an advanced monitoring system with a predictive and informed resource management approach using ML models.

In this study, we have established that there is much potential for the IoT ecosystem, along with AI, to enhance groundwater management practices by refining monitoring processes and enabling proactive water resource management. Ultimately, we are trying to conserve drinking water and provide advanced technological solutions for sustainable practices in the future.

The next section of this document provides some urgent groundwater and drinking water quality issues, some of which relate to their over-exploitation. We describe our research and prototype system that integrates IoT and ML to acquire water quality parameters and groundwater level with respectable accuracy, to foster sustainable and proactive groundwater and drinking water resource management. This article travels beyond only embodying the present and future capabilities of these emerging technologies because it presents a pathway for practical solutions to drinking water issues.

2 Context

Groundwater is essential for humanity, agriculture, and industry, and is one of the largest sources of freshwater on Earth after glaciers and ice caps. Groundwater is located in aquifers, which are permeable geologic formations such as sand, gravel, or rock. Groundwater is critical for global water security, but there are many challenges to managing groundwater resources in a sustainable manner, including issues of access to groundwater resources, resource quantity, and issues of water quality, all of which are getting worse due to pollution, depletion and climate change.

Water quality is a serious concern for groundwater supplies because the water can be contaminated from agriculture, industry and domestic demands, rendering the water unsafe to drink and unfit for use. Additives in agriculture, industrial chemicals, and other contaminants such as nitrates and pesticides can remain in aquifers for long periods and create risks to human and environmental health and safety.

This section will introduce the context of our system by focusing on water quality monitoring and the international potability standards and recommendations within the guidelines of the World Health Organization (WHO). Water quality can be impacted from over-extraction of groundwater and the effects of lowered groundwater levels on quality will also be examined. It is critical to discuss the use of Internet of Things (IoT) technologies for water quality monitoring (IoT-WQM) and groundwater monitoring including groundwater levels.

2.1 Monitoring of Drinking Water Quality

The method used in this study follows World Health Organization guidelines for water quality monitoring. Table 1 shows acceptable levels of parameters for the quality of drinking water. In our case, we are trying to use technology, such as almost real-time water quality monitors (WQM) and an Internet of Things (IoT) application to develop an integrated system called the IoT-WQM System (IoT-WQMS). Using an IoT-WQMS to monitor the quality of drinking water can be quite complex, given the multiple physicochemical and biological parameters that may impact human health and the safety of drinking water.

Table 2 provides a correlation matrix for WQM parameters and indicates many strong correlations among several parameters. The interdependence between parameters could form the basis for the streamlining of monitoring systems through using the most applicable parameters relating to drinking water quality. Past studies have highlighted the relevance of using these parameters within IoT-WQMS [9–12].

Table 1. WHO Recommended Thresholds for Drinking Water Quality Parameters

Parameter	WHO Recommended Threshold
Pathogenic microorganisms	Absent in 100 mL
Turbidity	≤ 5 NTU (ideally < NTU)
pH	6.5–8.5
Total Dissolved Solids (TDS)	≤ 1000 mg/L
Electrical Conductivity	≤ 1500 μS/cm at 25 °C
Temperature	Below 25 °C
Nitrates (NO_3^-)	≤ 50 mg/L
Fluorides (F^-)	≤ 1.5 mg/L
Lead (Pb)	≤ 0.01 mg/L
Arsenic (As)	≤ 0.01 mg/L
Cadmium (Cd)	≤ 0.003 mg/L
Nitrites (NO_2^-)	≤ 3 mg/L
Chlorides (Cl^-)	≤ 250 mg/L
Sulfates (SO_4^{2-})	≤ 250 mg/L

Source: http://www.who.int

Table 2. WQI parameters correlation-matrix

	TC	FC	TDS	TSS	TS	TH	DO	EC	Cl	T	Turb	pH	ORP	T-Cl	F-Cl	C-Cl
TC		■[1]					■				■		■	■	■	
FC	■												■	■	■	
TDS				■	■	■		■	■		■					
TSS			■		■							■				
TS			■	■		■	■	■		■	■					
TH			■		■				■		■	■		■		
DO					■			■								
EC			■		■		■		■	■	■	■				
Cl			■			■		■			■	■				
T					■			■					■			
Turb			■		■	■		■								
pH				■		■		■	■				■			
ORP	■	■								■		■		■	■	■
T-Cl	■	■										■	■			
F-Cl	■	■											■			
C-Cl													■			

Source: [7] Farmanullah Jan, Nasro Min-Allah, Dilek Düştegör.

The primary variables identified for their critical importance to evaluating groundwater quality include:

Potential Hydrogen (pH): Which informs about the water's acidity or alkalinity, is crucial for biological and chemical processes.

Turbidity: An indicator of water clarity, affects disinfection and aesthetic aspects of the water.

Temperature: Plays a role in chemical reactions and the biological activity of the water.

Electrical Conductivity: Reflecting the concentration of dissolved salts, impacts the taste and corrosion of infrastructure.

2.2 Overexploitation of Groundwater

Over-extraction of groundwater occurs when water is extracted from aquifers at a greater rate than it can be replenished naturally. This has resulted in economic and environmental problems, through significant declines in groundwater levels in many locations and has directly affected both the quantity and quality of water available. Figure 1 illustrates the direct relationship between over-extraction of groundwater and declining groundwater levels. A positive correlation between pumpage intensity and declining groundwater levels was noted, backed up by much previous literature as well. Groundwater resources are declining in size, and groundwater only storage facilities will yield water at varying rates without adequate replenishment, particularly in arid and semi-arid regions, at an immeasurable rate. Excessive groundwater use has lowered groundwater levels significantly, sometimes as many as 25 m lower than mean sea level [14, 15].

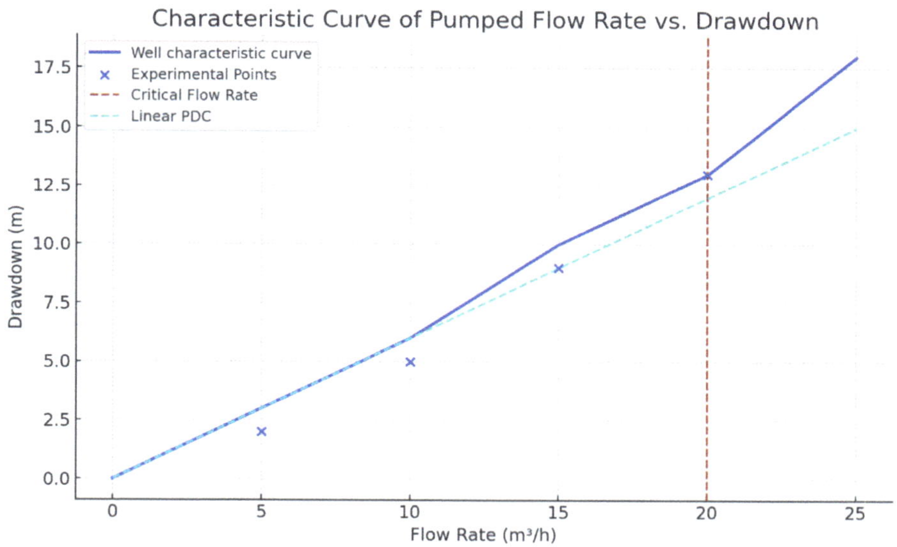

Fig. 1. Characteristic curve of the flow rate pumped from a well as a function of drawdow Source: SIGES Bretagne - Groundwater Management Information System in Brittany

Interpretation:

- As the rate of pumping (x-axis in m^3/h) increases, the water level in the well decreases and reflects increased drawdown (y-axis in m).
- The characteristic curve for wells shows that this is not a linear relationship; there is only limited drawdown associated with low flows, but as the flow rate increases, the drawdown increases at an increasing rate indicating that the aquifer is being over-exploited.
- The point at which the flow rate is critical indicates a threshold where the effect of pumping on drawdown increases at an increasing rate indicating an unsustainable operating point.
- Our experimental points support this trend by showing actual measurements that closely follow this curve, thus underpinning the argument that an increase in pumping rate is directly associated with a deeper lowering of the groundwater table.

The relationship of overexploitation on water quality is also a point of concern. A drop in water level can cause certain contaminants, such as dissolved salts, to be more concentrated leading to a rise in electrical conductivity and potential quality problems. Also, less aquifer recharge can cause less dilution of pollutants, ultimately constituting water quality issues [13].

The installation of a groundwater level monitoring system is integral to any water management plan aimed at avoiding over-exploitation of subsurface resources. These systems provide early detection of any significant lowering in groundwater levels, allowing for timely action to reduce the harmful impacts of over-pumping. Piezometric probes using IoT technology provide an efficient means of continuously monitoring groundwater levels in real-time and would facilitate timely mitigation of depletion risk.

3 The Proposed System

In the above mentioned context, particularly in Sect. 2 which identifies both qualitatively, with increased concentrations of pollutants, and quantitatively with decreasing groundwater levels, the system that we are developing takes a fresh look at the issues mentioned. By using a combination of Internet of Things (IoT) techniques and machine learning (ML).

The system aims to gather critical real-time data, including electrical conductivity, pH, turbidity, and temperature, regarding water quality with an emphasis on potability, along with groundwater levels. This data will be collected by IoT devices festooned with water quality monitoring (IoT-WQM) and groundwater level monitoring (IoT-SP). The data then processed, using ML models including Recurrent Neural Networks (RNN), time series, and random forest analysis to observe and extrapolate trends in terms of water quality and executive state of aquifers to optimize water resource management and consider the level of exploitation.

The goal of our approach is to facilitate a proactive management of water resources that has the ability to identify the risk of contamination or overexploitation efficiently. This represents a huge leap in the management of water resources and provides a highly relevant solution to their specific challenges of groundwater and drinking water quality, while also helping to ameliorate the pertinent problem of overexploitation (Fig. 2).

3.1 General Systems Architecture

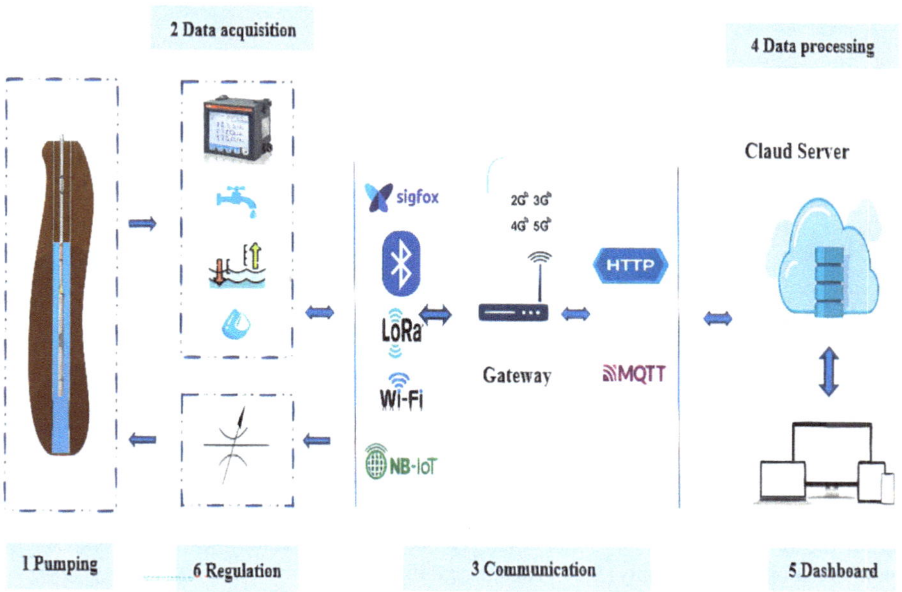

Fig. 2. Architectural diagram of the proposed system

Interpretations:

- **Pumping:** This is the starting point where water is extracted from the ground. It constitutes the first step in gathering information about the quality and level of the groundwater.
- **Data Acquisition:** IoT sensors are deployed to collect real-time data related to conductivity, pH, turbidity, temperature, and groundwater levels. These pieces of information are crucial for assessing the condition of the groundwater resource (Fig. 3).

- **Communication:** The captured data are transmitted using a range of communication technologies, including Sigfox, NB-IoT, cellular networks, and Wi-Fi. This allows the data to reliably reach the central server to be processed.
- **Data Processing:** The data arrives as inputs on a cloud server and are processed using ML algorithms. The ML model generates two outcomes, predicting the flow rate so that it can control the pump to adjust the exploitation rate, and also classifying the water quality of the pumped water. If the quality is acceptable, it allows the exploitation to continue, while if it (the quality) is deemed unacceptable, the pump would be stopped, as shown in Fig. 4.

– **Dashboard:** The results of the processing, flow rate prediction and water quality classification, are shown on an interactive dashboard for users. Users can see in real-time the status of the water resource, to make decisions and adjust the pumping operation.

Model	Picture	Description
IOT-485-PH pH sensor		1. Measuring range PH:0.0~14.0 Temp: (0~50.0) °C 2. Power supply: 12V or 24V DC 3. Power dissipation: 1W 4. Communication: Modbus RS485 5. Cable length: 5 meters
IOT-485-DD Conductivity sensor		1.Measuring range: Conductivity: 0-5000us/cm Temp:(0~50.0)°C 2. Power Supply: 12V or 24V DC 3. Power dissipation: 1W 4.Communication: Modbus RS485 5. Cable length: 5 meters

Fig. 3. WQM Probes for pH, Conductivity: Technical Specifications (Measurement Range, Communication, Power Supply) for System Integration.

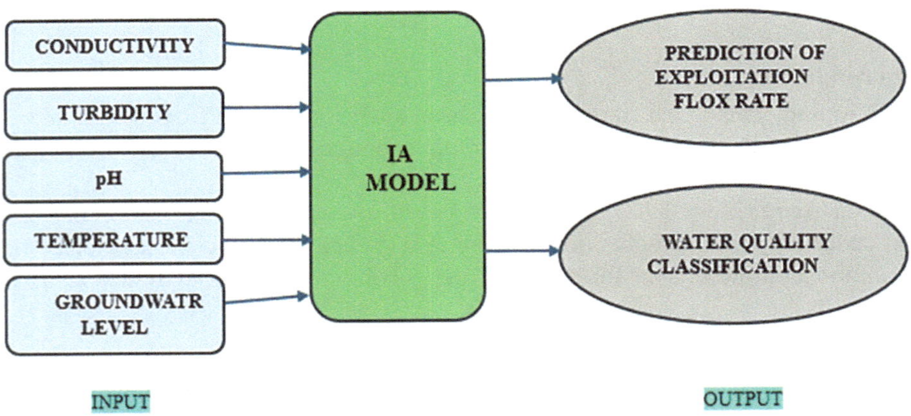

Fig. 4. Operational Diagram of the AI Model with System Inputs and Outputs.

4 Conclusion

As we begin a new chapter in the management of groundwater, our developing systems is an important step forward to enable sustainable and responsible management of these vital resources.

By leveraging the Internet of Things (IoT) and Artificial Intelligence (AI), we aim to deliver real-time data monitoring and predictive capability of groundwater quality and levels for the first time. As development continues, we will continue to prioritize the technical development of the system, particularly in relation to the accuracy of measurements and efficiency of data communication. Moving forward, we will aim to improve the accuracy of the AI and ML models we use. Greater accuracy will enable predictions and analyses to better inform more responsive and anticipatory management of water resources, by identifying risk of contamination or overexploitation cases faster. Equally, the communication of the IoT system had to improve to allow signal transmission with reliability and in real-time measures an important factor to allow better pro-active decision making.

We also want to extend our research to new AI methods and algorithms in order to refine predictive models even further. This includes a more in-depth study of deep learning and time series analysis of the original database to better equip the system to adapt to more environmental conditions and respond accurately to the intrinsic complex dynamics of the aquifers.

The possibility of our system being applied in other geographical and socio-economic contexts also encourages adaptive studies that will reflect the regional variability of groundwater resources. This will expand the portability and flexibility of our solution and promote international cooperative framework for developing universal standards of groundwater management.

References

1. Nature Reviews Earth & Environment: Global Water Resources and the Role of Groundwater in a Resilient Water Future, Nat. Rev. Earth Environ., 2019., https://www.nature.com/articles/s43017-019-0006-8
2. World Bank: The hidden wealth of nations: groundwater's critical role in a changing climate. World Bank. (2023) https://www.worldbank.org/en/news/feature/2023/06/14/the-hiddenwealth-of-nations-groundwater-s-critical-role-in-a-changing-climate
3. UN-Water: Water and Climate Change, UN-Water (n.d). https://www.unwater.org/waterfacts/climate-change/
4. UN-Water: UN World Water Development Report 2022 'Groundwater: making the invisible visible', UN-Water (2022). https://www.unwater.org/publications/un-world-waterdevelopment-report-2022/
5. OHCHR: World Must End Overexploitation of Groundwater, Says UN Expert, OHCHR (2022). https://www.ohchr.org/en/statements/2022/03/world-must-end-overexploitationgroundwater-says-un-expert
6. UN-Water: Groundwater: making the invisible visible: UN-Water (n.d). https://www.unwater.org/publications/world-water-development-report-202
7. Farmanullah Jan, F., Min-Allah, N., Düştegör, D.: IoT based smart water quality monitoring: recent techniques, trends and challenges for domestic applications. Water. **13**(13), 1729 (2021). https://doi.org/10.3390/w13131729
8. IoT-Based Water Monitoring Systems: a Systematic Review: Water. **13**(11), 1473 (2021) https://www.mdpi.com/2073-4441/13/11/1473
9. Geetha, S., Gouthami, S.: Internet of Things Enabled Real-Time Water Quality Monitoring System. Smart Water. **2**(1) (2016). https://doi.org/10.1186/s40713-017-0005-y

10. Manoharan, S., Sathiyaraj, G., Thiruvenkadakrishnan, K., Vetriselvan, G.V., Praveenkishor: Water quality analyzer using IoT. Int. J. Innov. Technol. Explor. Eng. (IJITEE). **8**(8S) (2019)
11. Lambrou, T.P., Anastasiou, C.C., Panayiotou, C.G., Polycarpou, M.M.: A low-cost sensor network for real-time monitoring and contamination detection in drinking water distribution systems. IEEE Sensors J. **14**(8), 2765–2772 (2014). https://doi.org/10.1109/JSEN.2014.231641
12. Priya, S.K., Shenbagalakshmi, G., Revathi, T.: IoT based automation of real-time in-pipe contamination detection system in drinking water. In: International Conference on Communication and Signal Processing (ICCSP), pp. 1014–1018, Chennai, India (2018). https://doi.org/10.1109/ICCSP.2018.8524255
13. US Geological Survey: "Groundwater Decline and Depletion," Water Science School, 2018., https://www.usgs.gov/centers/whcmsc/science/groundwater-decline-and-depletion
14. Nature Communications: South-to-North water diversion stabilizing Beijing's groundwater levels. Nat. Commun. (2019) https://www.nature.com/articles/s41598-019-46865-2
15. Groundwater overexploitation and seawater intrusion in coastal areas of arid and semi-arid regions: Water. (2020). https://doi.org/10.3390/w12081921

Overcoming Bottlenecks in DataOps: A Theory of Constraints Approach

Aymen Fannouch(✉), Jihane Gharib, and Youssef Gahi

Laboratory of Engineering Science, Ibn Tofail university, Kenitra, Morocco
{aymen.fannouch,jihane.gharib,youssef.gahi}@uit.ac.ma

Abstract. In today's data-driven landscape, organizations face unprecedented challenges in managing vast and complex information streams that overwhelm traditional infrastructures. DataOps integrating DevOps, agile methodologies, and lean principles offers a promising pathway to streamline data pipelines and enhance decision-making. Yet, its potential is often hampered by persistent bottlenecks. This paper employs the Theory of Constraints (ToC) as a strategic framework to systematically identify, categorize, and mitigate these bottlenecks. By segmenting constraints into three core areas hardware limitations, software inefficiencies, and organizational challenges we introduce a comprehensive taxonomy that captures the multifaceted nature of DataOps hurdles. Our analysis reveals that scalable cloud solutions, optimized query execution, containerization, and the cultivation of a collaborative culture are critical for transforming these constraints into opportunities for performance enhancement. Through an extensive literature review and empirical evaluation, our framework not only pinpoints the most restrictive elements within data pipelines but also offers targeted mitigation strategies to boost throughput and operational agility. The insights provided in this study serve as a robust blueprint for researchers and practitioners striving to overcome inherent DataOps challenges, ultimately enabling organizations to fully leverage Big Data for strategic innovation.

Keywords: DataOps · Data workflow automation · Agile methodologies · Data pipeline bottlenecks · Performance optimization in DataOps · Theory of Constraints

1 Introduction

The exponential growth of Big Data (BD) has transformed industries, creating vast opportunities for enhanced decision-making and operational efficiency. However, this rapid expansion introduces significant challenges, including data integration complexities, scalability issues, security risks, and governance constraints [1]. Traditional data management infrastructures, often struggle to handle the increasing velocity, volume, and variety of modern data streams including structured, semi-structured, and unstructured data [2], which originate from diverse sources [3, 4]. To address these limitations, organizations have adopted data pipelines, which orchestrate the extraction, transformation, and loading of data into centralized repositories, ensuring higher data quality,

B.-E. Boudriki Semlali et al. (Eds.): DATA 2025, LNNS 1908, pp. 111–123, 2026.
https://doi.org/10.1007/978-3-032-21200-9_10

consistency, and accessibility. These pipelines accelerate the generation of actionable insights which is crucial for timely decision-making [5].

This is what is referred to as the DataOps methodology, which emerges as a transformative approach integrating DevOps principles, agile methodologies, and lean manufacturing to enhance collaboration, automation, and continuous delivery. This approach significantly reduces operational bottlenecks, enabling enterprises to bridge the gap between data generation, ultimately maximizing the business value of Big Data [6]. in spite of its transformative potential, DataOps faces several obstacles that limit its effectiveness, posing challenges for organizations striving to fully capitalize on data-driven decision-making. These are referred to as bottlenecks [8].

Many efforts have focused on contributing innovative advancements to DataOps methodology in sectors like healthcare industry and others [9], the majority of them emphasizes enhancing DataOps practices, contributions specifically addressing bottlenecks in DataOps remain a significant gap. For this reason, this study aims to provide a comprehensive systematic methodology, from the identification of bottlenecks to their mitigation using the Theory of Constraints. To achieve this, the study is guided by the following research questions:

- RQ1: What are the main bottlenecks affecting data pipelines in DataOps?
- RQ2: How can we effectively identify and categorize bottlenecks in data pipelines?
- RQ3: What solutions contribute to mitigating bottlenecks in data pipelines?

The paper is structured as follows: Sect. 2 outlines the evolution of DataOps from traditional data management to the moderns and provides an overview of the theory of constraints. Section 3 reviews the literature on DataOps bottlenecks, highlighting key contributions and research gaps. Section 4 analyzes bottlenecks and mitigation strategies using the Theory of Constraints, while Sect. 4.4 focuses on the discussion, addressing the research questions, and outlining future directions. Finally, the conclusion.

2 Background

The evolution of DataOps has reshaped data management, progressing from early Big Data integration challenges (2004–2010) [7], to DevOps-inspired data operations (2011–2015). By 2016–2020, DataOps emerged as a formal discipline with structured frameworks, leading to automation advancements and sector-specific applications (2021–2024) [6]. As illustrated in Fig. 1, this transformation evolved from manual ad hoc analysis to semi-automated pipelines, culminating in the integration of DevOps, CI/CD, and automated monitoring for enhanced data quality and agility. This holistic approach eliminates silos, fosters innovation, and streamlines modern data ecosystems [10].

Fig. 1. Evolution of DataOps

However, despite its transformative potential, DataOps workflows remain vulnerable to bottlenecks critical constraints that slow performance and restrict throughput. These can arise from hardware and software limitations or human factors [12]. Effectively managing these obstacles is essential for ensuring seamless, high-performing data pipelines. To address these constraints the Theory of Constraints, introduced by Eliyahu Goldratt, provides a structured improvement method focused on identifying and eliminating the most critical bottleneck. The Process of On-Going Improvement (POOGI), as illustrated in Fig. 2 [11], follows a five-step cycle: identify, exploit, subordinate, elevate, and repeat. The case of Alex Rogo, depicted in The Goal, demonstrates that focusing on the primary constraint is the key to boosting productivity and reducing delays. In the context of DataOps analytics, bottlenecks can be diagnosed through key indicators, such as growing backlogs (WIP), frequent expedite requests and long cycle times. Addressing these constraints is crucial to optimizing workflows and improving efficiency within modern enterprises.

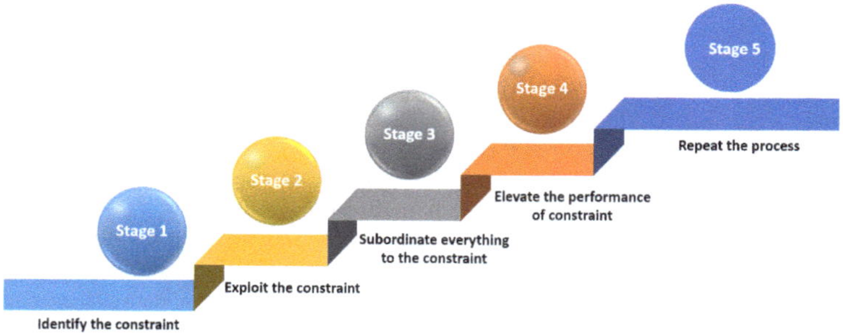

Fig. 2. Process of On-Going Improvement (POOGI)

3 Literature Review

Recent advances in DataOps have revealed a complex set of bottlenecks that impact data management efficiency across various workflows, including traditional ETL systems, machine learning (ML) and CI/CD pipelines. Performance limitations often stem from hardware and software constraints, necessitating innovations such as stream-based architectures, containerization to enhance throughput and scalability.

As shown in Table 1 the recent papers address the key bottlenecks in DataOps and propose various tools and techniques to improve performance. Each of these contributions has its strengths and weaknesses.

Table 1. Comparative of DataOps Contributions: Strength and Weakness

Contribution	Strength	Weaknesses
[12] Plumber Tool	- **High Impact:** Achieves up to 47× speedups in ML training pipelines. - **Automated Optimization:** Uses execution tracing and linear programming to dynamically adjust parallelization, caching, and prefetching. - **Performance Improvement:** Outperforms existing tools by 50%, significantly reducing model wait times.	- **Applicability Constraints**: limited to specific ML pipeline structures. - **Integration Complexity**: Requires careful tuning and may demand specialized knowledge for effective deployment. - **Scope**: Focused primarily on ML pipelines, which might restrict broader applicability.
[13] FFCV Library	- **Ease of Integration:** Acts as a drop-in replacement for standard data loaders. - **Efficient Data Handling:** Optimizes file formats, caching, asynchronous transfers, and JIT compilation. - **Performance Gains:** Doubles training speeds while maintaining accuracy; supports both vision and non-vision tasks.	- **Framework Dependency:** May be tightly coupled with specific ML frameworks or data formats. - **Flexibility Issues:** Potentially less adaptable to non-standard tasks or unconventional data structures. - **Resource Overheads:** Optimizations like JIT compilation might introduce trade-offs in memory usage or require further adjustments for diverse workloads.
[14] PRESTO for Preprocessing	- **Flexible Methodology**: Balances trade-offs between offline and online preprocessing, optimizing storage, time, and throughput. - **Performance Enhancement**: Demonstrates throughput improvements between 3× and 13 × . - **Automated Profiling**: The PRESTO tool profiles pipelines and recommends optimal strategies, enhancing decision-making.	- **Dataset Dependency:** Performance improvements (e.g., with compression) can vary significantly depending on the dataset. - **Tuning Requirements:** May require extensive adjustments to fit different data characteristics. - **Niche Focus:** Primarily addresses deep learning preprocessing, which might limit generalization to other domains.

(continued)

Table 1. (*continued*)

Contribution	Strength	Weaknesses
[15] Scientific Workflow Bottlenecks	- **Detailed Identification:** Clearly outlines seven critical bottlenecks (e.g., service availability, rigidity, interoperability, poor documentation) that hinder workflow reuse. - **Emphasis on Standardization:** Stresses the need for better documentation, discovery, and standardization practices. - **Broad Impact:** Highlights challenges that affect collaboration and automation in scientific research.	- **High-Level Analysis:** Provides conceptual insights without offering concrete technical solutions. - **Domain Specificity:** Focused on scientific workflows, which may not directly translate to other industries or general DataOps practices. - **Implementation Gap:** Lacks empirical validation or detailed case studies demonstrating the effectiveness of proposed remedies.
[16] Hierarchical Bottleneck Detection	- **Innovative Approach:** Utilizes canopy graphs to rapidly classify bottlenecks into latency, cycle-time, and occupancy categories. - **Continuous Optimization:** Offers a framework aligned with CI/CD practices, enabling ongoing performance enhancements. - **Analytical Rigor:** Provides a structured method to identify delays and inefficiencies in asynchronous pipelined systems.	- **Complex Implementation:** The canopy graph approach may be challenging to implement in diverse or highly heterogeneous systems. - **Specialized Scope:** Focused on asynchronous pipelined environments, which might limit broader applicability. - **Adaptability:** May require significant customization to suit different pipeline configurations or industries.
[17] ML-based CI/CD Optimization	- **Data-Driven Insights:** Leverages machine learning models (Random Forest, Logistic Regression, LightGBM) to predict and mitigate bottlenecks. - **Measurable Gains:** Demonstrates up to 81% accuracy in predictions and improves project efficiency by 7–30%. - **Automation:** Streamlines CI/CD processes by automating bottleneck detection across build, test, and deployment stages.	- **Data Dependency:** Effectiveness hinges on the availability and quality of historical data; sparse or noisy data can limit performance. - **Implementation Complexity:** Integrating ML-based solutions may be challenging in rapidly evolving or small-scale projects. - **Sensitivity:** The model may require regular retraining or adjustment to adapt to changes in pipeline structure.

Despite the powerful benefits each of these contributions offers addressing specific hardware or software bottlenecks, it's critical to develop an integrated approach that tackles hardware, software, and organizational challenges to deliver the most effective solution.

4 Bottleneck Analysis Using the Theory of Constraints

Bottleneck analysis is an essential part of DataOps workflow optimisation, enabling organisations to identify and address any factors limiting overall performance. By applying the theory of constraints, teams can improve efficiency, reduce delays and improve data pipeline throughput.

4.1 Identification of Bottleneck

Identifying bottlenecks in DataOps is essential for targeted improvements. The process begins by monitoring key performance metrics such as throughput, latency, cycle time, and work-in-progress to identify recurring delays. Value stream mapping provides a holistic view of the data pipeline, helping teams visualize accumulation points and potential inefficiencies. Complementing these observations with automated monitoring tools and direct team feedback further refines the diagnostic process, ensuring a comprehensive understanding of the system's weak points. Additionally, anomaly detection algorithms and predictive analytics can proactively identify emerging constraints before they cause significant disruptions.

4.2 Exploit the Constraint

After identifying the primary bottleneck, the next step is to exploit it that to maximize the throughput of the constraint using existing resources before considering any major changes. In a DataOps context, this means fine-tuning the specific stage that limits performance. For example, if a data ingestion module is identified as the bottleneck, teams might optimize batch sizes or refine query logic to reduce processing delays. By adjusting operational parameters and streamlining workflows at the constraint, organizations can extract the maximum performance from the current system, achieving immediate improvements without the need for costly infrastructure upgrades. This real-world, pragmatic approach ensures that the bottleneck is fully utilized.

4.3 Subordinate Everything to the Constraint

After optimizing the bottleneck with existing resources, the next step is to subordinate all other processes to it. In practical DataOps settings, this means adjusting upstream and downstream activities so that they align with the pace of the constraint. For example, if a specific data transformation module is the limiting factor, you might throttle the rate of data ingestion or reschedule less critical tasks to avoid overwhelming that module. This coordinated approach ensures that no process produces work faster than the constraint can handle, thereby maintaining a balanced and efficient pipeline. Additionally, implementing dynamic workload balancing and adaptive queue management can further refine this alignment, ensuring smooth data flow without unnecessary backlogs. Continuous feedback loops and performance monitoring allow teams to make real-time adjustments, preventing bottlenecks from shifting to other parts of the pipeline.

4.4 Elevate the Performance of the Constraint

Having established that elevating the constraint's performance is key through investments such as hardware upgrades, algorithm refinements, or workflow reengineering it becomes equally vital to address the underlying causes that impede the entire DataOps pipeline. In this context, a holistic approach is required, one that not only boosts the capacity of the bottleneck but also systematically mitigates hardware limitations, software inefficiencies, and organizational challenges.

- **Overcoming Hardware Bottlenecks**:

  Hardware constraints such as limited processing power, insufficient storage, and network latency pose significant obstacles to efficient DataOps workflows. To mitigate these challenges, organizations are increasingly adopting scalable solutions As illustrated in Fig. 3 like cloud computing, which leverages platforms such as AWS, Google Cloud, and Azure to provide on-demand computing power and eliminate the need for overprovisioning. Simultaneously, the implementation of parallel processing and distributed computing frameworks, including Apache Spark and Kubernetes-based systems, enables distributed task execution that significantly reduces overall processing time. Optimized storage solutions also play a critical role by employing advanced technologies such as SSDs, NVMe storage, and in-memory computing systems, organizations can minimize storage latency and enhance data throughput. Moreover, bolstering network infrastructure with Content Delivery Networks and bandwidth optimization tools like Cloudflare and AWS Global Accelerator further mitigates the impact of network delays. these integrated strategies collectively form a comprehensive approach to overcoming hardware-related bottlenecks.

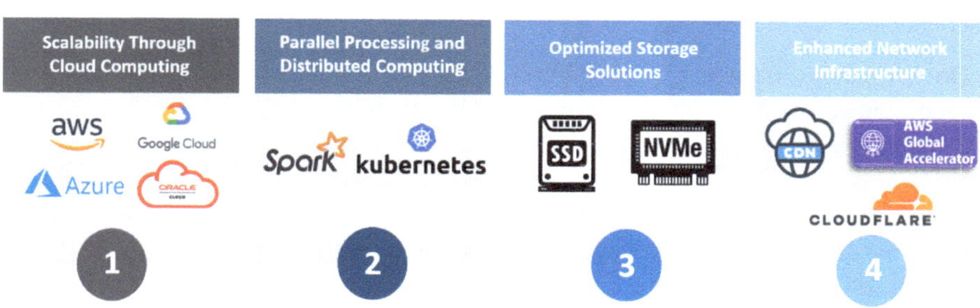

Fig. 3. Mitigating hardware Bottlenecks in DataOps

- **Addressing Software Bottlenecks**

  Software inefficiencies in DataOps, such as slow data processing, workflow execution delays, and query inefficiencies, present significant challenges, often restricting agility and scalability. Addressing these bottlenecks requires a structured, multi-layered approach as illustrated in Fig. 4 that involve:

Query Optimization
Poorly optimized SQL statements lead to excessive data scans, high latency, and resource contention. Modern query engines such as Apache Hive, Oracle, and Snowflake integrate advanced query optimizers to dynamically adjust and minimize processing overhead.

- Apache Hive leverages a cost-based optimizer (CBO) powered by Apache Calcite to analyze table statistics, optimize join ordering, and apply filter pushdown.
- Oracle employs adaptive execution techniques like partition pruning and runtime query transformations to dynamically adjust queries based on real-time data patterns and resource availability.
- Snowflake optimizes queries through micro-partitioning and dynamic pruning, automatically scaling compute resources to distribute workloads across clusters.

Workflow Orchestration
Data pipelines consist of interdependent tasks, where delays in one step propagate system-wide inefficiencies. Workflow orchestration tools like:

- Apache Airflow structures workflows as Directed Acyclic Graphs (DAGs), enabling parallel execution and automated retries to minimize idle time.
- Prefect introduces a hybrid execution model, allowing workflows to adapt dynamically to fluctuating workloads.
- Dagster integrates built-in data validation and lineage tracking, ensuring data integrity and traceability at every pipeline stage.

Containerization and Microservices
Mitigates software bottlenecks by decoupling tightly integrated processes into modular, independently scalable services.

- Docker enables lightweight, portable application environments, ensuring consistent execution across development, testing, and production.
- Docker Swarm and Kubernetes provide automated scaling, load balancing, and self-healing, dynamically allocating resources based on workload fluctuations.
- The modular architecture eliminates inefficiencies associated with monolithic systems, preventing a single failure from impacting the entire pipeline.
- Microservices-based architectures allow rapid iteration, minimal downtime for updates, and optimized resource allocation at a granular level.

Caching and Preprocessing Strategies
Even with optimized queries and orchestrated workflows, redundant computations and frequent database queries introduce latency and inefficiencies.

- Redis and Memcached store frequently accessed data in memory, significantly reducing disk I/O and query execution times.
- Apache Arrow introduces a high-performance, columnar in-memory format, eliminating serialization overhead and accelerating data transfers between distributed processing engines.

- Strategic caching mechanisms reduce database load, accelerate query responses, and ensure smoother, high-speed data processing within DataOps workflows.
- The modular architecture eliminates inefficiencies associated with monolithic systems, preventing a single failure from impacting the entire pipeline.
- Microservices-based architectures allow rapid iteration, minimal downtime for updates, and optimized resource allocation.

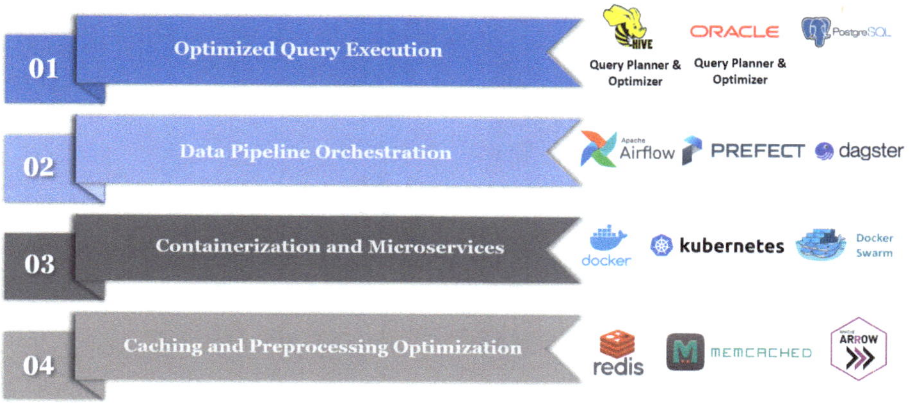

Fig. 4. Mitigating software Bottlenecks in DataOps

- **Mitigating Organizational Bottlenecks**

  A robust strategy to overcome organizational bottlenecks in DataOps, as illustrated in Fig. 5, begins with cultivating a culture where analytics are seamlessly embedded into daily operations and silos are dismantled through empowered "Data Champions" who lead workshops, curate tailored dashboards, and host cross-functional roundtables. This cultural shift is further reinforced by establishing a dynamic learning ecosystem that transforms upskilling into a continuous process featuring a dedicated DataOps Learning Hub, virtual bootcamps, hackathons, and an AI-driven mentoring platform that personalizes development pathways. In tandem, redefining role clarity through agile, cloud-based RACI systems provides real-time updates on responsibilities and decision paths, streamlining communication and accountability with automated alerts and regular review sessions. Underpinning these initiatives is a digitally integrated governance framework that embeds compliance into everyday operations through centralized digital portals, AI-powered monitoring, blockchain-based audit trails, and periodic virtual drills, fostering proactive oversight and continuous improvement.

By integrating these interconnected initiatives, organizations not only streamline their data workflows but also build a resilient infrastructure capable of adapting to rapid changes. After mitigating organizational bottlenecks, it's essential to recognize that the Theory of Constraints is inherently cyclical. Once you elevate the performance of the identified constraint, new limitations may emerge as a result of the improvements made. This final step repeating the cycle ensures continuous system optimization.

Fig. 5. Mitigating organizational Bottlenecks in DataOps

4.5 Discussion

Our analysis reveals that the main bottlenecks affecting DataOps (RQ1) are concentrated in three primary categories—hardware, software, and organizational constraints Fig. 6.

- **Hardware Bottlenecks**

  Hardware bottlenecks stem from limitations in computing resources, storage, and network infrastructure, which directly impact the speed and reliability of data processing. Insufficient storage capacity, high disk latency, and limited IOPS (Input/Output Operations Per Second) restrict efficient data retrieval and storage operations. Additionally, network and bandwidth constraints slow down data transfers. Computational power remains another critical challenge, where underpowered CPUs/GPUs, limited RAM, and energy constraints hinder performance, especially for large-scale data workloads. Addressing these limitations requires strategic resource allocation, optimized infrastructure, and cloud-based solutions to enhance scalability.
- **Software Bottlenecks**

  Software bottlenecks in DataOps can arise from various issues related to storage, data access, and processing algorithms. Challenges such as heterogeneous and incompatible formats, delayed updates, and corrupted or incomplete data can slow down workflows and impact data reliability. In terms of processing and algorithms, inefficiencies like suboptimal parallelization, computationally expensive algorithms, and inefficient queries can create bottlenecks. Additionally, issues like workflow delays, lack of data versioning and insufficient monitoring hinder automation and real-time decision-making.

- **Organizational Bottlenecks**
  Beyond technical limitations, organizational challenges play a significant role in constraining DataOps efficiency. Talent gaps, inadequate training and development, and skill mismatches prevent teams from fully leveraging DataOps methodologies. Moreover, siloed departments, resistance to change, and poor governance structures create friction in cross-functional collaboration. A lack of aligned goals, effective leadership, and clear communication channels further slows down workflow execution, leading to inefficiencies in data-driven initiatives.

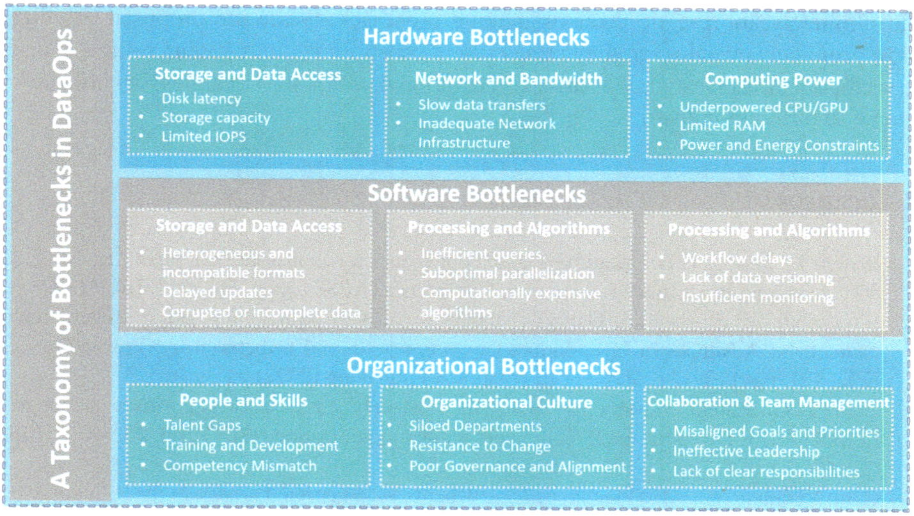

Fig. 6. A Taxonomy of Bottlenecks in DataOps: Hardware, Software, and Organizational Perspectives.

Additionally, organizational challenges, including siloed teams, resistance to change, and inadequate agile training, further impede the smooth implementation of automated data pipelines. To effectively identify and categorize these issues (RQ2), we integrated the Theory of Constraints with a DataOps-specific taxonomy, enabling us to systematically pinpoint the critical choke points across hardware, software, and organizational facets. In addressing the solutions (RQ3), our approach recommends leveraging scalable cloud infrastructures and distributed computing frameworks to overcome hardware limitations, optimizing orchestration tools and real-time monitoring to mitigate software inefficiencies, and fostering agile, cross-functional teams through continuous training and collaborative governance to surmount organizational barriers. Collectively, these targeted strategies not only streamline the data pipeline but also enhance the overall efficiency of DataOps in a specialized, coherent manner.

5 Conclusion

In conclusion, this paper offers a comprehensive exploration of the multifaceted bottlenecks hindering the full potential of DataOps in today's Big Data landscape. By systematically categorizing challenges into hardware, software, and organizational domains, we have illustrated how limitations from inadequate processing power and inefficient algorithms to siloed teams and misaligned goals can significantly disrupt data pipeline efficiency. Leveraging frameworks like the Theory of Constraints and Taxonomy of Bottlenecks in DataOps, the study not only identifies these critical choke points but also presents targeted mitigation strategies such as scalable cloud solutions, optimized data processing techniques, and the cultivation of a collaborative, data-driven culture. These integrated approaches pave the way for smoother data workflows, enhanced automation and ultimately, more agile decision-making processes. While the proposed solutions show promising avenues for alleviating current constraints, future research should focus on sector-specific investigations that explore tailored mitigation strategies for bottlenecks unique to different industries.

References

1. Bosch, J., Olsson, H.H.: Maturity assessment model for industrial data pipelines. In: 2023 30th Asia-Pacific Software Engineering Conference (APSEC), pp. 503–513. IEEE (2023)
2. Elouataoui, W., El Alaoui, I., Gahi, Y.: Metadata quality in the era of big data and unstructured content. In: Advances in Information, Communication and Cybersecurity: Proceedings of ICI2C'21, pp. 110–121. Springer International Publishing (2022)
3. Gahi, Y., Guennoun, M., Mouftah, H.T.: Big data analytics: security and privacy challenges. In: 2016 IEEE Symposium on Computers and Communication (ISCC), pp. 952–957. IEEE (2016)
4. Gahi, Y., Guennoun, M., Guennoun, Z., El-Khatib, K.: Encrypted processes for oblivious data retrieval. In: 2011 International Conference for Internet Technology and Secured Transactions, pp. 514–518. IEEE (2011, December)
5. Fannouch, A., Gahi, Y., Gharib, J.: Unified data framework for enhanced data management, consumption, provisioning, processing and movement. In: Proceedings of the 7th International Conference on Networking, Intelligent Systems and Security, pp. 1–7 (2024)
6. Fannouch, A., Gharib, J., Gahi, Y.: Enhancing DataOps practices through innovative collaborative models: a systematic review. Int. J. Inf. Manage. Data Insights. **5**(1), 100321 (2025)
7. Elouataoui, W.: AI-driven frameworks for enhancing data quality in big data ecosystems: error_detection, correction, and metadata integration. arXiv preprint arXiv:2405.03870. (2024)
8. Bergh, C., Benghiat, G., Strod, E.: The dataOps Cookbook. DataKitchen Hqrs (2019)
9. Bahaa, S., Ghalwash, A.Z., Harb, H.: DataOps lifecycle with a case study in healthcare. Int. J. Adv. Comput. Sci. Appl. **14**(1), 136 (2023)
10. Munappy, A.R., Mattos, D.I., Bosch, J., Olsson, H.H., Dakkak, A.: From ad-hoc data analytics to dataops. In: Proceedings of the International Conference on Software and System Processes, pp. 165–174 (2020)
11. Mulyono, D.P.I.: Model to Implement Theory of Constraint in Sea Transportation System (2020)

12. Kuchnik, M., Klimovic, A., Simsa, J., Smith, V., Amvrosiadis, G.: Plumber: diagnosing and removing performance bottlenecks in machine learning data pipelines. Proc. Mach. Learn. Syst. **4**, 33–51 (2022)
13. Leclerc, G., Ilyas, A., Engstrom, L., Park, S.M., Salman, H., Mądry, A.: FFCV: accelerating training by removing data bottlenecks. In: Proceedings of the IEEE/CVF Conference on Computer Vision and Pattern Recognition, pp. 12011–12020 (2023)
14. Isenko, A., Mayer, R., Jedele, J., Jacobsen, H.A.: Where is my training bottleneck? Hidden trade-offs in deep learning preprocessing pipelines. In: Proceedings of the 2022 International Conference on Management of Data, pp. 1825–1839 (2022)
15. Goderis, A., Sattler, U., Lord, P., Goble, C.: Seven bottlenecks to workflow reuse and repurposing. In: The Semantic Web–ISWC 2005: 4th International Semantic Web Conference, ISWC 2005, pp. 323–337. Springer Berlin Heidelberg, Galway, Ireland. Proceedings 4 (2005)
16. Gill, G., Singh, M.: Bottleneck analysis and alleviation in pipelined systems: a fast hierarchical approach. In: 2009 15th IEEE Symposium on Asynchronous Circuits and Systems, pp. 195–205. IEEE (2009)
17. Paul, S.P., Khan, S.T., Tasnim, Z., Haque, R., Shafikul Islam, Md.: System design and analysis for machine learning models using continuous integration and continuous Development (CI/CD). In: Swaroop, A., Kansal, V., Hassanien, A.E. (eds.) Proceedings of Sixth Doctoral Symposium on Computational Intelligence, pp 745–756. Springer Nature, Singapore (2026)

Digital Health

Digital Dining Redefined: Impacts of Food Delivery Apps in Croatia

Ernest Dudaković, Alen Orbanić, and Tihomir Orehovački(✉)

Faculty of Informatics, Juraj Dobrila University of Pula, Zagrebačka 30, 52100 Pula, Croatia
{edudakovic,aorbanic}@student.unipu.hr,
tihomir.orehovacki@unipu.hr

Abstract. Food delivery applications (FDAs) have become an integral part of modern dining habits, reshaping consumer behavior and everyday lifestyles. This study examines the multidimensional impacts of FDA usage within the Croatian socio-cultural context, addressing ten key dimensions: health impact, behavioral changes, emotional well-being, decision difficulty, food quality perceptions, food choices and variety, convenience and efficiency, financial implications, social influence, and data privacy and security. Using a survey-based approach, the findings reveal both benefits—such as convenience and access to diverse meal options—and challenges, including adverse health outcomes, emotional distress, financial strain, decision fatigue, and privacy concerns. By integrating these dimensions into a single framework, the study highlights their interconnected nature and provides actionable insights for app developers and policymakers. Recommendations include the implementation of user-centric features, such as nutritional transparency and robust data protection, alongside regulatory measures aimed at promoting fairness and ethical practices. These findings contribute to a broader understanding of FDAs and offer a foundation for future research seeking to maximize their benefits while mitigating potential drawbacks.

Keywords: Food Delivery Applications (FDAs) · User Behavior · Emotional and Physical Health · Financial and Social Factors · Privacy Concerns · Croatia

1 Introduction

The widespread adoption of food delivery applications (FDAs) has reshaped modern consumer behavior, particularly in how individuals interact with food and dining services [4]. Enabled by smartphone ubiquity and increased internet access, these platforms have catalyzed a global shift in eating habits, emphasizing convenience and instant gratification [2].

Within this broader trend, the Croatian context presents a distinctive case. Croatia is traditionally rooted in strong culinary heritage, home-cooked meals, and regionally sourced ingredients. However, rapid urbanization, shifting work patterns, and rising digital literacy—especially among younger populations—have led to growing reliance on FDAs. This shift is unfolding within a society that continues to value communal

B.-E. Boudriki Semlali et al. (Eds.): DATA 2025, LNNS 1908, pp. 127–139, 2026.
https://doi.org/10.1007/978-3-032-21200-9_11

eating and local food identity, making the tension between tradition and convenience particularly salient. Moreover, Croatia's relatively moderate-income levels introduce distinct financial trade-offs when using platforms that often charge additional delivery and service fees.

FDAs offer immediacy and access to a wide variety of meals, yet they also pose notable challenges. These platforms frequently promote calorie-dense, nutritionally imbalanced foods through persuasive design, encouraging impulse purchases and over-consumption [7]. Over time, such patterns may impact users' physical health and contribute to more sedentary lifestyles.

The user experience is further shaped by interface elements that reward repetitive use and may even foster compulsive ordering behaviors [11]. Beyond convenience, sustained engagement with FDAs is often driven by psychological factors—such as emotional eating, social disconnection, and perceived time scarcity—which, as Kaur et al. [9] observe, reflect a complex interplay of emotional, social, and utilitarian values.

While FDAs may seem cost-effective on the surface, the cumulative financial burden—particularly when accounting for hidden fees and habitual usage—can significantly strain individual budgets.

Given these intersecting concerns, there is a clear need to examine the broader lifestyle, behavioral, and financial implications of FDA usage in the Croatian context. This study explores these multidimensional impacts, offering insights into how such platforms shape not only eating habits, but also everyday routines and attitudes toward food.

The remainder of this paper is organized as follows. Section 2 provides a brief overview of the relevant literature. Section 3 outlines the methodological framework, including the development of constructs as well as procedures for data collection and analysis. Section 4 presents the main findings across ten key dimensions. Section 5 discusses their theoretical and practical implications, followed by Sect. 6, which addresses the study's limitations and identifies avenues for future research. Section 7 concludes with a reflection on the study's contributions and its broader significance.

2 Related Work

Food delivery applications (FDAs) have emerged as a critical component of modern consumer behavior, driven by rapid advancements in mobile technology and changing lifestyles. Several studies have investigated different aspects of FDA usage, identifying key factors influencing adoption, satisfaction, and continuous engagement.

Buettner et al. [1] examined the frequency and sociodemographic correlates of FDA usage among young adults aged 18–25 in the United States. Their findings revealed that young adults used these applications approximately twice per week. Higher usage was observed among individuals identifying as Hispanic/Latinx or Black, those with higher perceived subjective social status, and full-time students with greater financial responsibility. In contrast, living with others was associated with less frequent usage. These results provide a foundational understanding of factors that may influence dietary habits and health outcomes.

Similarly, Poon and Tung [15] explored consumer intentions during the COVID-19 pandemic, applying the Model of Goal-Directed Behavior (MGB). Their study identified that emotions, perceived behavioral control, and subjective norms significantly influenced users' desires to adopt FDAs, while perceived risks, including privacy and financial concerns, acted as barriers. The findings emphasized the importance of mitigating risks to enhance consumer trust and engagement.

Expanding on theoretical frameworks, Stephens et al. [17] highlighted the convenience offered by FDAs, identifying ease of use, promotional offers, and time efficiency as key adoption drivers. However, issues such as high delivery fees, inconsistent food quality, and a lack of nutritious options were also reported, prompting concerns about consumer well-being. The study emphasized the need for greater transparency and the integration of features that encourage more balanced eating habits.

Swothi et al. [18] focused on consumer preferences in Kerala, India, using a structured questionnaire administered to a sample of 50 respondents. The findings revealed that Uber Eats was the most frequently used app, followed by Swiggy and Zomato. Major factors influencing preferences included offers, pricing, and food taste, while challenges such as delivery delays and low food quality were prominent. The study underscored the need to enhance service reliability and customer satisfaction.

Kaur et al. [8] explored consumer resistance to FDAs using Innovation Resistance Theory (IRT). The research found that tradition-related barriers, particularly trust issues, negatively influenced adoption, whereas value and image barriers positively impacted word-of-mouth recommendations. The findings emphasized the importance of addressing trust and improving customer service to enhance adoption.

In another study, Kaur et al. [9] applied the Theory of Consumption Values (TCV) to examine the influence of consumption-related values on purchase intentions. The findings revealed that epistemic value (visibility) and price value were critical drivers. The study introduced the O2O TCV model, which separates attributes related to the app from those related to the product, offering actionable insights for service providers.

Choi [3] extended the Technology Acceptance Model (TAM) to examine how user familiarity, perceived ease of use, and perceived usefulness influence satisfaction and continued use intentions in the context of food delivery applications. The results indicated that familiarity and perceived usefulness were stronger predictors of satisfaction than ease of use, underscoring the relevance of intuitive platform design and habitual user engagement. These findings are consistent with previous research on digital service platforms that examined predictors of user satisfaction and continued usage intentions [12–14].

Ray et al. [16] utilized the Uses and Gratifications (U&G) theory to identify core motivations for FDA adoption, including convenience, customer experience, and intuitive app design. The study emphasized the significance of personalized engagement mechanisms to enhance user satisfaction and adoption rates.

Handayani et al. [6] applied the Pleasure–Arousal–Dominance (PAD) and Stimulus–Organism–Response (SOR) frameworks to explore how emotional responses shape user behavior within food delivery platforms. Their findings showed that convenience and price transparency significantly influenced emotional satisfaction, which in turn affected behavioral intentions.

Cho et al. [2] examined the influence of household size on FDA usage in China. Single-person households prioritized food variety and pricing, whereas multi-person households placed greater importance on design and convenience. Regardless of household composition, trustworthiness remained a consistently important factor, reinforcing its centrality in user decision-making. The study highlighted the need for tailored strategies that address the distinct preferences of different demographic segments.

Finally, Lin et al. [10] investigated the impact of food delivery app service quality on user satisfaction and behavioral intentions across different consumption stages. Key findings reveal that dimensions such as efficiency, fulfillment, contact, and billing significantly enhance satisfaction with the platform, which mediates the relationship between app quality and both food satisfaction and repurchase intentions. A positive pre-consumption experience was found to improve perceptions of food quality and increase the likelihood of continued usage, underscoring the critical spillover effect of app service quality on customer loyalty. These insights emphasize the importance of optimizing FDA functionality to enhance user satisfaction and foster retention within the online food delivery ecosystem.

Existing research on FDAs highlights their convenience, variety, and ability to cater to fast-paced lifestyles, particularly among younger users. However, critical gaps remain in understanding their broader health, emotional, behavioral, and financial impacts. While many studies emphasize functional attributes such as ease of use and promotional offers, they often overlook issues like sedentary lifestyles, decision fatigue, and reliance on calorie-dense foods—factors that raise important public health concerns.

Emotional dimensions, including feelings of guilt, loneliness, and social disconnection, have similarly received limited attention, as have long-term financial implications such as impulse spending and hidden costs. In addition, the cognitive burden associated with making frequent choices in apps offering dozens of food options—often resulting in decision fatigue—has not been widely explored. Nor has the role of social influence, including peer behavior and social media portrayals, in shaping ordering habits and app engagement.

Moreover, most existing studies focus on large, global markets, providing limited insights into culturally specific contexts like Croatia. These settings, shaped by distinctive culinary traditions and socio-economic dynamics, require deeper exploration to understand how FDAs influence local behaviors and food-related practices.

This study seeks to address these gaps by offering a comprehensive analysis of the multidimensional impacts of FDA usage, including health, emotional, behavioral, financial, cognitive, and social aspects, with a specific focus on the Croatian context. It aims to provide a deeper understanding of how FDAs shape users' daily lives within regional and cultural specificities often overlooked in existing literature.

3 Methodology

Data were collected using a post-use questionnaire administered via Google Forms. The instrument included seven demographic items: age, gender, frequency of using food delivery applications, frequency of cooking meals at home, frequency of dining out at restaurants, household size, and current employment status.

In addition to demographic information, the questionnaire assessed ten constructs central to the research framework. Health Impact (HI) was measured through 12 items examining the perceived effects of app usage on overall health, including changes in physical activity, weight, energy levels, and diet quality. Behavioral Changes (BC) were evaluated using 7 items focused on shifts in daily routines, such as cooking motivation, grocery shopping habits, and food ordering behavior patterns. Emotional Well-Being (EWB) was measured through 8 items assessing emotional consequences associated with food delivery app usage, including feelings of guilt, loneliness, uselessness, and fluctuations in mood. Decision Difficulty (DD) was captured using 4 items related to the cognitive burden of selecting meals due to the large number of options available within the apps. Food Quality Perceptions (FQP) were assessed through 9 items that addressed the perceived reliability, presentation, portion size, and overall quality of delivered food, including issues such as incorrect orders and late or damaged deliveries. Food Choices and Variety (FCV) were measured through 8 items reflecting users' selection tendencies, openness to trying new dishes, and the extent to which apps support or hinder dietary variety. Convenience and Efficiency (CE) was evaluated using 4 items that focused on the perceived usefulness of food delivery apps as a practical and time-saving alternative to home cooking. Financial Implications (FI) were explored through 6 items addressing users' financial experiences, including perceived increases in food-related expenses, cost-saving behaviors, and the influence of promotions and delivery fees. Social Influence (SI) was assessed through 5 items measuring the extent to which users' decisions are shaped by family, friends, ratings, media portrayals, and important others. Finally, Data Privacy and Security (DPS) was evaluated through 9 items measuring users' concerns and trust in the handling of personal data by food delivery platforms, including attitudes toward targeted advertising, data sharing, and transaction safety.

All items were rated on a five-point Likert scale ranging from 1 (strongly disagree) to 5 (strongly agree). The internal consistency of each construct was assessed using Cronbach's alpha coefficient.

4 Results

4.1 Participant Profile and Food Delivery Preferences

The study included 43 participants. Of these, 55.8% identified as male, 34.9% as female, and 9.3% as other. Participants ranged in age from 18 to 79 (M = 31.88, SD = 12.07), with 22 being the most common response (18.6%). With respect to occupational status, 44.2% were employed, 32.6% were students, 20.9% were not working, and 2.3% were retired.

In terms of food delivery app usage, Glovo emerged as the most widely used platform, with 81.4% of participants reporting its use. Additionally, 48.8% used Wolt and 20.9% used Bolt Food. One respondent indicated using a different application not listed among the predefined options.

As for household composition, 30.2% of participants lived in four-member households, followed by 23.3% in three-member households. Two- and five-member households each accounted for 18.6% of the sample, while 4.7% lived in six-member households. Single-person and seven-member households were each reported by 2.3% of participants.

Dining-out habits varied: 23.3% of respondents ate at restaurants two to three times per week, 14% dined out four to six times weekly, and 11.6% visited restaurants either once a week or two to three times per month. Additionally, 16.3% reported eating out once a month, while 9.3% did so less than once per month.

Regarding the preparation of home-cooked meals, 25.6% of participants cooked at home two to three times per week. Among the rest, 11.6% cooked multiple times daily, 14% four to six times per week, 7% every day, 20.9% once a week, and 9.3% only once a month.

4.2 Findings

Cronbach's alpha values for all ten constructs are presented in Table 1. All values exceed the threshold of 0.6, which is considered acceptable for exploratory research [5]. This indicates that the measurement items used to assess each construct demonstrated adequate internal reliability.

Table 1. Internal consistency of constructs.

Constructs	Cronbach's Alpha
Health Impact (HI)	0.845
Behavioral Changes (BC)	0.820
Emotional Well-Being (EWB)	0.858
Decision Difficulty (DD)	0.605
Food Quality Perceptions (FQP)	0.709
Food Choices and Variety (FCV)	0.640
Convenience and Efficiency (CE)	0.648
Financial Implications (FI)	0.691
Social Influence (SI)	0.645
Data Privacy and Security (DPS)	0.700

A considerable portion of participants perceived negative health consequences associated with FDA usage. Specifically, 34.88% believed that ordering meals via FDAs adversely affects their overall health. Physical inactivity appears to be a concern, with 41.86% reporting that FDAs increase their sedentary time and 44.19% stating they reduce physical movement. Furthermore, 32.56% claimed they gained weight due to FDA usage, and 55.81% reported feeling lazier since they started using such platforms. Although 46.51% noted no decline in their energy levels, 41.86% felt less physically active and fit. Concerns regarding nutritional quality were evident, as 48.84% indicated that the meals they order contain excessive salt, sugar, and unhealthy fats. Additionally, 37.21% noticed a decline in their overall fitness, while 51.16% believed that the use of FDAs negatively influenced their dietary habits. Portion control and caloric intake were also problematic, with 39.53% stating difficulty in estimating portion sizes and 37.21% admitting they often consume more calories than necessary.

The use of FDAs has altered users' daily routines and food preparation behaviors. A total of 41.86% felt that these platforms diminished their motivation to cook at home, while 39.53% ordered food due to time constraints. Other reasons for using FDAs include a lack of cooking skills (37.21%) and disinterest in learning to cook (39.53%). Moreover, 48.84% indicated that they no longer feel the need to shop for groceries. Disruptions in regular eating patterns were noted by 46.51% of users, and 37.21% felt that FDAs often lead them to order more food than they actually need, though the same proportion disagreed.

FDA usage also appears to have emotional ramifications. Feelings of guilt were reported by 39.53% of participants, and 41.86% said they felt lonely due to their ordering habits. Other negative emotions included feelings of uselessness (46.51%), lack of motivation (48.84%), and social detachment (44.19%). Irritation was common; 34.88% said they often felt frustrated when using FDAs, although 37.21% disagreed. Furthermore, 37.21% stated that FDAs negatively affected their mood, and 58.14% reported a general decline in emotional well-being linked to their use.

The vast array of options available through FDAs contributed to decision fatigue. Nearly half (48.84%) of respondents found the variety of food options overwhelming, and 41.86% felt similarly about the abundance of restaurants. Difficulties in decision-making were reported, with 48.84% unsure about what to order and 41.86% taking a long time to reach a decision.

Perceptions of food quality were mixed, yet numerous participants expressed dissatisfaction. 39.53% were not satisfied with the meals ordered, and 44.19% noted that the food did not match its presentation in the app. Portion size was also problematic for 41.86% who found meals smaller than expected. Errors in delivery were common: 32.56% frequently received the wrong order, 46.51% sometimes received incomplete deliveries, and 39.53% reported that food often arrived cold. Additionally, 37.21% reported receiving damaged meals, while 48.84% encountered frequent delays. Overall, 46.51% found the quality of delivered meals unacceptable.

Participants' responses suggest a complex relationship with food diversity. While 48.84% admitted to preferring unhealthier options when using FDAs, 41.86% noted that available meals rarely meet their dietary needs. FDAs were also seen as discouraging novelty: 32.56% felt dissuaded from trying new dishes, 53.49% typically ordered from the same restaurants, and 46.51% usually ordered the same meals. Nonetheless, a majority perceived a benefit in access and variety, with 48.84% saying FDAs facilitate access to diverse cuisines, 51.16% indicating they enable trying different meals, and 53.49% appreciating the ability to sample meals from various restaurants.

The appeal of FDAs is clearly linked to convenience. 53.49% of respondents agreed that these services are a practical solution to satisfy meal cravings, and 34.88% found them efficient. Enjoyment without effort was a factor, as 53.49% valued being able to consume meals without having to cook. Time-saving was also crucial, with 44.19% turning to FDAs when lacking time to prepare food.

Economic concerns were prominent. 46.51% of participants said their monthly food expenses increased due to FDA usage, while 55.81% considered delivery fees and surcharges when ordering. At the same time, 51.16% acknowledged that promotions and discounts help them save money. Notably, 53.49% believed home cooking is more economical, and an equal percentage reported actively seeking cheaper alternatives within FDAs. To reduce costs, 46.51% often added extra meals to a single order to offset fees.

Social context influenced user behavior. Peer influence was mixed: 41.86% agreed they used FDAs because of friends, while the same share disagreed. Similarly, 39.53% were influenced by family use. Social proof via ratings mattered for 44.19%, and media portrayals affected the opinions of 46.51%. Moreover, 23.26% considered the opinions of significant others before ordering, though 27.91% did not.

Participants reported differing levels of concern and trust in how FDAs use their personal data. 48.84% stated they take measures to protect their personal data. At the same time, 46.51% believed their data is used for targeted ads and promotional content, and 48.84% thought it is used to offer personalized recommendations. Prudence in sharing data was expressed by 39.53%, while 53.49% trusted that their data is not misused. Additionally, 46.51% believed FDAs do not share their data with third parties, and the same proportion felt that their data was securely stored. Finally, 44.19% expressed confidence in the safety of financial transactions carried out through these platforms.

Table 2 summarizes the main empirical insights for each construct, offering a descriptive synthesis of user responses across the measured dimensions.

The results provide a comprehensive overview of both the perceived advantages and disadvantages of FDA usage. While users appreciated the convenience, access to a wide variety of cuisines, and cost-saving opportunities through promotional offers, significant concerns were raised regarding health-related outcomes, emotional well-being, financial strain, decision fatigue, perceived food quality, and data privacy. The findings also underscore shifts in behavioral routines and the influence of social and environmental factors on app engagement. These insights point to critical areas for improvement and emphasize the importance of user-centered design, transparent data practices, and supportive policy frameworks to enhance user satisfaction, promote healthier habits, and ensure digital well-being.

Table 2. Summary of key findings by construct.

Constructs	Summary of Findings
Health Impact (HI)	FDA usage is associated with perceived weight gain, reduced physical activity, and concerns regarding the nutritional quality of delivered meals.
Behavioral Changes (BC)	Notable changes include decreased motivation to cook, reduced grocery shopping, and diminished control over eating routines.
Emotional Well-Being (EWB)	Users reported emotional effects such as guilt, listlessness, and social detachment, indicating a perceived decline in overall emotional state.
Decision Difficulty (DD)	The abundance of options within FDAs contributes to cognitive overload and decision fatigue.
Food Quality Perceptions (FQP)	Perceptions were mixed, with dissatisfaction related to portion size, temperature, presentation inconsistencies, and delivery errors.
Food Choices and Variety (FCV)	While diversity of options is appreciated, habitual ordering and reluctance to try new dishes remain prevalent.
Convenience and Efficiency (CE)	FDAs are widely regarded as practical and time-saving, especially when cooking is not feasible.
Financial Implications (FI)	Increased spending due to delivery fees and impulsive ordering was reported, though discount strategies were also valued.
Social Influence (SI)	Usage patterns were shaped by the behaviors of peers and family, social ratings, media representations, and opinions of significant others.
Data Privacy and Security (DPS)	While platforms are generally trusted, users expressed concern over data sharing, targeted advertising, and the handling of personal information.

5 Implications

The findings of this study offer both practical and theoretical implications for food delivery platforms, policymakers, and researchers. From a practical standpoint, FDAs should adopt strategies that directly address user concerns related to health, emotional well-being, financial burden, and data privacy. Integrating features such as nutritional labels, calorie-aware meal recommendations, and incentives for healthier food choices could help mitigate some of the adverse health effects identified in this research. Furthermore, simplifying interface design to reduce decision fatigue and offering personalized meal planning could enhance user satisfaction and engagement. Reducing the cognitive overload associated with excessive meal options—through streamlined categorization, filtering tools, or AI-powered suggestions—may also help users make quicker, less stressful decisions.

Emotional well-being issues—such as feelings of guilt, loneliness, and listlessness—indicate a need to foster more positive user experiences. Platforms might consider incorporating socially oriented features, including shared ordering options or group-based discounts, to help users feel more connected. Similarly, promoting balanced lifestyle habits through in-app content or optional wellness prompts could support emotional resilience. In addition, app features that reflect peer activity or social validation (e.g., trending meals among friends, social recommendations) could be leveraged to better align with how users are influenced by their social circles.

The financial strain reported by many users underscores the importance of transparent pricing and affordability. Delivery platforms could improve retention by offering personalized discounts, flexible fee structures, and loyalty programs aimed at cost-sensitive users. Policymakers, in turn, should consider regulatory measures to ensure pricing fairness and equitable access, particularly for socioeconomically disadvantaged populations such as students, unemployed individuals, and low-income groups.

Concerns related to data privacy and security highlight the need for rigorous data protection policies and user education. Platforms should communicate clearly how user data are collected, stored, and used, while implementing state-of-the-art security protocols. Acquiring recognized data protection certifications could further enhance user trust. Policymakers may also strengthen regulatory frameworks to safeguard personal information and uphold users' privacy rights industry-wide.

For researchers, the study points to the importance of exploring the broader societal implications of FDA usage, including shifts in dining behaviors and effects on local food systems. Cross-cultural and geographically diverse studies are needed to gain a more comprehensive understanding of how these platforms operate in varied contexts. Future research could also benefit from incorporating qualitative methods, such as in-depth interviews or focus groups, to explore users' underlying emotions, motivations, and behavioral patterns in greater depth. Moreover, longitudinal designs may offer valuable insights into the long-term health and lifestyle consequences associated with frequent use of food delivery applications.

This study calls for a collaborative, multi-stakeholder approach involving app developers, regulators, and researchers to amplify the benefits of food delivery services while minimizing their potential risks. Such efforts may enhance user satisfaction, encourage healthier behaviors, and promote greater trust in the expanding ecosystem of digital food access.

6 Limitations

Despite offering valuable insights, this study has several limitations. The relatively small sample size ($N = 43$) limits the generalizability of the findings, particularly given the overrepresentation of younger participants. With the majority aged between 21 and 24, the results may not adequately reflect the experiences of older age groups or capture broader population trends.

Additionally, the focus on Croatian respondents constrains the applicability of the results to other cultural or geographic contexts, where dining customs, food delivery practices, and technology use may differ substantially.

The sample was also skewed toward users of a single platform—Glovo—which was the most frequently used app among participants. Although some respondents reported using alternatives such as Wolt and Bolt Food, the data may be disproportionately influenced by Glovo-specific features and policies, thereby limiting the representativeness of the overall food delivery ecosystem.

The reliance on self-reported data introduces potential biases, including recall inaccuracies and social desirability effects. Participants' accounts of meal preparation or dining frequency may not fully align with their actual behaviors. Future studies could benefit from incorporating objective measures such as app usage logs, order histories, or screen-time data to enhance the validity of reported behaviors.

While household size and living arrangements were considered, the modest sample size precluded more detailed subgroup analyses of how these factors affect food delivery usage. Similarly, the high proportion of students in the sample may have influenced perceptions of time pressure and financial burden—factors that could vary meaningfully across different demographic profiles.

These limitations highlight the importance of conducting larger, more demographically diverse studies to better understand the nuanced dynamics of food delivery app usage.

7 Conclusion

This study offers a comprehensive analysis of food delivery application (FDA) usage across ten key dimensions: health, behavior, finances, emotions, decision difficulty, food quality, convenience, variety, social influence, and data privacy. It integrates these aspects into a single framework to provide a detailed understanding of FDA-related impacts within the Croatian context.

In contrast to previous studies that typically examine isolated factors—such as health outcomes or financial expenditures—this research emphasizes the interconnectedness of these domains and their cumulative influence on users' everyday lives. Moreover, while much of the existing literature centers on user satisfaction or adoption behavior, this study extends the analytical scope by exploring shifts in behavior, emotional and cognitive responses, and concerns related to personal data, offering a more comprehensive view of how FDAs shape modern food-related practices.

The findings underscore the importance of addressing user concerns holistically. For app developers, this includes implementing features that promote healthier choices, ensure price transparency, alleviate cognitive overload through improved interface design, and enhance trust via robust data protection. Policymakers are encouraged to establish regulatory frameworks that uphold ethical standards, particularly in the domains of consumer privacy, pricing transparency, and public health.

Future research should examine the long-term consequences of FDA usage, particularly its sustained impact on physical well-being, emotional resilience, and financial habits. Expanding the cultural and geographic scope of inquiry would also provide valuable insight into how local values and contexts shape engagement with food delivery services. Such efforts are essential to inform the development of more inclusive, responsible, and health-oriented digital food ecosystems.

Acknowledgments. This research was supported by the "European Digital Innovation Hub Adriatic Croatia (EDIH Adria) (project no. 101083838)" under the European Commission's Digital Europe Programme.

References

1. Buettner, S.A., Pasch, K.E., Poulos, N.S.: Factors associated with food delivery app use among young adults. J. Community Health. **48**, 1–7 (2023)
2. Cho, M., Bonn, M.A., Li, J.: Differences in perceptions about food delivery apps between single-person and multi-person households. Int. J. Hosp. Manag. **77**, 108–116 (2019)
3. Choi, J.-C.: User familiarity and satisfaction with food delivery mobile apps. SAGE Open. **10**, 2158244020970563 (2020)
4. Dannenberg, P., Fuchs, M., Riedler, T., Wiedemann, C.: Digital transition by COVID-19 pandemic? The German food online retail. Tijd. voor. Econ. Soc. Geog. **111**, 543–560 (2020)
5. Hair, J.F., Hult, G.T.M., Ringle, C.M., Sarstedt, M.: A Primer on Partial Least Squares Structural Equation Modeling (PLS-SEM), 3rd edn. Sage, Thousand Oaks (2022)
6. Handayani, P.W., Azzizah, S.F., Annisa, A.: The impact of user emotions on intentions to continue using online food delivery applications: the influence of application quality attributes. Cogent Bus. Manag. **9**, 2133797 (2022)
7. Jaworowska, A., Blackham, T., Davies, I.G., Stevenson, L.: Nutritional challenges and health implications of takeaway and fast food. Nutr. Rev. **71**, 310–318 (2013)
8. Kaur, P., Dhir, A., Ray, A., Bala, P.K., Khalil, A.: Innovation resistance theory perspective on the use of food delivery applications. J. Enterp. Inf. Manag. **34**, 1746–1768 (2021)
9. Kaur, P., Dhir, A., Talwar, S., Ghuman, K.: The value proposition of food delivery apps from the perspective of theory of consumption value. Int. J. Contemp. Hosp. Manag. **33**, 1129–1159 (2021)
10. Lin, P.M.C., Au, W.C.W., Baum, T.: Service quality of online food delivery mobile application: an examination of the spillover effects of mobile app satisfaction. Int. J. Contemp. Hosp. Manag. **36**, 906–926 (2023)
11. Moser, C., Schoenebeck, S.Y., Resnick, P.: Impulse buying: design practices and consumer needs. In: Proceedings of the 2019 CHI Conference on Human Factors in Computing Systems, pp. 1–15. Association for Computing Machinery, New York, NY, USA (2019)
12. Orehovački, T., Babić, S.: Predicting students' continuance intention related to the use of collaborative web 2.0 applications. In: Proceedings of the 23rd International Conference on Information Systems Development, pp. 112–122. University of Zagreb, Faculty of Organization and Informatics, Varaždin, Croatia (2014)
13. Orehovački, T., Babić, S., Jadrić, M.: Exploring the validity of an instrument to measure the perceived quality in use of web 2.0 applications with educational potential. In: Zaphiris, P., Ioannou, A. (eds.) Learning and Collaboration Technologies. Designing and Developing Novel Learning Experiences, pp. 192–203. Springer International Publishing, Cham (2014)
14. Orehovački, T., Cappiello, C., Matera, M.: Identifying relevant dimensions for the quality of web mashups: an empirical study. In: Kurosu, M. (ed.) Human-Computer Interaction. Theory, Design, Development and Practice, pp. 396–407. Springer International Publishing, Cham (2016)

15. Poon, W.C., Tung, S.E.H.: The rise of online food delivery culture during the COVID-19 pandemic: an analysis of intention and its associated risk. Eur. J. Manag. Bus. Econ. **33**, 54–73 (2022)
16. Ray, A., Dhir, A., Bala, P.K., Kaur, P.: Why do people use food delivery apps (FDA)? A uses and gratification theory perspective. J. Retail. Consum. Serv. **51**, 221–230 (2019)
17. Stephens, J., Miller, H., Militello, L.: Food delivery apps and the negative health impacts for Americans. Front. Nutr. **7**, 14 (2020)
18. Swothi, S., Bertle, T.G., Shehnaz, S.R.: A study on customer preference towards food delivery applications. Int. J. Adv. Innov. Res. **7**, 25–30 (2020)

A Deep Ensemble Transfer Learning Framework for Brain Tumor Classification Using MRI Scans

Abdullah Mateen Janjua, Muhammad Hassaan Ashraf(✉), Muhammad Esham Qureshi, Haider Ali, and Ahmed Khan

Faculty of Computing, Riphah International University, Islamabad 46000, Pakistan
abdullahjanjua.cs@gmail.com, mhassaan.scholar@gmail.com, mequreshi.cs@gmail.com, haiderchaudhary112@gmail.com, akhan.scholar@gmail.com

Abstract. The automated diagnosis of brain tumors from Magnetic Resonance Imaging (MRI) scans presents significant challenges due to tumor heterogeneity, dataset limitations, and imaging modality variability. In this research, we propose a novel deep ensemble transfer learning framework that integrates the strengths of two high-performing Convolutional Neural Networks (CNNs), HFF-Net and DenseNet-201, to overcome these challenges. The framework utilizes advanced preprocessing techniques, including bicubic interpolation for image resizing and Contrast Limited Adaptive Histogram Equalization (CLAHE), to improve image contrast and standardize inputs. We evaluate the performance of six state-of-the-art CNN models AlexNet, VGG-16, ResNet-50, InceptionV3, DenseNet-201, and HFF-Net on two publicly available datasets, focused on Multi-Cancer and Brain Tumor classification, respectively. The proposed ensemble model, which combines feature-level and decision-level fusion, demonstrates remarkable performance, achieving state-of-the-art results with 99.15% accuracy on the Multi-Cancer dataset, and 99.96% accuracy on the Brain Tumor MRI dataset.

Keywords: Magnetic Resonance Imaging (MRI) · Tumor Classification · Convolutional Neural Networks · Ensemble Learning · Transfer Learning · HFF-Net · DenseNet-201 · Contrast Limited Adaptive Histogram Equalization

1 Introduction

Brain tumors pose a serious global health challenge, with the World Health Organization (WHO) estimating over 250,000 deaths from cancers of the brain or central nervous system in 2020 [1]. The effective and timely categorization of brain tumors is essential for treatment, survival improvement and therapeutic approach optimization. MRI is the diagnostic modality of choice because of its excellent soft-tissue contrast that provides detailed morphology of tumors. Nonetheless, manual scanning of MRI data is labor-intensive, subjective and may result in inconsistencies, which underscores the need for automated and efficient classification frameworks [2]. Brain tumors demonstrate wide

B.-E. Boudriki Semlali et al. (Eds.): DATA 2025, LNNS 1908, pp. 140–154, 2026.
https://doi.org/10.1007/978-3-032-21200-9_12

ranges of morphology, which makes accurate classification quite difficult. Accurate categorization is essential to differentiate between benign and malignant tumors so that a suitable treatment can be administered. While aggressive treatment is mandated for high-grade tumors such as glioblastomas, benign ones may only need conservative treatments [3]. Conventional approaches which depend on the judgement of a human expert are subject to inter-observer variance, which makes the case for AI driven solutions even more compelling.

CNN-based frameworks have significantly advanced automated brain tumor classification in MRI scans. Classical architectures like AlexNet, VGG16, ResNet-50, and InceptionV3 demonstrate high accuracy due to their powerful feature extraction capabilities. However, variations in receptive field design and convolutional strategies lead to diverse feature representations. CNNs can be categorized into feedforward, multilevel, and multiscale models based on their spatial hierarchies and abstraction methods [4]. Using a single CNN model to classify brain tumors is inefficient because simple feedforward models do not easily capture low level texture features and high-level semantic features simultaneously enhancing generalization ability. Advanced architectures such as DenseNet-201 [5] and HFF-Net [4] alleviate these limitations through employing features of dense connected feature propagation, multiscale context learning, and hierarchical feature fusion. Incorporating the various feature representations from these CNNs models enhances the classification's robustness, discriminability, and the model interpretability which makes it more reliable [6].

In spite of the progress made on deep learning tumor classification models, the existing systems still suffer from critical dataset issues, poor performance on specific computational tasks or imaging modalities, and generalization problems. Moreover, a single CNN architecture usually has a feature loss problem at deeper layers, resulting in failure to accurately diagnose complicated cases. To address these problems, we developed a new ensemble transfer learning framework that incorporates HFF-Net and DenseNet-201 for brain tumor classification. This framework utilizes hierarchical and global feature extraction techniques, improving classification performance, feature loss, and generalization. Two publicly available MRI datasets, the Multi-Cancer Dataset and the Brain Tumor MRI Dataset comprising various tumor types such as glioma, meningioma, and pituitary tumors, were utilized to evaluate the proposed model. Extensive validation on these datasets demonstrates the method's effectiveness in addressing key challenges in automated brain tumor diagnosis.

1.1 Contributions

Following the introduction of an ensemble deep learning framework for brain tumor classification, HFF-Net and DenseNet-201 were integrated via transfer learning. The key contributions of this study include:

- Development of an ensemble framework that integrates HFF-Net and DenseNet-201 for improved brain tumor classification.
- Incorporation of transfer learning through pre-trained DenseNet-201 to accelerate convergence and enhance generalization.
- Application of CLAHE-based contrast enhancement and data augmentation to refine tumor boundaries and address class imbalance.

- Design of a multiscale feature fusion mechanism using HFF-Net to boost tumor localization and classification accuracy.
- Extensive evaluation on the Multi-Cancer and Brain Tumor MRI datasets, demonstrating superior performance compared to existing deep learning models.
- Validation of the proposed model's potential for real-world clinical deployment in automated brain tumor diagnosis.

These contributions establish the proposed ensemble learning framework as a benchmark for automated brain tumor detection, advancing the field of deep learning-based medical image analysis while ensuring practical applicability in real-world clinical settings.

2 Current State-of-the-Art

Guder and Cetin-Kaya [7] introduced three lightweight CNN models: ChannelNet, SpatialNet, and CbamNet optimized with attention mechanisms for brain tumor classification. Trained on the Kaggle Brain Tumor MRI dataset with four tumor classes: glioma, meningioma, pituitary, no tumor. The preprocessing involved resizing images and applying augmentation techniques, including rotations and flips. Robustness was evaluated under salt-and-pepper noise conditions. Particle Swarm Optimization (PSO) was utilized for hyperparameter tuning. The models featured softmax activation in their final layers and categorical cross-entropy loss. Their lightweight nature and resilience to noisy inputs make them suitable for resource-constrained environments.

Wang et al. [8] developed HQNet, a Hybrid Quantum-Classical Convolutional Network, used advanced attention mechanisms together with quantum computing for brain-tumor analysis. The design consists of four modules: Hybrid Residual (HR) for spatial feature extraction, Feature Extraction and Recombination (FER) with dual attention mechanisms, Random Quantum Circuit (RQC) for quantum-enhanced feature extraction, and Classification (CL). Evaluations were performed on BT-large-4c, BT-large-2c, and the Cheng dataset, covering glioma, meningioma, pituitary, and no-tumor classes. Preprocessing was performed with min-max normalization and resizing to 224x224 with class-imbalance-issues addressed by augmentation methods including translation and rotation. ReLU and softmax activation with a categorical cross-entropy loss demonstrated the ability of the framework to have diverse tumor morphologies.

Chandni et al. [9] with the aid of Dense-CNN for effective feature extraction coupled with classifier optimization using XGBoost (eXtreme Gradient Boosting), proposed a hybrid framework called BO-DenseXGB. They utilized the Figshare dataset, comprising 3,064 MRI images categorized into three tumor classes: meningioma, glioma, and pituitary. The preprocessing steps included resizing images to 224 x 224, normalizing the images, and applying data augmentation techniques such as horizontal and vertical flips, rotations, and zooms. Fine-tuning the hyperparameters of the XGB classifier was performed using Bayesian Optimization, whereas Dense-CNN utilized the softmax activation function in its last layer. Hence, the discriminative performance of the model demonstrates how deep learning can be more effective when combined with boosting techniques for tumor classification in clinical environments.

Subba and Sunaniya [10] addressed a computationally optimized CNN framework inspired by GoogLeNet integrated with an attention-based inception module (ABIM). Trained it on various datasets such as Figshare, Harvard Medical Repository, and Kaggle-merged dataset, preprocessing included min-max normalization and resized images of 224x224. Augmentation techniques are noise addition to test robustness. ABIM's channel and spatial attention mechanisms improved the feature learning concentration on specific tumor regions. This model demonstrated efficient computation using the ReLU activation and cross-entropy loss and generalized well for different datasets.

A. Karthik et al. [11] introduced a hybrid framework for brain tumor classification and segmentation, combining Attention-Augmented CNN, Random Forest, and U-Net. The dataset comprised 3100 MRI images, split into training and testing sets, with tumor annotations. Preprocessing involved Gaussian smoothing, intensity normalization, and contrast enhancement, alongside data augmentation techniques such as rotation and flipping. The U-Net ensured precise segmentation, while the Attention-Augmented CNN enhanced spatial and channel-wise features. Binary cross-entropy with an attention loss term was employed. This framework effectively addressed challenges like tumor heterogeneity and variations in imaging protocols.

Malik et al. proposed DEF-SwinE2NET [12], a hybrid model integrating Efficient-NetV2S, Swin Transformer, and a Dual Enhanced Features (DEF) scheme for brain tumor classification. Dilated convolutions and dual attention mechanisms captured spatial and channel features, addressing intra-class variability. Evaluations were performed on three multi-class datasets, including two from Kaggle and one from Figshare. Preprocessing included median filtering, CLAHE, Laplacian edge enhancement, and cropping, with augmentation techniques like rotation and flipping enhancing robustness. Swish activation and categorical cross-entropy loss were employed, showcasing the framework's effectiveness on small datasets and complex tumor patterns.

Remzan et al. [13] integrated RadImageNet-pretrained CNNs with ensemble learning and machine learning classifiers for brain tumor classification. Using a dataset of 7023 MRI images categorized into four tumor classes: glioma, meningioma, pituitary tumor, and normal. The preprocessing included resizing, bilateral filtering, gamma correction, and normalization. Two ensemble approaches were adopted: Feature Ensemble which combine features from ResNet-50 and DenseNet-121 and Stacking Ensemble using Logistic Regression as the meta-learner. Tailored preprocessing, ensemble strategies, and hyperparameter optimization through Randomized Search CV addressed challenges like overfitting and dataset limitations, demonstrating robust classification performance.

Sharma et al. [14] introduced a modified ResNet-50 framework for brain tumor classification using MRI images. The model leveraged transfer learning, replacing the ResNet-50's final layer with additional dropout, smooth, and dense layers, and employed a sigmoid activation function for probability output. The Kaggle brain MRI dataset was utilized, with images resized to 224x224 and augmented using rotation, shearing, and cropping. These preprocessing steps ensured balanced and normalized input. Comparative analysis highlighted the superior performance of the modified ResNet-50, underscoring the adaptability of transfer learning in medical imaging tasks.

2.1 Research Gap Analysis

Despite significant advancements in deep learning for medical image analysis, accurate and clinically applicable brain tumor classification remains challenging. Key limitations include poor generalization across diverse MRI modalities, class imbalance affecting minority tumor types, and the opaque nature of deep models that hinders clinical trust. Existing transformer-based ensembles often rely on homogeneous datasets, leading to reduced performance in real-world scenarios. Additionally, high computational demands restrict the use of advanced models like HFF-Net and DenseNet-201 in resource-limited settings, while lightweight alternatives remain underexplored. Inconsistencies in preprocessing techniques further affect model reproducibility. To bridge these gaps, this study introduces an ensemble transfer learning framework that combines HFF-Net and DenseNet-201, leveraging multimodal MRI inputs, optimized computation, and explainable AI to enhance accuracy, interpretability, and clinical readiness.

3 Brain Tumor Diagnosis Framework

This study presents a robust Brain Tumor Classification (BTC) framework leveraging MRI scans to enhance detection accuracy and reliability. The proposed model integrates HFF-Net with DenseNet-201 through features ensemble technique. Experiments conducted on benchmark datasets including the Multi-Cancer Dataset and the Brain Tumor MRI Dataset ensure comprehensive evaluation. The preprocessing pipeline employs CLAHE for contrast enhancement, image resizing to 224 × 224 pixels, and data augmentation to address class imbalance and improve generalization. Multiscale hierarchical feature fusion captures complementary deep features from HFF-Net and DenseNet-201, while transfer learning with ImageNet-pretrained weights accelerates convergence and strengthens representation learning. Fig. 1 illustrates the end-to-end workflow of the proposed framework, from MRI preprocessing to classification.

3.1 Data Preprocessing and Enhancement

To optimize feature extraction and model performance, an integrated preprocessing pipeline was developed combining contrast enhancement and image resizing. All MRI scans were resized to 224 × 224 pixels using bicubic interpolation, which preserves anatomical structures through smooth intensity transitions, ensuring architecture compatibility with HFF-Net and DenseNet-201. Given the limitations of resizing alone in low-contrast images, CLAHE was employed to enhance local contrast and delineate tumor boundaries. Unlike global histogram equalization, CLAHE divides images into 8 × 8 tiles and applies localized enhancement with a clip limit of 2.0, balancing contrast amplification and noise suppression. This preprocessing strategy standardizes input data and accentuates critical tumor features, ultimately improving classification accuracy within the proposed ensemble transfer learning framework.

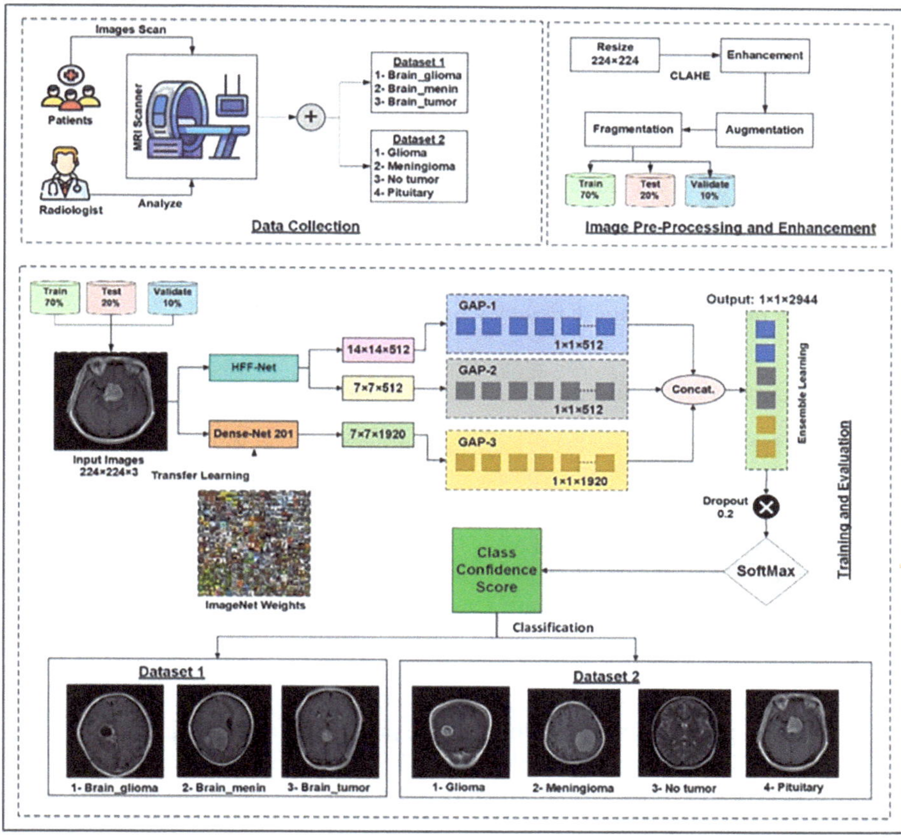

Fig. 1. The proposed framework for brain tumor classification.

3.2 Ensemble Learning

Ensemble learning enhances classification accuracy and robustness by integrating multiple models. The proposed framework leverages the complementary strengths of HFF-Net and DenseNet-201 to address tumor heterogeneity, interclass variation, and class imbalance in brain MRI datasets. Feature representations from both networks are aggregated and refined using Global Average Pooling (GAP), then concatenated into a unified descriptor. This fusion combines HFF-Net's multiscale hierarchical attention features with DenseNet-201's densely connected representations, improving tumor discrimination and generalization. Beyond accuracy, the ensemble approach enhances interpretability, promoting clinical applicability. By integrating hierarchical learning with computational efficiency, the framework minimizes overfitting and supports reliable, real-time brain tumor classification in clinical settings.

3.3 Transfer Learning

Transfer learning is a powerful technique in deep learning that improves accuracy and reduces training time by leveraging knowledge from pre-trained models on large datasets.

In the proposed framework, DenseNet-201 pretrained on ImageNet is employed to extract rich, generic features that are fine-tuned for brain tumor classification from MRI scans. This significantly reduces computational overhead and accelerates convergence, as DenseNet-201 provides well-initialized weights for capturing structural patterns in medical images. In parallel, HFF-Net is trained from scratch to focus on domain-specific, hierarchical, and multiscale feature extraction tailored to tumor morphology and heterogeneity. This approach combines transfer learning with training from scratch, combining globally transferable features from DenseNet-201 with specialized features from HFF-Net enables robust and accurate classification by capturing both coarse contextual structures and fine-grained tumor details while minimizing overfitting.

3.4 The Proposed Deep Ensemble Transfer Learning Architecture

The proposed framework introduces a hybrid ensemble-transfer learning method that combines DenseNet-201 and HFF-Net to accurately and effectively classify brain tumors from MRI images. DenseNet-201 guarantees deep feature propagation and gradient efficiency through densely linked layers, while HFF-Net is optimized for hierarchical multiscale feature extraction and excels at capturing complicated tumor shape. DenseNet-201 uses transfer learning to improve feature generalization and speed up convergence by utilizing pre-trained ImageNet weights. It is a reliable approach for automated brain tumor identification because of this integration, which guarantees the best possible balance between feature richness, computing efficiency, and classification accuracy.

Standardized preprocessing, which starts the processing pipeline, involves resizing MRI images to $224 \times 224 \times 3$ in order to preserve consistency between the two networks. DenseNet-201 collects deep structural representations at $7 \times 7 \times 1920$, while HFF-Net extracts hierarchical and multiscale features at $14 \times 14 \times 512$ and $7 \times 7 \times 512$ resolutions. The retrieved feature maps are refined by GAP layers, which eliminate redundancy while maintaining important information to guarantee efficient feature consolidation.

The essential component of the ensemble model is the feature fusion step, which creates a single high-dimensional feature vector by concatenating the spatially enriched features of HFF-Net with the thoroughly learned contextual representations of DenseNet-201. By combining macro-level structural patterns with micro-level tumor textures, this fusion improves the network's classification capabilities. In order to avoid overfitting, the fused feature vector is run through a fully connected classification layer that is tuned with dropout regularization (0.2). Glioma, Meningioma, Pituitary, and No Tumor can all be distinguished with a Softmax activation function, which guarantees precise multi-class classification.

The system uses sparse categorical cross-entropy loss to improve robustness and alleviate class imbalance, allowing for efficient learning across tumor kinds that are underrepresented. By dynamically modifying weight updates based on validation performance, the Adam optimizer and an adaptive learning rate scheduler stabilize convergence. Furthermore, by enhancing gradient flow, the Swish activation function maximizes learning effectiveness across deep layers.

Pre-trained weights speed up training convergence, and the framework uses a variety of feature representations for improved generalization by combining ensemble learning with transfer learning in a seamless manner. This collaboration overcomes issues with scalability, computing efficiency, and dataset variability to produce state-of-the-art results in brain tumor classification. DenseNet-201's densely linked architecture and HFF-Net's hierarchical feature extraction are designed to create a highly optimized, clinically feasible system that raises the bar for automated brain tumor identification.

3.5 Activation and Loss Function

In this research, Swish activation function used in CNN backbone. The Swish activation function, defined as:

$$Swish\,(x) = x.\sigma\,(x) \quad (1)$$

M where σ(x) is the sigmoid function is chosen because it can balance linear and non-linear transformations and smooth gradients. Swish is more effective than ReLU at capturing subtle tumor characteristics in MRI images because it enhances gradient flow and avoids dead neurons. This improves classification accuracy by allowing the model to catch intricate tumor borders and morphological differences.

Moreover, to handle multi-class classification, the sparse categorical cross-entropy loss function is applied, defined as:

$$L = -\frac{1}{N}\sum\nolimits_{i=1}^{N} \log\left(p_{i,yi}\right) \quad (2)$$

where:

- L: Average loss across all samples
- N: Total number of samples
- p_i: Predicted probability of the true class
- y_i: True class label for the sample

Class imbalance in brain tumor datasets is successfully addressed by this loss function, which encourages balanced learning across all tumor classes. The suggested framework ensures a strong and dependable brain tumor detection system by combining improved activation and loss functions, which improves learning efficiency, speeds up convergence, and increases classification accuracy.

4 Experiments

4.1 Model Training

To ensure computational performance needed to analyze large-scale MRI datasets and carry out deep learning computations with high precision, the suggested brain tumor classification framework was trained and evaluated on a Tesla T4 GPU. The GPU's high-speed memory and parallel processing capabilities made model training smooth and enabled effective management of intricate feature extractions and iterative optimizations.

A batch size of 32 was selected in order to strike the best possible balance between model performance and computational efficiency, guaranteeing seamless gradient updates without using excessive amounts of memory. With an initial learning rate of 0.001, the Adam optimizer was used, utilizing its adaptive learning mechanism to accelerate convergence. Furthermore, a learning rate scheduler effectively mitigated overfitting and ensured steady optimization throughout the training phase by dynamically adjusting the learning rate based on validation results.

The depth and ability of the suggested model to learn complex tumor traits while preserving computational viability are demonstrated by its 19,264,067 trainable parameters. In order to determine the best parameter combinations, the model was trained for 30 epochs, which allowed for ample learning time and ongoing validation performance monitoring. A checkpointing technique was used to maintain the best-performing weights, guaranteeing that the model kept its most useful parameters for assessment. The framework's resilience, dependability, and practicality in automated brain tumor diagnosis were demonstrated in the final evaluation step, which used the saved best model weights to evaluate classification accuracy.

4.2 Benchmark Dataset

This study utilizes two publicly available datasets for training, validation, and evaluation of the proposed ensemble transfer learning framework, chosen for their diversity and relevance to brain tumor classification. The first, the Multi-Cancer Dataset, includes 15,000 images equally distributed across three classes: Brain Glioma, Brain Meningioma, and Brain Tumor. It was partitioned into training, validation, and testing sets using a 70:10:20 ratio. The second, the Brain Tumor MRI Dataset by Masoud Nickparvar, comprises 7,023 images categorized into four classes: Glioma, Meningioma, Pituitary, and No Tumor, and follows the same split strategy as summarized in Table 1.

Table 1. Distribution of both brain tumor datasets.

Dataset	Total Images	Classes	Train (70%)	Test (20%)	Validate (10%)	Total
Multi-Cancer Dataset	15,000	Brain Glioma	3500	1000	500	5000
		Brain Meningioma	3500	1000	500	5000
		Brain Tumor	3500	1000	500	5000
Brain Tumor MRI Dataset	7,023	Glioma	1,134	325	162	1621
		Meningioma	1151	330	164	1645
		No Tumor	1400	400	200	2000
		Pituitary	1229	353	175	1757

4.3 Performance Evaluation Metrics

To ensure a comprehensive evaluation of the proposed model's performance across different brain tumor types, multiple metrics were employed. Accuracy in Eq. (3) measures the proportion of correctly classified samples among all samples, offering an overall assessment of effectiveness.

$$\text{Accuracy} = \frac{\text{TP} + \text{TN}}{\text{TP} + \text{FN} + \text{FP} + \text{TN}} \tag{3}$$

Precision in Eq. (4) evaluates the proportion of true positives among predicted positives, helping to minimize false alarms.

$$\text{Precision} = \frac{\text{TP}}{\text{TP} + \text{FP}} \tag{4}$$

Recall in Eq. (5) indicates the model's ability to detect actual tumor cases, reducing false negatives.

$$\text{Recall} = \frac{\text{TP}}{\text{TP} + \text{FN}} \tag{5}$$

F1-Score in Eq. (6), the harmonic mean of precision and recall, provides a balanced evaluation, especially valuable in the presence of class imbalance.

$$\text{F1} - \text{Score} = 2 * \frac{\text{Precision} \times \text{Recall}}{\text{Precision} + \text{Recall}} \tag{6}$$

These metrics are particularly useful for handling class imbalance, ensuring a comprehensive evaluation of classification performance.

4.4 Performance Evaluation Model

Table 2 presents the performance of a proposed model on two different datasets: the Multi-Cancer dataset and the Brain Tumor MRI Images dataset. For the Brain Tumor MRI images dataset, the model correctly identified most cases across four classes, including Glioma, Meningioma, No Tumor, and Pituitary, with very few misclassifications. The highest number of false negatives was observed in the Meningioma class, where eight cases were misclassified, while other classes had minimal errors. Precision, recall, and F1 scores remained consistently high, reflecting strong classification performance. In the Brain Tumor Multi-Cancer Dataset, the model demonstrated out-standing results, with nearly perfect classification across all three classes. Only one sample was misclassified in both the Meningioma and Tumor categories. The results in Table 2 highlight the model's effectiveness in brain tumor classification, with only minor misclassifications across both datasets.

Table 2. Performance of Proposed Model on each class

Dataset	Class	TP	FN	FP	TN	Precision	Recall	F1 Score
Brain Tumor MRI images dataset	Glioma	324	1	4	1079	99%	100%	99%
	Meningioma	322	8	2	1079	99%	98%	98%
	No Tumor	399	1	5	1003	99%	100%	99%
	Pituitary	351	2	1	1054	100%	99%	100%
Multi-Cancer Dataset	Brain Glioma	1000	0	0	2000	100%	100%	100%
	Brain Meningioma	1000	0	1	1999	100%	100%	100%
	Brain Tumor	999	1	0	2000	100%	100%	100%

The confusion matrix for the proposed model in brain tumor classification, as depicted in Fig. 2, illustrates the model's capability in distinguishing between different tumor categories. The results highlight the distribution of predictions across four classes: Glioma, Meningioma, No Tumor, and Pituitary.

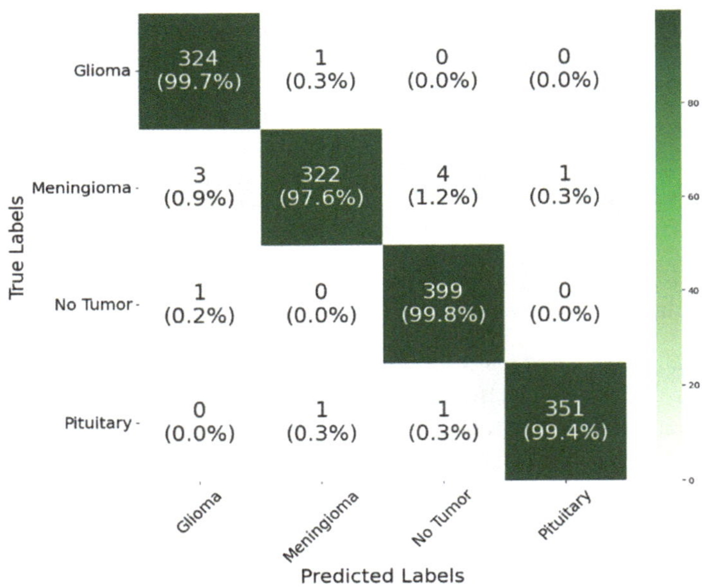

Fig. 2. Confusion matrix of Proposed Model for Brain Tumor MRI dataset

For the Glioma category, 324 out of 325 samples were correctly identified, with only 1 sample being misclassified as Meningioma. No misclassification was observed in the No Tumor or Pituitary categories. In the Meningioma category, the model correctly classifies 322 out of 330 samples. However, 3 samples were misclassified as Glioma,

4 samples as No Tumor, and 1 sample as Pituitary, indicating some overlap in feature representation between these categories.

The No Tumor class had 399 out of 400 samples correctly classified. Only 1 sample was misclassified as Glioma category, while no misclassification occurred in the Meningioma or Pituitary classes. For the Pituitary tumor category, 351 out of 353 samples were classified into their correct category. Only 2 samples were misclassified as Meningioma and No Tumor categories, with 1 sample in each.

The distribution of predictions suggests that while the model performs well across all categories, the Meningioma class exhibits more variation in misclassification. This could indicate similarities in features with Glioma and No Tumor, requiring further refinement in distinguishing these cases.

The confusion matrix for Brain Tumor MRI Images Dataset highlights the classification performance of the proposed model in distinguishing between Brain Glioma, Brain Meningioma, and Brain Tumor as depicted in Fig. 3. The distribution of predictions across these categories demonstrates the model's ability to correctly identify each class with minimal errors.

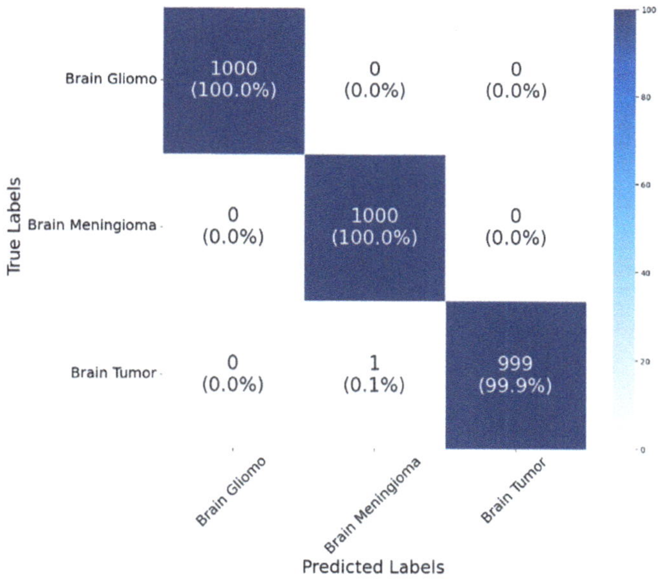

Fig. 3. Confusion Matrix of the Proposed Model for the Multi-Cancer Dataset.

For the Brain Glioma category, all 1000 samples were classified correctly, with no misclassification into other categories. In the Brain Meningioma category, the model also correctly identified 1000 out of 1000 samples, ensuring a perfect classification with no observed errors. For the Brain Tumor category, 999 out of 1000 samples were assigned correctly. However, 1 sample was misclassified as Brain Meningioma, indicating a rare case where the model confused a Brain Tumor with Meningioma.

Overall, the matrix suggests that the model maintains a high level of classification precision, with only a single misclassified sample in the Brain Tumor category. This minor confusion could be attributed to subtle similarities in imaging characteristics between Brain Tumors and Meningiomas, necessitating further refinement in distinguishing these classes.

4.5 Performance Comparative Analysis

The comparative analysis is presented in Tables 3 and 4 for brain tumor classification. The evaluation is based on key metrics such as Precision, Recall, F1-Score, and Accuracy for each architecture.

Table 3 shows the performance comparison on the Multi-Cancer dataset. AlexNet achieves a Precision of 88.40%, Recall of 87.92%, F1-Score of 87.45%, and Accuracy of 87.92%. ResNet-50 improves slightly with a Precision of 91.00%, Recall of 89.91%, F1-Score of 89.82%, and Accuracy of 89.91%. Inception V3 further enhances classification, reaching 92.70% Precision, 91.97% Recall, 92.02% F1-Score, and 91.97% Accuracy. VGG-16 shows significant improvement, achieving 94.52% Precision, 94.53% Recall, 94.51% F1-Score, and 94.53% Accuracy. DenseNet-201 performs even better, achieving 95.83% Precision, 95.66% Recall, 95.66% F1-Score, and 95.66% Accuracy. The HFF-Net model slightly surpasses DenseNet-201 with 96.88% across all metrices. However, the proposed model outperforms all architectures, achieving the highest Precision, Recall, F1-Score, and Accuracy of 99.15%.

Table 3. Proposed model comparison with state-of-the-art classifiers using Multi-Cancer Dataset

CNN Architecture	Precision %	Recall %	F1 Score%	Accuracy %
AlexNet	88.40	87.92	87.45	87.92
ResNet-50	91.00	89.91	89.82	89.91
InceptionV3	92.70	91.97	92.02	91.97
VGG-16	94.52	94.53	94.51	94.53
DenseNet-201	95.83	95.66	95.66	95.66
HFF-Net	95.88	95.88	95.88	95.88
Proposed	99.15	99.15	99.15	99.15

Table 4 presents the performance comparison on the Brain Tumor MRI Images dataset. InceptionV3 achieves a Precision of 87.18%, Recall of 79.96%, F1-Score of 79.75%, and Accuracy of 79.96%. AlexNet improves on this with 91.38% Precision, 89.99% Recall, 89.67% F1-Score, and 89.99% Accuracy. VGG-16 and ResNet 50 demonstrate further enhancements, achieving 93.53% and 95.54% Accuracy, respectively. DenseNet-201 achieves a Precision of 96.73%, Recall of 96.53%, F1-Score of 96.50%, and Accuracy of 96.53%. The HFF-Net model performs exceptionally well, reaching 98.65% Precision, 98.63% Recall, 98.63% F1-Score, and 98.63% Accuracy. Finally, the proposed model outperforms all state-of-the-art architectures, achieving an outstanding Precision, Recall, F1-Score, and Accuracy of 99.96%.

These results demonstrate the superior performance of the proposed model across both datasets. Its ability to significantly reduce misclassifications and achieve near-perfect classification highlights its effectiveness as a reliable and efficient tool for brain tumor detection and classification.

Table 4. Proposed model comparison with state-of-the-art classifiers using Brain Tumor MRI Dataset

CNN Architecture	Precision %	Recall %	F1 Score%	Accuracy%
InceptionV3	87.18	79.96	79.75	79.96
AlexNet	91.38	89.99	89.67	89.99
VGG-16	93.58	93.56	93.56	93.56
ResNet-50	95.54	95.26	95.18	95.26
DenseNet-201	96.73	96.53	96.50	96.53
HFF-Net	98.65	98.63	98.63	98.63
Proposed	99.96	99.96	99.96	99.96

5 Conclusions

In conclusion, this research introduces a novel deep ensemble transfer learning framework for brain tumor diagnosis using MRI scans, which combines the strengths of HFF-Net and DenseNet-201. The proposed framework addresses the challenges of tumor heterogeneity, class imbalance, and dataset limitations by leveraging advanced contrast adjustment techniques such as CLAHE. By incorporating both feature-level and decision-level fusion in the ensemble learning model, it achieves remarkable classification accuracy 99.15% on the Multi-Cancer dataset and 99.96% on the Brain Tumor MRI dataset. Compared to existing state-of-the-art methods, the proposed approach demonstrates superior performance, particularly in terms of precision, recall, F1-score, and overall accuracy. Additionally, the integration of transfer learning, through the use of pre-trained DenseNet-201 weights, accelerates model convergence and improves generalization, while HFF-Net's hierarchical feature fusion ensures robust tumor classification. Future work will focus on further refining the model's ability to handle diverse MRI modalities and enhancing its robustness under various clinical conditions. This work advances the field of automated brain tumor classification, offering a reliable tool for assisting medical professionals in early diagnosis and treatment planning.

Acknowledgments. No funding was received for this study.

Disclosure of Interests The authors declare no conflicts of interest relevant of this article.

References

1. Schwehr, Z., Achanta, S.: Brain tumor segmentation based on deep learning, attention mechanisms, and energy-based uncertainty predictions. Multimed. Tools Appl., 1–20 (2025). https://doi.org/10.1007/S11042-024-20443-0/FIGURES/8
2. Chauhan, P., et al.: Analyzing brain tumour classification techniques: a comprehensive survey. IEEE Access. (2024). https://doi.org/10.1109/ACCESS.2024.3460380
3. Das, S., Goswami, R.S.: Review, limitations, and future prospects of neural network approaches for brain tumor classification. Multimed. Tools Appl. **83**(15), 45799–45841 (Oct. 2023). https://doi.org/10.1007/S11042-023-17215-7
4. Ashraf, M.H., Alghamdi, H.: HFF-net: a hybrid convolutional neural network for diabetic retinopathy screening and grading. Biomed. Technol. **8**, 50–64 (2024). https://doi.org/10.1016/J.BMT.2024.09.004
5. Huang, G., Liu, Z., van der Maaten, L., Weinberger, K.Q.: Densely connected convolutional networks (2017). [Online]. Available: https://github.com/liuzhuang13/DenseNet
6. Ashraf, M.H., Jabeen, F., Alghamdi, H., Zia, M.S., Almutairi, M.S.: HVD-net: a hybrid vehicle detection network for vision-based vehicle tracking and speed estimation. J. King Saud Univ. Comput. Inf. Sci. **35**(8), 101657 (2023). https://doi.org/10.1016/J.JKSUCI.2023.101657
7. Guder, O., Cetin-Kaya, Y.: Optimized attention-based lightweight CNN using particle swarm optimization for brain tumor classification. Biomed. Signal Process. Control. **100** (2025). https://doi.org/10.1016/j.bspc.2024.107126
8. Wang, A., Mao, D., Li, X., Li, T., Li, L.: HQNet: a hybrid quantum network for multi-class MRI brain classification via quantum computing. Expert Syst. Appl. **261** (2025). https://doi.org/10.1016/j.eswa.2024.125537
9. Chandni, M.S., Kushwaha, A.K.S.: AI-based intelligent hybrid framework (BO-DenseXGB) for multi- classification of brain tumor using MRI. Image Vis. Comput. **154** (2025). https://doi.org/10.1016/j.imavis.2025.105417
10. Subba, A.B., Sunaniya, A.K.: Computationally optimized brain tumor classification using attention based GoogLeNet-style CNN. Expert Syst. Appl. **260** (2025). https://doi.org/10.1016/j.eswa.2024.125443
11. Karthik, A., et al.: Unified approach for accurate brain tumor multi-classification and segmentation through fusion of advanced methodologies. Biomed. Signal Process. Control. **100**, 106872 (2025). https://doi.org/10.1016/j.bspc.2024.106872
12. Abbas Malik, M.G., Saeed, A., Shehzad, K., Iqbal, M.: DEF-SwinE2NET: dual enhanced features guided with multi-model fusion for brain tumor classification using preprocessing optimization. Biomed. Signal Process. Control. **100** (2025). https://doi.org/10.1016/j.bspc.2024.107079
13. Remzan, N., Tahiry, K., Farchi, A.: Advancing brain tumor classification accuracy through deep learning: harnessing radimagenet pre-trained convolutional neural networks, ensemble learning, and machine learning classifiers on MRI brain images. Multimed. Tools Appl. (2024). https://doi.org/10.1007/s11042-024-18780-1
14. Sharma, A.K., et al.: Brain tumor classification using the modified ResNet50 model based on transfer learning. Biomed. Signal Process. Control. **86** (2023). https://doi.org/10.1016/j.bspc.2023.105299

Instance Segmentation of Brain Tumors in MRI Using Mask R-CNN with BrainSeg and PointRend Enhancements

Mourad Hikki[1,2](✉), Aziz Darouichi[1], and Abdessamad Elboushaki[1]

[1] L2IS, FST, Cadi Ayyad University, Marrakech, Morocco
hikki.mourad@gmail.com
[2] LAMIGEP, EMSI, Marrakech, Morocco

Abstract. Brain tumor segmentation is an essential step in the early diagnosis and treatment evaluation of brain tumor, but manual segmentation is often time-consuming, and the quality of the segmentation varies based on the operator's experience. This leads to the requirement of a fully automatic method for brain tumor segmentation. In this work, we explore the use of the Mask R-CNN framework to achieve this goal, with the idea of concentrating the segmentation on the detected region of interest, unlike conventional methods that perform the segmentation on the whole brain scan image, resulting in possibly reduced segmentation performance due to the disturbance of background regions. We tested different configurations to achieve optimal performance. Our method is trained and evaluated on the BraTS 2020 dataset containing 369 brain tumor cases, and the results show promising performance.

Keywords: MRI · Brain tumor · Segmentation · Mask R-CNN · CNN · BraTS 2020

1 Introduction

Over the past few years, the development of deep learning in medical imaging has led to significant advancements in brain tumor segmentation, a crucial step in brain function analysis and disease diagnosis [1]. Initially, brain tumor segmentation was approached as a patch classification problem using typical convolutional neural networks (CNNs) architectures in visual classification tasks. Sliding window and post-processing methods were then employed to achieve the final segmentation result. However, these methods are known to have drawbacks such as redundant calculations and loss of global information. To address these limitations, fully convolutional networks (FCNs) were introduced to provide a pixel-to-pixel solution for brain tumor segmentation, which effectively expanded the receptive fields, leading to superior segmentation accuracy and efficient calculation reduction [2, 3]. One of the most popular revolutionary versions of FCN is U-Net [4], which has integrated high-level and low-level features of medical images and has shown significant improvement in various medical segmentation tasks, making it the mainstream method in brain tumor segmentation. To enhance segmentation performance

B.-E. Boudriki Semlali et al. (Eds.): DATA 2025, LNNS 1908, pp. 155–164, 2026.
https://doi.org/10.1007/978-3-032-21200-9_13

further, several techniques, including attention mechanisms [5] and multi-scale fusion cascade ideology [6], have been integrated into the baseline model, leading to the development of advanced brain tumor segmentation methods. Despite these advancements, existing brain tumor segmentation networks directly operate on the whole image, which may result in reduced segmentation performance due to the interference of background regions.

Our goal is to explore state-of-the-art deep learning techniques for image segmentation and develop a system that achieves strong results for automatic brain tumor segmentation. We use the Mask R-CNN framework [7], which provides flexibility to modify or improve different parts of the model. To enhance feature extraction, we incorporate the BrainSeg model [8], and we further polish the generated segmentation masks using the PointRend model [9]. Our research leverages the Detectron2 library provided by the Facebook AI research team, which grants access to various state-of-the-art models and tools for development. Additionally, utilizing the library allows for easy integration with other projects and tools within it. Furthermore, we had to develop a method to convert the BraTS dataset from the original semantic segmentation task to the instance segmentation task required for our work.

In the rest of the paper, we will provide more details about our approach, including the implementation of our system, the experimental results we obtained, and a discussion of our findings.

2 Related Work

2.1 Mask R-CNN

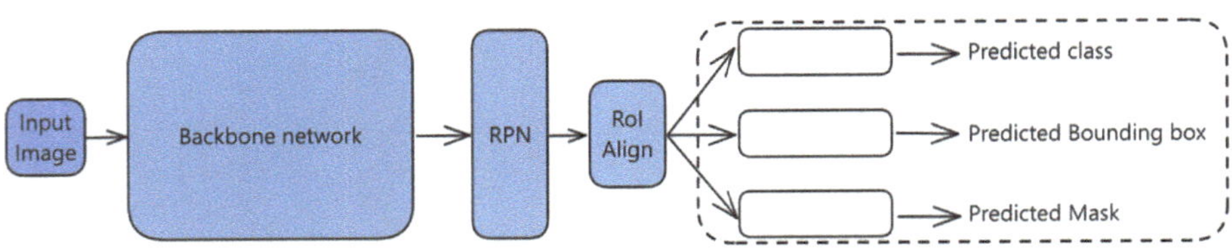

Fig. 1. Overview of the Mask R-CNN pipeline. It consists of a backbone CNN for feature extraction, an RPN for region proposal, the RoIAlign layer for accurately aligned feature pooling, and parallel heads for classification, bounding box regression, and pixel-wise mask prediction.

Mask R-CNN [7] is a popular state-of-the-art deep learning framework for instance segmentation, which extends Faster R-CNN [10], a widely used object detection framework, by adding a branch to predict segmentation masks for each detected object (see Fig. 1). Mask R-CNN has been widely used in various computer vision tasks, including object detection, instance segmentation, and human pose estimation. The main advantage of Mask R-CNN is its ability to accurately segment objects in an image, which is crucial for tasks such as medical image analysis.

The Mask R-CNN framework consists of two main components: a backbone network and a region proposal network (RPN). The backbone network is responsible for feature extraction from the input image, while the RPN proposes candidate objects bounding boxes for further processing. In Faster R-CNN, the RoIPool layer is used to extract features from each region of interest (RoI) by dividing the RoI into a fixed number of bins and then applying max pooling to each bin. However, this coarse spatial quantization can result in misalignments between the RoI and the feature map, which can lead to suboptimal performance. To address this issue, Mask R-CNN proposes the RoIAlign layer, which uses bilinear interpolation to allow for sub-pixel level alignment between the RoI and the feature map. By using RoIAlign instead of RoIPool, Mask R-CNN can more accurately extract features from the RoI, which leads to improved segmentation performance. After the RPN, the network's 3 heads generate the final predictions for the bounding box, class, and segmentation mask.

2.2 BrainSeg R-CNN

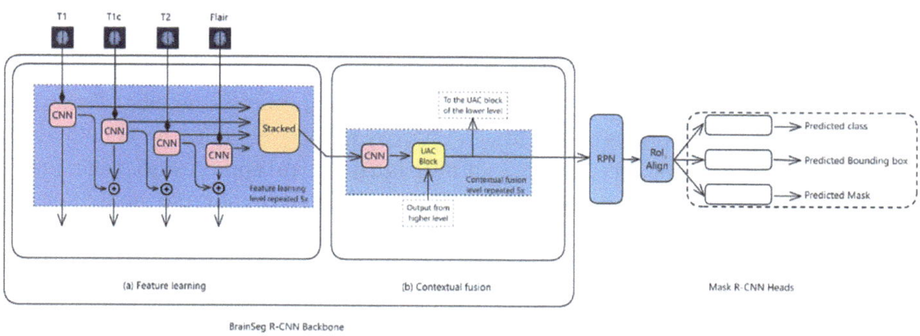

Fig. 2. Architecture BrainSeg R-CNN model. The backbone consists of two stages: (a) Multi-path feature learning, where four MRI modalities (T1, T1c, T2, Flair) are processed by separate CNN streams with cross-modality additive connections, repeated across multiple levels. (b) Contextual fusion, utilizing an FPN-like structure where features from different levels are combined using UAC (Upsample, Add, Concatenate) blocks. The resulting feature map feeds into a Region Proposal Network (RPN). Features for proposed regions are then extracted using RoIAlign and passed to the Mask R-CNN heads for parallel prediction of class, bounding box, and segmentation mask.

Inspired by Mask R-CNN, BrainSeg R-CNN [8] was developed for brain tumor segmentation, achieving results comparable to state-of-the-art methods. A key contribution is its custom-built backbone, specifically designed to utilize the different modalities available in MRI scans.

This backbone accepts four MRI modalities as input and is structured into two parts: one for feature learning and another for contextual fusion (see Fig. 2).

The feature learning part employs a Multi-path and Cross-Modality Feature Learning method specifically designed to exploit the information from the multiple imaging modalities available in the BraTS dataset. This initial stage of the backbone utilizes four separate CNN blocks to learn features from each modality (T1, T1c, T2, and Flair) individually. The features extracted from these different modalities are then combined at various levels through element-wise addition. This approach enables the model to learn not only modality-specific features but also multi-scale and cross-modality features, effectively considering the interactions among modalities to obtain discriminative representations of brain tumors.

The second part of the method, inspired by the Feature Pyramid Network [11], merges the semantic features from higher levels with the spatial features from lower levels to get better global contextual information. This is achieved through an improved feature pyramid structure that fuses features gained from feature learning under different pyramid resolutions. The UAC (Upsampling, Add, and Concatenation) block is used to fuse the features, which involve element-wise addition and concatenation operations. The resulting features contain global and local information with stronger semantic and finer spatial resolution, which are particularly helpful for segmentation. The final fused features go through vanilla RPN to generate Regions of Interest (RoIs) of the brain tumor, which are fed into the network heads for bounding-box recognition and mask prediction.

The network heads are structured similarly to Mask R-CNN. However, due to the considerable similarity between tumors and healthy tissues, their different shapes, and tiny size, the loss function used in Mask R-CNN may not pay enough attention to desirable tumor regions, potentially resulting in poor segmentation performance. Therefore, BrainSeg R-CNN incorporates a Dice loss to improve the segmentation branch. Overall, Multi-path and Cross-Modality Feature Learning is a powerful approach for brain tumor segmentation that exploits the complementary information from multiple modalities and combines it in a sophisticated way to obtain accurate results.

2.3 PointRend: Image Segmentation as Rendering

Reference [9] presents a new method for efficient high-quality image segmentation of objects and scenes. It gives a fresh viewpoint on image segmentation as a rendering problem by drawing an analogy to traditional computer graphics approaches.

For effective rendering with the over- and under-sampling challenges experienced in pixel labeling tasks. PointRend (Point-based Rendering) performs point-based segmentation predictions at adaptively selected sites using an iterative subdivision technique. By building on top of current state-of-the-art models, PointRend can be applied flexibly to both instance and semantic segmentation problems. PointRend produces crisp object borders in locations where prior approaches have over-smoothed them.

This proposed model was introduced to be integrated as a segmentation head for any model, and it was used as the mask head in Mask R-CNN.

PoinRend involves the training of a multi-layer perceptron (MLP) model to predict the class of a singular point. The model utilizes features extracted from the feature maps generated by ROIAlign and a coarse mask predicted by a compact convolutional network. Specifically, the coarse mask is predicted atop the feature maps that have been extracted by ROIAlign.

During the inference stage, the predicted coarse mask is subjected to an up-sampling procedure using bilinear interpolation. Following this, a selected subset of points exhibiting ambiguity in their class prediction are identified, and the MLP model, which was previously trained, is employed to predict their class. This process is repeated 5 times as recommended by the authors or until the desired mask resolution is achieved.

3 Proposed Method

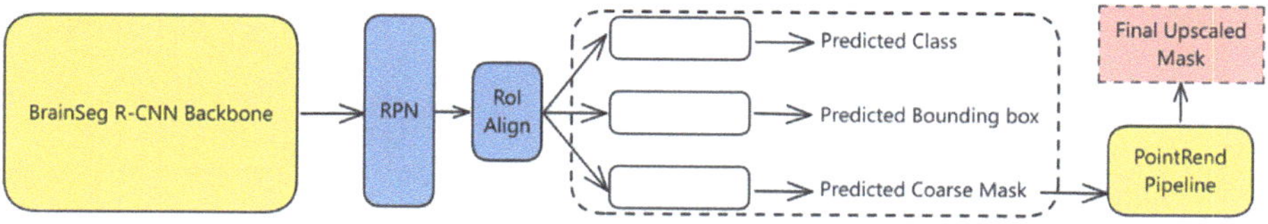

Fig. 3. Overview of the proposed model architecture. It integrates the BrainSeg R-CNN backbone for feature extraction and the PointRend pipeline for segmentation refinement within the Mask R-CNN framework. Features from the backbone feed into an RPN and RoIAlign layer. Parallel heads then predict class, bounding box, and a coarse mask, which is subsequently refined by PointRend.

After presenting the different components, we can now discuss the idea of our approach (see Fig. 3). As we presented in the Mask R-CNN section, this framework offers flexibility in modifying its various components, ranging from its backbone to the multi-head utilized by the model, so our idea is to improve some of these parts. Starting with the backbone, BrainSeg is well suited for our case, as it was designed for brain tumor segmentation, it takes advantage of the different MRI modalities provided by the BraTS dataset, by learning from each modality on its own, before mixing the learned features. Then we have PointRend, which gives a detailed segmentation mask at no additional computational cost, by adopting an efficient method to upscale the segmentation masks and at the same time refine the borders of the detected object. Therefore, we integrated these approaches into the Mask R-CNN framework, utilizing BrainSeg as the backbone and PointRend as the segmentation head.

3.1 BrainSeg R-CNN Implementation

Since the original authors of BrainSeg R-CNN did not provide their implementation code, we had to implement it ourselves, but we ran into some issues, that is because the authors did not mention many of the necessary implementation details such as the kernel size of the convolution filters, the number of channels for the convolution layers, the number of training epochs, learning rate and many other essential details and hyperparameters for a successful reproducibility.

In the feature learning part of the backbone, the input images are inserted into a residual block taken from ResNet [12], it was used because the authors used it in other parts of the model, then we proceeded to sum the outputs element-wise, repeated throughout the 5 levels as described by the authors. The number of filters for the residual blocks increases with each level, we start with 64 filters for the 4 residual blocks of the first level, then 128 filters in the second, and so on. Finally, in the fifth level the residual blocks have 1024 filters.

For each level we concatenate its outputs into one single tensor, which is then fed into another residual block with 128 filters due to the graphics card video memory limitations, we end up with 5 tensors with the same channel depth each from a different level.

For the context fusion stage, we adhered to the described architecture, despite the ambiguity regarding the number of channels. Following the paper's methodology, the feature tensors from the different levels are ultimately concatenated and fed into the region proposal network of the Mask R-CNN framework.

3.2 Preprocessing of BraTS Dataset

Our work utilized the BraTS 2020 dataset consisting of 369 patient scans, each comprising native (T1), post-contrast T1-weighted (T1Gd), T2-weighted (T2), and T2 Fluid Attenuated Inversion Recovery (T2-FLAIR) volumes, acquired from different clinical protocols and various scanners from multiple (n = 19) institutions. Each patient is also provided with a segmentation volume describing the segmentation of each slice of the brain, which includes the GD-enhancing tumor (ET — label 4), the peritumoral edema (ED — label 2), and the necrotic and non-enhancing tumor core (NCR/NET — label 1). All the images in the dataset were manually segmented by one to four raters using the same annotation technique and were reviewed and approved by professional neuroradiologists. The data was pre-processed, including co-registration to the same anatomical template, interpolation to the same resolution (1 mm3), and skull-stripping. The images have a resolution of 240x240 pixels, and for each modality, we are provided with 155 slices.

To make the BraTS dataset compatible with the Mask R-CNN framework, we converted it into the COCO format. The COCO format comprises a directory containing the images and a corresponding JSON file that provides annotations for each image, including object classes, bounding boxes, and segmentation bitmasks. This conversion was achieved by iterating through each image, processing its segmentation mask, and extracting the required information to generate annotations for every identified object using the following steps:

- Extracting class submask: Since the provided mask in the dataset includes masks for all classes, we needed to select only the corresponding mask for the desired class. This can be achieved by simply choosing the pixels of that class.

  Extracting polygons: After obtaining a class submask (which is a Boolean type), we extracted all the contours present in the mask. These contours were then transformed into polygons for easier processing and manipulation.
- Merge overlapping polygons: Objects with inner rings result in two polygons from the previous step. To address this, we identify the inner rings of a polygon and merge them into a single polygon. If one polygon is inside another, they are merged. Additionally, we merge polygons with overlapping bounding boxes if their Intersection over Union (IoU) score greater than 0.5.
- Generate COCO annotations: With our polygons in hand, we treat each polygon as a single instance and easily generate its bounding box, class label, and segmentation mask.

4 Results and Discussion

4.1 Experimental Setup

As we have 369 patients in our dataset we divided the dataset into two sets, a training set with 80% of the patients and a testing set with 20%.

All the tested models were trained for 8 epochs, mainly due to training hardware limitations on the Kaggle platform, which restricted the number of experiments and the training duration. As recommended in the Mask R-CNN paper [7], the learning rate was initially set to 0.02. However, after two epochs, we decreased it to 0.002 based on several experiments that showed it to be the optimal time to reduce the learning rate.

Models with a ResNet [12] backbone were trained with a batch size of 32, and models with BrainSeg backbone were trained with 8 images per batch because the models take up a large amount of memory, and since the original authors did not specify the video RAM that the model takes up, we cannot determine if our implementation suffers from memory leakage, requires further optimization, or if this memory usage is typical for the model

To facilitate the reuse of our code, we have made it available at our GitHub repository. Our code consists of two Jupyter notebooks, one for preprocessing the BraTs dataset and the other for training and evaluating the models using the Detectron2 library. We have also developed functions for the Detectron2 library that take our generated dataset and plug it into the library to be compatible with the different detection models offered by the library. This means that our dataset can be easily used to train a variety of models such as Cascade R-CNN, rotated bounding boxes, PointRend, DeepLab, and ViTDet, by simply modifying the name of the model to be used. In addition, the preprocessing code converts the BraTs dataset from a semantic segmentation task to an instance segmentation task and generates an annotation file under the COCO format. Overall, our code is designed to be easily reusable, with a variety of models provided for training and evaluation.

4.2 Results

As shown in Table 1, the initial model tested was a Mask R-CNN using a ResNet-50 backbone and a Feature Pyramid Network, maintaining default settings for the remaining components as per the original paper [7]. This model yielded good results with a Dice score of 0.86.

Then it was time to introduce the enhancements to the backbone and the segmentation head, starting with the BrainSeg backbone only, we observed an increase of 1 percentage point over the standard ResNet with 50 layers, but only 0.03 percentage points compared to ResNet-101, but keep in mind that we are using BrainSeg with only 4 layers out of the recommended 5 layers.

Next, we introduced the PointRend head to the network,with a ResNet-50 backbone, it achieved results comparable to the BrainSeg (4 layers) model. Finally, we combined the BrainSeg backbone this time with the full 5 layers, and the PointRend segmentation head, this yielded the best score in terms of Dice score.

Table 1. Results for whole tumor class

		COCO metrics									
		Object detection metrics					Segmentation metrics				
Models	Dice score	AP	AP50	AP75	APS	APM	AP	AP50	AP75	APS	APM
ResNet50 +FPN	0.867	36.15	59.46	39.3	16.44	68.94	36.72	59.74	40.06	15.26	70.1
ResNet101 +FPN	0.872	36.74	60	40.4	17.42	68.96	37.07	59.94	39.58	15.87	70.16
BrainSeg 4 levels	0.876	34.96	56.69	39.28	15.9	66.53	37.04	57.38	41.15	15.64	71.18
ResNet50 + PointRend	0.877	36.43	59.27	39.67	16.84	68.89	37.63	59.6	41.34	15.71	71.82
BrainSeg 5 levels + PointRend	0.88	35.89	59.45	38.94	18.39	65.02	38.6	60.4	42.62	17.46	71.49

4.3 Discussion

While our results are promising, they fall short of the 91.54% Dice score reported in the original BrainSeg paper [8] for whole tumor segmentation. This discrepancy may stem from several factors, including the use of a different dataset version (BraTS 2020 in our work versus BraTS 2017), the lack of a shared dataset conversion methodology from the original authors for Mask R-CNN compatibility (unlike our approach detailed in Sect. 3.2), and the lack of sufficient implementation details or code provided by the original authors, as discussed in Sect. 3.1, hindering exact replication.

An analysis of the COCO metrics reveals a consistent pattern across all models: lower average precision for small objects (APS metric, area < 32x32 pixels) in both detection and segmentation tasks. This indicates a general difficulty in accurately processing smaller tumor regions, suggesting a potential focus for future improvements. Notably, these APS values are comparable to those reported in the original Mask R-CNN paper on the COCO test-dev dataset (APS of 16.9), suggesting this is an inherent challenge rather than an anomaly specific to our setup.

4.4 Comparison with Other Approaches

For comparison, we trained and tested a U-Net model [4] (our implementation) on the same dataset. As shown in Table 2, our best-performing model (BrainSeg 5 levels + PointRend) demonstrates superiority over the standard U-Net model for this task, achieving a 10% higher Dice score.

We hope to submit our model to the official BraTS challenge to be tested on their private validation dataset, then our model would be tested against all the other submitted models, and we could get a clearer view of where our method stands.

Table 2. Comparison with other approaches.

Model	Dice score (Whole tumor)
U-Net	0.8
BrainSeg 5 levels + PointRend	0.88

5 Conclusion and Future Work

Brain tumor segmentation remains one of many challenging tasks in the medical imaging field, due to the various shapes and sizes tumors exhibit. Even though constrained to the brain, distinguishing tumor tissues from healthy tissues is difficult, especially in the borders between the two, such a task requires a specialist with a trained eye to make the best segmentation, consuming significant time and effort.

In this work we explored the use of Mask R-CNN in this task, as the use of object detection models limits the segmentation prediction only to the region of the tumor, avoiding invalid segmentation calculation in the background area. Our main contribution is the combination of a tailored backbone for MRI scans of the brain, called BrainSeg, and an enhanced segmentation head called PointRend. Furthermore, we worked on a method to convert the BraTS dataset from its original form of semantic segmentation to instance segmentation, making it compatible with Mask R-CNN and all other instance segmentation models. The results we got were very promising, encouraging further work in the direction of using instance segmentation models in the brain tumor segmentation task. Also, the use of the Detectron2 library in this work makes our code reusable, and with the variety of models provided by the library, future work should be much easier.

In future work, we would work on solving the problem of the detection of small regions of the tumors, mostly it would be through the use of other object detection models, as in this work we laid the groundwork for such experiments. During this work we were limited by time and hardware provided by the Kaggle platform, access to greater computational resources would allow for longer training times, and the application of sophisticated hyperparameter optimization techniques.

Disclosure of Interests. The authors have no competing interests to declare that are relevant to the content of this article.

References

1. Biratu, E.S., Schwenker, F., Ayano, Y.M., Debelee, T.G.: A Survey of brain tumor segmentation and classification algorithms. J. Imaging. **7**, 179 (2021)
2. Long, J., Shelhamer, E., Darrell, T.: Fully Convolutional Networks for Semantic Segmentation. arXiv. (2014)
3. Shen, H., Wang, R., Zhang, J., et al.: Multi-task fully convolutional network for brain tumour segmentation. In: Annual Conference on Medical Image Understanding and Analysis (MIUA), pp. 239–248 (2017)

4. Ronneberger, O., Fischer, P., Brox, T.: U-Net: convolutional networks for biomedical image segmentation. arXiv. (2015)
5. Zhang, J., Jiang, Z., Dong, J., Hou, Y., Liu, B.: Attention gate ResU-net for automatic MRI brain tumor segmentation. IEEE Access. **8**, 58533–58545 (2020)
6. Zhou, C., Ding, C., Wang, X., Lu, Z., Tao, D.: One-pass Multi-task Networks with Cross-task Guided Attention for Brain Tumor Segmentation. arXiv. (2019)
7. He, K., Gkioxari, G., Doll'ar, P., Girshick, R.: Mask R-CNN. arXiv. (2017)
8. Zhang, J., Cheng, X., He, T., Liu, D.: BrainSeg R-CNN for brain tumor segmentation. In: Proceedings of Image and Graphics Technologies and Applications, pp. 217–226 (2021)
9. Kirillov, A., Wu, Y., He, K., Girshick, R.: PointRend: image segmentation as rendering. arXiv. (2019)
10. Ren, S., He, K., Girshick, R., Sun, J.: Faster R-CNN: towards real-time object detection with region proposal networks. arXiv. (2015)
11. Lin, T., Doll'ar, P., Girshick, R., He, K., Hariharan, B., Belongie, S.: Feature pyramid networks for object detection. arXiv. (2016)
12. He, K., Zhang, X., Ren, S., Sun, J.: Deep residual learning for image recognition. arXiv. (2015)

Acute Lymphoblastic Leukemia (ALL) Detection Using Gene Expression Data (GEO Dataset) with Machine Learning Algorithms

Nusrat Jahan Ashika[1], Jobair Ahammed Tusher[1], Mamunur Rashid Alex[2], and Al-Imtiaz[1](✉)

[1] University of Information Technology and Sciences (UITS), Dhaka, Bangladesh
{2114951010,jobair.ahammed,al.imtiaz}@uits.edu.bd
[2] North South University, Dhaka, Bangladesh
mamunur.alex@northsouth.edu

Abstract. The heterogeneous blood cancer Acute Lymphoblastic Leukemia (ALL) needs immediate classification through specialized methods to create successful treatment strategies. This research employs current machine learning solutions to develop an adequate system which analyzes gene expression information for precise ALL subtype identification. The analysis process involved proper preprocessing strategies which included class balancing with SMOTE paired with dimensionality reduction using PCA. In this work, K-Nearest Neighbors (KNN), Random Forest, Logistic Regression and Decision Tree were used for an evaluation of subtype classification algorithms to determine the most effective solution. Logistic Regression succeeds over other examined models with its 97.20% accuracy. Random Forest maintained strong performance standards while KNN and Decision Tree struggled to properly classify subtypes that were similar to each other.

Strong preprocessing together with thorough model evaluation stands as a necessary requirement for dealing with biological data of high dimensions. The developed approach enables both better subtype discrimination accuracy alongside capability to bring machine learning methods into precision medical practice. The study advances individualized ALL patient care through its capability to help clinicians refine therapeutic decisions.

Keywords: Acute Lymphoblastic Leukemia (ALL) · GEO Dataset · Gene Expression Data · Machine Learning · Principal Component Analysis · Logistic Regression · Random Forest · KNN · Decision Tree

1 Introduction

Fast-growing immature lymphoid cells predominate in bone marrow and peripheral blood and other organs to cause the severe blood cancer Acute Lymphoblastic Leukemia (ALL) [8]. Public health faces substantial challenges from this sickness which mostly manifests in young patients causing extensive medical treatment needs. The classification

B.-E. Boudriki Semlali et al. (Eds.): DATA 2025, LNNS 1908, pp. 165–176, 2026.
https://doi.org/10.1007/978-3-032-21200-9_14

of ALL subtypes needs to happen early along with high accuracy to choose optimum treatments that lead to enhanced patient outcomes because various ALL forms show unique treatment reactions and disease behaviors [9].

Leukemia diagnostic methods of the past have primarily depended on morphological examination together with immunophenotyping and cytogenetic analyses. Traditional diagnostic methods functioning from clinical settings have delivered important disease information throughout various clinical applications [10]. Traditional diagnostic methods require extended processing times as well as limited sensitivity for discriminating between different forms of leukemia. Hand interpretation of data causes variations in results along with restrictions in spotting new and unrecognizable leukemia subtypes [11].

Biomedical research has experienced a major revolution because of high-throughput technologies that emerged specifically from transcriptomics and genomic research despite specific advancements. These technologies achieve deep insights into leukemia molecular foundations through their advanced method of gene expression pattern analysis which produced substantial data sets. Connectivity analysis has proven to become a promising tool when machine learning processes gene expression patterns for leukemia classification [12, 13].

Based on the gene expression data it developed a machine learning system that can accurately classify subtypes of acute lymphoblastic leukemia (ALL). Leveraging different models in the context of noise reduction and class balancing after data processing, results were obtained. Random Forest, KNN, Decision Tree each achieved a slightly lower accuracy of 97.2%, except Logistic Regression that achieved the highest one. Our results demonstrate that with ML oriented identification of leukemia can be done faster and more accurately, so can aid in personalized treatment.

Key contributions of this study are as follow:

1. We preprocess and convert the GEO dataset into a suitable format for efficient utilization in machine learning-based classification models for Acute Lymphoblastic Leukemia (ALL) subtypes.
2. As far as we know, no previous study has been conducted using the GSE184692 dataset for ALL subtype classification, making this the first work to explore its predictive potential [14].
3. We evaluate and compare systematically a variety of machine learning techniques, show that our proposed approach reaches the best, comprising accuracy, precision, recall, F1 score in comparison to the methods currently available in ALL detection.
4. The results of our study show the scope of ML models in the improvement of the diagnosis of leukemia and is structured to assist in futures consolidation in automatic clinical decision support systems.

For ease of reference, we hereby present the outline of the work:

- Abstract: Provides a summary of the study, including key findings and contributions.
- Introduction: Discusses the background of Acute Lymphoblastic Leukemia (ALL), the importance of precise classification, and the role of machine learning in leukemia detection.

- Related Work: Reviews existing literature on leukemia classification using machine learning and highlights the research gap addressed by this study.
- Methodology: Describes the data pre-processing methods (SMOTE-based balancing, PCA for dimensionality reduction), machine learning models and assessment framework.
- Results Analysis: Demonstrates the efficacy of multiple machine learning models, including Logistic Regression, Random Forest, K-Nearest Neighbors, and Decision Tree, alongside confusion matrices and a comparison analysis.
- Conclusion: Summarizes key findings and suggests potential improvements, such as integrating feature selection techniques and deep learning models.

2 Related Works

Researchers have conducted multiple investigations into the combination of machine learning with genomic data for detecting leukemia as well as classifying its variations.

The authors in Kashef et al. [1] applied both clinical and medical data to forecast outcomes of pediatric ALL treatment. The research utilized Support Vector Machines (SVM) and reached a 94.9% correct rate while determining fever frequency as the essential factor.

MD-ALL represents an integrative platform which combines machine learning algorithms with RNA-Seq data analysis for B-ALL subtype classification according to Hu et al. [2]. The system presented exceptional accuracy above current tools while verifying RNA-Seq could be executed as a routine diagnostic method. The research group led by Schmidt et al. [3] presented ALLSorts which utilizes hierarchical RNA-Seq analysis to classify B-ALL subtypes in their paper of 2022. The platform achieved 93% accuracy while offering its programming to the opensource community.

According to Walter et al. (2021) [4], whole transcriptome sequencing (WTS) served as an effective method for classifying ALL patients. The combination of WTS with WGS led the study to find novel fusion events while maintaining high diagnostic accuracy.

Pölönen et al. [5] studied both genomic and epigenomic data of T-lineage ALL. Ten investigators examined different genomic subtypes and established significant genomic risk markers for subtype-specific risk assessments. Beder and co-authors developed ALLCatchR which functions as an RNA-Seq classifier for B-ALL subtypes with a 95.7% accuracy level in their publication [6]. Developments stages of cells became linked to expression profiles through this research which generated understanding about how classification works between subtypes.

Ilyas et al. [7] developed a combination technique for microarray data-based leukemia type prediction. The model demonstrated 96.92% prediction accuracy after reducing the genes from 22,000 to five key determinants which exceeded current method achievements.

3 Methodology

Our methodology consists of two main parts: dataset preprocessing and model training and validation. Since we selected gene expression data extracted from RNA sequencing [14], it cannot be used directly for training without proper preprocessing. This involves transforming the gene expression dataset into a format that includes labeled data. Once the dataset is properly processed, it can then be used for prediction using machine learning models.

3.1 Dataset Preprocessing

The dataset requires thorough preprocessing before analysis, as it initially does not contain any labels. We needed to map the labels from the metadata corresponding to each gene expression sample, as shown in Fig. 1 and described below:

- Raw Data Collection: We chose GSE184692 as our dataset from the NCBI GEO database, comprising gene expression values derived from RNA sequencing of 372 samples [14].
- Transpose of Raw Gene Expression Data: In the collected data from the GEO dataset, there was a column named ID_REF containing probe IDs, and the sample identifiers along with their expression values were in columns that were not suitable for prediction in machine learning models. Therefore, we transposed the dataset, which converted the probe IDs into features of the dataset, and the sample identifiers became entries in a column where each row represented the expression values of all probes for a specific sample.

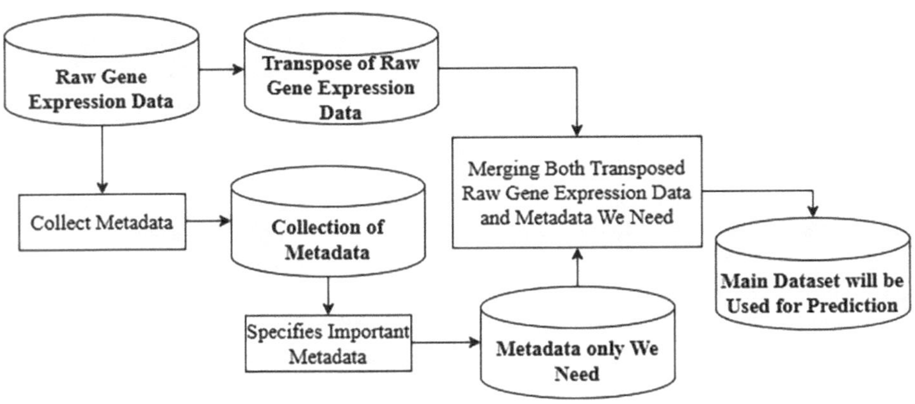

Fig. 1. Dataset Preprocessing.

- Meta Data Collection: After transposing, we had our dataset without metadata. So, using R Studio, we collected the metadata from the NCBI database and mapped it to the sample identifiers.
- Specification of Important Metadata: The metadata contains various information, such as when the sample was collected, who collected it, when the data was uploaded to NCBI, and so on—most of which we didn't need. Therefore, we filtered the metadata to retain only the relevant information, specifically the ALL subtype.
- Merging the Metadata and Transposed Data: To obtain our final dataset for prediction, we merged the transposed data and the metadata based on the sample identifiers.

After completing these processes, we obtained a dataset suitable for building predictive models using machine learning algorithms.

3.2 Model Training and Validation

Before training our models, we identified several remaining issues within the dataset that needed to be addressed. These issues, along with their solutions and the subsequent processes, are described in the following sections and illustrated in Fig. 5.

Label Encoding: Some columns in the dataset contained non-numeric values, particularly the label column, which contained string values. To simplify model training, we used a Label Encoder to convert the string labels into numeric values.
SMOTE: One major concern was the significant class imbalance in the dataset, as shown in Fig. 2, which could lead to biased or inaccurate predictions by the machine learning models.

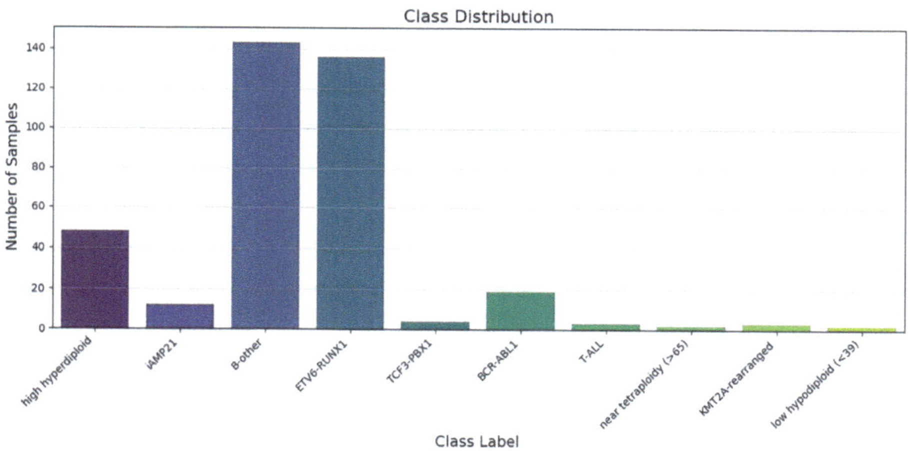

Fig. 2. Initial Class Distribution of the Dataset.

Therefore, we used SMOTE to create a balanced class distribution, as illustrated in Fig. 3. As a result, the issue of faulty predictions due to class imbalance will be mitigated.

After obtaining our balanced dataset, we proceeded to train our model using machine learning algorithms.

PCA for Dimensionality Reduction: PCA was employed to reduce the dataset to 100 principal components, capturing 66% of the variance. By retaining only the most important components, it removed less informative noise. Figure 4(a) illustrates that the original variables (e.g., gene expression values) are combined linearly to form the main components, with each dot representing a biological sample. The samples are color-coded according to their class labels, which correspond to different subtypes of acute lymphoblastic leukemia (ALL). Dots that are positioned far from the main cluster represent samples from distinctly different conditions or subtypes.

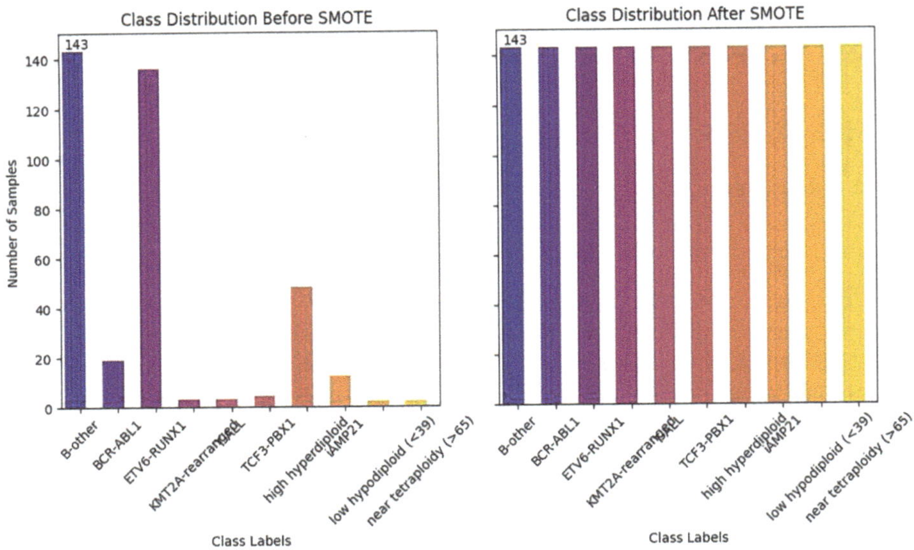

Fig. 3. Distribution of Classes Before and After SMOTE.

The bright yellow diagonal in Fig. 4(b)'s correlation heatmap of principle components, on the other hand, indicates that every principal component has a perfect connection with itself. Since the off-diagonal elements are close to zero, there is no correlation between the main components. This confirms that PCA successfully creates orthogonal (uncorrelated) components by design. Therefore, we can conclude that our PCA effectively reduced noise and multicollinearity, while compressing the data without losing its essential structure.

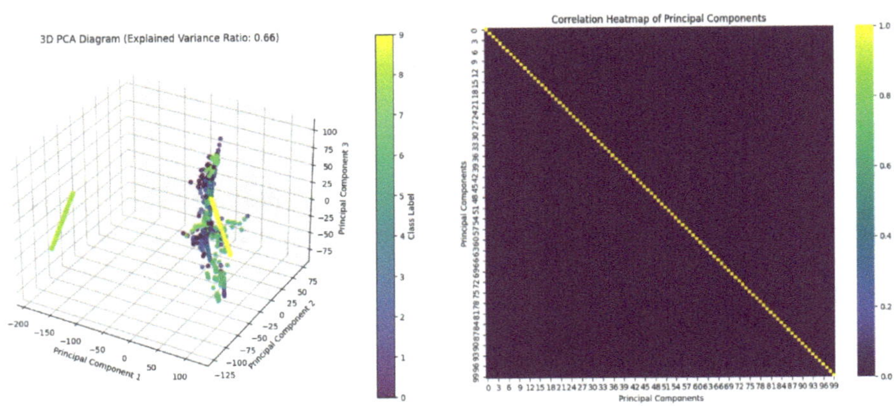

(a) 3D PCA Representation (b) Correlation Heatmap

Fig. 4. (a) 3D PCA representation of the dataset, with each dot representing a sample color-coded by leukemia subtype. (b) Correlation heatmap of principal components.

Stratified Train Test Split: The resulting dataset, after applying SMOTE and PCA, was divided via stratified sampling into a training set (80%) and a testing set (20%). Stratified sampling ensured that the class distribution ratio remained consistent in both the training and testing sets. As a result, during validation, all classes are predicted, ensuring the efficacy of the model.

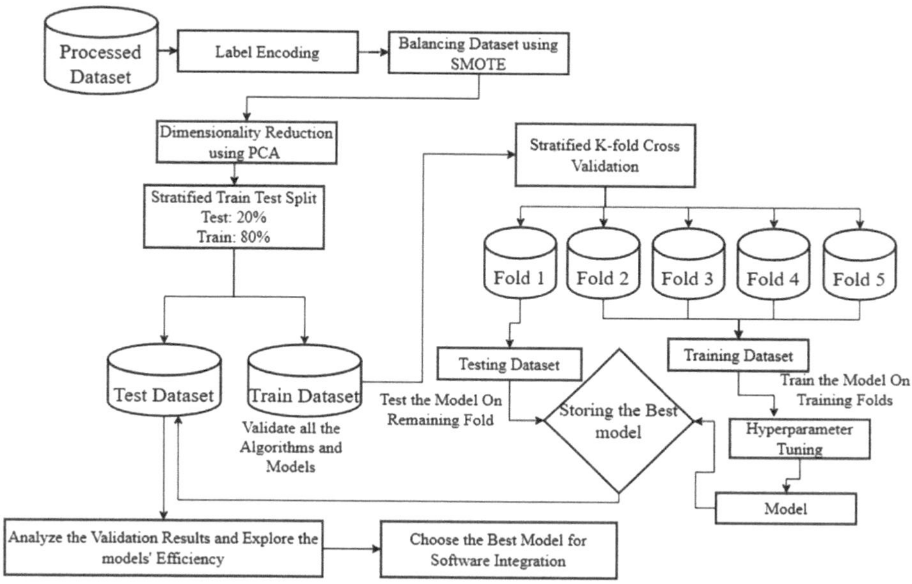

Fig. 5. Model Training and Validation Workflow.

Selection of the Best Model: The dataset was divided into training and testing sets, and the training data was then subjected to stratified K-Fold crossvalidation.

The following of ML Algorithms were considered:

- Logistic Regression
- Random Forest
- K-Nearest Neighbors (KNN)
- Decision Tree

We used five folds for each model, where in every iteration, one fold served as the validation set while the remaining four were used for training. In each fold, to get the ideal set of configuration parameters, we used Grid Search to do hyperparameter tuning, as shown in Table 1:

Table 1. Hyperparameter Grid for Model Tuning

Model	Hyperparameter	Values
Logistic Regression	C	{0.01, 0.1, 1}
	solver	{'lbfgs', 'liblinear'}
Random Forest	n_estimators	{50, 100, 200}
	max_depth	{None, 10, 20}
	min_samples_split	{2, 5}
K-Nearest Neighbors (KNN)	n_neighbors	{3, 5, 7}
	p (distance metric)	{1 (Manhattan), 2 (Euclidean)}
Decision Tree	max_depth	{None, 10, 20}
	min_samples_split	{2, 5}

We then validated our models using the test dataset, where Logistic Regression achieved the best performance. A detailed discussion of the results is provided in the Result Analysis section.

4 Result Analysis

The performances of each machine learning algorithm was evaluated using precision, recall and F1 score measures. That these metrics give as a quantitative measure how well these classifying leukemia subtypes classifying model is.

4.1 Performance Metrics

Table 2 presents a comparison of the performance metrics for LR, RF, KNN, and Decision Tree.

Table 2. Performance Metrics Comparison

Algorithm	Accuracy	Precision	Recall	F1-Score
Logistic Regression	97.2%	96.9%	97.4%	97.1%
Random Forest	94.5%	94.3%	94.7%	94.5%
KNN	91.3%	91.7%	90.8%	91.2%
Decision Tree	88.7%	87.9%	88.5%	88.2%

Figure 6 visualizes the comparative performance of these algorithms, highlighting Logistic Regression as the top performer across all metrics.

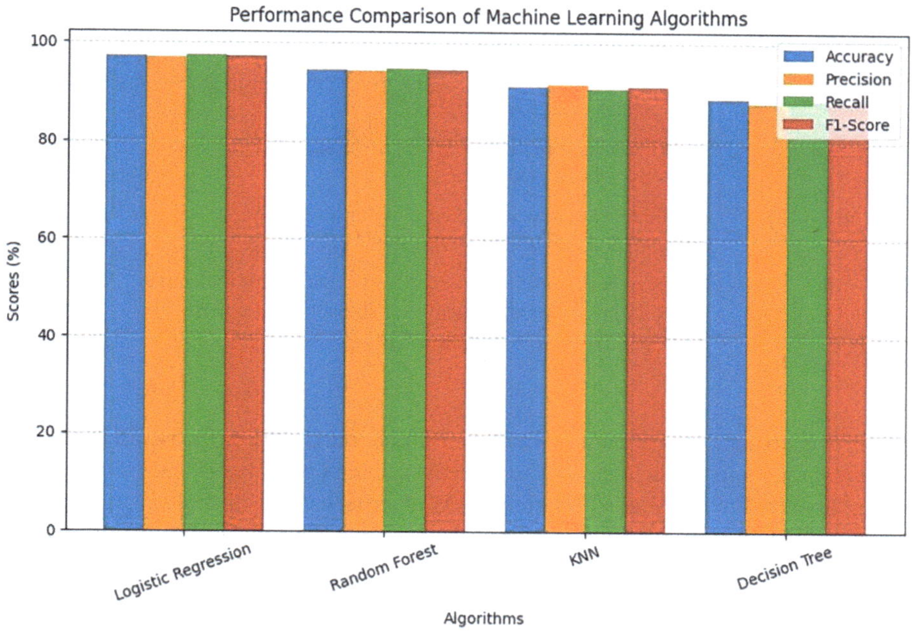

Fig. 6. Result Analysis: Comparison of Accuracy, Precision, Recall, and F1-Score for Each Algorithm.

Based on the results presented in Table 2 and Fig. 6, Logistic Regression achieved the best result at 97.2%, compared to Random Forest (94.5%), KNearest Neighbors (91.3%), and Decision Tree (88.7%). Notably, Logistic Regression also outperformed the other models in terms of precision (96.9%), recall (97.4%), and F1-score (97.1%). Random Forest secured the second-highest scores with a precision of 94.3%, recall of 94.7%, and an F1-score of 94.5%. The remaining models, KNN and Decision Tree, ranked third and fourth, respectively, across all evaluated performance metrics.

Table 3. Performance Comparison of Machine Learning Models in Leukemia Classification

Study	Methodology	Accuracy
Kashef et al. [1]	SVM-based classification of pediatric ALL using clinical and medical data	94.9%
Schmidt et al. [3]	ALLSorts: Hierarchical RNA-Seq-based model for B-ALL subtype classification	93.0%
Beder et al. [6]	ALLCatchR: RNA-Seq classifier mapping gene expression to developmental trajectories	95.7%
Ilyas et al. [13]	Hybrid feature selection technique for leukemia classification using microarray data	96.92%
Our Work	Dimensionality reduction and class balancing for leukemia prediction using machine learning algorithms (Logistic Regression) on a preprocessed GEO dataset	97.2%

Table 3 presents a comparative analysis of mentioned studies focused on leukemia classification, focusing their approaches. and the accuracy achieved by each. Kashef et al. [8] employed a Support Vector Machine (SVM) approach using clinical and medical data to classify pediatric Acute Lymphoblastic Leukemia (ALL), achieving an accuracy of 94.9%. Schmidt et al. [14] developed ALLSorts, a hierarchical RNA-Seq-based model tailored for B-ALL subtype classification, with an accuracy of 93.0%. Beder et al. [1] introduced ALLCatchR, an RNA-Seq classifier that maps gene expression data to developmental trajectories, achieving 95.7% accuracy. Ilyas et al. [6] proposed a hybrid feature selection technique using microarray data, which yielded an accuracy of 96.92%. In comparison, our work focused on dimensionality reduction using PCA and class balancing using SMOTE, followed by classification with Logistic Regression on a preprocessed GEO dataset. This approach achieved an accuracy of 97.2%, the highest among all the compared studies.

4.2 Confusion Matrices

An analysis of confusion matrices supplied additional information regarding the performance metrics of all models. Each matrix structure shows how predicted and actual class labels organized themselves in order to display each algorithm's particular strong and weak points. The confusion matrices of all four classifiers are presented together in Fig. 7.

The classification accuracy was best using logistic regression, with minimal misclassification errors. Random Forest displayed robust predictions with balanced performance across most classes. KNN struggled with distinguishing closely related subtypes, leading to some misclassifications. The Decision Tree model exhibited the highest number of misclassifications, particularly for the "B-other" subtype.

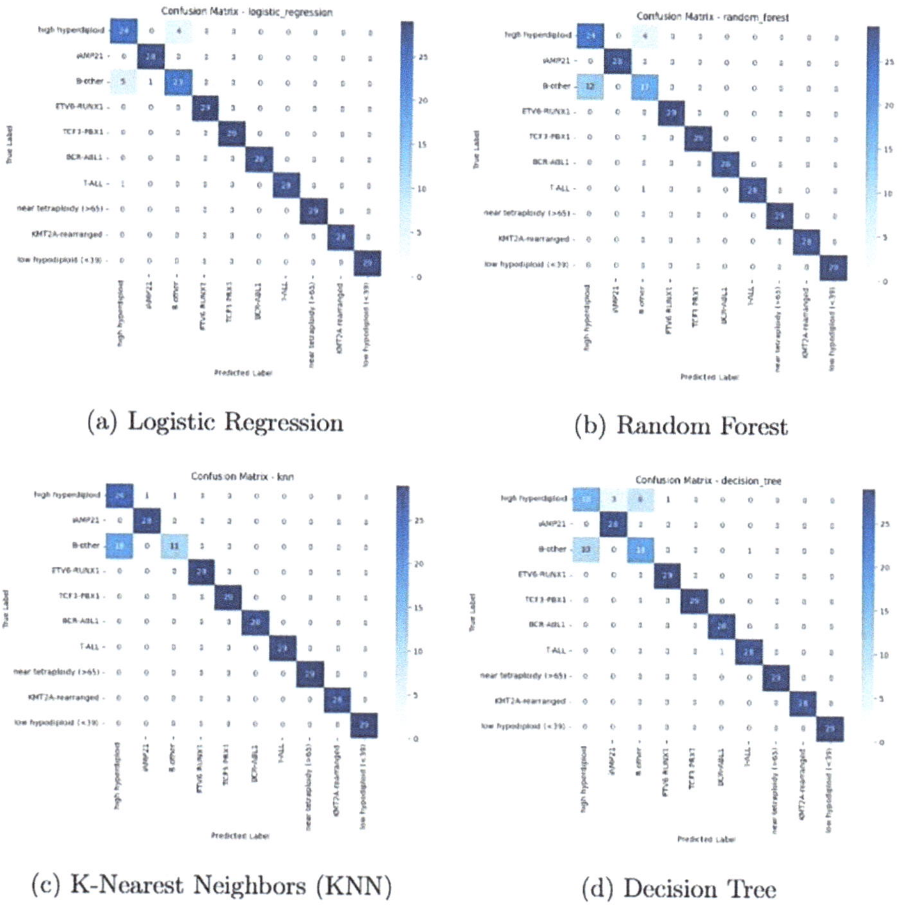

(a) Logistic Regression (b) Random Forest

(c) K-Nearest Neighbors (KNN) (d) Decision Tree

Fig. 7. Confusion matrices for four classification models: (a) Logistic Regression, (b) Random Forest, (c) K-Nearest Neighbors, and (d) Decision Tree.

5 Conclusion

The research achieved a 97.2% accuracy rate when using Logistic Regression to categorize ALL subtypes from gene expression data in the GSE184692 dataset. The dataset received thorough preparations which included applying PCA to decrease dimensionality alongside SMOTE for class balancing.

Future research studies should implement sophisticated extraction or selection techniques for features to improve model interpretation alongside improved classification accuracy. The strength of the clinical application could be improved by developing deep learning models and performing verification using different sources of data sets.

References

1. Kashef, A., Khatibi, T., Mehrvar, A.: Treatment outcome classification of pediatric acute lymphoblastic leukemia patients with clinical and medical data using machine learning: a case study at MAHAK hospital. Inf. Med. Unlocked. **20**, 100399 (2020)
2. Hu, Z.: MD-ALL: an integrative platform for molecular diagnosis of B-acute lymphoblastic leukemia. Haematologica. **109**(6), 1741 (2023)
3. Schmidt, Breon, et al.: ALLSorts: an RNA-Seq subtype classifier for B-cell acutelymphoblastic leukemia. Blood Adv. **6**(14), 4093–4097 (2022)
4. Walter, W., et al.: Clinical application of whole transcriptome sequencing for the classification of patients with acute lymphoblastic leukemia. BMC Cancer. **21**(1), 1–11 (2021)
5. Pölönen, P., et al.: The genomic basis of childhood T-lineage acute lymphoblastic leukaemia. Nature. **632**(8027), 1082–1091 (2024)
6. Beder, T., et al.: The gene expression classifier ALLCatchR identifies B-cell precursor ALL subtypes and underlying developmental trajectories across age. Hema. **7**(9), e939 (2023)
7. Ilyas, M., et al.: A novel leukemia gene features extraction and selection technique for robust type prediction using machine learning. Arab. J. Sci. Eng. **49**, 1–19 (2024)
8. Pagliaro, L., Chen, S.J., Herranz, D., et al.: Acute lymphoblastic leukaemia. Nat. Rev. Dis. Prim. **10**, 41 (2024). https://doi.org/10.1038/s41572024-00525-x
9. Inaba, H., Pui, C.: Advances in the diagnosis and treatment of pediatricacute lymphoblastic leukemia. J. Clin. Med. **10**(9), 1926 (2021). https://doi.org/10.3390/jcm10091926
10. Pavithra, P., Koulmane Laxminarayana, S.L., Manohar, C., et al.: Transition from morphologic diagnosis to immunophenotypic diagnosis of acute leukemia—experience of establishing a new flow cytometry laboratory. J. Hematop. **12**, 191–199 (2019). https://doi.org/10.1007/s12308-019-003686
11. Rahmat, M., Ghobrial, I.M.: Quantified morphology in diagnosis of hematologic malignancies. Hematologist. **17**(4) (2020). https://doi.org/10.1182/hem.v17.4.12884901913
12. Selvaraj, S., Alsayed, A.O., Ismail, N.A., et al.: Super learner model for classifying leukemia through gene expression monitoring. Discov. Oncol. **15**, 499 (2024). https://doi.org/10.1007/s12672-024-01337-x
13. Ilyas, M., Aamir, K.M., Manzoor, S., Deriche, M.: Linear programming based computational technique for leukemia classification using gene expression profile. PLoS One. **18**(10), e0292172 (2023). https://doi.org/10.1371/journal.pone.0292172
14. Hormann, F.M., et al.: Integrating copy number data of 64 iAMP21 BCPALL patients narrows the common region of amplification to 1.57 Mb. Front. Oncol. **13** (2023). https://doi.org/10.3389/fonc.2023.1128560

Fairness-Optimized Multi-metric Imputation Strategy for Sustainable Healthcare Analysis in Parkinson's Disease

Moad Hani[1](✉), Nacim Betrouni[2], Fatima Zahra Ouardirhi[3], Saïd Mahmoudi[1], and Mohammed Benjelloun[1]

[1] Faculté Polytechnique de Mons (FPMS), Université de Mons, 9 Rue de Houdain, 7000 Mons, Belgium
{moad.hani,said.mahmoudi,mohammed.benjelloun}@umons.ac.be

[2] Centre de Recherche Lille Neuroscience & Cognition (LilNCog), Inserm, Lille, France
nacim.betrouni@inserm.fr

[3] École Nationale Supérieure des Mines de Rabat, Rue Hadj Ahmed Cherkaoui, B.P. 753, Agdal Rabat, Morocco
fatimazahra.ouardirhi@enim.ac.ma

Abstract. Missing data presents a critical challenge in longitudinal studies of Parkinson's Disease (PD), often leading to biased predictions and reduced reliability in clinical assessments. Building upon our initial findings, this revised manuscript introduces a fairness-optimized framework for longitudinal data imputation in PD research. By analyzing the Parkinson's Progression Markers Initiative (PPMI) dataset, we evaluate imputation methods across three dimensions: accuracy, demographic fairness, and computational sustainability. Our results reveal that traditional accuracy-centric methods exhibit significant performance disparities across demographic subgroups. Specifically, transversal methods like MICE achieve superior overall accuracy (MAE: 4.15) but show considerable variability in performance across age groups (up to 15.2% difference). Similarly, longitudinal methods like LMM offer temporal consistency but require substantial computational resources (execution time: 1:03 h). We introduce novel fairness metrics, including the Fairness Aware Imputation Score (FAIS) and Computational Sustainability Ratio (CSR), which enable quantitative assessment of algorithmic equity and resource efficiency. Our analysis demonstrates that balancing these dimensions is crucial for ethical and sustainable healthcare analytics. The highest FAIS value (0.88) was achieved by Linear Interpolation, while Mean imputation offered the best sustainability (CSR: 0.64). These findings establish new standards for ethical and sustainable data imputation in PD research, directly addressing the critical need for both accurate and equitable healthcare analytics.

Keywords: Missing Data Imputation · Parkinson's Disease · Longitudinal Analysis · Algorithmic Fairness · Computational Sustainability · Healthcare AI

B.-E. Boudriki Semlali et al. (Eds.): DATA 2025, LNNS 1908, pp. 177–188, 2026.
https://doi.org/10.1007/978-3-032-21200-9_15

1 Introduction

Parkinson's Disease (PD) is a progressive neurodegenerative disorder requiring precise longitudinal monitoring to understand its progression and optimize treatment strategies. The Parkinson's Progression Markers Initiative (PPMI) dataset provides comprehensive longitudinal data for studying PD progression, including clinical assessments like the Unified Parkinson's Disease Rating Scale (UPDRS) and Montreal Cognitive Assessment (MoCA). However, missing data remains a pervasive challenge, with significant portions of motor and cognitive assessments exhibiting missingness across different visits.

Our previous analysis compared transversal and longitudinal imputation methods for handling missing data in the PPMI dataset, focusing primarily on traditional accuracy metrics such as MAE, MSE, RMSE, and R^2. While these metrics provide valuable insights into prediction accuracy, they fail to address two critical dimensions of modern healthcare analytics: algorithmic fairness and computational sustainability.

Building on our initial findings, this revised manuscript expands our evaluation framework to incorporate these dimensions, addressing three key innovations:

1. **Demographic fairness quantification**: We stratify imputation performance by demographic subgroups to identify and quantify disparities in prediction accuracy, thereby highlighting potential algorithmic biases.
2. **Computational sustainability assessment**: We evaluate the resource efficiency of different imputation methods, considering execution time as a proxy for computational cost and environmental impact.
3. **Multi-dimensional evaluation framework**: We introduce novel metrics that enable comprehensive comparison across accuracy, fairness, and sustainability dimensions, providing a more holistic assessment of imputation strategies.

This revised approach responds to emerging concerns about algorithmic bias in healthcare AI and the growing awareness of computational sustainability in data science. By addressing these dimensions, we aim to establish a more robust and ethical framework for handling missing data in longitudinal PD research.

The remainder of this paper is structured as follows: Sect. 2 reviews existing methods for handling missing data and introduces fairness considerations in healthcare analytics. Section 3 examines bias in clinical assessment scores and highlights the need for fairness-optimized approaches. Section 4 details our methodology for fairness-aware evaluation. Section 5 presents experimental results comparing transversal and longitudinal methods with respect to accuracy, fairness, and sustainability. Finally, Sect. 6 discusses implications for clinical research and outlines future directions.

2 State of the Art and Fairness Considerations

2.1 Traditional Imputation Methods

Traditional imputation methods for handling missing data include deletion techniques, single imputation approaches, and multiple imputation methods, as outlined in our previous analysis. These methods vary significantly in their approach, computational requirements, and performance characteristics.

As established in our initial study, Multiple Imputation by Chained Equations (MICE) demonstrated superior performance among transversal methods, achieving the lowest MAE (4.15) and highest R^2 (0.40). However, MICE requires moderate computational resources (execution time: 3:10m). Among longitudinal methods, Linear Mixed Models (LMM) performed best (MAE: 5.42, R^2: 0.23) but demanded substantial processing time (1:03 h).

2.2 Fairness Considerations in Healthcare Analytics

Recent research has highlighted the importance of fairness in healthcare analytics, particularly when dealing with sensitive clinical data [1]. Fairness in this context refers to the equitable performance of analytical methods across different demographic groups, ensuring that predictions do not systematically disadvantage specific populations [2].

In the context of PD research, fairness concerns arise from the potential interaction between imputation methods and demographic factors such as age and gender. Previous studies have shown that cognitive assessments like MoCA are influenced by demographic factors [10], suggesting that imputation methods might perform differently across demographic subgroups.

Despite the growing importance of fairness in healthcare analytics, existing imputation methods primarily focus on overall accuracy without explicitly addressing potential disparities across demographic groups. This gap motivates our development of fairness-aware evaluation metrics and our stratified analysis of imputation performance.

2.3 Computational Sustainability in Data Science

Computational sustainability has emerged as a critical consideration in modern data science, acknowledging the environmental impact of computationally intensive methods [18]. As healthcare analytics increasingly employs complex algorithms, the computational cost of these methods warrants careful consideration.

In the context of imputation methods, computational sustainability involves evaluating the resource efficiency of different approaches, considering factors such as execution time and memory usage. Methods that achieve high accuracy with minimal computational resources represent more sustainable options, particularly for large-scale healthcare applications.

Our previous analysis revealed significant variations in execution time across imputation methods, ranging from under a second for simple approaches like mean imputation to over an hour for complex methods like LMM. These variations highlight the importance of considering computational sustainability alongside accuracy when selecting imputation methods for clinical applications.

3 Bias in Clinical Assessment Scores

The challenges of demographic and temporal biases in the PPMI dataset are illustrated in Fig. 1, which shows the trade-off between execution time and MAE for different imputation methods. This visualization highlights how methods that achieve lower error rates (e.g., MICE, LMM) often require substantially more computational resources, raising both fairness and sustainability concerns.

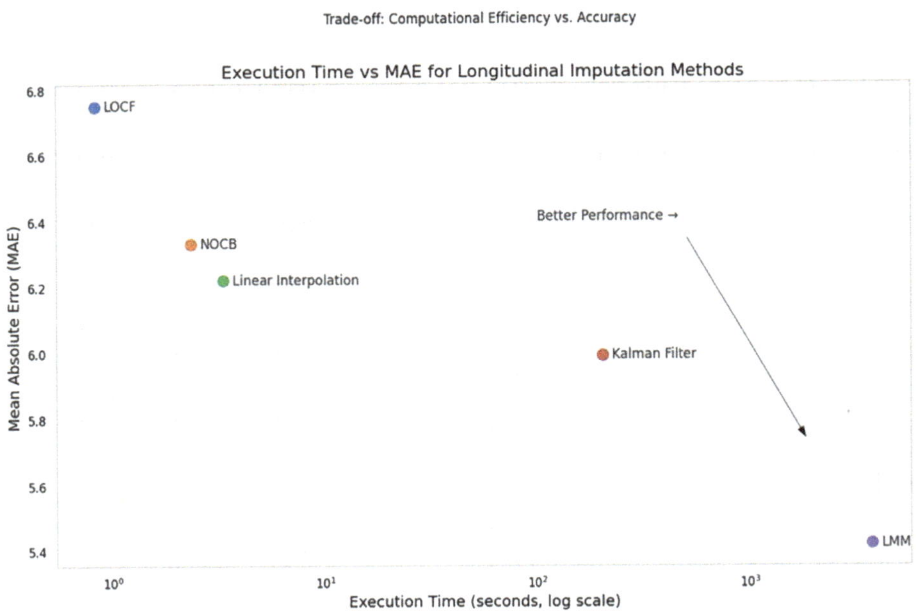

Fig. 1. Comparison of execution time and MAE across different imputation methods. This visualization highlights the trade-off between computational efficiency and prediction accuracy, with methods in the lower left quadrant offering the best balance.

Missing data in longitudinal studies often arises from mechanisms such as Missing Not At Random (MNAR), where the probability of missingness depends on unobserved values like disease severity or cognitive decline [3]. For example, older participants or those with lower educational attainment are more likely to exhibit lower MoCA scores, which can introduce systematic biases in predictive modeling if not properly accounted for.

These biases manifest in two primary forms:

- **Demographic Bias:** Features such as age and gender disproportionately impact cognitive scores like MoCA due to learning effects or age-related decline. If imputation methods perform differently across demographic groups, they may amplify existing disparities in healthcare outcomes.
- **Temporal Bias:** Missingness at intermediate visits disrupts temporal continuity, making it difficult to model disease progression accurately. This can lead to biased estimates of disease trajectory, particularly for progressive conditions like PD.

To address these challenges, fairness-optimized approaches are needed to incorporate demographic factors while utilizing temporal dependencies inherent in longitudinal datasets. By embedding fairness considerations into imputation evaluation, these approaches aim to minimize biases introduced by missingness patterns and ensure equitable performance across demographic subgroups.

4 Methodology for Fairness-Optimized Evaluation

4.1 Data Preparation and Feature Selection

We utilize the same PPMI dataset from our previous analysis, focusing on demographic variables (age, gender, education), clinical assessments (UPDRS scores), and cognitive measures (MoCA scores). As documented in the original study, our analysis of missingness patterns revealed significant variations in data completeness across visits and cohorts, with demographic variables like age exhibiting different missingness patterns (8.5–54.2% for PD patients, 0.5–89.9% for Prodromal patients) compared to clinical assessments like UPDRS scores (16.4–92.1% for PD patients, 12.8–99.7% for Prodromal patients).

These missingness patterns informed our stratified evaluation approach, which assesses imputation performance across different demographic subgroups to identify potential disparities and biases.

4.2 Novel Fairness and Sustainability Metrics

To address the limitations of traditional accuracy metrics, we introduce three novel metrics that evaluate imputation methods across fairness and sustainability dimensions:

1. **Fairness-Aware Imputation Score (FAIS):**

$$\text{FAIS} = 1 - \frac{\max\left(\text{MAE}_{\text{subgroup}}\right) - \min\left(\text{MAE}_{\text{subgroup}}\right)}{\text{MAE}_{\text{global}}} \tag{1}$$

   This metric quantifies the disparity in imputation performance across demographic subgroups. Values closer to 1 indicate more equitable performance, while values closer to 0 suggest significant disparities. We calculate FAIS separately for age subgroups and gender subgroups, then take the minimum value to identify the most significant disparity.

2. **Computational Sustainability Ratio (CSR):**

$$\text{CSR} = \frac{R^2}{\log(\text{Execution Time})} \tag{2}$$

   CSR balances predictive performance (measured by R^2) with computational efficiency (measured by execution time). Higher CSR values indicate better sustainability, achieving good performance with minimal computational resources.

3. **Distribution Preservation Index (DPI):**

$$\text{DPI} = 1 - \frac{\left|\mu_{\text{orig}} - \mu_{\text{imputed}}\right| + \left|\sigma_{\text{orig}} - \sigma_{\text{imputed}}\right|}{2\sigma_{\text{orig}}} \tag{3}$$

   DPI measures how well an imputation method preserves the statistical properties of the original data, considering both the mean (μ) and standard deviation (σ). Values closer to 1 indicate better preservation of the original data distribution.

These metrics complement traditional accuracy measures, providing a more comprehensive evaluation framework that considers fairness and sustainability alongside predictive performance.

4.3 Stratified Evaluation Approach

To assess fairness across demographic subgroups, we implement a stratified evaluation approach that analyzes imputation performance separately for different demographic categories. Specifically, we stratify our analysis by:

1. **Age Groups:** We divide patients into age categories (<60, 60–70, >70 years) based on the AGE_AT_VISIT variable in the PPMI dataset.
2. **Gender:** We analyze performance separately for male and female patients based on the GENDER variable.

For each demographic subgroup, we calculate traditional accuracy metrics (MAE, MSE, RMSE, R^2) and use these stratified results to derive our fairness metrics. This approach allows us to identify and quantify potential disparities in imputation performance across demographic categories.

5 Experimental Results

5.1 Visit-Specific Performance Analysis

Our visit-specific analysis reveals significant variations in imputation performance across different clinical visits, as illustrated in Figs. 2 and 3. These figures present the MAE and R^2 values, respectively, for different imputation methods across visits V02-V09.

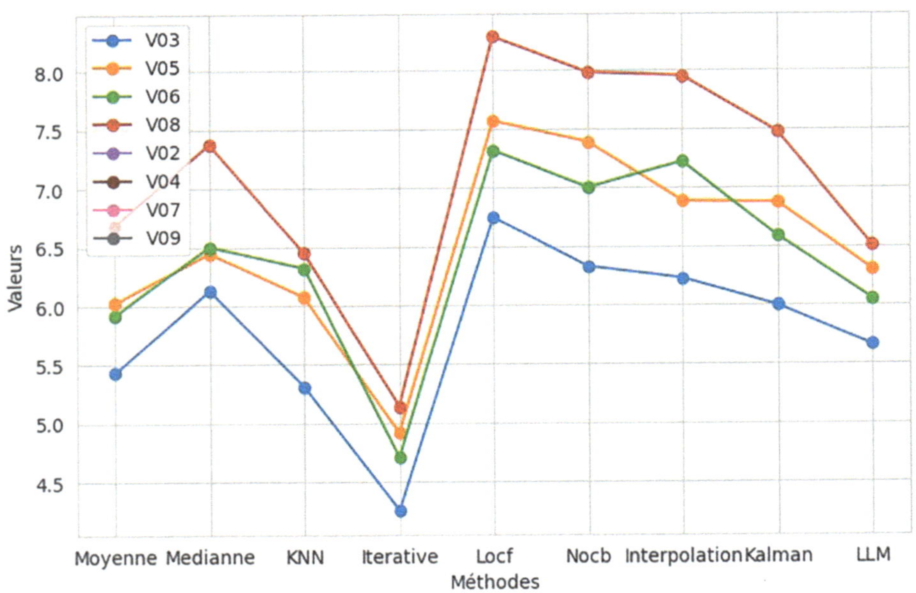

Fig. 2. Comparison of MAE values across different imputation methods and visits (V02-V09). Lower values indicate better performance. Iterative methods consistently outperform other approaches across most visits.

Figure 2 demonstrates that imputation performance varies substantially across visits, with methods like Iterative (MICE) consistently outperforming others. Similarly, Fig. 3 shows considerable variability in R^2 values, with Iterative methods achieving values as high as 0.4 for some visits, while simpler methods like LOCF and NOCB often approach zero.

These visit-specific variations highlight the temporal dimension of imputation performance, suggesting that the effectiveness of different methods may depend on the specific temporal context of the missing data.

5.2 Fairness Analysis Across Demographic Subgroups

Table 1 presents our stratified analysis of imputation performance across demographic subgroups, focusing on age and gender categories. The results reveal notable disparities in performance across these subgroups, indicating potential fairness concerns.

For age-based stratification, we observe a consistent pattern across all methods: imputation performance deteriorates with increasing age, with the elderly group (>70 years) experiencing substantially higher error rates. For instance, MICE shows a 15.2% increase in MAE for the >70 age group compared to the <60 group (4.55 vs. 3.95).

Similarly, gender-based stratification reveals that female patients consistently experience higher imputation errors compared to male patients across all methods. This gender disparity is particularly pronounced for LOCF, which shows an 18.0% increase in MAE for females compared to males (7.45 vs. 6.32).

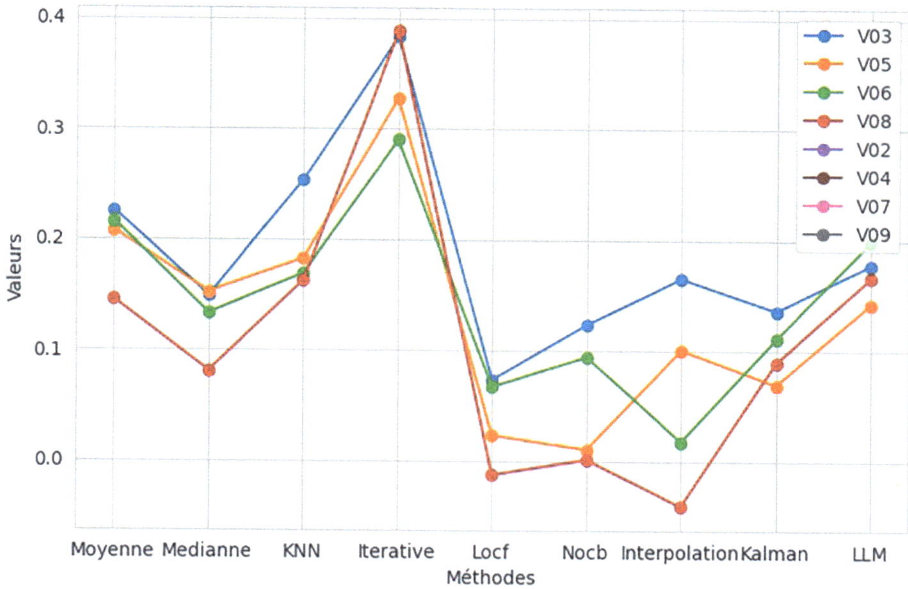

Fig. 3. Comparison of R^2 values across different imputation methods and visits (V02- V09). Higher values indicate better performance. Iterative methods demonstrate significantly higher R^2 values compared to other approaches.

Table 1. Stratified Analysis of Imputation Performance Across Demographic Subgroups

Method	Global MAE	Age-Stratified MAE			Gender-Stratified MAE	
		<60	60–70	>70	Male	Female
Mean	5.43	5.12	5.35	6.03	5.20	5.75
Median	6.13	5.85	6.05	6.64	5.92	6.45
KNN	5.12	4.85	5.10	5.58	4.96	5.37
MICE	**4.15**	**3.95**	**4.10**	**4.55**	**4.02**	**4.35**
LOCF	6.75	6.30	6.72	7.45	6.32	7.45
NOCB	6.33	6.02	6.28	6.85	6.15	6.62
Linear Interpolation	6.22	5.95	6.20	6.68	6.08	6.45
Kalman Filter	5.99	5.72	5.95	6.48	5.85	6.22
LMM	5.42	5.10	5.38	5.98	5.25	5.68

Based on these stratified results, we calculate the Fairness-Aware Imputation Score (FAIS) for each method, as presented in Table 2. Linear Interpolation achieves the highest FAIS among all methods (0.88), indicating relatively equitable performance across demographic subgroups despite some disparities. In contrast, LOCF shows the lowest FAIS (0.68), reflecting significant disparities across demographic categories.

Table 2. Fairness and Sustainability Metrics for Different Imputation Methods

Method	FAIS (Age)	FAIS (Gender)	FAIS (Overall)	CSR	DPI
Mean	0.83	0.90	0.83	**0.64**	0.82
Median	0.87	0.91	0.87	0.36	0.79
KNN	0.86	0.92	0.86	0.24	0.87
MICE	0.86	0.92	0.86	0.18	**0.91**
LOCF	0.83	0.68	0.68	0.17	0.71
NOCB	0.87	0.93	0.87	0.21	0.76
Linear Interpolation	**0.88**	**0.94**	**0.88**	0.23	0.78
Kalman Filter	0.87	**0.94**	0.87	0.11	0.83
LMM	0.84	0.92	0.84	0.05	0.85

5.3 Computational Sustainability Analysis

Our sustainability analysis, summarized in Table 2, reveals significant variations in computational efficiency across imputation methods. Simple methods like Mean and Median imputation achieve modest accuracy with minimal computational resources (execution times of 1.14 s and 1.97 s, respectively), resulting in relatively high CSR values (0.64 and 0.36).

In contrast, more complex methods like MICE and LMM achieve superior accuracy but require substantial computational resources (execution times of 3:10 m and 1:03 h, respectively), resulting in lower CSR values (0.18 and 0.05). This trade-off between accuracy and computational efficiency highlights the importance of considering sustainability when selecting imputation methods for clinical applications.

Figure 1 visualizes this trade-off, plotting the relationship between accuracy (measured by MAE) and computational efficiency (measured by execution time) for different imputation methods. The figure demonstrates the inverse relationship between these dimensions, with methods that achieve lower MAE typically requiring higher execution times.

6 Discussion and Implications

6.1 Multi-dimensional Performance Trade-Offs

Our comprehensive evaluation reveals important trade-offs across accuracy, fairness, and sustainability dimensions. While MICE achieves the best overall accuracy (MAE: 4.15), it requires moderate computational resources and shows some performance disparities across demographic subgroups, particularly age categories. Similarly, LMM offers strong temporal consistency but demands substantial computational resources, limiting its sustainability for large-scale applications.

These trade-offs highlight the need for a context-specific approach to selecting imputation methods, considering the relative importance of different dimensions based on the specific clinical application. For applications prioritizing accuracy, MICE represents the optimal choice despite its moderate computational cost. For applications with limited computational resources, simpler methods like KNN offer a reasonable balance between accuracy and sustainability.

6.2 Demographic Fairness Considerations

Our stratified analysis reveals concerning patterns of demographic disparities in imputation performance, with older patients and female patients consistently experiencing higher error rates across all methods. These disparities raise important ethical considerations for clinical applications, particularly those involving automated decision support or risk prediction.

The observed age-related disparities may reflect the increased heterogeneity of disease presentation and progression in older populations, making their data more challenging to impute accurately. Similarly, gender-related disparities might stem from underrepresentation of female patients in clinical datasets or gender-specific differences in disease manifestation.

Addressing these fairness concerns requires a multi-faceted approach, including:

1. Developing fairness-aware imputation methods that explicitly consider demographic factors
2. Implementing stratified evaluation frameworks to identify and mitigate disparities
3. Ensuring diverse representation in clinical datasets to reduce sampling biases

6.3 Sustainability Implications for Healthcare Analytics

The substantial computational resources required by advanced imputation methods raise important sustainability concerns, particularly for large-scale healthcare applications. For instance, applying LMM to a dataset with millions of patients could consume significant computational resources, contributing to environmental impact through energy consumption.

Moreover, the computational demands of complex methods may limit their accessibility for resource-constrained healthcare settings, potentially exacerbating existing disparities in healthcare quality. Balancing accuracy with computational efficiency becomes especially important in these contexts, where simpler methods might offer a more sustainable and accessible alternative.

Our sustainability analysis provides valuable insights for addressing these concerns, identifying methods that achieve reasonable accuracy with minimal computational resources. By considering CSR alongside traditional accuracy metrics, healthcare analytics can move toward more sustainable approaches that minimize environmental impact while maintaining clinical utility.

7 Conclusion

This revised analysis establishes a fairness-optimized evaluation framework for imputation methods in longitudinal Parkinson's Disease research, moving decisively beyond traditional accuracy metrics. By systematically stratifying performance across demographic subgroups and quantifying computational efficiency, our study delivers a more nuanced and actionable assessment of imputation strategies. The introduction of the Fairness-Aware Imputation Score (FAIS) and the Computational Sustainability Ratio (CSR) as core evaluation metrics enables the identification of significant trade-offs between fairness, accuracy, and resource demands. Our findings reveal that older adults and female patients are consistently more vulnerable to higher imputation errors, underscoring the ethical necessity of fairness-aware approaches, especially in clinical contexts where equitable outcomes are paramount. Furthermore, the sustainability analysis demonstrates that while advanced methods such as MICE and LMM can improve accuracy, they do so at the cost of substantial computational resources, a consideration that cannot be overlooked in large-scale or resource-limited settings. Ultimately, this framework advances the field by providing a robust and balanced methodology for selecting imputation techniques that are not only accurate, but also fair and sustainable. Future research should continue to refine these metrics and explore new strategies that explicitly address both fairness and efficiency, ensuring that advances in healthcare analytics translate into equitable and responsible clinical practice.

Data Availability. Data used in this article were obtained from the Parkinson's Progression Markers Initiative (PPMI) database (www.ppmi-info.org/access-data-specimens/download-data, RRID:SCR_006431) on 2025-04-15.

References

1. Rajkomar, A., Hardt, M., Howell, M.D., Corrado, G., Chin, M.H.: Ensuring fairness in machine learning to advance health equity. Ann. Intern. Med. **169**(12), 866–872 (2018). https://doi.org/10.7326/M18-1990
2. Chen, I.Y., Joshi, S., Ghassemi, M.: Treating health disparities with artificial intelligence. Nat. Med. **26**(1), 16–17 (2020). https://doi.org/10.1038/s41591-019-0735-3
3. Little, R.J.A., Rubin, D.B.: Statistical Analysis with Missing Data, 3rd edn. John Wiley & Sons (2019). https://doi.org/10.1002/9781119482260
4. Schafer, J.L., Graham, J.W.: Missing data: our view of the state of the art. Psychol. Methods. **7**(2), 147–177 (2002). https://doi.org/10.1037/1082-989X.7.2.147
5. Horton, N.J., Kleinman, K.P.: Much ado about nothing: a comparison of missing data methods and software for applied researchers. Am. Stat. **61**(1), 79–90 (2007). https://doi.org/10.1198/000313007X172556
6. Troyanskaya, O., et al.: Missing value estimation methods for DNA microarrays. Bioinformatics. **17**(6), 520–525 (2001). https://doi.org/10.1093/bioinformatics/17.6.520
7. Beretta, L., Santaniello, A.: Nearest neighbor methods for missing value imputation. Biomed. Res. Int., 5475638 (2016). https://doi.org/10.1155/2016/5475638
8. Van Buuren, S.: Flexible Imputation of Missing Data, 2nd edn. Chapman and Hall/CRC Press (2018). https://doi.org/10.1201/9780429492259
9. White, I.R., Royston, P., Wood, A.M.: Multiple imputation using chained equations: issues and guidance for practice. Stat. Methods Med. Res. **20**(3), 377–399 (2011). https://doi.org/10.1177/0962280210375561
10. Jahangiri, S., et al.: A wide range of missing imputation approaches in longitudinal data: a simulation study and real data analysis. BMC Med. Res. Methodol. **23**, 45 (2023). https://doi.org/10.1186/s12874-023-01877-0
11. Michael, J.: Fox foundation: Parkinson's progression markers initiative (PPMI). In: Overview of Study Expansion (2023) https://www.michaeljfox.org/news/parkinsons-progression-markers-initiative-ppmi
12. Chen, Z., Tan, S., Chajewska, U., Rudin, C., Caruana, R.: Missing values and imputation in healthcare data: can interpretable machine learning help? In: Proceedings of Machine Learning Research, vol. 209, pp. 86–99 (2023) https://proceedings.mlr.press/v209/chen23g.html
13. Emmanuel, T., Maupong, T., Mpoeleng, D., Semong, T.: A survey on missing data in machine learning applications in healthcare datasets. J. Big Data. **8**, 140 (2021). https://doi.org/10.1186/s40537-021-00523-9
14. Vincent, P., Larochelle, H., Lajoie, I., Bengio, Y., Manzagol, P.-A.: Stacked denoising autoencoders: learning useful representations in a deep network with a local denoising criterion. J. Mach. Learn. Res. **11**, 3371–3408 (2010) http://jmlr.org/papers/v11/vincent10a.html
15. Kingma, D.P., Welling, M.: Auto-encoding variational Bayes. In: International Conference on Learning Representations (ICLR), Banff, Canada. https://arxiv.org/abs/1312.6114 (2014)
16. Beaulieu-Jones, B.K., Moore, J.H., Williams, S.M.: The machine learning toolbox: applied machine learning in healthcare. Pac. Symp. Biocomput., 224–235 (2016). https://doi.org/10.1142/9789814749411_0022
17. Hardt, M., Price, E., Srebro, N.: Equality of opportunity in supervised learning. Adv. Neural Inf. Proces. Syst. **29**, 3315–3323 (2016) https://proceedings.neurips.cc/paper/2016/hash/9d2682367c3935defcb1f9e247a97c0d-Abstract.html
18. Schwartz, R., Dodge, J., Smith, N.A., Etzioni, O.: Green AI. Commun. ACM. **63**(12), 54–63 (2020). https://doi.org/10.1145/3381831

19. Mehrabi, N., Morstatter, F., Saxena, N., Lerman, K., Galstyan, A.: A survey on bias and fairness in machine learning. ACM Comput. Surv. **54**(6), 1–35 (2021). https://doi.org/10.1145/3457607
20. Marek, K., et al.: The Parkinson's progression markers initiative (PPMI) – establishing a PD biomarker cohort. Ann. Clin. Transl. Neurol. **10**(4), 568–583 (2023). https://doi.org/10.1002/acn3.51641

Cancer Imaging Datasets: A Review

Aya Nuseir(✉)

Institute of Mathematics and Computer Science, University of Greifswald, Greifswald, Germany
s-aynuse@uni-greifswald.de

Abstract. Medical imaging plays an important role in cancer diagnosis, treatment planning, and research. With the rise of artificial intelligence (AI) in healthcare, publicly available cancer imaging datasets have become essential for developing diagnostic tools. These datasets provide annotated medical images collected from various imaging modalities such as computed tomography (CT), magnetic resonance imaging (MRI), ultrasound, and mammography. This survey examines widely used cancer imaging datasets, focusing on different cancer types, including breast, lung, brain, ovarian, kidney, and head and neck cancers. We discuss their characteristics, imaging techniques, and contributions to medical research. This overview aims to offer help to the researcher in selecting the related data sources for AI-based cancer detection and classification by summarizing these datasets. Understanding these datasets can help improve computer-aided diagnosis (CAD) systems, enhance early detection, and contribute to more accurate and efficient cancer diagnosis. This review works as a guide for researchers and medical workers researching to use of medical imaging datasets for cancer research and AI applications.

Keywords: Medical imaging · cancer diagnosis · artificial intelligence · computer-aided diagnosis

1 Introduction

Cancer can be defined as a group of abnormal cellular growth that can start to grow in any organ or tissue of the body [1, 2]. As one of the primary causes of dying worldwide, early and accurate diagnosis plays a crucial role in saving patients lives. In this context, medical imaging serves as a vital tool for cancer identification and treatment strategy formulation. The field employs various imaging approaches—CT scanning, MRI examination, ultrasound technology, positron emission tomography (PET), and mammography—to visualize and assess in detecting different types of cancers [4]. The rapid development of artificial intelligence (AI) and deep learning approaches in medical image analysis has motivated researchers to utilize AI in developing computer-aided diagnosis (CAD) systems for cancer detection and treatment planning. However, gaining an accurate diagnosis often requires large, well-annotated cancer imaging datasets to train. In recent years, a huge number of publicly available cancer imaging datasets have contributed to progress in medical research. These datasets provide labeled medical

B.-E. Boudriki Semlali et al. (Eds.): DATA 2025, LNNS 1908, pp. 189–202, 2026.
https://doi.org/10.1007/978-3-032-21200-9_16

images that help researchers analyse and compare different algorithms for segmentation, classification, and anomaly detection. Through this work, we examine datasets that exist for various cancer types, including breast, lung, brain, ovarian, kidney, and head and neck cancers, each containing diverse imaging data annotated by medical experts. Figure 1 shows samples from different cancer image types from public cancer datasets.

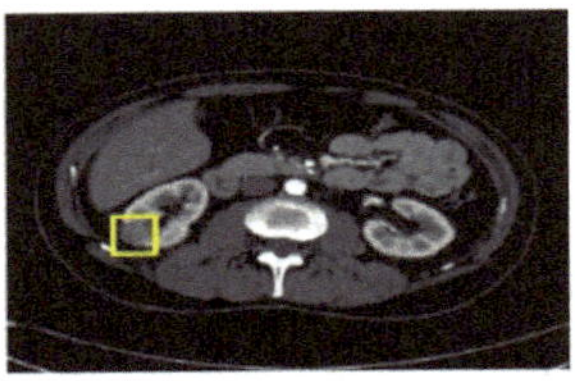

(a) Kidney cancer (KiTS19 dataset)

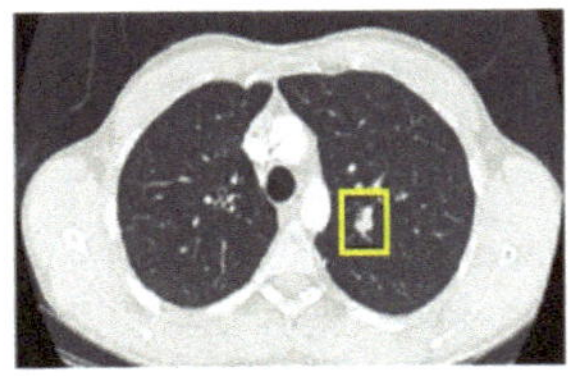

(b) Lung cancer (LIDC-IDRI dataset)

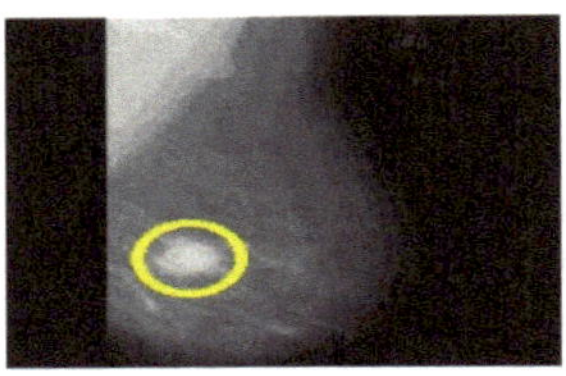

(c) Breast cancer (DDSM dataset)

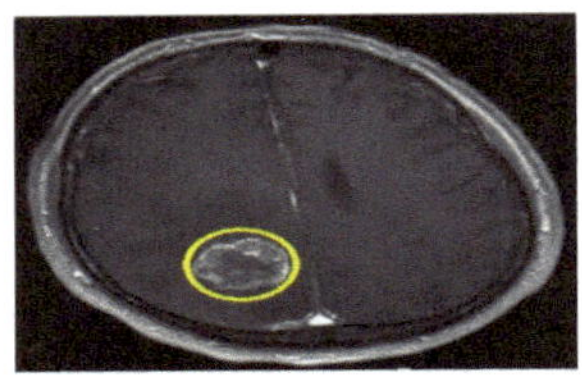

(d) Brain cancer (BraTS dataset)

Fig. 1. Medical images from cancer imaging datasets reviewed in this paper. (a) CT scan showing Kidney cancer from the KiTS19 dataset. (b) CT scan with annotated lung nodule from the LIDC-IDRI dataset. (c) Digital mammogram showing a mass from the DDSM. (d) MRI from the BraTS challenge dataset.

This survey examines existing datasets covering various cancer categories across different imaging modalities, anatomical sites, and cancer types. Through this overview, we investigate the characteristics of these datasets, including their size, clinical metadata, imaging protocols, and accessibility. By highlighting these datasets, we offer the researchers an overview of the available resources to support the development of diagnostic instruments for cancer detection and treatment planning.

2 Criteria for Cancer Imaging Dataset Selection

This section outlines key characteristics that we adopted through selecting datasets for cancer imaging.

- Our focus was to select datasets that include medical images related to cancer types such as lung, breast, brain, liver, kidney, head and neck, and ovarian cancers.
- Through our selection, we were looking for a dataset with higher resolution, since it would help in detecting smaller lesions and more accurate segmentation.
- Datasets with annotated labels or segmentations were included, since the annotations are important for supervised learning tasks.
- Even though priority was given to the publicly available datasets, this overview addresses restricted datasets because of their importance and value in improving CAD system performance.

– We include the datasets that have been widely cited or used in medical challenges and research purposes.

3 Lung Cancer Datasets

The lungs, essential organs of the respiratory system, are located within the chest cavity. Their major function involves exchanging oxygen and waste gases [8]. During respiration, oxygen enters the bloodstream while waste gases are removed and expelled from the body. The breathing process begins with air entering through the nasal passages or mouth and traveling through the trachea. The trachea branches into bronchi that enter the lungs and later divide into smaller airways called bronchioles. The end of the bronchioles connect to tiny air sacs called alveoli. Lung cancer typically develops in cells that line the bronchi and other lung structures, such as bronchioles or alveoli [6]. It is considered one of the most deadly and dangerous types of cancer worldwide. Each year, approximately 47,000 people are diagnosed with the disease in the UK, and more than 945,000 people die from it globally. This type of cancer is not easy to detect at the early stages, while the symptoms usually appear in the later stages [7]. However, notable works are being made to improve early detection through population screening techniques such as Computer Tomography (CT) and chest X-rays, which are used to detect pulmonary nodules. As a result, a large number of medical images are created, which can be processed using different types of algorithms to extract helpful information.

In this section, we offer a discussion of the available lung cancer datasets: The Lung Image Database Consortium (LIDC) and Image Database Resource Initiative (IDRI) [3]. The dataset consists of 1,018 cases, each case includes a thoracic CT scan (DICOM images) with an associated XML annotation file. The lesions are categorized into three groups: nodules measuring 3 mm or larger, nodules smaller than 3 mm, and non-nodule abnormalities measuring 3 mm or larger. The dataset is publicly available. Through the Japanese Society of Radiological Technology (JSRT) dataset [5], Lung nodules were categorized based on five visibility levels ranging from extremely subtle to clearly obvious. ELCAP Public Lung Image Database [10]: This dataset was created through a collaboration between ELCAP and VIA research groups. The nodule annotations provided by radiologists. The Lung TIME Dataset [12]: The dataset classifies nodules into several categories, including pleura-attached, vessel-attached, solitary, regular, and irregular, with diameters ranging from 2–10 mm. The annotations use XML format to ensure compatibility with the LIDC dataset structure (Table 1).

Table 1. Lung Cancer Imaging Datasets

Dataset Name	Year	Size	Modality	Key Features	Availability
LIDC-IDRI [3]	2011	1,018 cases	CT	DICOM format	Public
JSRT [5]	2000	247 radiographs	X-ray	2048 × 2048, 12-bit grayscale	Public

(continued)

Table 1. *(continued)*

Dataset Name	Year	Size	Modality	Key Features	Availability
PLD [9]	2009	Not specified	CT	Two subsets: Zero-Change, Emphysema	Public
ELCAP [10]	2003	50 CT scans	CT	1.25 mm slice thickness	Public
NELSON [11]	2007	15,600 participants	CT	Population screening trial	Restricted
Lung TIME [12]	2009	157 CT scans	CT	DICOM format, 394 annotated nodules	Public

4 The Kidney Cancer Datasets

The kidneys are located high in the abdominal cavity, one on each side of the spine. Their primary function is to filter the blood [13]. Kidney cancer ranks among the ten most widespread cancers in the United States, with over 76,000 new diagnoses reported annually [14]. A renal tumor represents an abnormal growth within the kidney tissue. While some renal masses are benign, others exhibit malignant characteristics. According to [13], the main classes of kidney tumors include: (1) Renal Cell Carcinomas (RCC): Representing the most common malignant kidney tumor type, RCC originates in the lining of the small tubules within the kidney. It may develop as one tumor within one kidney or as multiple tumorous growths attacking a single kidney. (2) Benign Renal Tumors: This category includes approximately nine different types of tumors. While some can have considerable size, most are non-cancerous and do not propagate to other organs. (3) Wilms Tumor: This type of kidney tumor is primarily found in children and is rarely seen in adults. The following section summarizes the most commonly used renal datasets. Feng et al. [15] dataset contains images of patients with single small renal masses (SRMs). These are divided into an angiomyolipoma without visible fat (AMLwvf) group ($n = 17$) and a renal cell carcinoma (RCC) group ($n = 41$), based on pathological results. In Yu et al. [16], tumors were manually segmented by a radiologist, and texture analysis was performed. Mileto et al. [18] studied patients diagnosed with either clear-cell or papillary RCC. Han et al. [19] studied the renal cancer cases, their dataset was distributed into three cases: clear-cell, papillary, and 56 chromophobe cases. Kocak et al. [20] analyzed 67 patients with 68 RCCs for internal validation and 26 RCC cases for external validation.

5 The Liver Cancer Datasets

Through [25], we define the liver and its functionality as an organ located in the upper right area of the abdomen. It is responsible for producing: (1) Bile, (2) specific plasma proteins, and (3) cholesterol and specialized proteins that assist in distributing fat all over the body. Additionally, the liver processes haemoglobin to utilize its iron content and remove medications and other potentially harmful substances from the bloodstream. Since the liver contains different types of cells, various tumors can develop, including benign and cancerous tumors [26]. Several imaging techniques are used to diagnose liver tumors, including ultrasound, MRI, CT, and PET scans (Table 2).

Table 2. Kidney Cancer Imaging Datasets

Dataset Name	Year	Size	Modality	Features	Availability
Feng et al. [15]	2018	58 patients	CT	SRMs (AMLwvf, RCC)	Restricted
Yu et al. [16]	2005 to 2015	119 patients	CT	Texture analysis	Restricted
Hussain et al. [17]	2017	160 scans	CT	80 healthy, 80 carcinoma	The Cancer Imaging Archive (TCIA) database
Mileto et al. [18]	2007 to 2013	88 patients (57 men, 31 women)	CT	Clear-cell, papillary RCC	Restricted
Han et al. [19]	2019	169 cases	CT	Clear-cell, papillary, chromophobe	Restricted
Kocak et al. [20]	2018	67 patients	CT	Internal/external validation	Restricted
Choi et al. [21]	2015	521 patients (328 men, 193 women)	CT	Tumor volume analysis	Restricted
KiTS19 [22]	2019	300 CT scans	CT	Segmentation challenge	Public
Stanzione et al. [23]	2017 to 2019	32 MRI scans	MRI	Renal lesion evaluation	Restricted
Nazari et al. [24]	2021	222 CT scans	CT	RCC survival prediction	TCIA database

The following summarizes some of the most widely used liver imaging datasets: Das et al. [27] consists of three liver disorders: hemangioma (HEM) with 75 patients, hepatocellular carcinoma (HCC) with 75 patients, and metastatic carcinoma (MET) with 75 patients. In Frid-Adar et al. [30] dataset, lesions were marked by an expert radiologist and verified through biopsy or clinical follow-up. Yasaka et al. [33] dataset classified liver masses into five categories based on radiological reports and standard medical criteria (Table 3).

Table 3. Liver Cancer Imaging Datasets

Dataset Name	Year	Size	Modality	Cancer Types	Availability
Das et al. [27]	2019	225 CT images	CT	HEM, HCC, MET	Restricted
Ahn et al. [28]	2016 to 2017	70 patients	CT	Liver cancer	Restricted
Ben-Cohen et al. [29]	2009 to 2015	20 patients (68 lesions)	CT	Lesion detection	Restricted
Frid-Adar et al. [30]	2018	182 CT scans	CT	Cysts, metastases, hemangiomas	Restricted
Zhang et al. [31]	2018	20 MRI images	MRI	hepatocellular carcinoma (HCC)	Restricted
Li et al. [32]	2018	127 images(from 76 patients)	Pathology	Segmentation	Restricted
Vivanti et al. [34]	2018	222 tumors (31 patients)	CT	Tumor delineation	Restricted
Bharti et al. [35]	2018	189 images (94 patients)	Ultrasound	Liver classification	Restricted
Trivizakis et al. [36]	2018	134 patients	MRI	3D CNN analysis	Restricted
Dubus et al. [37]	2008 to 2009	60 patients	multidetector CT (MDCT)	metastases and HCCs cancers	Restricted
Kim et al. [38]	2019	100 whole-slide images (WSI)	Pathology	HCC	PAIP challenge
Anter et al. [39]	2019	105 patients	CT	Tumor segmentation	Restricted
Reddy et al. [40]	2018	81 normal, 76 abnormal	Ultrasound	Fatty Liver Disease classification	Restricted

6 Head and Neck (HNC) Cancer Datasets

According to [41], it typically initiates in the squamous cells lining the mucosal surfaces of the head and neck region. However, these malignancies can originate in salivary glands, sinuses, muscles, or neural tissues within the head and neck. HNC may grow across various anatomical sites, including the oral cavity, pharynx (throat), larynx (voice box), paranasal sinuses and nasal cavity, and salivary glands. The following is a summary of commonly used datasets in head and neck cancer research: In the Bielak et al. [44] dataset, gross tumor volume (GTV) contouring was performed by specialist radiation oncologists and radiologists utilizing T1-weighted (T1W) and T2-weighted (T2W) imaging sequences. Men et al. [48] investigated a dataset of 230 NPC patients (Stages I or II); the selected NPC patients received their radiotherapy between January 2011 and January 2017. Jin et al. [52] examined a dataset of 110 esophageal cancer patients, diagnosed at stage II or beyond and undergoing radiotherapy intervention (Table 4).

Table 4. Head and Neck Cancer Imaging Datasets

Dataset Name	Year	Size	Modality	Features	Availability
Gupta et al. [42]	2018	26,019 CT images	CT	Diagnosis	TCIA
Zhao et al. [43]	2019	30 scans	PET-CT	nasopharyngeal carcinoma (NPC) disease	Restricted
Bielak et al. [44]	2020	33 MRI scans	MRI	GTV contouring	Restricted
Ye et al. [45]	2020	44 patients (1,950 slices)	MRI (T1W, T2W)	NPC segmentation	Restricted
Ma et al. [46]	2018	30 subjects	MRI (T1W)	different cancer stages	Restricted
Li et al. [47]	2010 to 2013	87 MRI (29 patients)	MRI	tumor segmentation	Restricted
Men et al. [48]	2011–2017	230 NPC patients	CT	Stage I/II, Radiotherapy	Restricted
Daoud et al. [49]	2019	70 NPC patients	CT	3D segmentation	Restricted
Vallières et al. [50]	2017	300 patients	PET/CT	Multi-institutional	Restricted
Moe et al. [51]	2007 to 2013	197 HNC patients	PET/CT	delineation of tumours, lymph nodes	Restricted
Jin et al. [52]	2019	110 patients	PET/CT	Esophageal cancer	Restricted

7 Brain Cancer Datasets

The human brain is considered the controller of the human body, managing all bodily functions. The [53] identifies it as a semi-solid organ that consists of a complex mass of neural tissue protected by the skull. Anatomically, the brain comprises three main regions: the cerebrum, cerebellum, and brainstem, which work together, each with distinct functions. The brain receives information from at least one of the five senses and responds immediately after interpreting the information. The cerebrum is divided into two right and left hemispheres. Each hemisphere comprises four lobes: frontal, parietal, temporal and occipital. A brain tumor could be benign, pre-carcinoma, or malignant. The following summarizes widely used datasets in brain tumor research:

Cheng et al. [54] dataset includes three types of brain tumors: meningioma (708 slices), glioma (1,426 slices), and pituitary tumor (930 slices). Repository of Molecular Brain Neoplasia Data (REMBRANDT) [57] dataset provides molecular and clinical data, including 130 pre-surgical magnetic resonance multi sequence images. Sahoo et al. [59] dataset includes malignant and benign brain tumors affecting the central nervous system (CNS), skull base, and pituitary gland (Table 5).

Table 5. Brain Cancer Imaging Datasets

Dataset Name	Year	Size	Modality	Features	Availability
Cheng et al. [54]	2005 to 2010	3,064 images (233 patients)	MRI (T1W)	Meningioma, glioma, pituitary	Public
BraTS [55]	2012 to 2020	65 MR scans(from glioma patients)	MRI (multi-seq)	Segmentation challenge	Public
DB-PET-MRI [56]	2020	22 studies	PET/MRI	neurodegenerative or neuro functional pathologies	Restricted
REMBRANDT [57]	2004 to 2006	671 patients	MRI (multi-seq)	130 pre-surgical scans	Public
Rajesh et al. [58]	2013	20 MRI images	MRI	10 with tumors, 10 without	Public
OASIS [60]	–	Varies by year	MRI	A broad demographic, cognitive, and genetic spectrum	Public

8 Breast Cancer Datasets

Breast cancer is characterized by uncontrolled cell growth and spread. While this type of cancer occurs in both women and men, it primarily affects women. In the UK, approximately 54,700 women and about 390 men are diagnosed with breast cancer annually

[61, 62]. According to [63], in the United States, breast cancer represents the second most common cancer affecting women, following skin cancer. The average diagnostic age for women in Western societies is over sixty, whereas in the Middle East, particularly in Jordan, the average age at diagnosis is approximately 51 years [64, 65]. Early detection techniques for breast cancer include mammography, ultrasound scans, and magnetic resonance imaging (MRI). Mammography is widely employed as a screening test, utilizing X-ray technology to capture breast tissue images. However, in cases where mammography and ultrasound fail to detect a mass or tumour, histopathological tissue analysis is conducted using biopsies. Grading systems such as the Nottingham Grading System are used to evaluate the aggressiveness of breast cancer [66].

The following datasets are widely used for breast cancer research: The Aksac et al. [68] dataset comprises 162 microscopic biopsy images in TIFF format with a resolution of 1360 × 1024 pixels. These images, collected for educational purposes, are categorized into six classes: mitosis, apoptosis, tumor nuclei, non-tumor nuclei, tubule, and non-tubule. The Nottingham Grading System was used to assign cancer grades. The Moreira et al. [69] dataset contains six categories: normal mammograms, mammograms with masses, calcifications, architectural distortions, asymmetries, and multiple findings. Huang et al. [70] selected 106 breast mammography images with masses and applied data augmentation, increasing the dataset size to 7,632 images (Table 6).

Table 6. Breast Cancer Imaging Datasets

Dataset Name	Year	Size	Modality	Features	Availability
Al-Dhabyani et al. [67]	2020	780 images	Ultrasound	Normal, benign, malignant	Public
Aksac et al. [68]	2019	162 images	Histopathology	6 classes, 1360 × 1024 pixels	Public
Moreira et al. [69]	2008 to 2010	115 cases (410 images)	Mammography	6 categories	Public
Huang et al. [70]	2020	7,632 images (augmented)	Mammography	Masses	Public
BreaKHis [71]	2014	7,909 images (82 patients)	Histopathology	8 tumor types	Public
DDSM [72]	1999	2,620 cases	Mammography	Normal, benign, malignant	Public
Nijmegen [73]	Not specified	40 images (21 patients)	Mammography	105 microcalcification clusters	Public
BancoWeb [74]	2010	1,400 images (320 patients)	Mammography	include a functional search system	Public
MIAS [75]	2015	161 cases (322 images)	Mammography	50-μm resolution, PGM format	Public

(continued)

Table 6. (*continued*)

Dataset Name	Year	Size	Modality	Features	Availability

9 Ovarian Cancer Datasets

The ovary is an organ in the female reproductive system responsible for producing ova [76]. The female reproductive system contains two ovaries, one located on each side of the uterus. In [77], it the clarified that each ovary is approximately the size of an almond and produces both eggs (ova) and the hormones estrogen and progesterone. This cancer type is often called a "silent killer" because, in most cases, it remains undetected until it has spread within the pelvis and abdominal cavity [77]. According to [78], ovarian cancer is the fifth leading cause of cancer-related deaths, with over 21,500 women diagnosed annually and nearly 15,000 fatalities each year. Ultrasound imaging represents the most commonly employed technique for ovarian cancer detection.

The following datasets are commonly used in ovarian cancer research: Khazendar et al. [81] compiled a dataset from the International Ovarian Tumor Analysis (IOTA) study, consisting of 187 static ultrasound images of ovarian masses collected between 2005 and 2013. The dataset includes 112 benign and 75 malignant tumor images. Akazawa et al. [82] developed a dataset that includes 202 patients with ovarian tumors: 53 with ovarian cancers, 23 with borderline malignant tumors, and 126 with benign ovarian tumors. The median age of patients was 51 years, with a median tumor size of 9.4 cm. Acharya et al. [84] compiled a dataset of 469 non-consecutive women, where 238 were classified as non-suspicious and 231 as suspicious. This dataset categorizes masses into unilocular, multilocular, unilocular-solid, multilocular-solid, or solid types, with all images standardized to 256 × 256 pixel resolution (Table 7).

Table 7. Ovarian Cancer Imaging Datasets

Dataset Name	Year	Size	Modality	Features	Availability
Acharya et al. [79]	2013	2,600 images (20 women)	Ultrasound	1,300 benign, 1,300 malignant	Restricted
Carvalho et al. [80]	2019	107 images	Ultrasound (BRF)	5 image types, 512 × 512 pixels	Restricted
Khazendar et al. [81]	2005–2013	187 images	Ultrasound	112 benign, 75 malignant	IOTA
Akazawa et al. [82]	2013–2019	202 patients	MRI	Cancer, borderline, benign	Restricted
Srivastava et al. [83]	2020	240 images	Ultrasound	Normal, cysts	Restricted

(*continued*)

Table 7. (*continued*)

Dataset Name	Year	Size	Modality	Features	Availability
Acharya et al. [84]	2018	469 women	Ultrasound	238 non-suspicious, 231 suspicious	Restricted

10 Conclusion

This survey has highlighted the publicly available cancer imaging datasets, discussing their characteristics. By summarizing these datasets, we aim to assist researchers in selecting the most suitable resources for their studies. As AI and deep learning continue to develop, future research should focus on improving dataset quality, expanding annotations, and ensuring diversity in data to enhance generalization and clinical applicability.

References

1. Cleveland Clinic: Cancer. Available at: https://my.clevelandclinic.org/health/diseases/12194-cancer
2. World Health Organization: Cancer. Available at: https://www.who.int/health-topics/cancer#tab=tab_1
3. Armato III, S.G., et al.: The lung image database consortium (LIDC) and image database resource initiative (IDRI): a completed reference database of lung nodules on CT scans. Med. Phys. **38**(2), 915–931 (2011) Wiley Online Library
4. Jin, J., Jiang, Y., Zhao, Y., Huang, P.: Radiomics-based machine learning to predict the recurrence of hepatocellular carcinoma: a systematic review and meta-analysis. Acad. Radiol. **31**, 467–479 (2023)
5. Shiraishi, J., et al.: Development of a digital image database for chest radiographs with and without a lung nodule: receiver operating characteristic analysis of radiologists' detection of pulmonary nodules. Am. J. Roentgenol. **174**(1), 71–74 (2000) American Roentgen Ray Society
6. Thorat, M.B., Jain, N., Bhosale, A. S., Pawar, B.R.: Lung Cancer Cytology: Potential Pitfalls – A Review. Available at: https://core.ac.uk/download/ 230742016.pdf (2019)
7. Gindi, A., Al Attiatalla, T.A., Sami, M.M.: A comparative study for comparing two feature extraction methods and two classifiers in classification of early-stage lung cancer diagnosis of chest X-ray images. J. Am. Sci. **10**(6), 13–22 (2014)
8. McGraw, M., Sherlock, L., Bailey, K., Abman, S.: Developmental origins of chronic lung diseases: mechanical stretch, micro-RNAs, and hydrogels. Am. J. Respir. Cell Mol. Biol.. (Online). **59**(2), 267–270 (2018)
9. Reeves, A.P. et al.: A public image database to support research in computer-aided diagnosis. In: 2009 Annual International Conference of the IEEE Engineering in Medicine and Biology Society, pp. 3715–3718. IEEE (2009)
10. Cornell Vision and Image Analysis Group: Lung Database. Available at: https://www.via.cornell.edu/databases/lungdb.html

11. Gietema, H.A., Schilham, A.M., van Ginneken, B., van Klaveren, R.J., Lammers, J.W.J., Prokop, M.: Monitoring of smoking-induced emphysema with CT in a lung cancer screening setting: detection of real increase in extent of emphysema. Radiology. **244**(3), 890–897 (2007) Radiological Society of North America
12. Dolejsi, M., Kybic, J., Polovincák, M., Tuma, S.: The lung TIME: annotated lung nodule dataset and nodule detection framework. In: Medical Imaging 2009: Computer-Aided Diagnosis, vol. 7260, pp. 538–545. SPIE (2009)
13. Urology Care Foundation: Kidney Cancer. Available at: https://www.urologyhealth.org/urology-a-z/k/kidney-cancer
14. Comprehensive Urology: Kidney Cancer – Comprehensive Urology. Available at: http://www.urologist.org/new/kidney-cancer/
15. Feng, Z., Rong, P., Cao, P., et al.: Machine learning-based quantitative texture analysis of CT images of small renal masses. Eur. Radiol. **28**(4), 1625–1633 (2018)
16. Yu, H., Scalera, J., Khalid, M., et al.: Texture analysis as a radiomic marker for differentiating renal tumors. Abdom. Radiol. **42**(10), 2470–2478 (2017)
17. Hussain, M.A., Amir-Khalili, A., Hamarneh, G., et al.: Collage CNN for renal cell carcinoma detection from CT. In: International Workshop on Machine Learning in Medical Imaging, pp. 229–237 (2017)
18. Mileto, A., Marin, D., Alfaro-Cordoba, M., et al.: Iodine quantification to distinguish clear cell from papillary renal cell carcinoma at dual-energy CT. Radiology. **273**(3), 813–820 (2014)
19. Han, S., Hwang, S.I., Lee, H.J.: The classification of renal cancer in 3-phase CT images using a deep learning method. J. Digit. Imaging. **32**(4), 638–643 (2019)
20. Kocak, B., Yardimci, A.H., Bektas, C.T., et al.: Textural differences between renal cell carcinoma subtypes. Eur. J. Radiol. **107**, 149–157 (2018)
21. Choi, S.M., Choi, D.K., Kim, T.H., et al.: A comparison of radiologic tumor volume and pathologic tumor volume in RCC. PLoS One. **10**(3), e0122019 (2015)
22. Heller, N., Sathianathen, N., Kalapara, A., et al.: The kits19 challenge data. arXiv preprint. (2019)
23. Stanzione, A., Ricciardi, C., Cuocolo, R., et al.: MRI radiomics for the prediction of Fuhrman grade in clear cell RCC. J. Digit. Imaging. **33**(4), 879–887 (2020)
24. Nazari, M., Shiri, I., Zaidi, H.: Radiomics-based machine learning model for RCC survival prediction. Comput. Biol. Med. **129**, 104135 (2021)
25. https://www.stanfordchildrens.org/en/topic/default?id= anatomy-and-function-of-the-liver-90-P03069
26. https://www.mayoclinic.org/diseases-conditions/liver-cancer/ symptoms-causes/syc-20353659
27. Das, A., Acharya, U.R., Panda, S.S., Sabut, S.: Deep learning-based liver cancer detection using watershed transform and Gaussian mixture model techniques. Cogn. Syst. Res. **54**, 165–175 (2019)
28. Ahn, S.H., Yeo, A.U., Kim, K.H., et al.: Comparative evaluation of atlas and deep learning-based auto-segmentation for liver cancer. Radiat. Oncol. **14**, 213 (2019)
29. Ben-Cohen, A., Klang, E., Kerpel, A., et al.: Fully convolutional network and dictionary learning for liver lesion detection. Neurocomputing. **275**, 1585–1594 (2018)
30. Frid-Adar, M., Diamant, I., Klang, E., et al.: GAN-based synthetic medical image augmentation for liver lesion classification. Neurocomputing. **321**, 321–331 (2018)
31. Zhang, F., Yang, J., Nezami, N., et al.: Liver tissue classification using an auto-context-based deep neural network. In: Lecture Notes Computer Science (2018)
32. Li, S., Jiang, H., Yao, Y., et al.: Structure convolutional extreme learning machine for HCC nucleus segmentation. Neurocomputing. **312**, 9–26 (2018)
33. Yasaka, K., Akai, H., Abe, O., Kiryu, S.: Deep learning for differentiation of liver masses. Radiology. (2018)

34. Vivanti, R., Joskowicz, L., Lev-Cohain, N., et al.: Patient-specific and global CNNs for liver tumor delineation. Med. Biol. Eng. Comput. (2018)
35. Bharti, P., Mittal, D., Ananthasivan, R.: Preliminary study of chronic liver classification. Ultrason. Imaging. (2018)
36. Trivizakis, E., Manikis, G.C., Nikiforaki, K., et al.: Extending 2D CNNs to 3D for cancer classification. IEEE J. Biomed. Health Inform. (2018)
37. Dubus, L., Gayet, M., Zappa, M., et al.: Comparison of semi-automated and manual liver tumor volume measurements. Eur. Radiol. (2010)
38. Kim, Y.J., Jang, H., Lee, K., et al.: PAIP 2019: liver cancer segmentation challenge. Med. Image Anal. **67**, 101854 (2021)
39. Anter, A.M., Hassenian, A.E.: CT liver tumor segmentation hybrid approach. Artif. Intell. Med. (2018)
40. Reddy, D.S., Bharath, R., Rajalakshmi, P.: A novel CAD framework using deep learning for FLD classification. IEEE Healthcom. (2018)
41. https://www.cancer.gov/types/head-and-neck/head-neck-fact-sheet
42. Gupta, P., Malhi, A.K.: Using deep learning to enhance head and neck cancer diagnosis and classification. In: 2018 IEEE International Conference on System, Computation, Automation and Networking (ICSCAN) (2018)
43. Zhao, L., Lu, Z., Jiang, J., et al.: Automatic nasopharyngeal carcinoma segmentation using fully convolutional networks. J. Digit. Imaging. (2019)
44. Bielak, L., Wiedenmann, N., Berlin, A., et al.: Convolutional neural networks for head and neck tumor segmentation. Radiat. Oncol. (2020)
45. Ye, Y., Cai, Z., Huang, B., et al.: Fully-automated segmentation of nasopharyngeal carcinoma using convolutional neural networks. Front. Oncol. (2020)
46. Ma, Z., Wu, X., Song, Q., et al.: Automated nasopharyngeal carcinoma segmentation using convolutional neural networks. Exp. Ther. Med. (2018)
47. Li, Q., Xu, Y., Chen, Z., et al.: Tumor segmentation in contrast-enhanced MRI for NPC: deep learning with CNN. Biomed. Res. Int. (2018)
48. Men, K., Chen, X., Zhang, Y., et al.: Deep deconvolutional neural network for target segmentation of nasopharyngeal cancer. Front. Oncol. (2017)
49. Daoud, B., Morooka, K.I., Kurazume, R., et al.: 3D segmentation of NPC using cascade deep learning. Comput. Med. Imaging Graph. (2019)
50. Vallières, M., et al.: Radiomics strategies for risk assessment of tumor failure in HNC. Sci. Rep. (2017)
51. Moe YM, et al. Deep learning for automatic tumor segmentation in PET/CT. arXiv preprint 2019.
52. Jin, D., Guo, D., et al.: Accurate esophageal tumor segmentation using 3D deep network fusion. In: International Conference on Medical Image Computing and Computer-Assisted Intervention (2019)
53. https://www.healthline.com/human-body-maps/brain
54. Cheng, J., Huang, W., Cao, S., Yang, R., et al.: Enhanced performance of brain tumor classification via tumor region augmentation and partition. PLoS One. **10**(10), e0140381 (2015)
55. Menze, B.H., Jakab, A., Bauer, S., et al.: The multimodal brain tumor image segmentation benchmark (BRATS). IEEE Trans. Med. Imaging. **34**(10), 1993–2024 (2015)
56. Buenamaizón, R., Quinteros, E., Graffigna, J.P., et al.: Database of PET/MRI studies of patients with brain pathologies. Revista Argentina de Bioingeniería. **24**(1), 55–59 (2020)
57. Scarpace, L., Flanders, A.E., Jain, R., et al.: Data from REMBRANDT. Cancer Imaging Arch. (2015)

58. Rajesh, T., Malar, R.S.: Rough set theory and feed forward neural network-based brain tumor detection in MRI. In: IEEE International Conference on Advanced Nanomaterials & Emerging Engineering Technologies (2013)
59. Sahoo, L., Sarangi, L., Dash, B.R., et al.: Detection and classification of brain tumor using MRI. In: Advances in Electrical Control and Signal Systems (2020)
60. Washington University in St. Louis.: OASIS Brains Project. Available at: https://sites.wustl.edu/oasisbrains/
61. https://www.cancerresearchuk.org/about-cancer/breast-cancer/about
62. https://www.nhs.uk/conditions/breast-cancer/
63. https://www.mayoclinic.org/diseases-conditions/breast-cancer/symptoms-causes/syc-20352470
64. Abdel-Razeq, H., Zaru, L., Badeeb, A., Hijjawi, S.: The application of Gail model to predict the risk of developing breast cancer among Jordanian women. J. Oncol. (2020)
65. Shulman, L.N., Willett, W., Sievers, A., Knaul, F.M.: Breast cancer in developing countries: opportunities for improved survival. J. Oncol. **2010**, 1–6 (2010)
66. Park, H.L., Hong, J.: Vacuum-assisted breast biopsy for breast cancer. Gland Surg. (2014)
67. Al-Dhabyani, W., Gomaa, M., Khaled, H., Fahmy, A.: Dataset of breast ultrasound images. Data Brief. **28**, 104863 (2020)
68. Aksac, A., Demetrick, D.J., Ozyer, T., Alhajj, R.: BreCaHAD: a dataset for breast cancer histopathological annotation and diagnosis. BMC Res. Notes. **12**, 82 (2019)
69. Moreira, I.C., Amaral, I., Domingues, I., Cardoso, A., Cardoso, M.J., Cardoso, J.S.: Inbreast: toward a full-field digital mammographic database. Acad. Radiol. (2012)
70. Huang, M.L., Lin, T.Y.: Dataset of breast mammography images with masses. Data Brief. **31**, 105928 (2020)
71. Spanhol, F.A., Oliveira, L.S., Petitjean, C., Heutte, L.: A dataset for breast cancer histopathological image classification. In: IEEE Transactions on Biomedical Engineering (2015)
72. Lee, R.S., Gimenez, F., Hoogi, A., Miyake, K.K., Gorovoy, M., Rubin, D.L.: A curated mammography dataset for computer-aided detection and diagnosis. Sci. Data. (2017)
73. Marrocco, C., Molinara, M., D'Elia, C., Tortorella, F.: A computer-aided detection system for clustered microcalcifications. Artif. Intell. Med. **50**, 23–32 (2010)
74. Matheus, B.R., Schiabel, H.: Online mammographic images database for CAD development. J. Digit. Imaging. (2011)
75. Suckling, J., et al.: Mammographic Image Analysis Society (MIAS) Database v1.21. (2015)
76. https://www.cancer.gov/publications/dictionaries/cancer-terms/def/ovary
77. https://www.mayoclinic.org/diseases-conditions/ovarian-cancer/symptoms-causes/syc-20375941
78. https://www.eurekalert.org/pub_releases/2010-01/acor-up011910.php
79. Acharya, U.R., Mookiah, M.R., Sree, S.V., et al.: Evolutionary algorithm-based classifier parameter tuning for ovarian cancer characterization. In: Ovarian Neoplasm Imaging. Springer (2013)
80. Carvalho, C., Marques, S., Peixoto, C., et al.: Deep learning approaches for gynecological ultrasound image segmentation. In: International Conference on Image Analysis and Recognition. Springer (2019)
81. Khazendar, S., Sayasneh, A., Al-Assam, H., et al.: Automated characterization of ultrasound images of ovarian tumors using machine learning. Facts Views Vision ObGyn. **7**, 7–15 (2015)
82. Akazawa, M., Hashimoto, K.: Artificial intelligence in ovarian cancer diagnosis. Anticancer Res. **40**, 4795–4800 (2020)
83. Srivastava, S., Kumar, P., Chaudhry, V., et al.: Detection of ovarian cysts in ultrasound images using a fine-tuned VGG-16 deep learning network. SN Comput. Sci. (2020)
84. Acharya, U.R., Akter, A., Chowriappa, P., et al.: Use of nonlinear features for automated characterization of ovarian tumors in ultrasound images. Int. J. Fuzzy Syst. (2018)

Improving MRI Brain Tumor Classification with Simulated Annealing in a Convolutional Neural Network

Sofia El Amoury[1,2], Youssef Smili[2](✉), and Youssef Fakhri[1]

[1] RI, Faculty of Sciences of Kenitra, Ibn Tofail University, Kenitra, Morocco
{sofia.elamoury,fakhri}@uit.ac.ma

[2] ENIC, Faculty of Sciences and Technologies of Settat, Hassan First University, Settat, Morocco
smili.fst@uhp.ac.ma

Abstract. The classification of brain tumors through MRI scans continues to present significant difficulties, primarily due to the intricate nature of tumors themselves and the subjective variations inherent in human-led analysis. Convolutional Neural Networks (CNNs) have advanced automated medical image analysis, but their performance relies heavily on computationally intensive hyperparameter optimization. In this study, we leverage Simulated Annealing (SA)—a probabilistic metaheuristic algorithm—to systematically refine the hyperparameters of a CNN architecture specifically tailored for brain tumor detection in MRI analyses. The proposed framework was rigorously tested on a curated dataset of 7023 clinical MRI scans spanning four diagnostic categories: glioma, meningioma, pituitary tumors, and non-tumor cases, achieved a validation accuracy exceeding 97%, demonstrating SA's efficacy in enhancing model performance. These results emphasize the critical role of advanced hyperparameter tuning in improving diagnostic accuracy, offering a pathway toward more reliable clinical tools for tumor classification.

Keywords: brain tumor classification · convolutional neural networks (CNNs) · simulated annealing (SA) · hyperparameter optimization · automated tumor diagnosis

1 Introduction

Brain tumors, marked by uncontrolled cellular growth within the brain or adjacent tissues, impose a significant global health burden due to their potential to impair neurological function, diminishing quality of life, and increasing mortality rates. These tumors are broadly categorized into three primary types—Meningiomas, Gliomas, and Pituitary tumors—each differing in origin, biological behavior, and clinical impact [1]. Meningiomas, typically benign lesions arising from the meninges, are the most frequently encountered primary brain tumors and often induce neurological deficits through mass effect on adjacent structures

B.-E. Boudriki Semlali et al. (Eds.): DATA 2025, LNNS 1908, pp. 203–213, 2026.
https://doi.org/10.1007/978-3-032-21200-9_17

[2]. In contrast, gliomas originate from glial cells and are typically malignant, with subtypes like glioblastoma multiforme demonstrating rapid, aggressive progression and notable resistance to treatment [3]. Although pituitary tumors are usually non-malignant, disrupt endocrine homeostasis by altering hormone secretion, thereby triggering systemic complications such as metabolic irregularities and reproductive disorders [4]. Accurate and timely diagnosis remains challenging, hindered by the intricate architecture of brain, the morphological diversity of tumors, and the subtlety of imaging differences that separate pathological from healthy regions. While magnetic resonance imaging (MRI) is the cornerstone of diagnostic evaluation, relying on advanced techniques like contrast-enhanced and diffusion-weighted imaging, the manual analysis of these scans remains time-consuming and prone to inconsistencies across radiologists due to inherent subjectivity in human-led assessments. This limitation underscores the critical need for robust, automated diagnostic solutions that leverage artificial intelligence, particularly deep learning and convolutional neural networks (CNNs), to enhance diagnostic accuracy, accelerate image analysis, and support clinical decision-making.

The transformative potential of CNNs in image-based tasks has been amplified by unprecedented advancements in computational infrastructure and information technology, catalyzing breakthroughs in artificial intelligence and machine learning (ML). Characterized by their hierarchical architecture and capacity for automated feature extraction, CNNs excel at discerning intricate patterns within visual data, making them indispensable for tasks requiring high-dimensional analysis and classification. In medical imaging, for instance, CNNs have been instrumental in the early detection and diagnosis of pathologies by effectively analyzing complex imaging modalities, which is critical for timely clinical interventions [5].

Despite their transformative potential, the design and optimization of CNNs are inherently constrained by their reliance on hyperparameters, which govern critical aspects of model architecture and training dynamics. These parameters—including the depth and configuration of convolutional layers, selection of activation functions (e.g., ReLU, sigmoid), learning rates, batch sizes, and dropout rates—remain static during training, yet their calibration profoundly impacts model generalizability, convergence speed, and computational efficiency. Manual tuning of these variables is often infeasible due to the combinatorial explosion of possible configurations, particularly as CNN architectures grow in complexity to address specialized tasks, such as medical image analysis. This challenge has spurred the development of automated Hyperparameter Optimization (HPO) frameworks, which employ systematic search strategies to iteratively explore the hyperparameter space and identify optimal settings. By minimizing human intervention, HPO not only accelerates model development but also enhances reproducibility and uncovers configurations that may elude manual experimentation [6].

In this study, we employ Simulated Annealing Method (SA), a metaheuristic optimization algorithm inspired by the thermodynamic annealing process, as our

HPO strategy to systematically calibrate a CNN for MRI-based brain tumor categorization. SA's stochastic search strategy uniquely addresses the non-convex, high-dimensional hyperparameter space of CNNs by probabilistically accepting suboptimal configurations during its exploration phase. This controlled randomization enables SA to escape local minima—a common pitfall in gradient-based methods—while systematically narrowing the search scope through an adaptive temperature parameter. Unlike population-based approaches, SA operates with a single candidate solution, significantly reducing computational overhead and memory demands. This non-population strategy is particularly advantageous in medical imaging tasks, where training complex CNNs on MRI data already imposes substantial resource constraints.

2 Related Work

To overcome limitations in manual tumor diagnosis, recent advancements leverage stochastic metaheuristic frameworks to automate CNN optimization, achieving higher accuracy in MRI scan interpretation. For instance, Anaraki et al. [7] utilized genetic algorithms to evolve CNN designs for glioma grading and multi-class tumor classification, achieving accuracies of 90.9% and 94.2%, respectively. While effective, their approach encountered challenges in managing the expansive search space, leading to overly complex models with high computational demands. Similarly, El Amoury et al. [8,9] applied particle swarm optimization to hyperparameter optimization, attaining validation accuracies of 96.64% and 96.8%. Despite these results, PSO-based methods often struggle with premature convergence in high-dimensional spaces, limiting their ability to balance exploration and exploitation.

Celik et al. [10] proposed a hybrid model combining a custom CNN with classical machine learning classifiers, optimized via Bayesian optimization, achieving a mean accuracy of 97.15%. While Bayesian optimization excels in sample efficiency, its reliance on probabilistic surrogate models can become computationally prohibitive for large-scale or non-differentiable search spaces. Bansal et al. [11] reported exceptional accuracies using a CNN-SVM framework, reporting 99% for binary and 98% for multi-class tasks, though their work did not emphasize hyperparameter optimization, leaving room for further refinement. Pre-trained models like InceptionV3, achieved 97.12% accuracy, yet their reliance on transfer learning may hinder adaptability to domain-specific MRI data [12]. Ait Amou et al. [13] further demonstrated the promise of Bayesian optimization in CNN tuning, reaching 98.70% validation accuracy.

3 Convolutional Neural Networks (CNNs)

CNNs have become indispensable tools in medical imaging, particularly for brain tumor classification. Their ability to discern intricate patterns surpasses that of alternative methods, significantly enhancing diagnostic accuracy [14]. A standard CNN framework comprises several distinct layers: convolutional layers, which extract spatial features from images; pooling layers, which reduce dimensionality without sacrificing critical details; and fully connected layers, which consolidate these features to make final classification decisions. Each layer is built from neurons associated with weights and biases; these neurons perform mathematical operations on their inputs by combining them with respective weights and subsequently applying an activation function. CNNs operations can be summarized as:

- The input image X, represented as a tensor, is fed into the network and sequentially processed through each layer.
- At each layer j, the operation computes Z_j by combining the output from the preceding layer with the corresponding weights W_j and biases b_j.
- Subsequently, the activation function $f_j(Z_j)$ is applied to Z_j to produce the layer's output O_j.

This process continues throughout the network until the final layer generates a prediction. Activation functions play a critical role by introducing nonlinearity into the network, which is essential for modeling and capturing the complex relationships among features. For multiclass classification tasks, the softmax cross-entropy loss serves as a critical error metric to measure the divergence between a model's probabilistic outputs and ground-truth label assignments, thereby directing the optimization process. During training, backpropagation is employed to iteratively adjust the network's weights with the objective of minimizing the loss function. To mitigate overfitting, regularization techniques such as dropout are commonly incorporated, which reduces the network's reliance on specific neurons [15]. Several hyperparameters [16], significantly influence CNNs performance:

- **Batch size:** This determines how many samples are processed in each training iteration, affecting both training efficiency and computational resource usage.
- **Learning rate:** Determines the magnitude of weight updates, impacting learning speed.
- **Optimizers:** Algorithms like SGD with momentum, Adam, and RMSProp manage weight updates during training.
- **Number of filters and filter sizes:** These hyperparameters dictate the extent and granularity of feature extraction in convolutional layers, with larger numbers and sizes capturing more complex patterns.
- **Dropout rates:** Specifies the proportion of neurons excluded during training to improve generalization.
- **Activation functions:** Functions like ReLU, LeakyReLU, and ELU introduce non-linearity.
- **Fully connected neurons:** The number of neurons within these layers directly influences the model's representational power, balancing its aptitude for discerning intricate spatial features, thereby impacting its overall performance.

4 Simulated Annealing Method (SA)

Simulated Annealing Method (SA) is a robust optimization technique modeled after the physical annealing process in metallurgy. In this process, a material is gradually heated and then slowly cooled to reduce defects and achieve a stable, low-energy state. Introduced by Kirkpatrick et al. (1983), SA has gained widespread recognition for its effectiveness in avoiding local minima and identifying near-optimal solutions to complex optimization challenges [17].

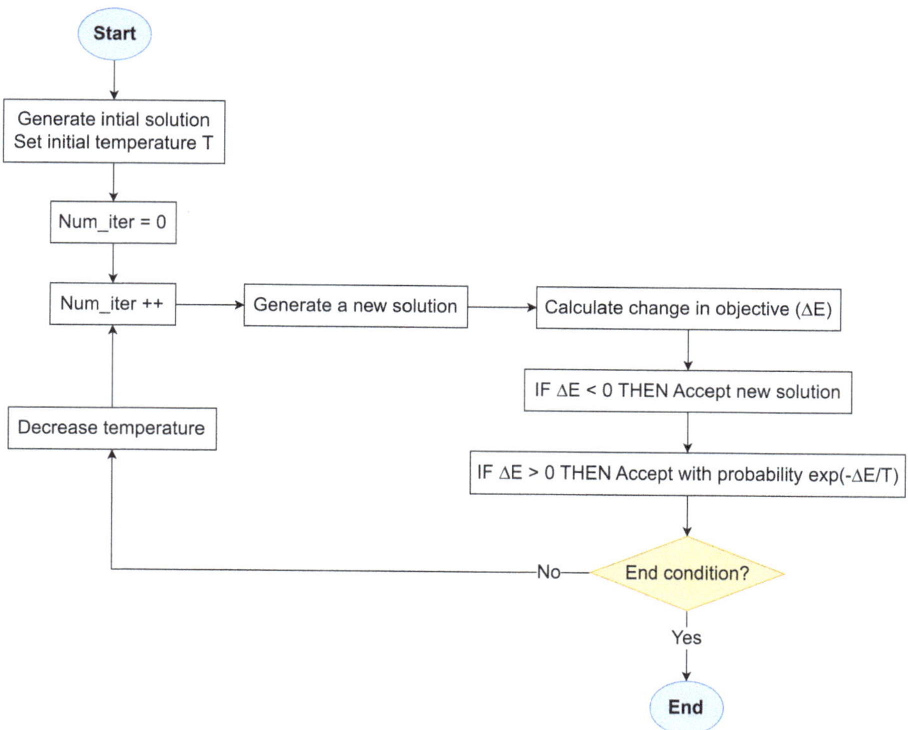

Fig. 1. Simulated Annealing Process Flowchart.

Figure 1 outlines the workflow of simulated annealing (SA). The algorithm begins with an initial solution and a high *temperature* parameter, which regulates the acceptance of suboptimal states. Unlike deterministic approaches, SA employs random elementary transformations (moves) to generate neighboring solutions. Each move induces an energy change $\Delta E = E_{\text{new}} - E_{\text{current}}$, where lower energy signifies improved fitness. The Metropolis criterion determines acceptance: moves with $\Delta E \leq 0$ are always adopted, while those increasing energy ($\Delta E \geq 0$) are accepted probabilistically via $e^{-\Delta E/T}$. This rule, rooted in statistical mechanics, ensures configurations asymptotically follow the

Maxwell-Boltzmann distribution, mirroring equilibrium thermodynamics. Early high-temperature phases favor exploratory behavior by tolerating unfavorable moves, enabling escape from local minima. As temperature decreases according to a predefined cooling schedule, the search transitions toward exploitation, increasingly prioritizing fitness improvements. The process terminates when further cooling yields negligible energy changes, ideally converging near the global optimum. While slower than basic iterative improvement, SA's stochastic acceptance mechanism enhances robustness by balancing exploration and exploitation, a trade-off governed by temperature dynamics [18].

5 SA Applied to CNNs HPO

For optimizing CNNs hyper-parameters, SA begins with an initial solution, representing specific values for parameters like learning rate, optimizer type, number of filters, filter sizes, dropout rate, activation functions, and the number of neurons. These hyperparameters directly represent the solution, with no need for additional encoding. In each move, every hyperparameter has a 50% chance to alter to the next entry in its cyclic list of permitted values. The energy function is inversely correlated with validation accuracy, meaning that lower energy indicates improved performance. An exponential cooling schedule progressively reduces exploration flexibility, with each temperature evaluating one neighbor. The process terminates after a maximum iteration threshold.

The proposed algorithm also excels at balancing model performance and complexity. It assesses solutions not only by their validation accuracy but also by the number of parameters they contain. A new solution is accepted if it offers greater validation accuracy. Additionally, if the validation accuracies are similar but the new solution has fewer parameters, it is still accepted, thereby fostering simplicity and efficiency. In cases where neither of these conditions is satisfied, the algorithm may still accept the new solution based on the Metropolis criterion.

6 Experimental Results

6.1 Dataset

For the purposes of this study, an open-access brain MRI dataset from Kaggle [19], was utilized. This dataset aggregates scans from three repositories: Figshare, SARTAJ, and Br35H. The composite dataset comprises 7023 grayscale images in JPG format, divided into four categories: glioma (1621 total images), meningioma (1645 total images), pituitary tumor (1757 total images), and no tumor (2000 total images). The data is partitioned into training and testing subsets following an 81.3:18.7 ratio. Specific class-wise splits include 1321 training and 300 testing samples for glioma, 1339 training and 306 testing samples for meningioma, 1457 training and 300 testing samples for pituitary tumors, and 1595 training and 405 testing samples for healthy scans. Figure 2 displays sample images from each class.

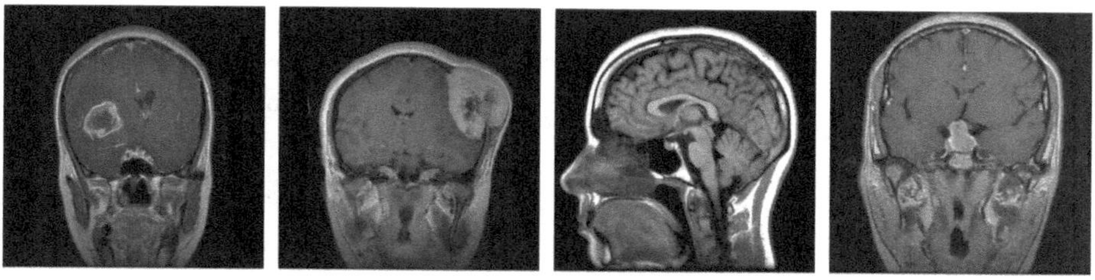

(a) Glioma (b) Meningioma (c) No tumor (d) Pituitary

Fig. 2. Sample images from each class in the dataset.

6.2 Algorithm Parameters

The parameters governing our proposed framework are detailed in Table 1. Our CNN model adopts a streamlined topology comprising two convolutional layers, succeeded by two fully connected (FC) layers. To enhance generalization, dropout regularization is strategically integrated prior to the final classification layer, effectively minimizing overfitting risks.

Table 1. Algorithm Parameters

Parameter	Values
Maximum Iterations	20
Initial Temperature	1.0
Cooling Rate	0.5
Epochs per Evaluation	Starts at 1, increases by 1 each iteration
Maximum Training Epochs	20
Number of Filters	8, 16, 32, 64
Filter Sizes	3, 5, 7
Number of Neurons (FC)	32, 64, 128
Dropout Rates	0.1 to 0.3 in increments of 0.01
Activation Functions	ReLU, LeakyReLU, ELU
Learning Rates	0.0001 to 0.001 in increments of 0.0001
Optimizers	SGD, Adam, RMSProp

6.3 Results and Discussion

Figure 3 depicts the evolution of validation accuracy achieved by the current CNN model over 20 iterations of the proposed algorithm. The results indicate a marked improvement, with accuracy increasing from approximately 65.67% in the initial iteration to around 97.25% by the 14th iteration, after which performance stabilizes. The optimal set of hyperparameters identified through this process is presented in Table 2. Training was performed using a batch size of 32 images.

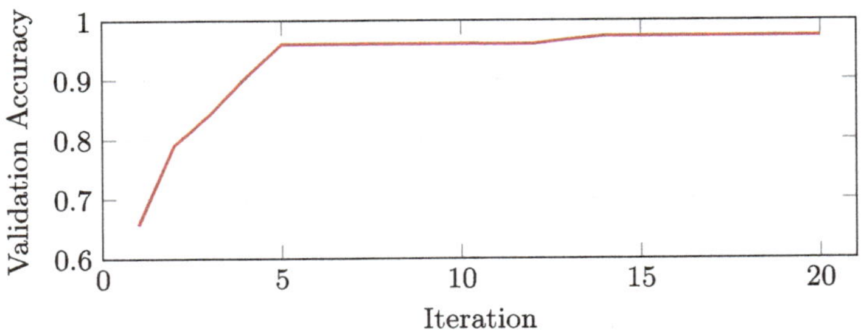

Fig. 3. Validation accuracies across iterations.

Table 2. Best Solution Found

Parameter	Values
Number of Filters	8, 16
Kernel Sizes	5, 3
Number of Neurons (FC)	64
Dropout Rates	0.2
Activation Functions	LeakyReLU
Learning Rate	0.0006
Optimizer	RMSProp
Number of Parameters	3213364

Figure 4 displays the loss progression for the optimal solution identified. The training loss exhibits a steady decrease over successive epochs, indicating that the model effectively reduces prediction errors as training advances. Concurrently, the validation loss follows a similar declining pattern, implying that the model maintains robust generalization capabilities when applied to unseen data.

Figure 5 illustrates the accuracy trends associated with the optimal model configuration. The training accuracy shows a consistent improvement, eventually reaching 98.25%. In parallel, the validation accuracy steadily rises, peaking at approximately 97.25%, which reflects the network's robust generalization capability. The narrow margin between the training and validation curves further highlights the network's robustness and reliability.

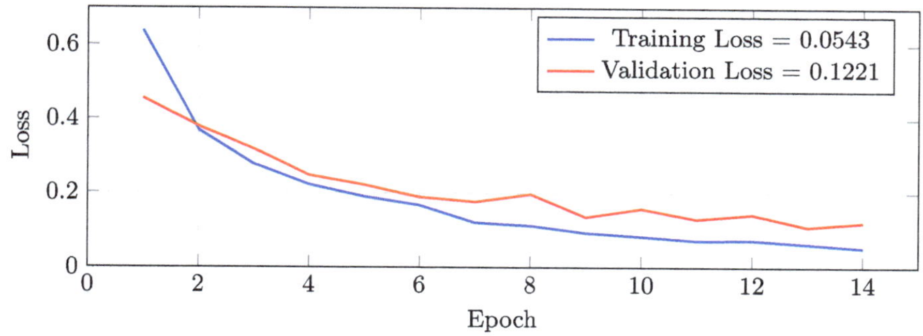

Fig. 4. Training and validation loss progression of the best-obtained solution.

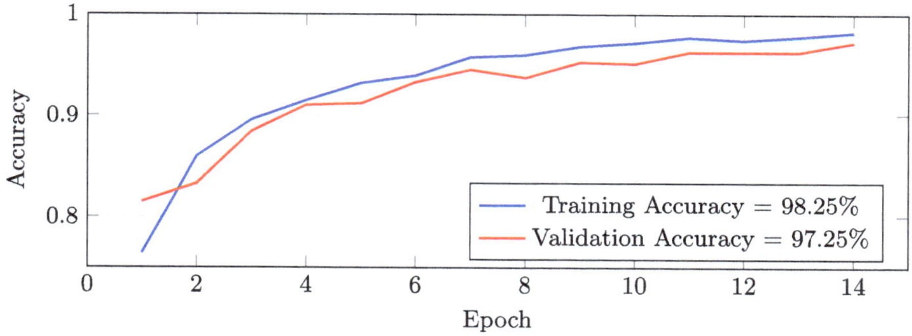

Fig. 5. Training and validation accuracy progression of the best-obtained solution.

6.4 Results Compared to Related Work

Table 3 presents a comparison of validation accuracy across different approaches. The developed framework achieved 97.25% validation accuracy, closely trailing the top-performing Bayesian-optimized convolutional frameworks (98.70% and 97.93%) while surpassing other contemporary strategies. This positions the proposed approach as a competitive alternative to existing neural network enhancement and hyperparameter tuning methodologies.

Table 3. Validation Accuracy Comparison

Approach	Validation Accuracy (%)
Genetic Algorithm-Enhanced CNNs [7]	94.2
Particle Swarm-Optimized CNNs (I) [8]	96.64
Particle Swarm-Optimized CNNs (II) [9]	96.8
Advanced CNN-Integrated ML Framework [10]	97.93
Transfer Learning-Integrated CNNs [12]	97.12
Bayesian Optimization-Enhanced CNNs [13]	98.70
Proposed Framework	**97.25**

7 Conclusion

The integration of Simulated Annealing Method (SA) into hyperparameter optimization for Convolutional Neural Networks offers a methodical strategy to enhance classification accuracy in MRI-based brain tumor diagnostics. By systematically navigating the hyperparameter landscape, SA efficiently identifies configurations that maximize model performance. Despite its efficacy, the approach incurs notable computational costs due to repeated training cycles, especially with complex architectures and high-volume imaging data. While SA showcases promise in refining diagnostic precision, its resource-intensive nature underscores the importance of exploring adaptive techniques—such as parallelization or early stopping mechanisms—to mitigate computational overhead. Future advancements could focus on hybrid methodologies that balance exploration and efficiency, ensuring scalability without sacrificing the rigor of hyperparameter optimization.

References

1. Tariq, A., Iqbal, M.M., Iqbal, M.J., Ahmad, I.: Transforming brain tumor detection empowering multi-class classification with vision transformers and efficientnetv2. IEEE Access 1–1 (2025). https://doi.org/10.1109/ACCESS.2025.3555638
2. Maggio, I., et al.: Meningioma: not always a benign tumor. a review of advances in the treatment of meningiomas. CNS Oncol. **10**(2), CNS72 (2021). https://doi.org/10.2217/cns-2021-0003
3. Grochans, S., et al.: Epidemiology of glioblastoma multiforme–literature review. Cancers **14**(10) (2022). https://doi.org/10.3390/cancers14102412
4. Tsukamoto, T., Miki, Y.: Imaging of pituitary tumors: an update with the 5th WHO classifications—part 1. pituitary neuroendocrine tumor (PitNET)/pituitary adenoma. Japan. J. Radiol. **41**(8), 789–806 (2023). https://doi.org/10.1007/s11604-023-01400-7
5. Abid, M.A., Munir, K.: A systematic review on deep learning implementation in brain tumor segmentation, classification and prediction. Multimedia Tools Appl. 1–40 (2025). https://doi.org/10.1007/s11042-025-20706-4
6. Bohmrah, M.K., Kaur, H.: Advanced hybridization and optimization of DNNs for medical imaging: a survey on disease detection techniques. Artif. Intell. Rev. **58**(4), 122 (2025). https://doi.org/10.1007/s10462-024-11049-x
7. Anaraki, A.K., Ayati, M., Kazemi, F.: Magnetic resonance imaging-based brain tumor grades classification and grading via convolutional neural networks and genetic algorithms. Biocybern. Biomed. Eng. **39**(1), 63–74 (2019). https://doi.org/10.1016/j.bbe.2018.10.004
8. El Amoury, S., Smili, Y., Fakhri, Y.: Optimization of convolutional neural network architecture by PSO algorithm for MRI brain tumor image classification. In: 2024 Sixth International Conference on Intelligent Computing in Data Sciences (ICDS), pp. 1–8 (2024). https://doi.org/10.1109/ICDS62089.2024.10756503
9. El Amoury, S., Smili, Y., Fakhri, Y.: Design of an optimal convolutional neural network architecture for MRI brain tumor classification by exploiting particle swarm optimization. J. Imaging **11**(2) (2025).https://doi.org/10.3390/jimaging11020031

10. Celik, M., Inik, O.: Development of hybrid models based on deep learning and optimized machine learning algorithms for brain tumor multi-classification. Expert Syst. Appl. **238**, 122159 (2024). https://doi.org/10.1016/j.eswa.2023.122159
11. Bansal, S., Jadon, R.S., Gupta, S.K.: A robust hybrid convolutional network for tumor classification using brain mri image datasets. Int. J. Adv. Comput. Sci. Appl. **15**(4) (2024). https://doi.org/10.14569/IJACSA.2024.0150459
12. Gómez-Guzmán, M.A., et al.: Classifying brain tumors on magnetic resonance imaging by using convolutional neural networks. Electronics **12**(4) (2023). https://doi.org/10.3390/electronics12040955
13. Ait Amou, M., Xia, K., Kamhi, S., Mouhafid, M.: A novel mri diagnosis method for brain tumor classification based on cnn and bayesian optimization. Healthcare **10**(3) (2022). https://doi.org/10.3390/healthcare10030494
14. Azeez, O., Abdulazeez, A.: Classification of brain tumor based on machine learning algorithms: a review. J. Appl. Sci. Technol. Trends **6**(1), 01–15 (2025). https://doi.org/10.38094/jastt61188
15. Igual, L., Seguí, S.: Deep Learning, pp. 211–232. Springer, Cham (2024). https://doi.org/10.1007/978-3-031-48956-3_11
16. Thakur, A., Biswas, S.K., Majumdar, S., Thounaojam, D.M.: A comprehensive review on different cnn architectures. In: 2025 3rd International Conference on Intelligent Data Communication Technologies and Internet of Things (IDCIoT), pp. 1997–2004 (2025). https://doi.org/10.1109/IDCIOT64235.2025.10914792
17. Chopard, B., Tomassini, M.: Simulated Annealing. In: An Introduction to Metaheuristics for Optimization. NCS, pp. 59–79. Springer, Cham (2018). https://doi.org/10.1007/978-3-319-93073-2_4
18. El Amoury, S., Smili, Y., Fakhri, Y.: Cnn hyper-parameter optimization using simulated annealing for mri brain tumor image classification. In: 2025 5th International Conference on Innovative Research in Applied Science, Engineering and Technology (IRASET) (2025)
19. Msoud, N.: Brain tumor MRI dataset. Kaggle (2021). https://www.kaggle.com/dsv/2645886. Accessed 9 Mar 2025

Caries-TransUNet: An Enhanced TransUNet Model for Dental Caries Segmentation in Panoramic Radiography

Manal Touahri(✉) and Aissam Berrahou

Mohammed V University, Rabat, Morocco
manal_touahri@um5.ac.ma

Abstract. This study aims to develop an enhanced version of the TransUNet architecture to improve the segmentation accuracy across various caries types in panoramic dental X-rays. The DC1000 dental caries dataset was used for this study. We propose Caries-TransUNet, an enhanced version of TransUNet, designed to improve the identification and segmentation of diverse caries types in dental panoramic radiography. The proposed model's performance was compared against several state-of-the-art models, including U-Net, Attention UNet, DeepLabv3+ and TransUNet, using standard metrics such as Accuracy, Dice score, Intersection over Union (IoU), Precision and Recall. The proposed model attained an Accuracy of 99.81%, Dice score of 77.88%, IoU of 63.85%, Precision at 81.37% and Recall reaching 76.48% surpassing TransUNet and other state-of-the-art models consistently in Dice score, IoU score and Recall while demonstrating robust segmentation of all three types of caries across multiple X-ray images.

Keywords: Dental caries segmentation · Panoramic radiography · Deep learning · Transformer · Convolutional neural network

1 Introduction

Dental caries is the leading condition affecting oral health in both children and adults [6] with a profound impact on overall well-being as it can damage tooth structure. It is mainly caused by oral bacteria that generate lactic acid, leading to enamel demineralization over time. Although dental caries tend to develop gradually, without timely and proper intervention they may develop into a significant health concern leading to infection, pain and tooth loss [16]. Therefore, preventing and detecting caries at an early stage is crucial in dental care. A thorough evaluation of the patient's oral condition is essential for establishing an accurate diagnosis and an effective treatment plan. This process largely relies on radiographic imaging and the clinical expertise of the dentist [12].

Various radiographic imaging techniques including periapical radiographs, cone beam computed tomography (CBCT), bitewing radiographs, panoramic radiography and digital radiography, are available and are chosen by dentists based on the patient's condition and diagnostic needs. Panoramic X-ray imaging is widely used [11] due to its

B.-E. Boudriki Semlali et al. (Eds.): DATA 2025, LNNS 1908, pp. 214–223, 2026.
https://doi.org/10.1007/978-3-032-21200-9_18

low radiation exposure, cost-effectiveness, and enhanced patient comfort. While it may not present detailed assessment of individual teeth it still offers a baseline oral health assessment. Moreover, the analysis of dental panoramic X-rays remains a demanding task due to issues like excessive noise, low contrast and resolution, overlapping structures and soft tissue artifacts.

In recent years, incorporating artificial intelligence into dental imaging has led to notable advancements in diagnostic accuracy and treatment planning [14], and various studies have focused on dental caries segmentation on panoramic X-rays. A study conducted by Asci et al. [1] utilized 6075 panoramic X-rays acquired from pediatric patients aged 4 to 14. The dataset was categorized into three main types: primary, mixed, and permanent dentition. The research employed a UNet model which demonstrated strong results across multiple evaluation metrics, including sensitivity, precision, and F1-score. Another study by Mărginean et al. [13] introduced CariSeg, an advanced intelligent system designed for caries detection in dental panoramic radiographs. The framework integrates four neural network architectures to enhance segmentation accuracy. Initially, the first component, built on the architecture of U-Net, locates the dental area and trims the radiograph to focus on it. Subsequently, caries lesions are segmented using an ensemble approach integrating U-Net, DeepLabV3 and Feature Pyramid Network. After training on 150 radiographic images, the model reached a dice score of 68.2%. Another technique was introduced by a Dayi et al. [7], who proposed Dental Caries Detection Network (DCDNet), a novel deep learning model developed to segment proximal, cervical, and occlusal caries. A unique aspect of this architecture is its Multi-Predicted Output structure, which enhances segmentation accuracy by channeling feature extraction through three dedicated paths, each specializing in a different caries type. The proposed model demonstrated strong segmentation capabilities, achieving an average F1-score of 62.79%.

In medical image segmentation, various deep learning techniques have been developed to improve segmentation accuracy. TransUNet [2], in particular, has proven to be highly effective and demonstrated strong performance across various medical imaging tasks, and over time several extensions and variations of TransUNet have been proposed [5, 19] to further optimize its segmentation capabilities. In our work we introduce Caries-TransUNet, an enhanced version of TransUNet developed to improve dental caries segmentation in panoramic radiography by incorporating atrous spatial pyramid pooling (ASPP) [3], squeeze and excitation (SE) [10] and attention gates. We evaluated our proposed model on the first publicly available dental caries dataset DC1000 [20] and compared its performance against state-of-the-art models namely U-Net [17], Attention UNet [15], TransUNet [2] and DeepLabv3+ [4].

The remainder of this paper is organized as follows. In Sect. 2 we describe the Materials and Methods, including the dataset, preprocessing techniques used, the TransUNet framework, and the proposed model architecture. The Experiments and Results in Sect. 3 follows, detailing the experimental setup, results and comparison, ablation study, and discussion. Finally, the conclusion is presented in Sect. 4.

2 Materials and Methods

2.1 Dataset Description

The DC1000 dataset is the first public dental panoramic dataset for caries segmentation. It was originally proposed in [20] and comprises 1000 dental panoramic radiographs with a resolution of 2943x1435 pixels, containing over 7500 caries lesions categorized into three classes: shallow, middle and deep caries. Data was obtained from the Zhejiang Provincial People's Hospital and was annotated by five dental experts. Figure 1 presents a sample of images from the DC1000 dataset along with their corresponding masks. Among the 1000 panoramic images, 593 panoramic images are well annotated, while the remaining 407 have rough annotations. The authors originally used this dataset for a semi-supervised learning task; however, the publicly available version includes only the 593 fully labeled panoramic radiographs, which we used for caries segmentation. This dataset contains a total of 4520 caries across all images with an average of 7.6 caries per image. Additionally, around 77% of patients have fewer than 11 caries, whereas around 6% are diagnosed with more than 15 caries. Furthermore, around 73.5% of the panoramic radiographs in the dataset have a caries region occupying less than 1.5% of the total image area.

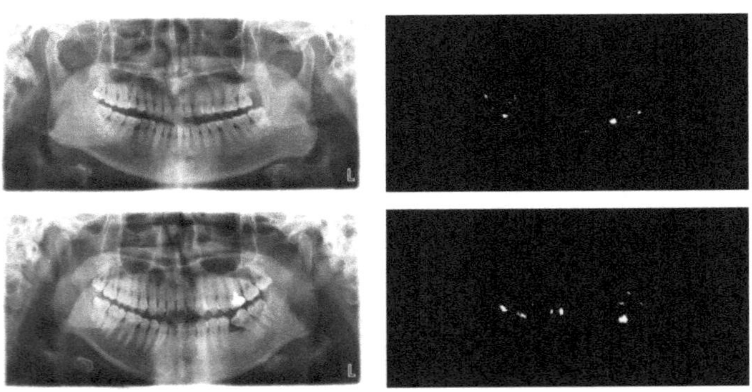

Fig. 1. Sample Images and Ground Truth Masks from the DC1000 Dataset.

2.2 Preprocessing

The DC1000 dataset, initially proposed in [20], was divided by the authors into 493 images for training and 100 images for testing. To reduce computational demands, we first cropped the images to focus on relevant regions and then resized them to a uniform resolution of 320×640 pixels. To enhance the diversity of the training data and increase the robustness of the model, we implemented several data augmentation techniques. These transformations included random rotations, zooming, translations, and horizontal flips, which expanded the training dataset by an additional 400 images. Additionally, we incorporated random adjustments of brightness and contrast, combined with advanced methods such as adaptive histogram equalization (CLAHE) and the addition of Gaussian noise. This approach not only optimized computational resources but also simulated a variety of imaging conditions, thereby enhancing the model's ability to generalize.

2.3 TransUNet

TransUNet [2] is the first medical image segmentation model to use a hybrid CNN-Transformer design. Initially, a CNN generates a feature map from the input image and a Transformer then encodes tokenized patches from the feature map to capture global context. The self-attentive features produced by the Transformer are then upsampled using a cascaded upsampler and merged with high-resolution features from the encoder's skip connections to achieve precise localization (Fig. 2).

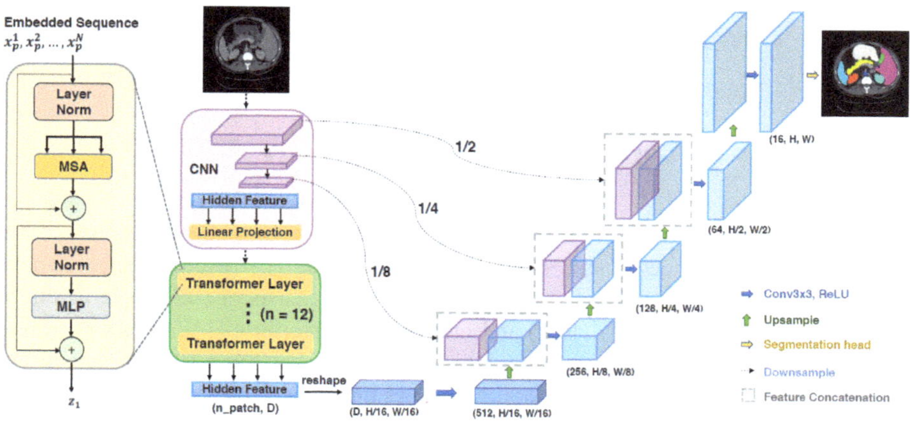

Fig. 2. TransUNet architecture from [2].

2.4 Proposed Model: Caries-TransUNet

Overview of the Architecture. We introduce Caries-TransUNet, an enhanced adaptation of the TransUNet architecture specifically tailored for dental caries segmentation in panoramic radiographs. The model leverages a hybrid design that combines CNNs for local feature extraction and transformers for global context modeling. This architecture is further augmented with advanced modules, including Atrous Spatial Pyramid Pooling (ASPP) [3] and Squeeze-and-Excitation (SE) [10] blocks, to improve multi-scale feature representation and channel-wise recalibration. Additionally, attention gates are integrated into skip connections to refine feature fusion during decoding. Below, we provide a comprehensive overview of the key components as well as their roles in enhancing the segmentation performance. The overall architecture of Caries-TransUNet is illustrated in Fig. 3.

CNN Encoder and Transformer Backbone. The architecture begins with a CNN encoder, where we employ a VGG19 network [18] pre-trained on ImageNet [8]. This

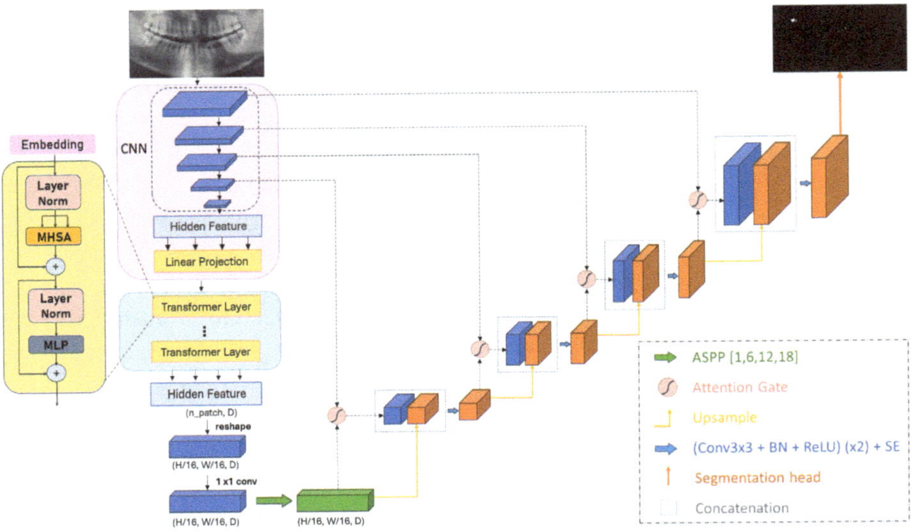

Fig. 3. Architecture of Caries-TransUNet.

choice ensures robust extraction of low-level features from input images, which are critical for capturing fine-grained details of dental caries. The resulting feature maps are then passed through a transformer module, which consists of successive layers of multi-headed self-attention (MHSA) and multi-layer perceptron (MLP) blocks. Layer normalization (LN) is applied throughout the transformer to stabilize training and ensure consistent gradient flow. This combination enables the model to learn both fine-grained and broader contextual features, laying a strong foundation for accurate segmentation.

Multi-scale Feature Capture with ASPP. Following the transformer, we integrate an Atrous Spatial Pyramid Pooling (ASPP) module immediately after the encoder to address the challenge of segmenting caries of varying sizes. The ASPP module captures multi-scale contextual information while preserving spatial resolution, ensuring precise localization of both small superficial caries and larger deep lesions.

As illustrated in Fig. 4 (a), the ASPP module operates through multiple parallel convolutional branches. The first branch applies a 1×1 convolution to capture fine-grained details. Three additional branches use 3×3 atrous convolutions with varying dilation rates to extract multi-scale spatial information. A final branch employs global average pooling to capture global context, followed by a 1×1 convolution to refine the extracted features. The outputs from all branches are concatenated along the channel dimension and processed through a final 1×1 convolution to reduce channel dimensionality while enhancing feature representation. This multi-scale strategy greatly enhances the model's ability to handle diverse caries sizes, particularly in challenging cases where lesions vary in scale and appearance.

Decoder with Attention Gates and SE Block. The feature map produced by the ASPP module is subsequently passed to the decoder, which consists of four blocks designed to progressively upsample and refine the feature maps. Each decoder block follows a structured pipeline:

1. **Transposed Convolution:** Upsamples the feature map to restore spatial resolution.
2. **Attention Gate:** Refines skip connections by focusing on relevant regions before concatenating them with the upsampled output.
3. **Convolutional Layers:** Two 3 × 3 convolutional layers are applied, each followed by batch normalization to stabilize training and a ReLU activation function to introduce non-linearity.

   To further enhance the discriminative power of the feature maps, we incorporate a Squeeze-and-Excitation (SE) block after the ASPP module. As shown in Fig. 4 (b), the SE block operates in two main steps:
4. **Squeeze Step:** Global average pooling compresses spatial dimensions into a single descriptor per channel.
5. **Excitation Step:** A lightweight two-layer fully connected network processes these descriptors, applying a ReLU activation followed by a sigmoid function to generate adaptive channel weights. These weights are then multiplied element-wise with the original feature maps, dynamically recalibrating their importance. By emphasizing the most informative channels and suppressing irrelevant ones, the SE block significantly improves the model's ability to distinguish between different types of dental caries, particularly subtle lesions that are challenging to detect.

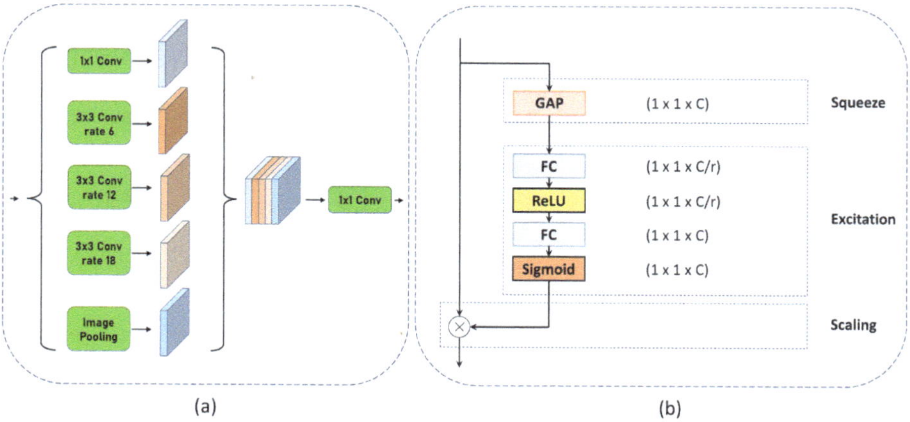

Fig. 4. Architecture of (a) ASPP and (b) SE.

Final Segmentation Output. After the last decoder block, a 1×1 convolutional layer with a sigmoid activation function is applied to produce the final segmentation mask. This step ensures that the model outputs pixel-wise predictions, enabling precise localization of carious regions across the entire image. Qualitative results of the segmentation are presented in Fig. 5, showcasing the model's accuracy in detecting shallow, middle, and deep caries compared to other state-of-the-art methods.

3 Experiments and Results

3.1 Experimental Setup and Evaluation

The model was implemented and trained using the TensorFlow framework on Google Colab Pro +, leveraging an NVIDIA A100 GPU with 40 GB of dedicated GPU RAM. During training, a batch size of 16 was used, the learning rate was initially configured to 0.001 and the Adamax optimizer was employed to update the model's parameters. All segmentation models underwent training for 50 epochs, with performance assessed using evaluation metrics including Accuracy, Dice coefficient, IoU, Precision and Recall. Finally, we used the IoU loss function to optimize the segmentation performance.

3.2 Results and Comparison

The performance of our proposed Caries-TransUNet model was evaluated and compared against several deep learning-based methods such as U-Net [17], Attention UNet [15], TransUNet [2] and DeepLabv3+ [4]. We used VGG19 [18] as CNN backbone with pretrained weights from ImageNet [8] for U-Net, Attention UNet, TransUNet and Caries-TransUNet and we employed ResNet50 [9] backbone, also initialized with ImageNet weights, for Deeplabv3+ model.

The results are reported in Table 1. The experimental results indicate that Caries-TransUNet achieves superior performance compared to the state-of-the-art methods consistently in terms of Dice coefficient, IoU score and Recall with a Dice score of 77.88%, an IoU score of 63.85% and a Recall of 76.48%. The qualitative results produced by our model and the models used for comparison are displayed in Fig. 5. We can see that our proposed model produces the most accurate segmentation relative to the ground truth as it rarely segments non-carious teeth and is more accurate in detecting the three categories of caries, particularly shallow caries.

Table 1. Evaluation results (%) of different methods on DC1000 dental caries dataset.

Model	Accuracy	Dice	IoU	Precision	Recall
U-Net	99.81	76.59 ± 0.25	62.11 ± 0.32	84.33 ± 1.65	71.91 ± 1.74
TransUNet	99.81	77.06 ± 0.22	62.74 ± 0.28	**84.33 ± 0.76**	72.49 ± 0.64
Attention UNet	99.81	76.20 ± 0.15	61.63 ± 0.20	81.72 ± 1.94	73.05 ± 1.58
DeepLabv3+	99.80	72.89 ± 0.75	57.46 ± 0.96	83.23 ± 1.11	66.01 ± 1.86
Caries-TransUNet (Ours)	99.81	**77.88 ± 0.23**	**63.85 ± 0.30**	81.37 ± 0.69	**76.48 ± 0.69**

3.3 Ablation Study

We conducted an ablation study to assess the contribution of different components of our model by systematically adding components to the baseline model TransUNet and measuring their effect on segmentation performance. The results of the ablation study are illustrated in Table 2. In the table, ASPP refers to atrous spatial pyramid pooling [3], SE refers to squeeze-and-excitation [10], and AG refers to attention gate.

Our ablation study begins with TransUNet as the baseline. Adding ASPP module improved the Dice score by 0.24% and IoU by 0.32%. Incorporating SE module further increased the Dice score by 0.42% and IoU by 0.57%. Finally, adding AG, which completes our proposed model, resulted in a total Dice score gain of 0.82% and an IoU gain of 1.11% over the baseline.

Table 2. Ablation study on network components.

Model	Dice	IoU	Precision	Recall
TransUNet	77.06 ± 0.22	62.74 ± 0.28	**84.33 ± 0.76**	72.49 ± 0.64
TransUNet + ASPP	77.30 ± 0.36	63.06 ± 0.47	83.12 ± 0.81	73.84 ± 1.16
TransUNet + ASPP + SE	77.48 ± 0.31	63.31 ± 0.41	84.01 ± 0.91	73.52 ± 0.72
TransUNet + ASPP + SE + AG (Ours)	**77.88 ± 0.23**	**63.85 ± 0.30**	81.37 ± 0.69	**76.48 ± 0.69**

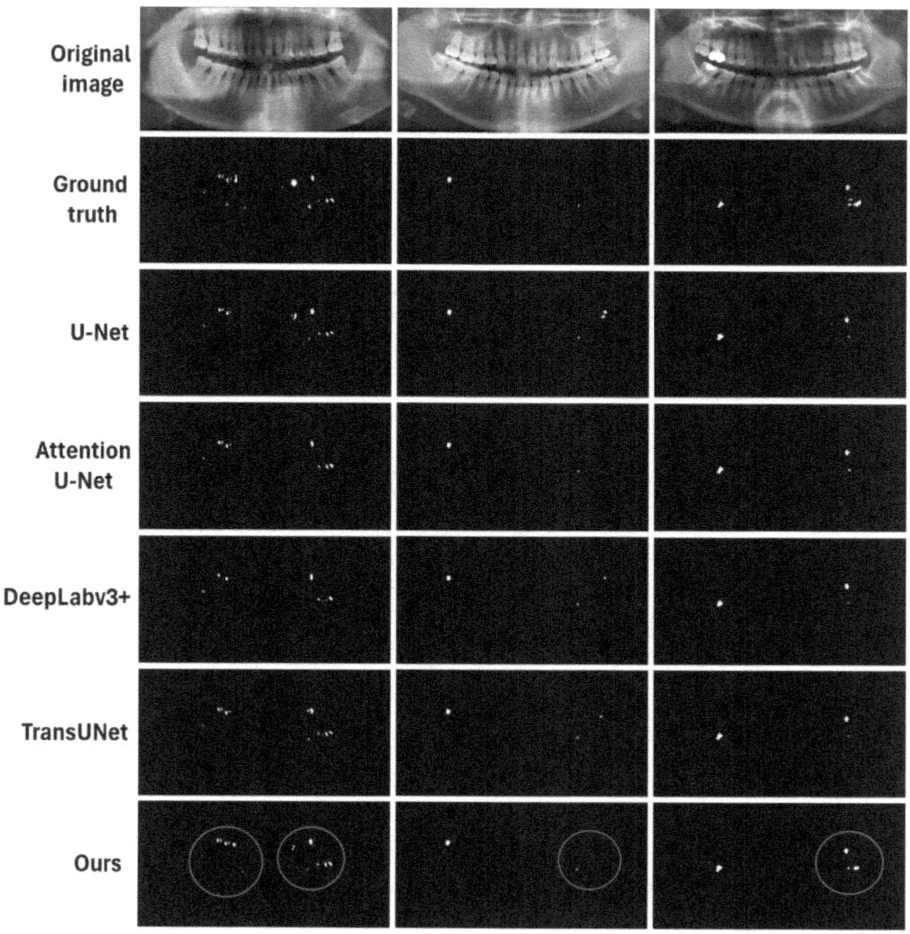

Fig. 5. Qualitative Results of Dental Caries Segmentation on the DC1000 dataset.

3.4 Discussion

Our proposed model Caries-TransUNet outperformed TransUNet and other state-of-the-art models, with Dice coefficient gains ranging from 0.82% to 5% and IoU improvements between 1.11% and 6.39%, demonstrating its effectiveness in segmenting dental caries in panoramic radiographs. The qualitative results further validate its accuracy in correctly identifying and segmenting the different categories of caries: shallow, middle and deep caries. Due to the relatively small size of shallow and medium caries, compared to the larger deep caries, detecting shallow or middle caries is more challenging and more crucial since early detection is essential for timely intervention and better prognosis. We can see from the prediction visualizations, that our model is more effective at segmenting shallow caries compared to other models.

The dataset used in this study contained only 593 images with 493 images used for training and 100 images used for testing. Without data augmentation, the model achieved a Dice score of 75.39% and an IoU score of 60.5%. Similarly, the other models also achieved lower IoU and Dice scores. Additionally, panoramic X-ray images contain a lot of background region, while we want our model to focus on the teeth region specially since caries occupy less than 1.5% of the total image area. Therefore, we applied cropping, which not only helps in removing non-essential parts, but also lowers computational cost and leads to faster training without losing important details.

In this study, we aimed to develop an enhanced version of TransUNet to improve dental caries segmentation in panoramic X-rays by integrating ASPP and SE modules and by adding attention gates to skip connections. These additions increased the model's parameter count from 31M to parameters 41M. The results from the ablation study confirm that every added component contributes significantly to the model's performance. For future work, we aim to employ instance segmentation to distinguish between different categories of caries more accurately.

4 Conclusion

In this paper, we proposed Caries-TransUNet model to enhance caries segmentation in panoramic dental X-rays. Building upon the TransUNet architecture, we introduce an enhanced version incorporating ASPP module to help segment caries of varying sizes by aggregating multi-scale features, SE module to help the network focus on more relevant channels and finally attention gates to refine skip connections. We evaluated the proposed model on the DC1000 dental caries dataset and compared its performance against several state-of-the-art models. The results indicate that our model consistently outperformed them across multiple metrics.

References

1. Asci, E., et al.: A deep learning approach to automatic tooth caries segmentation in panoramic radiographs of children in primary dentition, mixed dentition, and permanent dentition. Children. **11**(6), 690 (2024)
2. Chen, J., et al.: TransUnet: transformers make strong encoders for medical image segmentation. arXiv preprint https://arxiv.org/abs/2102.04306. (2021)

3. Chen, L.C.: Rethinking atrous convolution for semantic image segmentation. arXiv preprint arXiv:1706.05587. (2017)
4. Chen, L.C., Zhu, Y., Papandreou, G., Schroff, F., Adam, H.: Encoder-decoder with atrous separable convolution for semantic image segmentation. In: Proceedings of the European Conference on Computer Vision (ECCV), pp. 801–818 (2018)
5. Chong, Y., Xie, N., Liu, X., Pan, S.: P-TransUnet: an improved parallel network for medical image segmentation. BMC Bioinf. **24**(1), 285 (2023)
6. G.B.D. Collaborative Network, Bernabe, E., et al.: Global, regional, and national levels and trends in burden of oral conditions from 1990 to 2017: a systematic analysis for the global burden of disease 2017 study. J. Dent. Res. **99**(4), 362–373 (2020)
7. Dayı, B., Üzen, H., Çiçek, İ.B., Duman, Ş.B.: A novel deep learning-based approach for segmentation of different type caries lesions on panoramic radiographs. Diagnostics. **13**(2), 202 (2023)
8. Deng, J., Dong, W., Socher, R., Li, L.J., Li, K., Fei-Fei, L.: ImageNet: a large-scale hierarchical image database. In: 2009 IEEE conference on computer vision and pattern recognition, pp. 248–255. IEEE (2009)
9. He, K., Zhang, X., Ren, S., Sun, J.: Deep residual learning for image recognition. In: Proceedings of the IEEE Conference on Computer Vision and Pattern Recognition, pp. 770–778 (2016)
10. Hu, J., Shen, L., Sun, G.: Squeeze-and-excitation networks. In: Proceedings of the IEEE Conference on Computer Vision and Pattern Recognition, pp. 7132–7141 (2018)
11. Jeon, K.J., Han, S.S., Lee, C., Choi, Y.J., Jung, H.I., Kim, Y.H.: Application of panoramic radiography with a multilayer imaging program for detecting proximal caries: a preliminary clinical study. Dentomaxillofac. Radiol. **49**(8), 20190467 (2020)
12. Keenan, J.R., Keenan, A.V.: Accuracy of dental radiographs for caries detection. Evid. Based Dent. **17**(2), 43–43 (2016)
13. Mărginean, A.C., Mureşanu, S., Hedeşiu, M., Dioşan, L.: Teeth segmentation and carious lesions segmentation in panoramic x-ray images using CariSeg, a networks' ensemble. Heliyon. **10**(10), e30836 (2024)
14. Mohammad-Rahimi, H., et al.: Deep learning for caries detection: a systematic review. J. Dent. **122**, 104115 (2022)
15. Oktay, O., et al.: Attention u-net: learning where to look for the pancreas. arXiv preprint arXiv:1804.03999. (2018)
16. Pitts, N., et al.: Dental caries. Nat. Rev. Dis. Primers. **3**, 17030 (2017)
17. Ronneberger, O., Fischer, P., Brox, T.: U-net: convolutional networks for biomedical image segmentation. In: Medical Image Computing and Computer-Assisted Intervention–MICCAI 2015: 18th International Conference, Munich, Germany, October 5–9, 2015, Proceedings, Part III 18, pp. 234–241. Springer (2015)
18. Simonyan, K., Zisserman, A.: Very deep convolutional networks for large-scale image recognition. arXiv preprint https://arxiv.org/abs/1409.1556. (2014)
19. Sun, G., et al.: Da-TransUNet: integrating spatial and channel dual attention with transformer u-net for medical image segmentation. Front. Bioeng. Biotechnol. **12**, 1398237 (2024)
20. Wang, X., et al.: Multi-level uncertainty aware learning for semi-supervised dental panoramic caries segmentation. Neurocomputing. **540**, 126208 (2023)

Predicting Mutations and Metastases in Colorectal Tumors Using Artificial Intelligence

Sofyan El Idrissi[1](✉), Yassine Drider[1], Ikram Ben Abdel Ouahab[2], Mohammed Bouhorma[3], and Fatiha El Ouaai[3]

[1] Computer Science and Smart Systems (C3S) Laboratory, Tangier, Morocco
IDRISSIidrissisofyan3838@gmail.com
[2] Data and Intelligent Systems (DIS) Team, C3S Laboratory, Tangier, Morocco
[3] Smart Systems & Emerging Technologies (SSET) Team, C3S Laboratory Faculty of Sciences and Technologies of Tangier, University Abdelmalek Essaadi, Tangier, Morocco

Abstract. Treatment resistance poses a major obstacle in precision oncology, diminishing therapeutic efficacy and patient survival chances. Molecular genetic mutations play a crucial role in this phenomenon, altering the biological characteristics of tumor cells. In colorectal tumors, specific genetic mutations have been identified as markers of treatment resistance. However, detecting and characterizing these mutations is a technical challenge with limitations due to traditional laboratory techniques. Recent advances in deep learning allow for accurate prediction of genetic mutations from histological images, providing a rapid and cost-effective alternative to traditional methods. Furthermore, predicting colorectal metastatic sites based on molecular mutations shows promising potential to enhance clinical management of patients. This innovative approach could guide personalized treatments and improve clinical outcomes.

Keywords: Treatment resistance · Genetic mutations · Colorectal tumors · Deep learning · Mutation prediction · Metastatic sites · Metastasis site prediction

1 Introduction

Treatment resistance poses a major challenge in oncology, hampering the effectiveness of therapies and diminishing patients' survival prospects. Among the numerous factors contributing to this resistance, genetic mutations play a central role by altering the biological, molecular, and microenvironmental properties of cancer cells. Consequently, the detection and characterization of these mutations in tumors are of crucial importance for elucidating resistance mechanisms, predicting treatment response, and adjusting therapeutic strategies.

In the case of colorectal tumors, which represent the third leading cause of cancer-related deaths worldwide, several genetic mutations have been identified as markers of resistance to chemotherapy. Detecting and characterizing genetic mutations in colorectal

B.-E. Boudriki Semlali et al. (Eds.): DATA 2025, LNNS 1908, pp. 224–231, 2026.
https://doi.org/10.1007/978-3-032-21200-9_19

tumors pose a technical and methodological challenge. The technologies required to detect these mutations must be sensitive, specific, and cost-effective.

Laboratory genetic characterization techniques such as cytogenetics or anatomopathological and molecular genetic techniques have certain limitations, such as high cost, time-consuming result production, which can delay the diagnostic process and therapeutic decision-making. Some standard genetic characterization techniques require tissue samples of sufficient quality, and the reliability and reproducibility of results can vary from one laboratory to another depending on the techniques used and the expertise of the personnel, among several other limitations.

The prediction of molecular mutations in colorectal cancers (CRC) from histopathological images represents a significant advancement in patient treatment personalization. Recent studies have demonstrated that deep learning models can successfully predict genetic mutations directly from histological sections or other medical imaging. These methods offer a cost-effective and rapid alternative to traditional genomic tests.

2 Prediction of Molecular Status from Image Processing by Deep Learning Models and Their Performance

Mutations in the KRAS gene, encoding a protein involved in cellular signaling, are associated with resistance to anti-EGFR agents (epidermal growth factor receptor), such as cetuximab or panitumumab [1]. Similarly, mutations in the BRAF gene, which encode another protein in the same signaling pathway, are associated with resistance to anti-EGFR agents and poor survival [2], other genetic mutations, such as those in the TP53, PIK3CA, PTEN, or APC genes, can also influence the response to treatments for colorectal tumors [3].

To detect and analyze these genetic mutations, there are various methods based on the extraction and sequencing of tumor DNA, either from biopsies or from circulating DNA in the blood. However, these methods have limitations such as the low quantity and quality of tumor DNA, spatial and temporal variability of mutations, or the cost and complexity of sequencing techniques [4]. Therefore, new approaches are needed to improve the detection and characterization of genetic mutations in colorectal tumors. One of these approaches involves using image processing with Deep Learning models, which are artificial intelligence techniques capable of learning from large amounts of data and performing complex tasks, such as image recognition. Deep Learning can be applied to histological images of tumors obtained by optical or electron microscopy to extract information about the morphology, structure, composition, and distribution of tumor cells. This information can then be correlated with the genetic mutations present in the tumors using classification or regression algorithms [5].

Several studies have demonstrated the feasibility and utility of image processing with Deep Learning models for predicting genetic mutations in colorectal tumors. For instance, one study utilized a convolutional neural network to predict the presence of KRAS gene mutations from histological images of colorectal tumors, achieving an accuracy of 83% [6]. Another study employed a recurrent neural network to predict the presence of mutations in the KRAS, BRAF, and TP53 genes from histological images of colorectal tumors, achieving accuracies of 86%, 75%, and 69%, respectively [7].

The use of whole-slide images (WSIs) of CRC to predict common mutations in the APC, KRAS, PIK3CA, SMAD4, and TP53 genes has demonstrated promising performance using deep learning models, with ROC curves showing AUCs ranging from 0.693 to 0.809. Increasing the size of the datasets has enhanced prediction performance [8], A model called DeepHE, which is based on the residual network (ResNet), was put forth to forecast the tumor mutational burden (TMB), a crucial component of immunotherapy's effectiveness in colorectal cancer. This model showed a strong correlation between genetic mutations and histopathological images, outperforming several other models in the process [9], Lastly, unsupervised clustering techniques can also be used to optimize deep learning models for the prediction of gene mutations, enabling the identification of spatial regions associated with particular mutations and the exclusion of areas lacking predictive information [10] And tissue analysis based on deep learning can predict outcomes in patients with CRC, even outperforming histological visual evaluation conducted by human experts [11].

3 Prediction of Metastatic Sites of Colorectal Tumors by Molecular Mutations of Primary Tumors

Oncogenic mutations are key in colorectal cancer development and can predict metastatic relapse sites. Research in predicting CRC metastases is growing, with deep-learning models playing a significant role. Quantum biology studies suggest molecular mutations might affect distant sites via quantum entanglement, inspiring predictions of metastasis sites to monitor recurrence after treatments.

This predictive approach will improve the quality and efficiency of monitoring tumor recurrences after treatment by specialists. Study [12] is a promising research to demonstrate the relationship between molecular genetic mutations and their correlation with tumor metastasis sites. They have shown that Deep Learning models can provide relevant results for predicting metastases based on molecular mutations. Additionally, the ColoMet19, a model based on the expression of 19 genes to predict metastases in CRC patients at an early stage and with low clinical risk, has demonstrated an adjusted positive predictive value in the population of 0.67, suggesting that positively tested patients have a substantially higher risk of developing metastases [13].

The spatiotemporal stability of somatic mutations is crucial in metastatic CRCs with stable microsatellites. A study using next-generation sequencing of 24 primary CRCs and multiple metastases found that 40% of CRC-related genes had somatic mutations, with 100% genetic concordance for APC, KRAS, and other genes between primary and secondary sites. Except for new mutations in 4 cases, 84.4% shared genetic lesions across all metastases, regardless of location or time after tumor resection [14].

More precisely, the mutation of the KRAS gene is significantly more frequent in pulmonary and cerebral metastases compared to hepatic metastases. This increased prevalence of KRAS mutations in pulmonary and cerebral metastases suggests an association between KRAS mutation and metastasis at these specific sites [15].

A further finding of the study [16] was the correlation between PIK3CA mutations and pulmonary metastases. Individuals with mutated p53 had statistically more lesions than patients with wild-type p53, according to the correlation between p53 status and clinical parameters [17] More specifically, 41% of patients with mutated p53 had three or more metastases, while only 14% of patients with wild-type p53 met this requirement. Whereas only 29% of patients with wild-type p53 had synchronous metastases, 70% of patients with p53 mutations had them. In addition, patients with p53 mutations have a higher chance of experiencing recurrence (73%) than those with wild-type p53 (33%). According to the findings, p53 status may be a significant prognostic factor for forecasting the course and probability of treatment failure following hepatic resection [17], According to [18], KRAS, p53, SMAD4, and BRAF mutations play significant roles in CRC metastasis and can serve as both potential biomarkers of CRC metastasis and therapeutic targets. RAS mutational status clearly stratified survival in patients with inadequate response to preoperative chemotherapy. Therefore, RAS mutational status can be used to complement current prognostic indicators in patients undergoing curative resection of CLM after modern preoperative chemotherapy. All these findings were corroborated by [19] which demonstrated that KRAS, p53, SMAD4, and BRAF mutations play significant roles in CRC metastasis and can be both potential biomarkers of CRC metastasis and therapeutic targets.

These findings underscore the importance of somatic mutations as potential biomarkers for monitoring and personalized management of patients with metastatic colorectal cancer. Molecular mutations in colorectal tumors play a crucial role in determining distant metastasis sites. Understanding the genetic stability between primary tumors and metastases can influence the clinical management of patients with metastatic CRC and the selection of potential therapeutic targets.

4 Proposal for an Innovative Approach to Characterizing Colorectal Tumors

These studies suggest that image processing using Deep Learning models can be an effective tool for predicting genetic mutations in colorectal tumors, thereby stratifying patients and optimizing treatments. Furthermore, predicting metastatic sites after identifying molecular mutations could revolutionize the management of colorectal tumors and significantly improve the prognosis of these tumors, particularly for increasing the survival of patients with CRC, especially mobile colon tumors, which pose challenges for the application of radiotherapy.

Our innovative approach builds upon the aforementioned findings. By utilizing medical imaging of tumors, deep learning models can predict the molecular status of tumors with high efficiency, guiding personalized treatments against tumor resistance.

We propose a groundbreaking approach to analyze colorectal tumors, leveraging recent breakthroughs in the field of medical imaging and artificial intelligence. Our approach aims to overcome the limitations of traditional laboratory methods by providing a fast, cost-effective, and accurate solution for predicting the molecular characteristics of tumors, as well as their metastatic tendencies.

4.1 Background and Significance

Genetic mutations play a crucial role in colorectal tumors' ability to resist treatments, making it an essential area of study for oncology specialists. However, the precise detection of these mutations is often hindered by technical and financial constraints associated with conventional laboratory methods. Our approach offers a promising alternative by harnessing the capabilities of artificial intelligence models to analyze histological images of tumors, thus providing an innovative and effective means to predict genetic mutations associated with treatment resistance.

4.2 Potential Advantages

One of the main advantages of our approach is its ability to predict not only the genetic mutations of tumors but also their metastatic tendencies. By identifying these characteristics early on, we could enable physicians to tailor treatments more precisely, thereby improving outcomes for patients with colorectal cancer. Additionally, this method could lead to the identification of potential therapeutic targets, paving the way for new, more effective treatment strategies (Fig. 1).

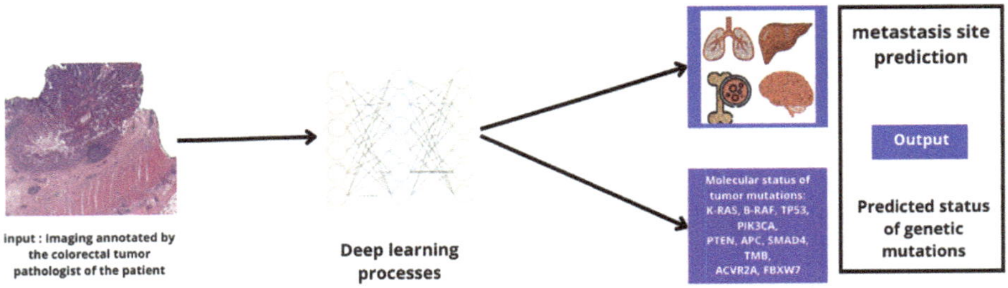

Fig. 1. Overview of the proposed computational framework, which takes the image profile of any primary cancer sample and predicts the molecular status of patients and predicted metastasis sites in the brain, bones, lungs, or liver based on a deep learning architecture.

4.3 Proposed Approach

Our approach relies on the use of advanced Deep Learning techniques, a branch of artificial intelligence, to analyze histological images of colorectal tumors. These models are capable of learning to recognize visual features associated with different genetic mutations, thus enabling them to accurately predict the molecular profile of tumors. To achieve this, we support the application of convolutional neural networks, an architecture specifically created for image analysis and proven successful in a number of domains. Furthermore, according to [20] the multi-modal deep learning model ICSDA, which integrates clinical data, gene expression data, and colored hematoxylin and eosin (H&E) histopathological images, produces promising results in predicting the risk of metastasis and recurrence in patients with breast cancer. The superior performance of this model over methods that only used sequencing, clinical, or H&E image data highlights the significance of multimodal integration in risk prediction.

The authors of [21] introduce a novel deep learning model to capture mutation information in multivariate time series predictions. This model utilizes an attention mechanism and an LSTM with a transformation mechanism to handle mutation information, demonstrating superior performance compared to baseline methods on five real datasets.

Continual adaptation of deep learning models is explored in [22], where the authors develop an online learning method via meta-learning for model-based reinforcement learning. This approach enables effective adaptation in non-stationary task distributions, which is crucial for control in changing environments.

Finally, the authors of [23] examine the application of deep learning in predicting cancer prognosis. The improvement in cancer prognosis prediction is attributed to increased computational power and the rapid advancement of artificial intelligence, particularly deep learning. The use of multi-omics data is suggested to potentially enhance the accuracy of cancer prognosis models.

4.4 Perspectives

After presenting our approach, we explore future research directions. Implementing this in clinical settings requires high-quality image databases for training deep learning models. Future studies could integrate clinical and genomic data for better predictions or explore new modeling methods. Our approach aims to enhance colorectal tumor characterization and guide personalized treatments by combining medical imaging and AI.

5 Discussion

The present study demonstrates the promising potential of using artificial intelligence, particularly Deep Learning models, for predicting genetic mutations and metastasis sites in colorectal tumors (CRC). Our findings confirm the conclusions of several previous studies, which have shown that Deep Learning models can successfully predict genetic mutations from histological images of colorectal tumors with high accuracy.

The clinical implications of this approach are significant. By identifying genetic mutations associated with treatment resistance and metastasis sites early on, clinicians may be able to tailor therapeutic strategies more personalized, thereby improving clinical outcomes for CRC patients. For example, early detection of KRAS or BRAF gene mutations could help avoid the use of EGFR-targeted therapies, which are ineffective in patients with these mutations.

However, despite the encouraging results obtained in this study, and in the context of North Africa, for example, several limitations must be considered. Firstly, the proposed approach relies on Deep Learning models that require large datasets to achieve optimal performance. Therefore, it is crucial to continue collecting and annotating high-quality histological data to effectively train these models.

Furthermore, the generalizability of Deep Learning models may be limited by the diversity of populations and CRC subtypes. Additional studies and investigations on larger and more diverse patient cohorts are needed to validate the effectiveness of this approach in real clinical settings.

In the Moroccan context, it is recommended that with the adoption of electronic medical records in university hospital complexes, a database of annotated images of the Moroccan population be created to increase the accuracy and relevance of models trained for specific population subtypes.

Furthermore, future research could focus on improving image processing techniques, combining clinical and genomic data for more precise predictions, as well as exploring new modeling approaches such as reinforcement learning or federated learning to overcome current challenges in this field.

6 Conclusion

In conclusion, genetic mutations play a key role in colorectal tumor resistance to chemotherapy, making detection and analysis crucial in precision oncology. Deep learning models provide a promising approach for predicting mutations from histological images, offering benefits in speed, simplicity, and cost. Challenges include validating models on independent data, generalizing results across tumor types and treatments, and interpreting visual features related to mutations.

References

1. Karapetis, C.S., et al.: K-ras mutations and benefit from cetuximab in advanced colorectal cancer. N Engl J Med. **359**, 1757–1765. https://www.nejm.org/doi/full/10.1056/nejmoa0804385 (2024)
2. De Roock, F., et al.: KRAS codon 61, 146 and BRAF mutations predict resistance to cetuximab plus irinotecan in KRAS codon 12 and 13 wild-type metastatic colorectal cancer. Br. J. Cancer. **101**, 715–721. https://www.nature.com/articles/6605177 (2024)
3. Dienstmann, R., Vermeulen, L., Guinney, J., Kopetz, S., Tejpar, S., Tabernero, J.: Consensus molecular subtypes and the evolution of precision medicine in colorectal cancer. Nat. Rev. Cancer. **17**(2), 2 (2017). https://doi.org/10.1038/nrc.2016.126
4. Siravegna, G., Marsoni, S., Siena, S., Bardelli, A.: Integrating liquid biopsies into the management of cancer. Nat. Rev. Clin. Oncol. **14**(9), 9 (2017). https://doi.org/10.1038/nrclinonc.2017.14
5. Kather, J.N., et al.: Deep learning can predict microsatellite instability directly from histology in gastrointestinal cancer. Nat. Med. **25**, 1054–1056. https://www.nature.com/articles/s41591-019-0462-y (2024)
6. Kather, J.N., et al.: Predicting survival from colorectal cancer histology slides using deep learning: a retrospective multicenter study. PLoS Med. **16**(1), e1002730 (2019). https://doi.org/10.1371/journal.pmed.1002730
7. Bychkov, D., et al.: Deep learning based tissue analysis predicts outcome in colorectal cancer. Sci. Rep. **8**, 3395. https://www.nature.com/articles/s41598-018-21758-3 (2024)
8. Bilal, M., et al.: Development and validation of a weakly supervised deep learning framework to predict the status of molecular pathways and key mutations in colorectal cancer from routine histology images: a retrospective study. Lancet Digit. Health. **3**, e763–e772. https://www.thelancet.com/journals/landig/article/PIIS2589-7500(2100180-1/fulltext (2024)

9. Liu, Y., Huang, K., Yang, Y., Wu, Y., Gao, W.: Prediction of tumor mutation load in colorectal cancer histopathological images based on deep learning. Front. Oncol. **12**, 906888 (2022) Consulté le: 3 mars 2024. [En ligne]. Disponible sur:https://www.frontiersin.org/journals/oncology/articles/10.3389/fonc.2022.906888
10. Chen, Z., Li, X., Yang, M., Zhang, H., Xu, X.S.: Optimization of deep learning models for the prediction of gene mutations using unsupervised clustering. J. Pathol. Clin. Res. **9**(1), 3–17 (2023). https://doi.org/10.1002/cjp2.302
11. Jang, H.-J., Lee, A., Kang, J., Song, I.H., Lee, S.H.: Prediction of clinically actionable genetic alterations from colorectal cancer histopathology images using deep learning. World J. Gastroenterol. **26**(40), 6207–6223 (2020). https://doi.org/10.3748/wjg.v26.i40.6207
12. Albaradei, S., et al.: MetastaSite: predicting metastasis to different sites using deep learning with gene expression data. Front. Mol. Biosci. **9**, 913602 (2022) Consulté le: 3 mars 2024. [En ligne]. Disponible sur: https://www.frontiersin.org/articles/10.3389/fmolb.2022.913602
13. Low, Y.S., et al.: A formalin-fixed paraffin-embedded (FFPE)-based prognostic signature to predict metastasis in clinically low risk stage I/II microsatellite stable colorectal cancer. ScienceDirect. https://www.sciencedirect.com/science/article/abs/pii/S030438351730383X. (2024)
14. Yu, J., et al.: Novel recurrently mutated genes and a prognostic mutation signature in colorectal cancer. Gut. **64**(4), 636–645 (2015). https://doi.org/10.1136/gutjnl-2013-306620
15. Tie, J., et al.: KRAS mutation is associated with lung metastasis in patients with curatively resected colorectal cancer. Clin. Cancer Res. **17**(5), 1122–1130 (2011). https://doi.org/10.1158/1078-0432.CCR-10-1720
16. Russo, A.L., et al.: Mutational analysis and clinical correlation of metastatic colorectal cancer. Cancer. **120**(10), 1482–1490 (2014). https://doi.org/10.1002/cncr.28599
17. Tullo, A., et al.: Characterization of p53 mutations in colorectal liver metastases and correlation with clinical parameters. Clin. Cancer Res. **5**(11), 3523–3528 (1999)
18. Mise, Y., et al.: RAS mutations predict radiologic and pathologic response in patients treated with chemotherapy before resection of colorectal liver metastases. Ann. Surg. Oncol. **22**(3), 834–842 (2015). https://doi.org/10.1245/s10434-014-4042-6
19. Huang, D., et al.: Mutations of key driver genes in colorectal cancer progression and metastasis. Cancer Metastasis Rev. **37**(1), 173–187 (2018). https://doi.org/10.1007/s10555-017-9726-5
20. Yao, Y., et al.: ICSDA: a multi-modal deep learning model to predict breast cancer recurrence and metastasis risk by integrating pathological, clinical and gene expression data. Brief. Bioinform./Oxford Acad. **23**, bbac448. https://academic.oup.com/bib/article/23/6/bbac448/6761046 (2024)
21. Hu, J., Zheng, W.: A deep learning model to effectively capture mutation information in multivariate time series prediction. Knowl.-Based Syst. **203**, 106139 (2020). https://doi.org/10.1016/j.knosys.2020.106139
22. Nagabandi, A., Finn, C., Levine, S.: Deep online learning via meta-learning: continual adaptation for model-based RL. arXiv.org, 18 décembre 2018. https://arxiv.org/abs/1812.07671v2. (2024)
23. Zhu, W., Xie, L., Han, J., Guo, X.: The application of deep learning in cancer prognosis prediction. Cancer. **12**(3), 3 (2020). https://doi.org/10.3390/cancers12030603

An Advanced Feistel Lap Enclosed Within Two Genetic Operators for the Medical Image Encryption

Mariem Jarjar[1(✉)], Abdellah Abid[1], Abdellhamid Benazzi[1], and Abdellatif Jarjar[2]

[1] MATSI laboratory Mohamed first university, Oujda, Morocco
jarjarmariem@gmail.com
[2] Hight School Moulay Rachid, Taza, Morocco

Abstract. In this paper, we will develop a new technique for medical image encryption. Based on chaos, this method relies on the implementation of an improvement of the Feistel scheme surrounded by two genetic operators widely adapted to color image encryption. After a partitioning of the image vector into blocks, a genetic mutation is applied on the two adjacent blocks, followed by a Feistel round ensured by a chaotic movement, a chain between the encrypted blocks and the clear blocks secured by a strong diffusion function is implemented to bridge any differential attack. This step is followed by a genetic crossing under the control of a pseudo-random crossbreeding table. Simulations performed on a large volume of images of different sizes and formats ensure that our approach is not exposed to known attacks.

Keywords: Chaotic map · Feistel-round diagram · genetic operator

1 Introduction

The rapid development of chaos theory in mathematics provides researchers with opportunities to further improve some classic encryption systems. In front of this great security focus, many techniques for color image encryption have flooded the digital world, mostly exploiting number theory and chaos [1–2]. Others are attempting to update their policies by improving some classical techniques, such as Hill [3–4], Cesar, Vignere [5–6], Feistel [7–8].

1.1 Feistel's Classical Technique

After separating the plain text into blocks of size 2 $h \in \mathbb{N}$, Feistel's classical method applied in the block ($U_k = (G_k D_k)$ resumed by the scheme of the figure below (Fig. 1)

.

B.-E. Boudriki Semlali et al. (Eds.): DATA 2025, LNNS 1908, pp. 232–243, 2026.
https://doi.org/10.1007/978-3-032-21200-9_20

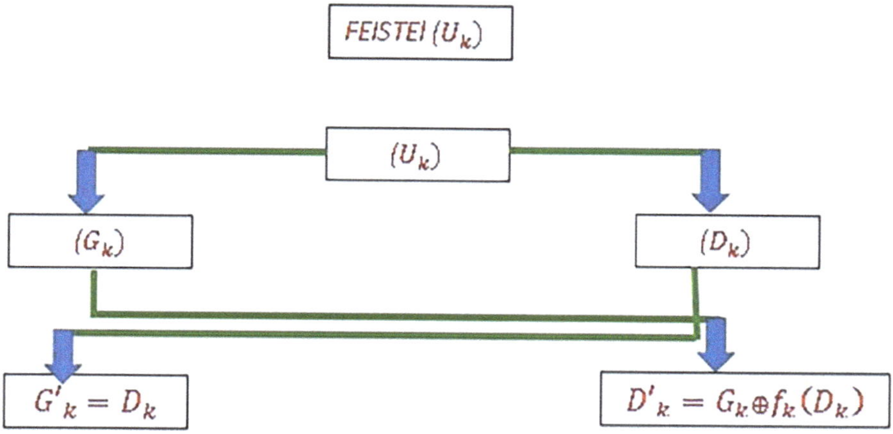

Fig. 1. Feistel's classic round scheme

This encryption can be explained by the following expression, which works on two blocks (,,).

$$g_k(G_{k,}, D_k) = \left(G'_{k,} D'_k\right)$$
$$\left\{ \begin{cases} G'_k = D_k \\ D'_k = G_k \oplus f_k(D_k) \end{cases} \right. \quad With\ k \in [\![1\ t]\!] \tag{1}$$

() Is a pre-defined pseudo-random. In the absence of any diffusion, these three systems remain exposed to differential attacks. Several techniques to improve Feistel's solution have invaded the network world [9–10]. Most of them are focused on increasing the number of turns.

1.2 Genetic Operator

Any genetic algorithm is based on particular functions. The most known and used in cryptography are:

- acting on DNA:
  - Relocation
  - Reversion
  - Mutation
- acting on Codon
  - Crossover

In our approach, we will apply a mutation at re-entry and a crossover at the exit of Feistel's turn.

2 The Proposed Method Description

Centered on the implementation of chaos and confused diffusion, our new method is a profound improvement of Feistel's scheme with the introduction of two genetic operators extremely adapted to color image encryption. This new technique is built around the following steps.

2.1 Chaotic Sequences Development

The choice of the two chaotic maps for the development of the keys and subkeys of our algorithm is related to the ease of their handling and to the high sensitivity to the initial conditions

PWLCM Map
It is a real linear sequence by pieces defined by the equation below

$$\left\{ w_n = f(w_{n-1}) \right.$$

$$= \left\{ \begin{array}{cc} w_0 \in]0\ \ 1[& d \in]0,5\ \ 1[\\ \frac{w_{n-1}}{d} & if\ 0 \leq w_{n-1} \leq d \\ \frac{w_{n-1}-d}{0.5-d} & if\ d \leq w_{n-1} \leq 0.5 \\ f(1 - w_{n-1}) & else \end{array} \right. \quad (2)$$

The Skew Tent Map (SKTM)
The Skew tent map will be redefined as the next equation

$$v_0 \in \left]0 \quad 1\right[\quad p \in]0,5 \quad 1]$$

$$\left\{ v_{n+1} = \left\{ \begin{array}{ll} \frac{v_n}{p} & if\ 0 \prec v_n \prec p \\ \frac{1-v_n}{1-p} & if\ p \prec v_n \end{array} \right. \right. \quad (3)$$

Chaotic Vector Design
Our work requires the construction of four chaotic vectors(), (), ()and () generated from the following algorithm

$$\left\{ \begin{array}{c} for\ i = 1\ to\ 3nm \\ CL(i) = \mathrm{mod}\left(E\left(\frac{w(i)+2v(i)}{3}{}^{*}10^{10}, 254\right) + 1\right) \\ KL(i) = \mathrm{mod}\left(E\left(\frac{Sup(w(i),v(i))+w(i)*v(i)}{2}{}^{*}10^{11}, 253\right) + 2\right) \end{array} \right. \quad \text{(Algorithm 1)}$$

$$KR(i) = E\left(\frac{2^{*}KL(i)+3CL(i)}{5}\right)$$
$$\left\{ \begin{array}{c} if \ \ w(i) \geq v(i) then \\ CR(i) = 0 \ \ else \ \ CR(i) = 1 \\ end \ \ if \\ Next \ \ i \end{array} \right.$$

Chaotic Translation Vector

$$\begin{array}{c} for \ \ i = 1 \ \ to \ \ m \\ VD(i) = \bmod\left(E\left(\frac{w(i)+3v(i)}{4}{}^{*}10^{12}, 3n - 15\right) + 15\right) \\ Next \ \ i \end{array} \quad \text{(Algorithm 2)}$$

2.2 Original Image Vectoring

After the three () color channels extraction and their conversion into size vectors (), (), () (,) each, a concatenation is established to generate a vector (, ,) of size (1,3). This operation is described by the following algorithm

$$\begin{array}{c} for \ \ i = 2 \ \ to \ \ nm \\ \left\{ X(3i-2) = Vr(i) : X(3i-1) = Vg(i) : X(3i) = Vb(i) \right. \\ Next \ \ i \end{array} \quad \text{(Algorithm 3)}$$

This step slightly reduces the high correlation between the pixels, then, the image and the chaotic vectors are proportioned in () blocks of size (3) (Fig. 2).

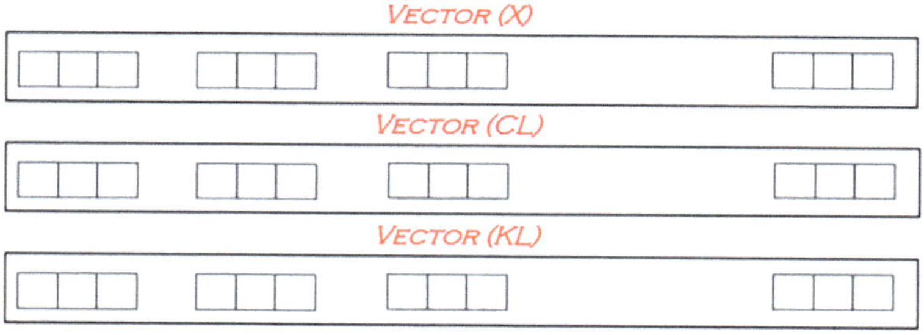

Fig. 2. block partitioning of (3n) pixels

Since the blocks will be taken by pair, we are in front of two possible cases $m \equiv 1[2]$

The, the last block will be treated separately. We will admit that $m \equiv 0\,[2]$

In our approach, we will encapsulate Feistel's scheme by a genetic mutation and a genetic crossover.

Adapted Genetic Mutation
Mutation is the exchange of two sub-blocks and the modification of the other sub- blocks (Fig. 3)

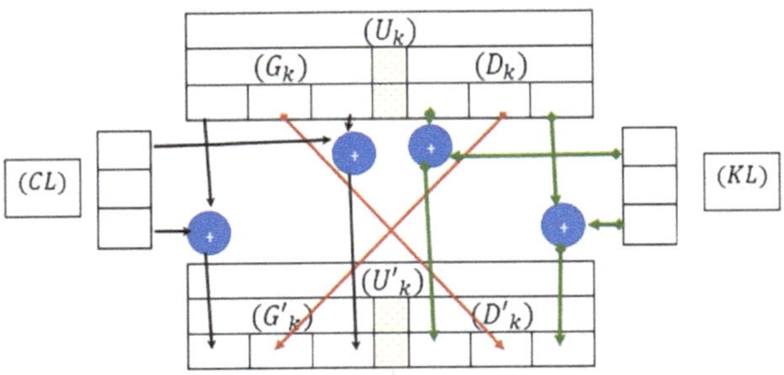

Fig. 3. Proposed Mutation

This scheme is interpreted by the algorithm below

$$\left\{\begin{array}{c} For\ i = 1\ to\ n \\ G'_k(i) = G_k(i) \oplus CL(i) \\ D'_k(i) = D_k(i) \oplus KL(i) \\ G'_k(i+n) = D_k(i+n) \\ D'_k(i+n) = G_k(i+n) \\ G'_k(i+2n) = G_k(i+2n) \oplus CL(i+2n) \\ D'_k(i+2n)\Big) = D_k(i+2n) \oplus KL(i+2n) \\ Next\ i \end{array}\right. \quad \text{(Algorithm 4)}$$

Feistel's Diagram
The output block of the mutation is exposed to a Feistel trick, illustrated by the diagram in the following figure (Fig. 4)

() is a chaotic displacement of the pixels of the block, of step (). This scheme is interpreted by the algorithm below

$$\left\{\begin{array}{c} For\ i = 1\ to\ 3n \\ G''_k(i) = D'_k(i) \\ D''_k(i) = G'_k(i) \oplus D'_k\Big(\mathrm{mod}(i + VD(k); 3n) \\ Next\ i \end{array}\right. \quad \text{(Algorithm 5)}$$

The output ($''$) block of the lathe will be subjected to a genetic crossing adapted to the encryption of medical and color images.

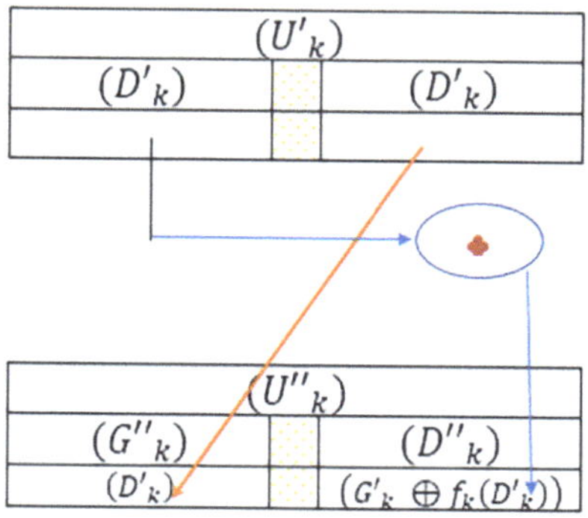

Fig. 4. New Feistel Scheme

Genetic Cross-over
The genetic crossover is an operation that will act at the pixel level by the intervention of three permutation vectors $({}_1)$, $({}_2)$, $({}_3)$ and a binary () control vector, and the two chaotic vectors () ().

Transformation of the Pixel
$(''())$

If () = 0 then it is the vector ()which will be taken for confusion, otherwise it is the vector (). This adapted crossing is interpreted by the following algorithm

$$\begin{cases} For\ i = 1\ to\ 3n \\ If\ \ CR(i) = 0 Then \\ Z(P_3(i)) = U''(P_1(i)) \oplus CL(P_2(i)) \\ Else \\ Z(P_3(i)) = U''(P_1(i)) \oplus KR(P_2(i)) \\ Next\ i \end{cases} \quad \text{(Algorithm 6)}$$

The vector () represents the encrypted image vector.

3 Examples and Simulations

In this section, a large number of medical images and reference colors are chosen to be tested later by our algorithm. First, a simulation study is described for testing a few images.

3.1 Brutal Assaults

They consist in reconstructing the encryption keys in a random manner.

Key-Space Analysis
The brute force attack is a technique used in cryptanalysis to find the encryption key. It

consists of reconstructing the encryption key using all possible combinations. However, the size of our secret key exceeds 128 , which guarantees a better protection against any brute force attack.

Secret Key's Sensitivity Analysis
All chaotic maps used in our algorithm are very sensitive to the initial conditions. Therefore, any perturbation of at least one parameter for the regeneration of the secret key, will produce a different key, this high sensitivity of the key of our new technology, can be seen in the following figure (Fig. 5)

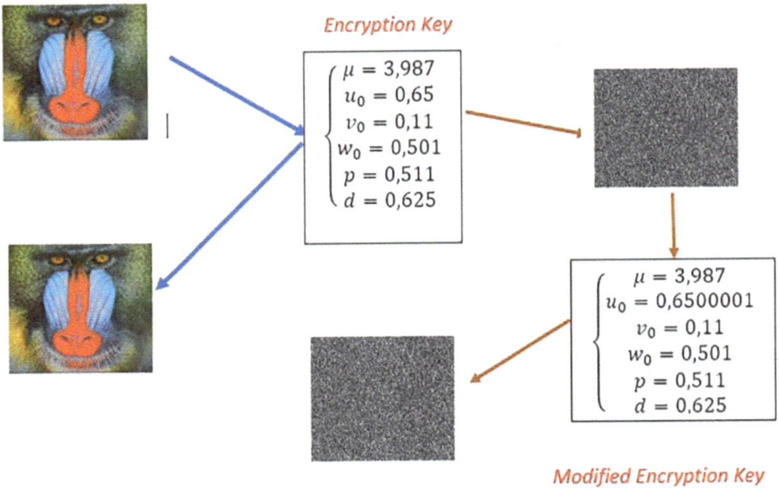

Fig. 5. Encryption key sensitivity

3.2 Statistics Attack Security

In order to prove the robustness of our new medical and color image encryption technique against statistical attacks, many tests on images have been performed, of which we cite the most interesting.

Histogram Analysis
A graphical representation of the number of pixels with the same gray level is called an image histogram. From a cryptography standpoint, an encrypted image's consistent color distribution is critical and provides good protection against statistical assaults. Table 1 shows some of our system's simulation outcomes.

The remarkable resilience to statistical attacks is demonstrated by the uniform distribution of the histograms of the numerical images. The diagram below exemplifies this.

Table 1. Histogram analisys

Name	Original Image	Size	Cipher Image
Img1	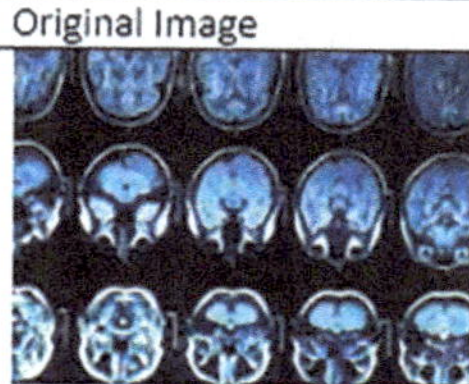	594x814	
Img2	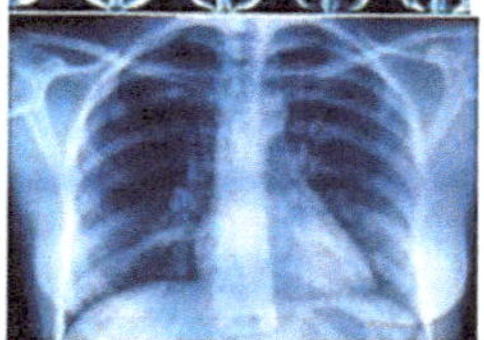	1057x1200	
Img3	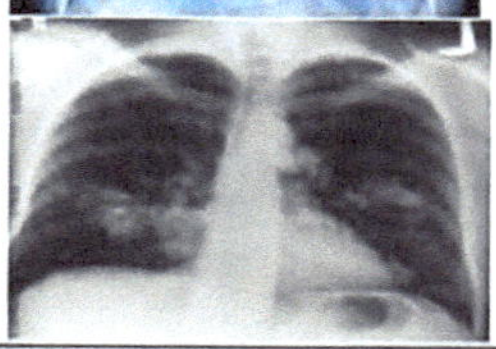	837x921	

Table 2. Histogram analisys

Name	Original Image	Size	Histogram original	Cipher Image	Histogram Cipher
Img4		512x512	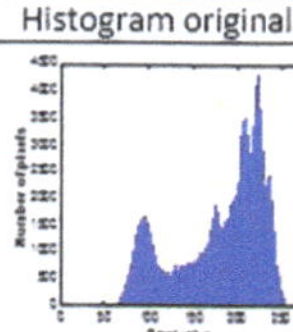	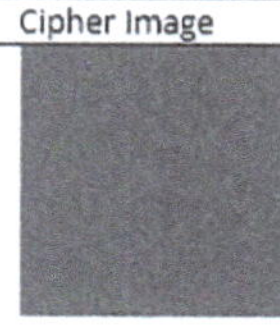	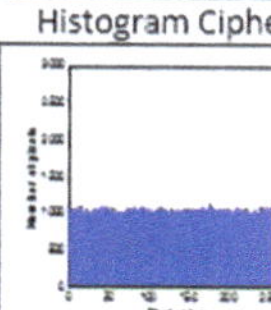
Img5	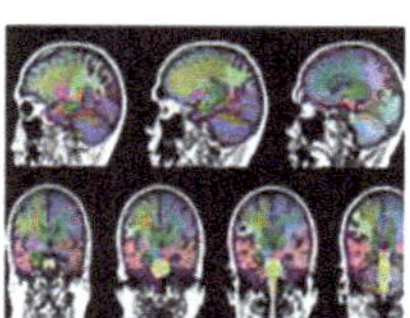	1057x1200	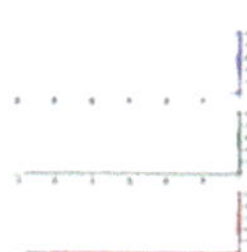		
Img6	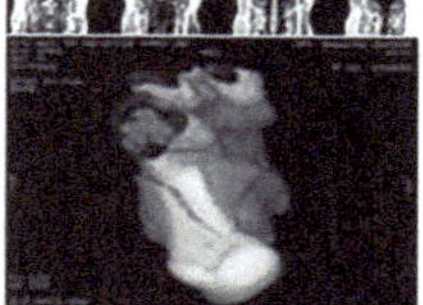	480x480	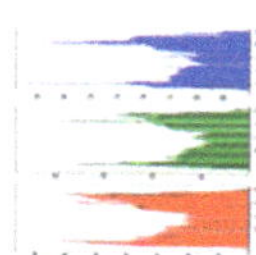		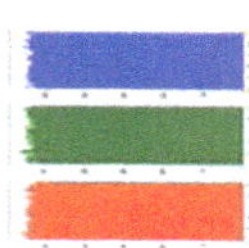

In this work, we notice that all the reference images tested by our algorithm, produce encrypted images whose histogram is flattened. This uniformity of the histogram, proves a great protection against any attack by histogram.

Entropy Analysis

Founded in 1949 by Claude E. Shannon. the modern theory of information is interested in the correction of errors, in the protection against attacks. The entropy of an image of size (,)is given by the following equation

$$\mathrm{H}(MC) = \frac{1}{t}\sum_{i=1}^{t} -p(i)\log_2(p(i)) \tag{4}$$

() is the probability of occurrence of level () in the original image. Table 2 shows the entropy of the images tested by our technique (Table 3)

Table 3. Entropy and correlation analisys

Name	Vertical Correlation			Horizontal Correlation			Diagonal Correlation			Entropy		
	Red	Green	Blue	Red	Green	Blue	Red	Green	Blue	Red	Green	Blue
Img1	–0.0042	–0.0008	–0.0029	–0.0004	–0.0060	–0.0025	–0.0056	–0.0009	–0.0057	7.9978	7.9981	7.9980
Img2	–0.0009	–0.0006	0.0031	–0.0005	0.0001	–0.0005	0.0016	–0.0004	–0.0002	7.9995	7.9995	7.9995
Img3	0.0003	0.0002	–0.0002	0.0003	0.0004	0.0001	–0.0002	0.0004	0.0003	7.9994	7.9996	7.9991

Correlation Analysis

Correlation is the search for a simple linear relationship between adjacent pixels [17]. The correlation of an image of size (,) is given by the equation below

$$r = \frac{\mathrm{cov}(x, y)}{\sqrt{V(x)}\sqrt{V(y)}} \tag{5}$$

The table below compares the original image's pixel correlations to the encrypted image's pixel correlations.

We can see that the correlation value is near to zero, which means that we are better protected from correlation assaults.

3.3 Performance Time

The following figure shows the encryption and decryption time (Table 4)

When compared to the size of the medical photos evaluated, the encryption and decryption times are encouraging.

3.4 Differential Analysis

Let be two encrypted images, whose corresponding free-to-air images differ by only one pixel, from ($\boldsymbol{C_1}$) and ($\boldsymbol{C_2}$), respectively. The mathematical analysis of an image is given by the equation below

$$\mathrm{NPCR} = \left(\frac{1}{nm}\sum_{i,j=1}^{nm} D(i,j)\right) * 100 \tag{6}$$

Table 4. Performance time

Name	Encryption time	Decryption time
Img1	0.04	0.023
Img2	0.032	0.014

$$With\ D(i,j) = \begin{cases} 1 & if \quad C_1(i,j) \neq C_2(i,j) \\ 0 & if \quad C_1(i,j) = C_2(i,j) \end{cases} \tag{7}$$

The mathematical analysis of an image is given by the below

$$\text{UACI} = \left(\frac{1}{nm} \sum_{i,j=1}^{nm} Abs\Big(C_1(i,j) - C_2\big(i,j\big)\Big) \right) * 100 \tag{8}$$

In almost all cryptographic hash functions and block coding algorithms, the avalanche effect is a required property. As the data propagates through the algorithm's structure, it generates increasingly significant modifications. The avalanche impact of the cryptographic structure in place is determined by this constant. The equation below is a good approximation.

$$\text{AE} = \left(\frac{\sum_i bit\ change}{\sum_i bit\ total} \right) * 100 \tag{9}$$

The differential constants found in our simulations performed on a set of reference images are shown in the table below (Table 5)

Table 5. Differential analisys

Name	Original Image	NPCR	UACI	AE (%)
Img7	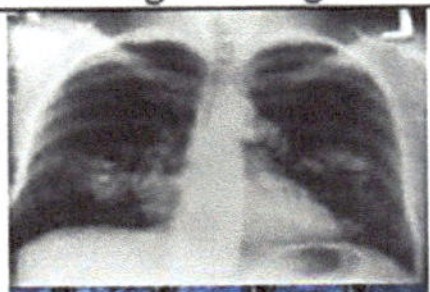	99 ,99	34,64	71
Img8	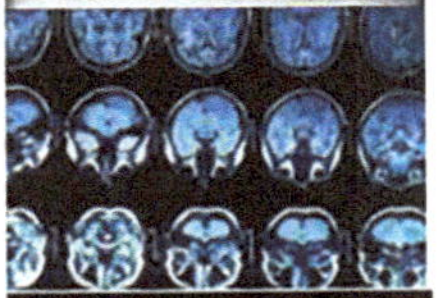	99,99	34,43	73
Img9	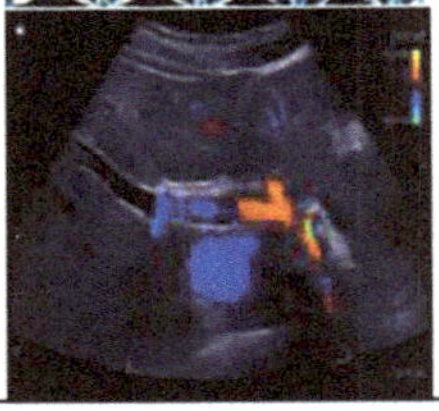	99,98	34,35	73

4 Math Security

The chaotic maps employed in this study are all extremely sensitive to beginning conditions. At the same time, the system's great security against brute force attacks is due to the big size of our secret key. The attack's complexity is increased by the permutation's randomness. Finally, the new encryption scheme is protected against any known attack thanks to the powerful chaining implemented in both towers.

5 Conclusion

The two genetic operators that surrounded the single Feistel trick produced a high-quality encrypted image when combined. This novel technique has a high level of complexity due to the mutation, which consists of a chaotic interchange of somatic blocks under the supervision of a decision vector. The most important role of the genetic crossover, which installs confusion with the chaotic maps used and the implementation of a diffusion to increase the impact of the avalanche effect and thus protect the algorithm against differential attacks, is that of Feistel's trick of moving the pixels. The results obtained support the method's robustness.

References

1. Adinaryana Reddy, K., Vishnuvardham, B., Madhu Viswanatham, M., Krishnaveni, N.: A modified hill chipper based on circulate matrices. Procedia Technol. **4**, 114–118 (2012)

2. Hraoui, S., Gmira, F., Jarar, A.O., Satori, K., Saaidi, A.: Benchmarking AES and chaos based logistic map for image encryption. In: Computer Systems and Applications (AICCSA), 2013 ACS International Conference
3. Hua, Z., Zhou, Y.: Image encryption using 2D logistic adjusted since map. Inf. Sci., 1–25 (2016)
4. Gayathri, J., Subashini, S.: An efficient statio temporel chatic image cypher with an emprovement scrambling algorithm driven by dynamic diffusion phase. Inf. Sci. **489**, 227–254 (2019)
5. Jarjar, A.: Improvement of hill's classical method in image cryptography. Int. J. Stat. Appl. Math. **2**(3, Part A) (2017)
6. Zhao, S.Y.: Image scrambling based on chaos theory and vigenere cypher. In: Computational Intelligence and Security (2011)
7. JarJar, A.: Improvement of Feistel method and the new encryption scheme. Optik. **157**, 1319–1324 (2018)
8. Zhang, G., Liu, Q.: A novel image encryption method based on total shuffling scheme. Opt. Commu., 284–2775 (2011)
9. Feng, X., Tian, X., Xia, S.: An improved image scrambling algorithm based on magic cube rotation and chaotic sequences. In: 4th International Congress on Image and Signal Processing, pp. 1021–1024. IEEE press (2011)
10. Patarin, J.: Security of random feistel schemes with 5 or more rounds. In: Annual International Cryptology Conference CRYPTO 2004.: Advances in Cryptology–CRYPTO 2004, pp. 106–122

Digital Learning

Sustainable Multi-class Classification for Enhancing Adaptive Learning Systems: A Deep Learning Approach

Ikram Amzil[1](✉), Hicham Er-Rradi[1], Souhaib Aammou[1], and Youssef Jdidou[2]

[1] TIMS Laboratory, Faculty of Science, Abdelmalek Essaadi University, Tétouan, Morocco
ikram.amzil@etu.uae.ac.ma

[2] Laboratory of Intelligent Systems and Applications (LSIA), Ecole Marocaine des Sciences de l'Ingénieur, Tangier, Morocco

Abstract. Sustainable education requires adaptive learning systems that efficiently personalize instructional content while ensuring equity, scalability, and resource efficiency. Traditional multi-class classification models, such as logistic regression and decision trees, struggle to capture the complexity of student learning behaviors and lack adaptability for large-scale deployment. Deep learning offers a promising solution, but computational cost, fairness, and interpretability remain key challenges in sustainable AI-driven education. This study proposes a sustainable deep learning-based classification model integrating Deep Neural Networks (DNNs) and transformers to improve multi-class classification in adaptive learning systems. Additionally, Explainable AI (XAI) techniques ensure fair, transparent, and interpretable model decisions. The model is trained using the EdNet dataset, leveraging features such as quiz performance, study duration, and engagement history to capture long-term learning dependencies. The proposed transformer-based model achieves 94% classification accuracy, outperforming traditional methods. XAI analysis (SHAP & LIME) confirms that quiz performance and study duration are the most influential factors in learner classification. Optimization techniques, including model pruning and quantization, reduce computational overhead, enabling resource-efficient deployment. This research highlights the role of sustainable AI in adaptive learning, ensuring accuracy, efficiency, and fairness. Future work should explore hybrid reinforcement learning, graph neural networks (GNNs), and federated learning to develop scalable, privacy-conscious adaptive learning systems that support ethical and accessible education.

Keywords: Adaptive learning · Multi-class classification · Deep learning · Educational data mining

1 Introduction

Adaptive learning systems represent a transformative approach to education, emphasizing personalization by dynamically tailoring learning resources and instructional pathways to individual student needs, abilities, and preferences [1]. The core principle behind

B.-E. Boudriki Semlali et al. (Eds.): DATA 2025, LNNS 1908, pp. 247–257, 2026.
https://doi.org/10.1007/978-3-032-21200-9_21

adaptive systems is the accurate classification of learners based on their learning behaviors, interactions, and performance metrics, enabling precise intervention and effective instructional guidance [2]. Multi-class classification methods are particularly valuable in this context, allowing educators and learning platforms to categorize students into meaningful groups or levels of competency. Effective classification is essential for delivering timely and contextually relevant content, thus enhancing student engagement, improving retention, and promoting sustainable learning outcomes [3].

However, traditional multi-class classification methods, including logistic regression, decision trees, and support vector machines (SVM), often encounter several critical limitations in adaptive educational environments. These limitations include handling complex and non-linear relationships, managing data sparsity and imbalance, and adequately capturing the dynamic, sequential nature of student interactions [4]. Specifically, issues such as class imbalance—where certain student categories are overrepresented—and the cold-start problem (difficulty in classifying students with limited historical data) further limit the effectiveness of these conventional methods [5].

Recent advances in deep learning, especially the integration of deep neural networks (DNNs) and transformer-based models, provide promising alternatives to address these challenges [6 and 7]. DNNs can effectively capture complex, nonlinear student-content interactions, facilitating more accurate and scalable classifications. Meanwhile, transformer architectures have shown notable success in modeling sequential data patterns, thus enabling more sophisticated modeling of student behaviors over time [8].

The primary objective of this research is to develop an efficient deep learning model for multi-class student classification within adaptive learning systems. Specifically, we propose combining DNNs with transformer architectures to model complex and dynamic educational interactions. Additionally, this study investigates the interpretability of the proposed model through explainable AI (XAI) methods to promote transparency and educator trust. Ultimately, the proposed approach seeks to enhance the precision and adaptability of classification processes in digital educational platforms, ensuring improved learning experiences and outcomes.

2 Related Work

2.1 Overview of Classification Techniques in Educational Recommendation Systems

Educational recommendation systems leverage classification techniques to categorize students based on their engagement, performance, and learning behaviors, thus enabling personalized learning pathways [4]. Various classification methods have been utilized, ranging from simple rule-based systems to sophisticated deep learning approaches.

Rule-based systems, historically among the earliest methods applied, use predefined heuristics to categorize learners based on explicit thresholds or conditions. While easy to implement and transparent, these approaches struggle with scalability and capturing nuanced learner behaviors, often resulting in oversimplified recommendations [9].

2.2 Comparison Between Rule-Based Models, Traditional ML, and Deep Learning Approaches

Traditional machine learning (ML) approaches, including **decision trees, logistic regression**, and **Support Vector Machines (SVM)**, have significantly improved upon rule-based methods by utilizing statistical patterns extracted from historical student data [10]. Decision trees are popular due to their interpretability, while logistic regression and SVM provide robust classification boundaries. However, these approaches have limitations, particularly their difficulty in modeling complex, high-dimensional, and sequential learning interactions [5].

In recent years, **deep learning (DL)** models, such as **Deep Neural Networks (DNNs), Convolutional Neural Networks (CNNs)**, and **Transformers**, have emerged as powerful alternatives due to their ability to model intricate patterns in large-scale educational datasets [7]. DNNs offer flexibility in modeling non-linear interactions, CNNs excel in feature extraction tasks from structured data (e.g., student responses or interaction logs), and transformer architectures have proven particularly effective in modeling sequential learning behaviors and interactions over time [6 and 8]. Recent empirical studies have highlighted the superior performance of deep learning approaches, especially in capturing fine-grained distinctions between learners, thereby improving recommendation accuracy and student engagement.

2.3 Emerging Trends in Explainable AI (XAI) and Reinforcement Learning for Personalized Education

Despite deep learning's advantages, interpretability remains a critical challenge, prompting growing interest in **Explainable Artificial Intelligence (XAI)**. XAI methods, such as **SHAP (SHapley Additive exPlanations)** and **LIME (Local Interpretable Model-agnostic Explanations)**, enhance model transparency by providing clear rationales behind classification outcomes, facilitating greater trust among educators and learners [11 and 12]. XAI's integration into educational contexts significantly impacts model adoption, enabling educators to better understand and utilize predictive insights from adaptive systems.

Additionally, **reinforcement learning (RL)** has gained attention as an emerging methodology in personalized education, primarily for dynamically optimizing learning pathways and recommendations. RL treats educational recommendation as a sequential decision-making problem, allowing systems to adaptively recommend content based on long-term learning goals and observed student responses [13]. Combining deep learning and reinforcement learning has resulted in novel adaptive recommendation approaches that significantly enhance personalization by continuously adjusting recommendations according to evolving student behaviors and performance [14].

In conclusion, the literature underscores the ongoing shift from traditional classification approaches toward advanced deep learning, reinforced by explainable AI methods and reinforcement learning techniques, highlighting the potential for transformative improvements in adaptive educational recommendations.

3 Methodology

3.1 Dataset Description

This study employs the EdNet dataset, a comprehensive and publicly available educational dataset, containing extensive records of student interactions with diverse educational content, including quizzes, instructional videos, assignments, and interactive exercises [15]. EdNet provides detailed interaction data such as timestamps, student responses, correctness indicators, and various engagement metrics, making it particularly suited for research into multi-class classification within adaptive learning systems. To strengthen the generalizability and robustness of the results, additional education-based datasets, such as the ASSISTments dataset [10] and the Open University Learning Analytics Dataset [16], were included for supplementary validation.

3.2 Feature Engineering

To facilitate accurate multi-class classification, systematic feature engineering was performed to transform raw interaction data into meaningful, structured representations. Three types of data were specifically addressed:

- **Categorical data** (e.g., student IDs, course identifiers, content types) were converted into numerical embeddings using embedding layers. This method effectively captures the latent semantic relationships among categories.
- **Numerical data** (e.g., interaction timestamps, session durations, quiz scores, response times) were standardized using Min-Max normalization to ensure uniform input ranges, thus improving model convergence and stability during training.
- **Behavioral data**, capturing the sequential patterns in student interactions, were represented using sequential embedding techniques and transformer architectures. This approach leverages self-attention mechanisms to accurately model temporal dependencies, sequence patterns, and evolving student behaviors.

3.3 Model Architecture

The proposed multi-class classification model combines **Deep Neural Networks (DNNs)** with **transformer-based architectures** to effectively analyze complex educational interaction data, particularly focusing on capturing both sequential and non-linear relationships in student behaviors and engagement patterns.

Input Layer and Embeddings
Initially, raw input features, consisting of categorical variables (e.g., student IDs, content types, course categories) and behavioral interaction sequences, are transformed into numerical vector representations through embedding layers. Embedding dimensions are selected via hyperparameter tuning, balancing representational richness and computational efficiency.

Transformer-Based Sequence Modeling
The embedded sequences of student interactions are subsequently processed by

transformer-based models. The transformer architecture incorporates self-attention mechanisms to capture the dependencies and sequential relationships among student interactions effectively. By applying multiple self-attention layers, the transformer generates contextually rich representations that reflect the temporal dynamics and evolving learning behaviors of students. Positional encoding is also integrated, enabling the model to explicitly account for the order of interactions.

Deep Neural Network (DNN) Layers
Outputs from the transformer architecture, which represent sequential learner behavior, are concatenated with additional numerical and categorical embeddings and passed through multiple fully connected neural network layers (dense layers). Each dense layer utilizes ReLU (Rectified Linear Unit) activation functions to capture complex and highly non-linear relationships within the aggregated feature space. Dropout regularization is employed between these layers to mitigate overfitting risks and enhance model generalization. Additionally, batch normalization is applied to stabilize and accelerate model convergence during training.

Multi-Class Output Layer (Softmax Activation)
The final output layer consists of neurons corresponding to the defined learner categories or proficiency levels. This layer utilizes a softmax activation function, converting outputs into probability distributions over the multiple classes. The softmax function ensures that the output probabilities sum to one, providing a clear indication of the predicted learner category or level of engagement. Learners are ultimately classified based on the highest probability output.

Hybrid Reinforcement Learning (RL) Component (Optional Extension)
Optionally, the model architecture can integrate a reinforcement learning-based component, specifically Deep Q-Networks (DQN). The RL extension dynamically refines and optimizes recommendations by treating adaptive recommendations as a sequential decision-making process. This component continuously updates recommendation strategies based on real-time student feedback, optimizing long-term engagement and learning outcomes.

By integrating transformer models, deep neural networks, and potentially reinforcement learning, the proposed architecture effectively addresses the limitations of traditional methods, capturing intricate patterns in educational data, enhancing prediction accuracy, and enabling truly adaptive educational experiences.

3.4 Model Training and Optimization

Model training utilized the Adam optimizer for efficient convergence, with hyperparameter tuning conducted via grid search over several parameters, including embedding dimensions, learning rates, and dropout rates. Cross-validation techniques (five-fold) ensured the robustness and reliability of the model. To address class imbalance, which commonly affects multi-class classification in educational datasets, synthetic oversampling methods, specifically SMOTE (Synthetic Minority Oversampling Technique), were employed. Additionally, early stopping was integrated into the training process, halting model training when validation loss ceased to improve after five consecutive epochs, thereby mitigating overfitting risks.

3.5 Evaluation Metrics

The performance of the multi-class classification model was comprehensively assessed using multiple evaluation metrics:

- **Accuracy**: measures overall correctness across all classification categories.
- **Precision and Recall**: evaluate the effectiveness of the model in correctly identifying learner categories, especially important for classes representing at-risk or high-performing learners.
- **F1-score**: provides a harmonic mean of precision and recall, particularly valuable in scenarios of class imbalance.
- **Area Under the ROC Curve (AUC-ROC)**: assesses the model's capability to discriminate effectively among the multiple learner categories.
- **Engagement Prediction Metrics**: assess the practical impact of classification-informed recommendations on actual student engagement outcomes, including average session duration and course completion rates, thus validating the educational utility of the model.

4 Results and Discussion

4.1 Performance Comparison of Classification Models

The evaluation results indicate that the proposed deep learning-based classification approach significantly outperforms traditional machine learning methods. As shown in Table 1, our transformer-based model achieved the highest classification accuracy (94%) compared to deep neural networks (DNNs) (91%), support vector machines (SVMs) (84%), logistic regression (82%), and decision trees (78%).

Table 1. Model performance comparison

Model	Accuracy	Precision	Recall	F1-Score	AUC-ROC
Decision Tree	0.78	0.76	0.75	0.75	0.80
Logistic Regression	0.82	0.80	0.79	0.80	0.83
SVM	0.84	0.83	0.82	0.83	0.86
Deep Neural Network (DNN)	0.91	0.89	0.88	0.89	0.92
Transformer-based Model	0.94	0.93	0.92	0.93	0.95

The significant improvement in accuracy and F1-score for deep learning models, particularly transformer-based models, highlights the effectiveness of leveraging sequential modeling in educational classification tasks. The superior performance of the transformer-based approach underscores its ability to capture temporal dependencies in student interactions, leading to better classification precision.

4.2 Visualizing Model Performance

To further illustrate the comparative performance of these models, Fig. 1 presents a bar chart depicting key classification metrics.

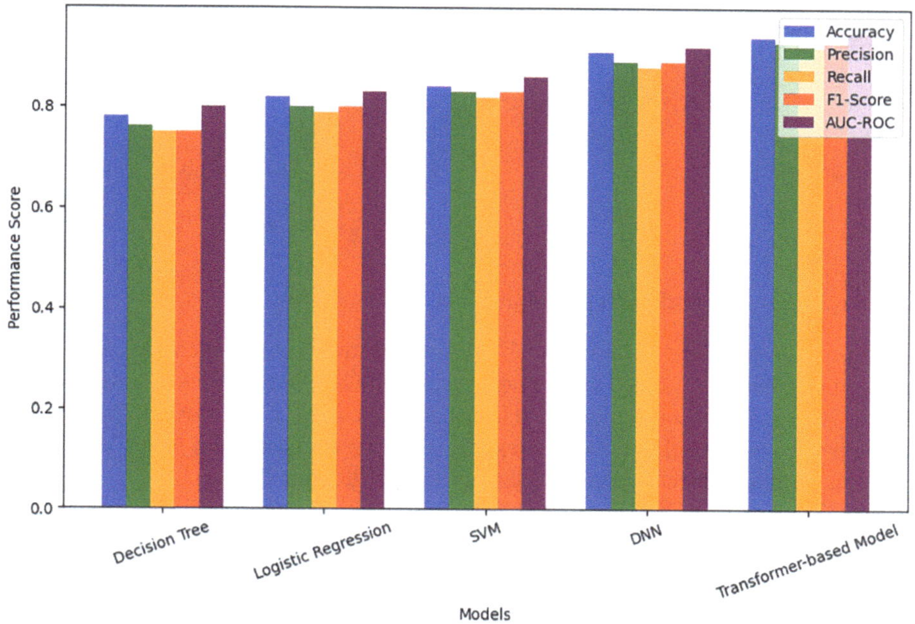

Fig. 1. Model performance comparison across metrics

As shown in the visualization, the transformer-based model consistently outperforms other approaches across all evaluated metrics. This result reinforces the need for deep learning models that capture long-term dependencies and evolving student behaviors in adaptive learning environments.

4.3 Impact on Learner Engagement Prediction

Beyond classification accuracy, the study also assessed the impact of different classification models on learner engagement. We analyzed engagement metrics such as average session duration, interaction frequency, and course completion rates to understand how well each model contributed to personalized learning experiences (Table 2).

The results indicate that higher classification accuracy correlates with improved student engagement, as learners received more accurate and relevant content recommendations. The transformer-based model led to the highest engagement levels, with an 80% course completion rate and the longest session duration (21.4 min per session). These findings demonstrate the importance of accurate classification models in driving personalized learning success.

Table 2. Engagement metrics comparison

Model	Avg. Session Duration (min)	Interaction Frequency	Course Completion Rate (%)
Decision Tree	12.4	5.2	45
Logistic Regression	14.1	6.1	51
SVM	15.3	6.8	58
Deep Neural Network (DNN)	18.7	8.4	72
Transformer-based Model	21.4	9.6	80

4.4 Explainability and Interpretability via XAI

To ensure transparency and interpretability in educational AI applications, this study applied Explainable AI (XAI) techniques, specifically SHAP (SHapley Additive Explanations) and LIME (Local Interpretable Model-agnostic Explanations). These methods help uncover how the model makes classification decisions and identify the most influential features affecting student engagement predictions (Fig. 2).

Key Findings from XAI Analysis

1. Feature Importance: The most significant factors in learner classification were quiz performance, time spent on content, and sequential engagement history.
2. SHAP Analysis: SHAP results indicated that students with higher quiz completion rates and consistent study habits were more likely to be classified as highly engaged learners. SHAP values provided quantitative insight into how much each feature contributed to classification outcomes, revealing that quiz performance had the strongest influence.
3. LIME Visualizations: LIME provided local explanations by showing how slight variations in study duration and response correctness impacted predictions. Students who slightly reduced their study time received lower engagement classifications, whereas those with consistently correct quiz responses had improved classification confidence.

These insights are crucial for educators, as they allow for data-driven interventions while maintaining trust in AI-driven adaptive learning systems. By understanding which factors most influence classification, instructors can personalize learning experiences, optimize content delivery, and provide targeted support to improve student outcomes.

4.5 Limitations and Future Directions

Despite the significant improvements achieved in classification accuracy and engagement predictions, the proposed deep learning approach presents several challenges that need to be addressed for broader adoption in real-world educational environments.

One of the primary limitations is the cold-start problem, where the model struggles to classify new learners who have minimal interaction history. Since adaptive learning

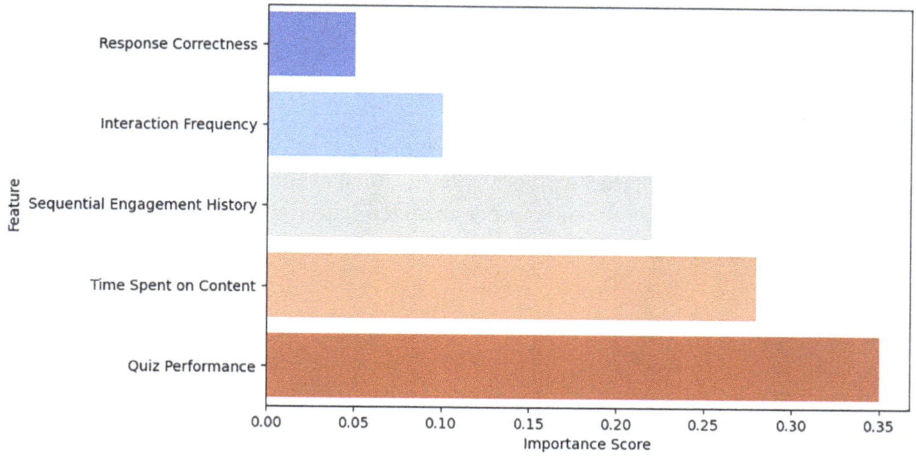

Fig. 2. SHAP Analysis: feature contributions to learner classification

systems rely on previous engagement patterns to make predictions, students with limited recorded activity may receive less accurate classifications. To overcome this, future research could explore meta-learning techniques that enable the model to generalize better to new learners with minimal data. Additionally, graph-based approaches, such as Graph Neural Networks (GNNs), could be utilized to model relationships between students, learning materials, and behavioral similarities, thus improving classification accuracy even in low-data scenarios.

Another critical limitation is computational complexity. Transformer-based models, while highly effective in capturing sequential dependencies and long-term learning behaviors, require substantial computational resources for both training and inference. This computational demand may pose challenges for deploying these models in real-time adaptive learning environments, especially in resource-constrained educational institutions. Future advancements should focus on optimization techniques, such as knowledge distillation, model quantization, and pruning, to reduce computational overhead while maintaining high classification performance.

Additionally, ethical considerations and potential biases in AI-driven educational recommendations require continuous monitoring. Since AI models learn from historical data, they may inadvertently reinforce existing biases in educational assessment and classification. If not carefully managed, the system could disproportionately favor certain learner groups while disadvantaging others. To mitigate this risk, future research should incorporate fairness-aware AI algorithms that ensure balanced and unbiased predictions, as well as mechanisms to audit and interpret AI decisions transparently.

To enhance the effectiveness of adaptive learning systems, several future research directions should be explored. Hybrid reinforcement learning models could be developed to dynamically adjust classification thresholds based on real-time student feedback, ensuring that learners receive personalized recommendations aligned with their evolving needs. Additionally, graph neural networks (GNNs) could be leveraged to improve representation learning, allowing the model to capture complex relationships between students and learning materials in a more structured way. Lastly, federated learning techniques should be investigated to enable privacy-preserving AI models that allow educational institutions to train models collaboratively while keeping student data secure and decentralized.

By addressing these challenges, future developments in deep learning and adaptive learning technologies will lead to more robust, efficient, and ethical AI-driven educational systems, ultimately enhancing personalized learning experiences for students worldwide.

5 Conclusion

This research presents a sustainable deep learning-based multi-class classification model for adaptive learning systems, integrating transformers and deep neural networks (DNNs) to improve classification accuracy and engagement prediction. The model significantly outperforms traditional approaches, achieving 94% accuracy while addressing key sustainability concerns, such as computational efficiency, fairness, and privacy.

Through Explainable AI (XAI) techniques such as SHAP and LIME, the study provides insights into the factors influencing learner classification, ensuring that educators can trust and interpret AI-driven recommendations. Findings confirm that quiz performance, study duration, and sequential engagement history are the most influential features for classification.

While deep learning improves adaptive learning, challenges remain, including the cold-start problem, high computational cost, and potential bias in AI-driven systems. Future research should focus on hybrid reinforcement learning models for real-time adaptive classification, graph neural networks (GNNs) for better representation learning, and federated learning to ensure privacy-preserving AI models.

By incorporating sustainable AI practices, this research contributes to ethical, efficient, and scalable adaptive learning technologies, supporting equitable and personalized education for diverse learning populations.

References

1. Taylor, D.L., Yeung, M., Bashet, A.Z.: Personalized and adaptive learning. In: Innovative Learning Environments in STEM Higher Education: Opportunities, Challenges, and Looking Forward, pp. 17–34 (2021)
2. Gligorea, I., Cioca, M., Oancea, R., Gorski, A.T., Gorski, H., Tudorache, P.: Adaptive learning using artificial intelligence in e-learning: a literature review. Educ. Sci. **13**(12), 1216 (2023)
3. UNESCO: The Future of Digital Education and Sustainable Learning, [Online]. (2021). Available: https://www.unesco.org
4. Romero, C., Ventura, S.: Educational data mining and learning analytics: an updated survey. Wiley Interdiscip. Rev. Data Min. Knowl. Discov. **10**(3), e1355 (2020)
5. Er-radi, H., Touis, B., Aammou, S.: Machine learning in adaptive online learning for enhanced learner engagement. In: Technological Tools for Innovative Teaching, pp. 43–63. IGI Global (2024)
6. Vaswani, A., et al.: Attention is all you need. Adv. Neural Inf. Process. Syst. **30** (2017)

7. Zhang, S., Yao, L., Sun, A., Tay, Y.: Deep learning based recommender system: a survey and new perspectives. ACM Comput. Surv. **52**(1), 1–38 (2019)
8. Sun, F., et al.: BERT4Rec: sequential recommendation with bidirectional encoder representations from transformer. In: Proceedings of the 28th ACM International Conference on Information and Knowledge Management, pp. 1441–1450 (2019)
9. Brusilovsky, P., Millán, E.: User models for adaptive hypermedia and adaptive educational systems. In: The Adaptive Web, pp. 3–53. Springer (2007) [Online]. Available: https://doi.org/10.1007/978-3-540-72079-9_1
10. Pardos, Z.A., Baker, R.S., San Pedro, M.O., Gowda, S.M., Gowda, S.M.: Affective states and state tests: investigating how affect and engagement during the school year predict end-of-year learning outcomes. J. Learn. Analytics. **1**(1), 107–128 (2014)
11. Lundberg, S.M., Lee, S.I.: A unified approach to interpreting model predictions. In: Advances in Neural Information Processing Systems, vol. 30, (2017)
12. Ribeiro, M.T., Singh, S., Guestrin, C.: Why should I trust you? Explaining the predictions of any classifier. In: Proceedings of the 22nd ACM SIGKDD International Conference on Knowledge Discovery and Data Mining, pp. 1135–1144 (2016)
13. Agrebi, M., Sendi, M., Abed, M.: Deep reinforcement learning for personalized recommendation of distance learning. In: New Knowledge in Information Systems and Technologies, vol. 2, pp. 597–606. Springer (2019)
14. Amzil, I., Aammou, S., Tagdimi, Z., Erradi, H.: Personalizing learning experiences with Q-learning in adaptive educational systems. In: Proceedings E-Learning and Smart Engineering Systems (ELSES 2023), pp. 70–77. Atlantis Press (2024)
15. Choi, Y., et al.: EdNet: A large-scale hierarchical dataset in education. In: Artificial Intelligence in Education, vol. 21, pp. 69–73. Springer (2020)
16. Kuzilek, J., Hlosta, M., Zdrahal, Z.: Open university learning analytics dataset. Sci. Data. **4**(1), 1–8 (2017)

Implementing Technological Change in an English Secondary School

Louise Hayes(✉)

Manchester Metropolitan University, Manchester, UK
l.hayes@ucl.ac.uk

Abstract. Computer Science in English schools is reported to have a lack of existing teacher confidence in expertise in teaching programming in its secondary schools. More recently, much progress is being made in pedagogical practice that supports pupil progress in the subject of Computing (Stoet & Geary, 2018). However, this presentation reflects an this presentation reflects on an early piece of writing that looked at the introduction of an e-learning system to support delivery programming from a curriculum intervention from the perspective of a teacher within the computing discipline. Through an analysis of the teachers lessons and reflections on classroom practice, the presentation considers the aim to understand how an online learning environment can support teachers with subject knowledge in the delivery of python programming to a class of Key Stage (KS) 4 (14 year old) pupils. In initial teacher education, having the appropriate digital skills and reflection on practice are integral parts of the course (DFE, 2025). The reflective data is then discussed using Bourdieu's frameworks specifically of field, habitus and capital in a desire to problematise teacher habitus-field with a deliberate attempt to improve pupil outcomes. It concludes that system outputs support the notion of pedagogical benefits of using learning platforms when correctly embedded into lesson planning, with the potential to provide more useful data to subject teachers for formative and summative assessment and planning to support pupil progress. Furthermore, for non-specialist teachers in the subject, where a shortage of computing teachers is reported to exist, the findings support the use of e-learning systems which have potential to support pupils in the delivery of programming, and give teachers in computing, a more stable field to develop a secure subject pedagogy.

Keywords: AI pedagogy · Bourdieu · Computing · Data · Education · English Secondary School · Intervention · Habitus-field · Python · Reflective Work · Subject Pedagogy

1 Computing Education in England

1.1 Introduction

The aim of this presentation is a reflective piece that investigates the changes in pedagogical practice, focusing on the implementation of an e-learning platform to support the teaching and learning of programming to secondary school students. The presentation is underpinned by a reflective paper was written prior to the COVID-19 pandemic.

B.-E. Boudriki Semlali et al. (Eds.): DATA 2025, LNNS 1908, pp. 258–270, 2026.
https://doi.org/10.1007/978-3-032-21200-9_22

The aim is to give the audience perspectives of a computing teacher, and how systems developed support the pedagogical practices of python programming to a mixed ability, and gender, pupils at Key Stage 4 (Year 10) class of pupils in a Secondary School in England.

1.2 Background

Firstly, in setting the scene, the computer science discipline has undergone a series of changes in recent years. A report in the United Kingdom (UK), written by The Royal Society (2017) stated that 'evidence gathering showed that pedagogies for computing in schools remain less developed than those for other subjects, as computing education research in the past has focused on higher education' The Royal Society (2017). As a result of this curriculum change, now nearly 10-years on, there continues to be a shortage of trained computing teachers. A post-pandemic longitudinal study would give a more in-depth understanding of the influence on pupil progress and uptake in the subject. The subject has changed significantly and the view taken was that the system would support teachers that have seen many changes, from the delivery of ICT qualifications, in reports written about what can be seen as 'good pedagogical practice'.

Furthermore, in more recent research, a large number of schools in the UK are reported to not offering computing, where in a more recent study by Hadwen-Bennett & Kemp (2024) 'the majority of teachers have taught students better at programming than themselves, with female teachers more likely to say this than male teachers' (Hadwen-Bennett & Kemp, 2024). It is not all bad news, however, and there are some positives, 'there is much to celebrate and there are many pockets of excellence' (Royal Society, 2017) but this is not reported to be the case nationally in England. Whilst it can be seen that there are many reasons for this, the concern remains for a large number of school pupils who are not being given the opportunity to study a subject or being taught in what constitutes 'good teaching', in a subject that gives them a fundamental understanding of how computers work, irrespective of whether they continue to study it beyond school. This view is shared by many others. The report goes on to say that this is a generation of pupils who will do jobs in industries that do not even exist today. I would argue that giving all pupils the opportunity to be educated in the discipline is as fundamental today as learning to read and write was a 100+ years ago.

1.3 Research Questions

As an educator it is important to challenge practice and therefore, in problematising the implementation of an e-learning platform., I wished to understand how a theoretical lens could give me an alternative perspective in teaching programming. Whilst is has been reported that there are many benefits of teaching through e-learning platforms, which offer problem-based learning programming activities, and data to track the progress of pupils, the use of them in classrooms is not without ethical considerations. This is worthy of further exploration (Jorgensen et al., 2013: 222) and I would argue that 'the political rhetoric in favour of the current policy direction suggests that testing, accountability, league tables are all strategies to improve standards' Jorgensen et al., (2013: 222). Furthermore, Ball (2012) who sees himself as a 'neoliberal academic', is more specific in this debate in the neoliberal education policy discourses that 'we come to know and value others by their outputs rather than by their individuality and humanity' (Ball, 2012). With all this being the case, Ozga (cited in Ball, 2012a, b) more succinctly puts it, as professionals we are 'literally governed by numbers' (Ozga, cited in Ball, 2012a, b); from this I mean that it can be seen that systems provides school with an easy measure of not only pupil progress, but it could be argued teacher output too. As an now ex- teacher, I am inclined to agree with Ball (2012) that at times I was required to deliver lessons that are seen as 'calculable rather than memorable' (Ball, 2012a, b), and I wonder if a system can ever be a replacement for good teaching. This is material for a further study, whilst the focus of this presentation is not the question of what constitutes good teaching (see Fig. 1.), but to reflect on the implementation of a system to support teaching programming into a classroom.

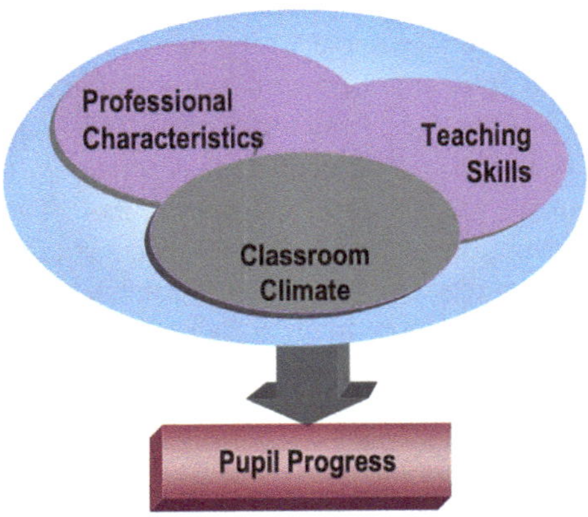

Fig. 1. What makes a good teacher Hay/McBer (2000)

At this point, it is important to acknowledge pupil progress, which holds particular significance for teachers and teacher educators in computing education. In a report by The CEM Centre a research centre at Durham University about 'what makes great teaching' (Coe et al., 2014) found that 'student progress is the yardstick by which teacher quality should be assessed' (Coe et al., 2014). The study outlines six common components that should be considered included: content knowledge; quality of instructions; classroom climate classroom management; teacher beliefs and professional behaviours (Appendix 1), where the very best teachers demonstrate all of the features, as believed by Coe et al., (2014). To further understand what makes good computing teaching, according to Gage (n.d), more recent research on the support of adaptive planning (Appendix 2), such as the application of Rosenshine (2012) in computing lessons, where the notion of hinge questioning will further support pedagogical practice.

1.4 E-Learning Pedagogy

As a classroom practitioner. the complexities of using learning environments in pedagogical practice in Secondary School classrooms adds an additional layer for deconstruction.

In a study written by The Open University (2015), with the Centre for Technology in Learning, explored new forms of teaching, learning and assessment for modern school education. For Massive Open Online Courses [MOOC] learning pedagogy is based around conversation and social networking. Much research is centred around their use in further education, and according to Dziabenko and Adorno (2017) are 'not used extensively in schools', who looked specifically at the 'micro-MOOC' (Dziabenko and Adorno, 2017) for remote scientific experiments, where a small part of a topic is taught 'because of the time restricted nature of a lesson in a school' (Dziabenko and Adorno 2017). Their research suggests that it supports teachers and learners based on their pedagogical preferences, and teachers are able to adapt the MOOC schedule with teaching and learning resources. For this paper, whilst this is of interest to a wider study, the focus will remain on the teacher reflections on the delivery of pre-prepared curriculum content within the system, delivered over the course of a school term.

1.5 Methodology

Firstly, I explore the notion of field through Bourdieu's concept of field theory, whereby 'a field is a key organiser' (Jorgensen et al., 2013: 224). In exploring these positions within the computing secondary school 'field', as with other fields, doxa exist, which are natural or normal taken from granted assumptions in a field 'contingent nature of these discourses are not questioned nor even recognised' Nolan (2012: 205). The practices are in early development in the subject and as stated earlier in the Royal Society Report (2017), the pedagogy is less-developed than other subjects. With the relative newness of the computing discipline, in comparison to more established subjects, the norms in the field are not yet fully entrenched. I would argue at this stage likely to differ from teacher to teacher, and in different school settings. If 'mathematics holds a privileged place in the school curriculum' (Jorgensen et al., 2013) it is difficult to say where computing teachers fit within the field as it exists in some schools, and has not been granted a place in the curriculum of others. In addition to this, not all pupils are able to access the subject, where some schools require pupils to have achieved a Level 6 in mathematics to be able to take computing at Key Stage 4. This clearly has a major impact upon the status and inclusive access to the pupils who are able to study the subject. It is worth pointing out that when writing this my 'field' norms may be very different to that of another computing teacher. Other teachers in the subject will draw their guidance from a different set of norms and pedagogical practices. New teachers of computing are more exposed to pedagogy that did not exist when the subject was newly developing in school curricular. This all contributes to a reported lack of pedagogical practice understanding, and subject knowledge in existing teachers.

Secondly, Bourdieu's theory of 'habitus', is defined by Jorgensen et al. (2013: 223) as a collection of informal skills and knowledge which participants have constructed over time. Whereby the influence of the home environment adds an additional layer. Bourdieu believes that informal skills and knowledge are 'rather more than mere preferences as they are operationalised by class positions' (Bourdieu, 1984). He believes that social class impacts upon a pupils access to the curriculum. Consequently, one pupil's access and use of using a computer at home is going to be different from another's. By this I mean pupils who live with families who are not only able financially to purchase the equipment, but also know how to use it, for example, they learn how to write computer programs, take them apart and rebuild them will enter the classroom with a better understanding, or 'habitus', thus ensuring they are better equipped in the subject. Pupils with different experiences in their home use of computers, without access to equipment, or have parents who use them, may be less able to engage with their adoption of the subject. As these factors go beyond the influence of the teacher in the classroom, habitus provides a real challenge to any teacher and researcher in the discipline of computing in particular. This influence has implications in the classroom. To pupils who have little, or no access to computers at home, will likely by ill at ease with the work to enable them to fully access the curriculum.

Another missing layer of interpretation as with other subjects, according to Jorgensen et al., (2013), social heritage converts to academic success, is an integral part of this heritage. I was interested to further explore this comparison to mathematics in teaching computing. Whereas, in society, taking an example of mathematics, this subject is traditionally held as important, according to Jorgensen et al. (2014). In making a comparison with computing as a discipline, the 'cultural capital' of this subject, whilst seen as important to those in the 'know', is clearly not so in a wider context. As a pupil of the subject in the 1980's, I was exposed to computers as a child in the 1980's, at home and at school. My parents decided it was a good subject for me to learn they both worked in the telecommunication industry, so it can be seen that my home 'habitus' led me to value the subject. My own school experience is likely to be very different from another practitioner in the subject.

A third concept 'one that plays an important role in the relationship between field, and habitus is capital' Nolan (2012: 204), whereby cultural capital accordingly is a synonym for status or position that one brings to the field, according to Grenfell (cited in Nolan, 2012). Here I consider the subject of a secondary school teacher, which plays a significant part in the capital they hold within the school; status in the subject that is taught, according to Day et al., (2006) will give a more dominant position (Day et al., 2006). Another important point to consider here is the school, which in itself has its own position within a field. Pupil performance is often seen as a reflection of the teacher's reputation in this field.

1.6 Data Collection and Small Data Sets

I now explore a small-scale reflective paper, which draws on data from a classroom of 10 pupils in an English secondary school. The pupils study 5*60-min lessons per week. The system was introduced to the pupils at the start of their GCSE course. The data collected for this study can be considered a small dataset. While much has been written about the analysis of large datasets, this study was intentionally limited in scope due to the initial research brief and was not designed as a large-scale investigation. In many cases, small datasets are used because of constraints such as cost, time, or ethical considerations. These factors also contributed to the limited size of the dataset in this study. The principles of statistics apply in the analysis of a small dataset where according to Faraway and Augustin, (2017) 'Big data is justifiably a major focus of research and public interest. Even so, small data is still with us' (Faraway and Augustin, 2017).

Although the group was small, the aim of the study was to gain a deeper understanding of the implementation of an online system—essentially prompting a shift in teachers' pedagogical practices, rather than simply aiming to improve efficiency or convenience 'claims about all teenagers' (Johnson, 2009). If more time had been available, a random sample or a comparison made with a control group could have been implemented. And/or even the experiences of another member of staff would enable me to widen or compare the data with the experiences of another member of staff, however, in the time frame it was not possible to extend the survey. For a larger study, which may afford a more comprehensive picture, the data collection method would need to follow ethical guidelines. For more rigorous analysis of the data, it is important for a researcher to consider guidelines advised by the UK Economic and Social Research Council [ESRC] and University Guidelines. Which were followed as part of this reflective piece of work, and additionally the pupils referred to in this study have been anonymised to ensure this is taken into consideration.

The first stage of the data collection was conducted through note taking and a reflective reflections at the end of a series of 60-min lessons. Reflective diaries or journals are an expectation of teachers undertaking initial teacher training and lessons were planned to include the use of the learning environment. The location of the lesson was a computing laboratory, which is arranged in a series of 4 banks of computers on desks, and the teacher has a desk at the front of the classroom. This aspect of the work influenced my future research in classroom design. Returning to them, I completed diary entries at the end of each lesson which was collected over a period of a school term. The second stage in the process was to reflect upon the data that was gathered. It would be easy to adopt a positive approach to the collection of the data, and whilst I have developed this aspect of the presentation to look at the progress of the pupils using quantitative data, the original paper was written as a reflective piece. Finally, the data is collected from the system was entered into a spreadsheet that enabled me to understand pupil progress through the modules of learning and compare them to examination board requirements, discussed in the next part of my presentation.

1.7 Data Findings

To reaffirm, the focus of the presentation is to reflect upon a paper written pre-COVID-19 to understand how the introduction of an e-learning system can support the teaching of programming.

In the first programming lesson the system was introduced to the pupils. Initially there was some resistance to it, and one of the pupils asked why they were not learning programming using the software Python (see Appendix 3). The pupils then made a number of sarcastic comments about the name of the product [anonymised for this study], this resulted the class then searching the internet to find out what it meant. It could be seen from this that the class already had a 'feel for the game' (Bourdieu, 1984), which is important, but it created some conflict as their expectations were not initially met as I had shifted the rules in asking them to use an alternative means of learning to program Python. This would emphasise a need for them to know the 'rules' of the game. I also had to learn how to use the system, and complete the modules ahead of the pupils in my care.

1.8 Modules and Pupil Progress

Lesson 1: Once the pupils were enrolled onto the system, the lessons were initially more manageable in terms of planning.

A specific example of 'habitus' can be seen from Pupil A who asked 'can we have more than one hour of programming a fortnight?'. I did not press him further on this matter, other than to state the GCSE AQA course has a higher content requirement of theoretical knowledge than practical programming study. This perhaps suggests this particular pupil's home 'habitus' of programming and Pupil A does not resonate or align with the practices of computer science in the school environment.

During the first lesson Pupil B then managed to delete an account. This resulted in an email to the organisation to get his account re-instated; a lesson was lost as he had not followed the instructions. This should have been an easy task in a familiar environment. Whilst it is not possible to comment with any certainty, it would appear that in presuming a 'habitus' for Pupil B, it was not considered that he would have a problem with logging onto a system.

Lesson 2: In the second lesson, the pupils were able to access the software and were happy to get straight on with their work. This left me feeling that I was surplus to requirements and the 'doxa' that usually exists were challenged; my usual mode of teaching instruction altered as the pupils worked independently on tasks. This did not last for long. I have often found problem solving in computer science often divides the more able in the class and it became evident that not all of the pupils sat within this more able boundary.

Lesson 3: The next lesson started to show more of this divide. The 'easy tracking' facility in the system [replicated in Fig. 2.0] showed the pupils who had a computational habitus (Appendix 2) in its progress chart. This then highlighted the competitive nature of the group. I had not seen that coming. An example of Bourdieu's 'social relationships' that are invisible to observation, and as an independent school this result driven ethos in the pupils is one that I frequently forgot. It is clear from this that my own habitus did not align with that of the school, demonstrating the competitive nature of pupils [capital] who have an expectation to see progress. This is an advantage for the pupils who were able and were achieving well. But is quite exposing for the pupils who were behind the rest of the class. I was once again uncomfortable with this.

Lesson 4: In the fourth lesson, one particularly interesting comment from Pupil B who is a highly able boy, was that he had managed to find a system fault, and was then trying to 'hack' the programming code behind the platform. He then stated 'if I get the answer wrong 3 times it will give me the right answer anyway'. The culture of having to get things right is reinforced with the system. This resulted in me having to follow school sanctions, for the 'hacking' offence. As a further example of habitus, it is not clear if the pupil's home use of computers had led him to being inquisitive as he had the confidence and understanding of what to do behind the scenes of the online environment. I would argue that his 'habitus' enabled him to behave as a programmer. If this was the case, this natural action was shaped by his experiences in his home environment, unfortunately for him, it did not sit within school sanctions, and he ended up in detention.

1.9 Key Findings

The presentation has made use of the Google Scholar and Researchgate databases, to further support the reflective piece of work and what could be argued as a small data set gathered in the classroom. Due to the nature of the assignment brief, the timeline of the assignment was short where a longer wider study was not possible. However, it became apparent that this needed to be extended using literature that had been written after a significant change in teaching had occurred due to COVID-19 pandemic (01/01/2020-01/07/2020). Therefore the search for literature was extended over the period since the original study to gain a better understanding of the uses of pedagogy of e-learning systems since this time.

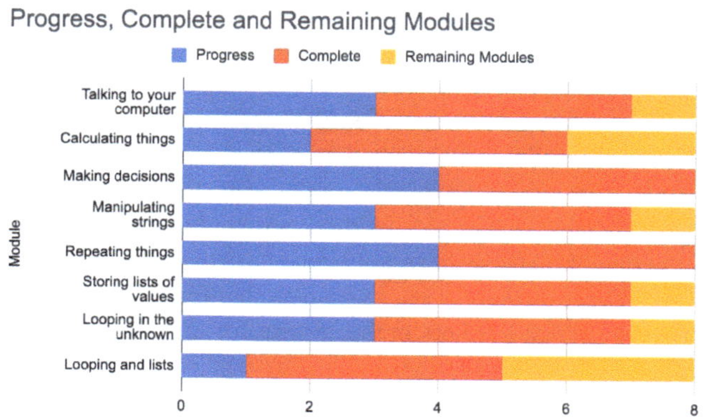

Fig. 2. Pupil data modules sample recreated as a bar chart.

The data is discussed in the next part of the presentation is analysed through two aspects; firstly using a positivist approach where a re-created example of a pupils achievement (see Fig. 2.). Secondly through a novel approach through the use of Bourdieu's frameworks which allows me to problematise the findings and shift any pre-set writers viewpoints.

1.10 Discussion

Teachers training as part of the PGCE Computing must prepare a lesson plan and an evaluation of the lesson. Assessment, as formative and summative supports planning for further lessons and tracking pupil progress (Table 1).

The example re-created (see Table 1.) what the pupils have learned, but not necessarily what they can recall from the lesson. Interventions in practice are required by teachers and trainees, and from the data shown (see Fig. 2). Lesson planning is supported through the use of the system enabling more time to be spent on adapting lessons, additionally using progress of modules completed for future lessons, allowing teachers to build in recall practice as part of planning. The key theme from the data collected shows the progress of the pupils through the modules. In the bar chart shown (see Fig. 2). the results from the data demonstrate progress towards the modules and give teachers a clear

Table 1. Pupil data extracted into a spreadsheet 2D Bar Chart Example of pupil progress.

Module	Progress	Complete	Remaining Modules
Talking to your computer	3	4	1
Calculating things	2	4	2
Making decisions	4	4	0
Manipulating strings	3	4	1
Repeating things	4	4	0
Storing lists of values	3	4	1
Looping in the unknown	3	4	1
Looping and lists	1	4	3

set of data on module progress, and the order of completion the system will support the learning and progress of pupils, and provide data of progression and engagement in work and lessons. The more specific benefits relating to using an online learning system outweigh the disadvantages in it's use. Whilst there are challenges in the implementation, and differing pedagogical methods have not been studied as part of the reflective piece, there are clearly benefits of using a system such as this for newly qualified teachers of computing and non-specialist teachers in the subject for that matter.

1.11 Applying Bourdieu to Data Findings

Reflective Diary from the Lessons

Within a Bourdieusian framework, the 'structuring practices in the field' (Jorgensen et al., 2012, 233), which in this case is ability within the class, has created a divide in the group particularly for the less able pupils. By lesson 4 this contrast was clearly seen as the more able pupils had sped ahead of the rest of the group, as shown in the data (see Fig. 2.), where three of the group had not completed modules as expected and 'were behind' expected progress as shown in the bar chart.

Identifying Misconceptions

In this example, the data supports pupil progress and evidences my need or 'doxa' as a 'good' teacher for intervention to take place. The e-learning system gives clear data on progress and engagement in learning activities of the pupils, that enable teachers to adjust planning for future lessons, identifying misconceptions in pupil understanding. Additionally, it became apparent that additional sessions were required to help pupils achievement in the group to make further progress.

Pupil Progress

From the data the evidence suggests, as it was clear that Pupil D [the only female], whilst outwardly saying everything was OK, things were clearly not so if her progress was to be believed. Through the use of the chart it was evident she was not where she should be. However, reviewing my diary notes, she was showing a disposition towards solving the problems set, and was doing so at her own pace. What was not clear as a teacher was that has she not felt the need to cover up her progress falling behind others, or she had not used or discovered, the system flaw that a pupil came across in a way to beat the system.

Adapting Practice

The final observation in viewing the pupil progress and that of pedagogy if the pupils are showing progress through the chart, I still felt an sense of discomfort with 2 of the pupils who could be seen through progress charts and were underachieving, whilst they were making progress, were not able to answer direct questions that were asked of them. By this I mean the pupils who fail to learn were being sorted within the class and the system was classifying them as able or not through the speed of their work. I would argue that the pedagogic mode of the online environment is limited, and teacher intervention was then required.

1.12 Conclusion

This presentation has re-visited a reflective paper which illustrated the application of Bourdieu's theory as an alternative perspective to the introduction of an e-learning system. On returning to the work, it is clear that whilst some themes have emerged from the data, there are a number that require further exploration to fully understand what constitutes good pedagogical practice of the use of an e-learning system in the computing discipline. Furthermore, the influence and potential benefits of the use of to support inclusion, pupil progress and motivation to achieve.

What the findings suggest is that the 'habitus' of the pupils, and the teacher, potentially play an important part in their access and predisposition to the curriculum. In choosing computing, assumptions had already been made by the pupils and their parents about the subject, but confusion exists about what is studied in the lessons as the evidence suggests that pupils wanted and expected to have more programming lessons. Without this clarity the subject identity and status will continue to be affected. It is essential that clarity is gained in the discipline to ensure it can gain a more commanding presence in the curriculum. This can be done, in part, with a clear link to curriculum mapping (see Appendix 2) and teaching schema such as through an examination board for example in the UK the AQA GCSE Specification.

A particular issue around adapting teaching and differentiation in lessons was found to be the case, and a benefit of the study afforded teacher intervention to address this. What is of concern is that without a clear view of good pedagogy in computing, teachers in the subject will not be confident in their delivery in lessons, and may use systems such as this as a method of subject delivery. Whilst the report found there were many benefits to using the system, a further study would be required to look at what constitutes good pedagogical practice in the use of e-learning systems within the subject.

This reflective piece of writing finds that pupil engagement was evident and the system was found to provide motivation, in addition to data on progress. Pupils who were better players in the game [habitus], made quicker progress, were more able and were afforded a more dominant position [capital] in the class. A further investigation of social status [habitus] and its impact upon successful access to the subject will give further insights into pupil progress. For example, the 'habitus' of the pupils who were underperforming was unknown. This is a concern, as clearly through introducing a system, presumptions have been made when they enter the classroom, and home 'habitus' can impact upon access and ability in the discipline.

1.13 Recommendations/Further to the Study

The initial intention of this paper was to write a paper as part of my early writing as a Doctoral student as a reflective piece. Needless to say Bourdieu was 'more than a little helpful' Ball (2015: 816) in my thinking about an intervention and I was able to make some adjustments to my practice throughout the data gathering period. Finally, returning to Ball (2015), once again, succinctly puts it, my individuality and humanity prevailed and the research agenda in the end was to improve my practice and ultimately support my pupils learning and attainment, through the implementation of a system that allows for data to be measured to support pupil progress.

In revisiting this work for the presentation, I recommend the following five areas for a future, more in-depth study to support teachers in initial teacher education:

1. Questioning techniques the use of a system will enable teachers to target questioning and plan lessons around the modules. For example, hinge questioning can be targeted in adapting lesson planning.
2. Statistical analysis further statistical analysis of the data to support a more rigorous study and gather a larger data set with a pilot and control study. This will support forecasting of potential pupil progress in the subject using historical data sets. Artificial Intelligence (AI) and Machine learning (ML) practices in the classroom are becoming more prevalent in their use since writing this paper where for example systems such as https://www.kaggle.com/ could be used to support dataset manipulation, adaptive planning and more approachable learning for computing teachers.
3. e-Learning pedagogy a more in-depth section and study of e-Learning pedagogical methods, with particular reference to the post-pandemic use of e-learning and MOOC systems in the classroom to further support inclusive teaching practices, pupil progress and critically engagement in the computing subject.
4. Habitus as part of lesson planning, teachers use data to support a wider collection of data of pupils for inclusion purposes, such as teacher intervention from system feedback, pupil premium data, home postcodes (as with any lesson preparation ensure that school policy is followed ensuring ethics are considered as part of this practice).
5. Data tracking pupil progress machine learning will help with reducing teacher workload and support more targeting planning using a system to track progress through the curriculum will enable teachers to measure pupil progress through modules and gain insights into areas for further support that may require pedagogy. Statistical analysis using machine learning will better predict pupil outcomes using datasets from existing classes to support future forecasting.

Acknowledgments. This paper was an early piece of reflective writing, with express thanks and gratitude as a student of the Phase A Education Doctorate at Manchester Metropolitan University.

Disclosure of Interests The author has no competing interests to declare that are relevant to the content of this article.

Appendix 1

What makes great teaching?
6 Common Teaching

Pedagocial content knowledge – strong evidence of impact on student outcomes
Quality of instruction strong evidence of impact on student outcomes
Classroom climate moderate evidence of impact on student outcomes
Classroom management moderate evidence of impact on student outcomes
Teacher beliefs some evidence of impact on student outcomes
Professional behaviours some evidence of impact on student outcomes

Appendix 2

GCSE Computer Science AQA Examination Board
What's assessed
Computational thinking, problem solving, code tracing and applied computing as well as theoretical knowledge of computer science from subject content 1–4 above.
How it's assessed
Written exam set in practically based scenarios: 1 h 30 min
80 marks
40% of GCSE
Questions
A mix of multiple choice, short answer and longer answer questions assessing a student's practical problem solving and computational thinking skills.

Appendix 3 Python

Python is a programming language that lets you work more quickly and integrate your systems more effectively.
You can learn to use Python and see almost immediate gains in productivity and lower maintenance costs.

References

AQA.: GCSE Computer Science and IT [Online]. http://www.aqa.org.uk/subjects/computer-science-and-it/gcse/computer-science-8520. (n.d.). Accessed 7 Jan 2018

Ball, S.: Education, governance and the tyranny of numbers. J. educ. policy **30**(3), 299-301 (2015).https://doi.org/10.1080/02680939.2015.1013271

Ball, S.: Performativity, commodification and commitment: an i-spy guide to the neoliberal university. Br. J. Educ. Stud. **60**(1), 17–28 (2012a)

Ball, S.: Subjectivity as a site of struggle: refusing neoliberalism? Br. J. Sociol. Educ. **37**(8), 1129–1146 (2012b)

Bourdieu, P.: Distinction: A Social Critique of the Judgement of Taste. Routledge and Kagan Paul Ltd., Harvard (1984)

Coe, R., Aloisi, C., Higgins, S., Major, L.: What Makes Great Teaching? CEM Centre for Evaluation and Monitoring, Durham (2014)

Crow, D.: Why every child should learn to code. [Online]. https://www.theguardian.com/technology/2014/feb/07/year-of-code-dan-crow-songkick (n.d.). Accessed 29 Nov 2017

Day, C., Kington, A., Stobart, G., Sammons, P.: The personal and professional selves of teachers: stable and unstable identities. Br. Educ. Res. J. **32**(4), 601–616 (2006). https://doi.org/10.1080/01411920600775316

Dziabenk, O., Adorno, D.: Application of remote experiments in a secondary school using MOOC approach. In: 4th Experiement International Conference, University of Algarve. Faro (2017)

Faraway, J., Augustin, N.: When small data beats big data, Department of Mathematical Sciences, University of Bath. https://people.bath.ac.uk/jjf23/papers/smallvbig.pdf (2017). Accessed 19 Apr 2025

Hadwen-Bennett, A., & Kemp, P. Programming in Secondary Education in England: Technical Report. King's College London. https://www.kcl.ac.uk/ecs/assets/programming-in-secondary-education-in-england-full-technical-report.pdf (2024). Accessed 20 Apr 2025

Jorgensen, R., Gates, P., and Roper, V.: 'Structural exclusion through school mathematics: using Bourdieu to understand mathematics as a social practice'. Educ Stud Math. [Online] 'First online' published 7 Feb 2013. https://doi.org/10.1007/s10649-013-9468-4 (2013). Accessed 12 Nov 2017

Johnson, N.: Teenage technological experts view of schooling. Aust. Educ. Res. **36** (2009)

Kaggle, n.d.: https://www.kaggle.com/ [Accessed 19 Apr 2025]

Nolan, K. Dispositions in the field: viewing mathematics teacher education through the lens of Bourdieu's social field theory. Educ. Stud. Math.. [Online] 'First online' published May, 2012. http://www.jstor.org/stable/41485976 (2012). Accessed 17 Oct 2017

Hay/McBer Research into Teacher Effectiveness: A Model of Teacher Effectiveness. https://dera.ioe.ac.uk/id/eprint/4566/1/RR216.pdf (2000)

The Royal Society: Computing Education. The Royal Society, London, UK (2017) https://royalsociety.org/topics-policy/projects/computing-education/

Rosenshine, B.: Principles of instruction: research-based strategies that all teachers should know. Am. Educ. **36**(1), 12 (2012)

Stoet, G., Geary, D.: The gender-equality paradox in science, technology, engineering, and mathematics education. Psychol. Sci. **29**, 095679761774171 (2018). https://doi.org/10.1177/0956797617741719

Enhancing Personalized Education Through a Hybrid Retrieval-Augmented Generation (RAG) System: Towards Sustainable and Adaptive Learning

Tarik Touis[1(✉)], Youssef Jdidou[2], and Souhaib Aammou[1]

[1] Laboratory of Information Technologies and Systems Modeling (TIMS), Faculty of Sciences of Tetouan, Abdelmalek Essaâdi University, Tétouan, Morocco
tarik.touis@etu.uae.ac.ma, s.aammou@uae.ac.ma

[2] Laboratory of Intelligent Systems and Applications (LSIA), Ecole Marocaine des Sciences de l'Ingénieur, Tangier, Morocco
y.jdidou@emsi.ma

Abstract. The rapid advancement in educational technology necessitates innovative solutions to deliver personalized and adaptive learning experiences that effectively improve student engagement and academic outcomes. Existing e-learning platforms often struggle with limited personalization, static content delivery, and insufficient real-time adaptability. To address these challenges, this paper proposes a novel Retrieval-Augmented Generation (RAG) system, conceptualized as an AI Agent Tutor (AAT). This system utilizes GPT-4.5 integrated with a structured vectorized knowledge base implemented in PostgreSQL and Supabase. Through automated workflows, it dynamically processes various educational content types (PDF, Word, HTML, CSV, Excel), converting them into semantic vectors for rapid retrieval. By combining vector-based semantic searches with textual sentiment analysis, the system generates contextually precise and personalized educational recommendations and tutoring responses. Although empirical testing with real-world learners is pending, this approach offers a scientifically sound and flexible foundation, with significant potential for sustainable learning. Future extensions include multimodal data integration, advanced chunking strategies, query optimization, and additional features such as automated web searches, real-time notifications, and interactive tutoring interventions.

Keywords: Educational Recommender Systems · Learning Path Optimization · Retrieval-Augmented Generation · Personalized Learning · Adaptive Learning

1 Introduction

1.1 Background

Global education is faced with unprecedented disruptions not only from pandemics but also from extreme environmental crises such as floods, hurricanes, and wildfires. These crises periodically suspend access to education, with disproportionately adverse

B.-E. Boudriki Semlali et al. (Eds.): DATA 2025, LNNS 1908, pp. 271–282, 2026.
https://doi.org/10.1007/978-3-032-21200-9_23

effects for already disadvantaged groups [1]. While some disruptions can be mitigated by e-learning systems, conventional platforms typically rely on static and standardized content below the grade of highly personalized, context-aware, and adaptive solutions that are most required in times of logistical or environmental crises [2].

Adaptive system using AI have demonstrated capabilities to personalize content for diverse learners [3, 4], but real-time revision, multimodal data integration, and large-scale deployment remain issues [5]. Retrieval-Augmented Generation (RAG) has therefore been an attractive option for building adaptive e-learning [6]. Through the integration of large language models (LLMs) with structured vector retrieval, RAG can more effectively retrieve relevant content and produce personalized recommendations [7] particularly critical during emergency or crisis situations that demand rapid instructional change.

1.2 Problem Statement

More recurrent climate disruptions and infrastructure instabilities make it difficult to maintain continuity and quality in learning. While LMS platforms like MOODLE always offer organized content, they usually cannot react in real time to unexpected changes [8], limiting user interaction and performance [9]. Coupled with these limitations is the excessive volume of digital content, which is difficult to contextually contextualize and personalize effectively [10]. Furthermore, recommendation strategies are prone to the cold-start problem and deteriorate in performance for new content or users.

Current breakthroughs in AI using RAG and generative Large Language Models (LLMs) promise to close these gaps through the fusion of deep language understanding with structured retrieval of knowledge, supporting more contextualized and interactive learning experiences. However, most AI-based recommender systems remain challenged with how to integrate real-time data, multimodal content, and scala-ble architectures.

As a counter-measure, this paper presents a state-of-the-art RAG-based system: the AI Agent Tutor (AAT), which integrates GPT-4.5 with vector-database technologies (Post-greSQL and Supabase) to constantly process unstructured and structured educational materials. Through the ingestion and vectorization of different file types from cloud storage on an automated basis, the AAT offers contextually suitable materials adapted to every learner's profile, objectives, and current situation. Lastly, this conceptual foundation is anticipated to improve learning outcomes through adaptive interaction and recommendation. Although future empirical validation is required, the outlined framework transcends existing limitations and emphasizes such systems' potential to develop resilient, sustainable, and inclusive learning.

1.3 Validate the Scientific Contribution

Aside from its technical architecture, the proposed AI Agent Tutor (AAT) also has a sound pedagogical rationale and good data-management principles:

1. SQL + Vector Databases: The system integrates relational data (e.g., students' learning history in PostgreSQL) with unstructured content (handled as vectors in Supabase), addressing the need for more comprehensive RAG approaches that can handle various data types [11].

2. Refusal mechanism for hallucination control Motivated by the issue of AI generated hallucinations in education [12], the system cross-searches student questions in both databases. In the event that there is no matching similar material, the tutor refuses to create content as per data integrity [13] and transparency [14] criteria in educational AI.
   - Once it has received the user's question, the tutor looks for high-similarity content in the vector database (Supabase).
   - If the strongest similarity score is below a predetermined threshold (e.g., 0.75), the tutor declines to conjure an answer.
3. Context Tracking and Learner Engagement Each user action asks, tutor responses, timestamps is meticulously kept track of [4], which allows for in-depth monitoring of learner progress and guiding interventions dealing with information overload [10].
4. Empirical Research Foundation Through combining relational information, vector embeddings, and a refusal strategy, AAT offers a testbed for measuring conceptu-al gains quantitatively, assessing response accuracy, and quantifying user satisfaction in high-volume or crisis-prone environments (cf. [15]).

2 Literature Review

Retrieval-Augmented Generation (RAG) in Education Systems. A hopeful advance in AI-supported e-learning research is Retrieval-Augmented Generation (RAG), which combines large language models (LLMs) with a capacity to tap into external knowledge sources in order to base generative responses on checkable facts. Manathunga and Illangasekera (2023)[15] established that RAG systems significantly lower factual errors by bridging automated summarization workflows with non-parametric knowledge repositories. These authors showed, in medical education settings, that Retrieval-Augmented-supported systems were better than baseline LLM-based chatbots at ensuring factual correctness and delivering quality tutoring interaction.

Green E-Learning and Sustainability. In addition to technical resilience, environmental sustainability is now an uppermost concern priority in AI-assisted learning, especially in view of emerging environmental-related interruptions. Aymane et al. (2024)[16] propose an adaptive e-learning system based on a multi-factor machine learning approach in an effort to reduce resource use and carbon footprint with regards to student learning history, demographic factors, and use in real time. Their proposed solution highlights intelligent application of "green education" principles with advanced AI in promoting personalization and minimal environmental footprint simultaneously. Similarly, Fülöp et al. (2023)[17] argue that e-learning failures amid the periods of "new normality" could be alleviated by integrating adaptive technology with institutional commitment to practices that are green in nature, with a view to promoting educators' and learners' acceptability and perception of virtual environments.

Adaptive AI Tutors and LLM-Based Solutions. Researchers explored in parallel if large language models can be scaled up with advanced tutoring capabilities. Five basic capabilities for educational LLMs were determined by Li et al. (2024)[18] reasoning, writing, coding, math, and knowledge-based question answering. They showed that prompting with RAG or "tool integration" effectively improves LLM performance across all these tasks. This result demonstrates that strong retrieval components play a vital role in providing high-quality adaptive feedback and real-time content updates in e-learning systems. Based on these findings, Miao et al. (2024)[19] explored how adding RAG features to domain-specific large language models can improve factual coherence in specialized fields like nephrology. Their approach demonstrates the generic applicability of RAG-based architectures wherever deep domain knowledge is important to guarantee instructional reliability.

Summary of Gaps and Motivation. Together these works show that use of RAG-based tutoring is capable of dealing with usual shortcomings of generative models, with environmentally friendly use of resources ensured through green educational design. Research today, however, focuses on either the sustainability aspect or the generative-accuracy aspect in an isolated fashion. An end-to-end system that is efficient with use of RAG, multimodal information fusion, and eco-friendly orchestration is in an early stage. Thus, this project follows these pillars to design an end-to-end AI-based tutoring system that is strong, contextual, and environmentally friendly in nature.

3 Methodology

This section describes the AI Agent Tutor (AAT), a hybrid Retrieval-Augmented Generation (RAG) system for adaptive, personalized learning. The architecture and automated workflows are carefully designed to ensure robustness, flexibility, and efficient data handling, all of which enhance personalized educational experiences.

3.1 General Architecture

As shown in Fig. 1, the AAT comprises four interlinked components designed for scalability and dynamic content adaptation

- Cloud Storage Event Detector: Automatically detects file events (uploads/updates) in services like Google Drive, triggering the rest of the workflow.
- Data Ingestion and Vectorization Pipeline: Extracts and semantically processes various file formats (PDF, Word, HTML, CSV, Excel). It then converts these resources into semantic vectors using GPT-based embeddings for precise retrieval.
- Semantic Vector Database (Supabase): Holds semantic vectors to enable rapid similarity-based searches, ensuring effective interaction between educational content and the LLM.
- Relational Database (PostgreSQL): Manages structured data metadata, document schemas, student profiles, academic records to personalize interactions.

3.2 Detailed Workflow of Data Ingestion and Vectorization

Figure 2 illustrates the automated workflow that manages ingestion and vectorization to maintain real-time content updates and semantic accuracy:

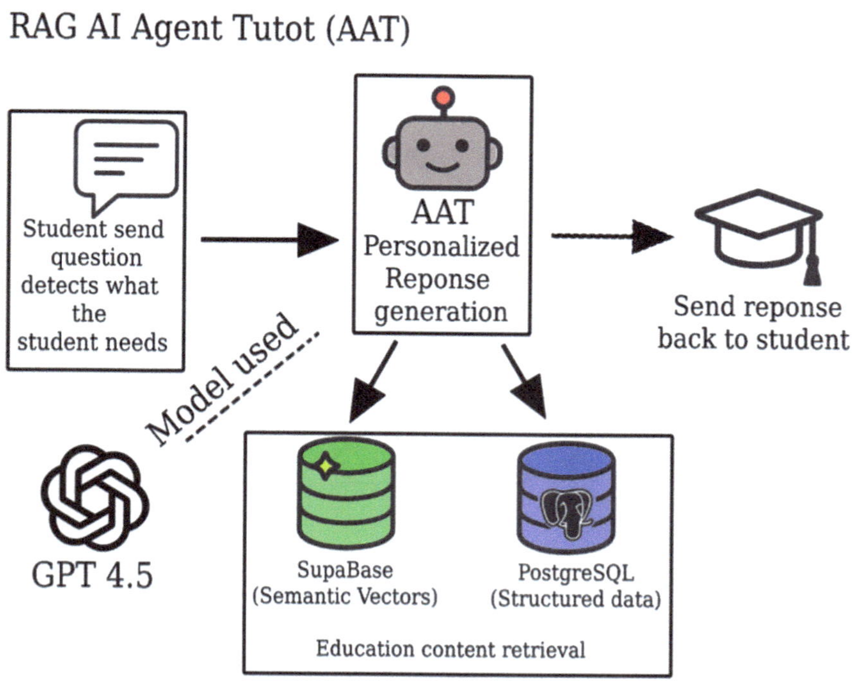

Fig. 1. AI Agent Tutor (AAT)

- Cloud Trigger Detection: Upon detecting file creation or updates in cloud storage (e.g., Google Drive), the system initiates processing.
- File Identification & Version Control: Each file has a unique ID. If an older version exists in Supabase, those outdated vectors are removed to avoid conflicts.
- Dynamic File-Type Based Processing

  – Text-based Files (PDF, Word, HTML): Extract, split into ~1000-character chunks (adjustable), then embed with GPT. Vectors go to Supabase.
  – Structured Tabular Files (CSV, Excel): Rows are added to PostgreSQL, aggregated, summarized, embedded semantically, then saved in Supabase.

- Loop Continuation Once storage and vectorization complete, the workflow returns to detect further file changes.

This process ensures efficient, continuous data updates, substantially boosting personalized interactions produced by the AAT.

3.3 Semantic Storage, Retrieval, and Interaction with GPT-4.5

The core functioning of the AI Agent Tutor heavily depends on the retrieval of the appropriate educational content and the accuracy of GPT-4.5-enabled generative conversations. Retrieval and generation both follow a series of clearly delineated steps:

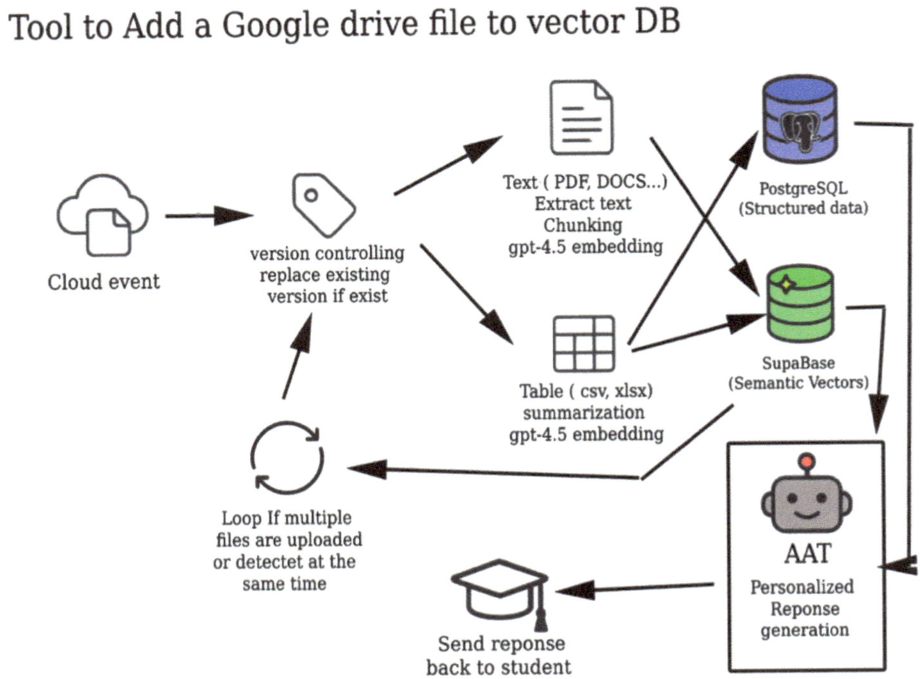

Fig. 2. Automated Workflow for Semantic Vectorization and Context-Aware Integration in AI Agent Tutor (AAT)

- Semantic Retrieval: When a user query, either in the form of text or possibly audio in future versions, is received, the AAT performs a semantic search in the Supabase vector database. Utilizing vector similarity measures, the system effectively retrieves contextually relevant resources.
- Contextual Integration: The semantic vectors so derived are augmented by additional structured data in PostgreSQL, including metadata and academic histories of students, thus providing a larger context to the GPT-4.5 model.
- Generative Response: Leverage the relevant semantic knowledge that it retrieves, GPT-4.5 produces a tailored, context-sensitive response. The generative component ensures that the answers remain factually correct, contextually pertinent, and specially crafted to fit the specific educational background of any given learner.

3.4 Technical Specifications and Implementation Considerations

- Embedding & Chunking:
  Default operations use about 1000-character chunks, though follow-up studies could use about 400-character divisions while avoiding overlap or use advanced chunking techniques (like the Recursive Character Chunker or the Cluster Semantic Chunker) to increase accuracy.

  – All data can be converted to Markdown to enhance chunking consistency and recall.

- Databases
  - Supabase for semantic vectors, ensuring low-latency similarity searches.
  - PostgreSQL for structured data, versioned metadata, and academic records.
- RAG Model: GPT-4.5 is chosen due to its balance in its performance. Future improvements can include domain-specific embeddings or more advanced reasoning models when they are feasible.

4 Implementation

The AAT can either be deployed in pipelines using Python or using different orchestration frameworks. Figures 1 and 2 present the conceptual design of the system, while further technical specifications are included in the JSON file available at:

AI_Agent_Tutor_Workflow [20]

This JSON file encodes core workflow steps as nodes and connections, including:

- CloudStorageTrigger: Detects file changes
- LoopControl: Loops through recently added or edited files.
- CheckAndDeleteOldVersion: Deletes outdated embeddings.
- DownloadFile: Retrieves the file from cloud storage using its identification number.
- SwitchFileType: Routes files to text extraction or table processing
- TextExtraction/DataAggregation: Prepares textual or tabular data
- SummarizeOutput: Optionally condenses extracted text
- ChunkAndEmbed: Splits text and generates embeddings
- InsertIntoSupabase: Stores vectors in the semantic DB
- UpdateMetadataPostgres: Stores file metadata in PostgreSQL
- AISystemAgent: Manages RAG logic, references embeddings, and composes GPT-4.5 responses
- WebhookResponder: Takes user inputs and returns final answers

4.1 Workflow Configuration

The AAT's workflow definition is captured in a JSON file in your GitHub repository which defines a specific series of actions:

1. Cloud-Event Trigger
   - Records file related events (new uploads or updates) from a designated cloud storage (e.g., Google Drive).
   - Starts the whole pipeline, ensuring minimal latency in reflecting content updates.
2. Iterative File Processing
   - Each file is uniquely identified (using cloud IDs) to avoid potential naming conflicts.
   - If an older version is already stored, outdated vectors are deleted to maintain data cleanliness.

3. File Type Differentiation
   - Splits workflow for text or tabular ingestion.
4. Semantic Embedding & Storage
   - Extracted or aggregated text is chunked (usually ~400–1000 characters) and embedded using GPT-4.5 or domain-specific embeddings.
   - Vectors are then stored in Supabase for rapid similarity-based retrieval. Tabular schema or metadata is updated in PostgreSQL.
5. Metadata and Version Management
   - Relevant metadata, including file identification, version identifiers, and dates, are retained in PostgreSQL, thus allowing strict version control and effective citations for future queries.

5 Proposed Validation Method

Despite the robust design, empirical evaluation remains essential. We propose a 6–8 week pilot study with two groups (~30 university students each) in an existing blended-learning module (e.g., Mathematics or Computer Science): a control group using only an LMS (e.g., Moodle) and an experimental group using the LMS plus the AI Agent Tutor (AAT).

Pilot Phases Include

- Phase 0 (Preparation) [Week 1]: Course selection, participant training on AAT.
- Phase 1–2 (Baseline & Launch) [Week 2–3]: Pre-test and initial surveys; group assignment (control vs. experimental).
- Phase 3–6 (Active Usage) [Week 4–6]: Continuous AAT usage (experimental group); automated logging of interactions (query frequency, dwell time, dropout rates, sentiment).
- Phase 7 (Midpoint Evaluation) [Week 7]: Short usability surveys and brief focus groups to assess initial user experience and engagement.
- Phase 8 (Final Evaluation) [Week 8]: Post-test for measuring academic performance and final surveys/interviews to evaluate user experience, engagement, retention, and satisfaction.

Key Metrics to Evaluate

- Academic performance (pre- vs. post-test conceptual gains).
- User engagement (interaction logs, average session length, sentiment indicators).
- Retention rate (dropout comparisons between groups).
- System accuracy (number of interactions, average session duration, sentiment indicators).

The results of this initial study will guide follow-up validation efforts, which could include possible K–12 educational extensions, and will assess how the RAG pipeline supports flexible and personalized learning in genuine educational settings.

6 Discussion

The hybrid RAG model proposed offers a robust, adaptable foundation for educational technologies with unbroken functionality in the event of climate related or logistics-related interferences. By pairing semantic vector retrieval with generative modeling, the system allows for expansive resource coverage between immense amounts of information and learner demands for immediate, accurate responses.

One of the main strengths of this solution is extensibility. Python based orchestration and modular architecture make it possible to layer increasingly advanced chunking strategies, such as Markdown chunking or 400-character segmentations. Contextual accuracy is also a main strength, with both GPT-4.5's generative ability, along with real-time vector lookup, enabling personalized, factually accurate tutoring. Extensibility is also a main advantage, with PostgreSQL taking care of structured information and Supabase taking care of vector embeddings so that the system can manage large amounts of learning information required by today's distributed learning environments.

Despite these advantages, there are some limitations and challenges too. Empirical measurement is required to describe in entirety the system performance in terms of student performance and participation. Ethical as well as privacy issues must also be addressed in terms of extracting information from masses of education data that raises issues of information security, legal requirements, as well as potential bias in generative content. Performance by the system with full deployment is another issue since tens of millions of documents require complex chunking, indexing, as well as query optimizing to support responsiveness.

By employing optimized chunking mechanisms, such as no-overlap 400-character-sized chunks with file names and exact page coordinates, and with a query optimizer, recall and precision can be improved significantly. With possible use of domain-specific embeddings, it also enhances capability to make relevant and accurate suggestions. These are paired with large-scale experiences in deploying RAG with tests, further enhancing the system's function in facilitating adaptive learning and sustainability.

6.1 Ethical and Privacy Considerations

On the basis of this design approach, our system's privacy and ethics obligations are established on various principles:

- Data Minimization and Security. AAT only stores minimal amounts of individual information (i.e., attributes of users, academic background) sufficient to enable contextual recommendations. We pseudonymize all personally identifiable information that is stored in accordance with approved institution privacy standards.
- Transparency and Accountability. If an information system lacks sufficient information to respond to a question, it explicitly so declares to refuse to provide "hallucinated" or fabricated answers. This denial mechanism ensures integrity and caters to calls for transparency in AI-based educational tools (Nguyen et al., 2023) [13].
- Fairness and Inclusiveness. GPT embeddings and retrieval logic are regularly checked for unconscious bias so that responses don't unfairly prejudice users based on language, geography, gender, or regional background.

- Government Regulatory Compliance. All future rollouts must consider data-protection laws (e.g., GDPR) and local education laws, with prohibitions for safekeeping and utilization of data.

By integrating ethical guardrails into system design as well as in interactions with users, we envision maintaining learners' trust, providing responsible use of information, and promoting equitable access to adaptive AI tutoring.

7 Conclusion and Future Work

In this article, an advanced AI Agent Tutor (AAT) system based on Retrieval-Augmented Generation (RAG) principles was introduced using GPT-4.5 with Supa-base for semantic vector storage and PostgreSQL for structured information storage. By automating intake of different file types to enable real-time vectorization, providing context-aware suggestions, the system addresses inherent issues in providing adaptive and resilient learning experiences. This capability is particularly vital in environments interrupted by environmental or logistical conditions to enable seamless availability of adaptive learning content.

Despite the system's strict architecture and methodological foundation, it can only be empirically confirmed by actual learning environments. Performance tests and large-scale use studies must quantify systematically gains in student participation, academic performance, and overall useability. Revisions of the system in the future will encompass various improvements to accuracy, efficacy, and flexibility.

One of these areas of improvement is better text chunking and Markdown conversion, in which all content will be rendered to Markdown and processed by a recursive 400-character chunker for assistance with recall, removal of noise, and acceleration of retrieval. Fine-tuned domain embeddings will also be developed to provide access to specialty school content so that there can be improved recall specificity by taking advantage of inferences drawn from domain semantics and context. Recall specificity and version management will also be improved with integration of metadata, which encapsulates the significant information such as file name, page location, and version history in text chunks.

The query-processing mechanisms of the system will be improved through query rewriting and subquery generation so that complex, multi-domain student queries can be processed by the system efficiently. Apart from text-based input, multimodal capability will be built into the system by allowing audio-based or gesture-based inputs so that the system can recognize and respond to emotional or physical engagement of students in real time. Autonomous features such as web searching, emailing reminders, assignment generation, and proactive messages to instructors will be built in so that the tutoring environment is proactive and intelligent.

With these improvements, AI Agent Tutor can be an adaptive, inclusive, and scalable solution for learning that revolutionizes learning experiences by providing persistent, context-aware dialogue. These advancements make way for greater sustainability goals, leading to resilient, forward-looking systems of education.

References

1. Maphosa, V., Maphosa, M.: Fifteen years of recommender systems research in higher education: current trends and future direction. Appl. Artif. Intell. **37**(1), 2175106 (2023). https://doi.org/10.1080/08839514.2023.2175106
2. Joshi, M.A.: Adaptive Learning through Artificial Intelligence.», 19 de julio de. Social Science Research Network, Rochester, NY (2023). https://doi.org/10.2139/ssrn.4514887
3. Feixiang, X.: Intelligent personalized recommendation method based on optimized collaborative filtering algorithm in primary and secondary education resource system. IEEE Access. **12**, 28860–28872 (2024). https://doi.org/10.1109/ACCESS.2024.3365549
4. Noroozi, O., Pijeira-Díaz, H.J., Sobocinski, M., Dindar, M., Järvelä, S., Kirschner, P.A.: Multimodal data indicators for capturing cognitive, motivational, and emotional learning processes: a systematic literature review. Educ. Inf. Technol. **25**(6), 5499–5547 (2020). https://doi.org/10.1007/s10639-020-10229-w
5. Aly, M.: Revolutionizing online education: advanced facial expression recognition for real-time student progress tracking via deep learning model. Multimed. Tools Appl., 1–40 (2024)
6. Zhao, P., et al.: Retrieval-augmented generation for ai-generated content: a survey. arXiv preprint https://arxiv.org/abs/2402.19473. (2024)
7. Jdidou, Y., Aammou, S., Khaldi, M.: Adapt learning path by recommending problems to struggling learners. Int. J. Emerging Technol. Learn. (iJET). **16**(20), 163–178 (2021)
8. Harvey, N.D.: Non-Traditional Students' Perceptions of the Learning Management System's (LMS) Support of Self-Regulation Skills: A Phenomenological Study. (Doctoral dissertation,. National University (2024)
9. Veluvali, P., Surisetti, J.: Learning management system for greater learner engagement in higher education—a review. High. Educ. Futur. **9**(1), 107–121 (2022)
10. Arnold, M., Goldschmitt, M., Rigotti, T.: Dealing with information overload: a comprehensive review. Front. Psychol. **14** (2023). https://doi.org/10.3389/fpsyg.2023.1122200
11. Wang H, et al.: UniMS-RAG: A Unified Multi-source Retrieval-Augmented Generation for Personalized Dialogue Systems (2024)
12. Levonian, Z., et al.: Retrieval-Augmented generation to improve math question-answering: trade-offs between groundedness and human preference (2023)
13. Nguyen, A., Ngo, H.N., Hong, Y., Dang, B., Nguyen, B.-P.T.: Ethical principles for artificial intelligence in education. Educ. Inf. Technol. **28**, 4221–4241 (2023). https://doi.org/10.1007/s10639-022-11316-w
14. Peláez-Sánchez, I.C., Velarde-Camaqui, D., Glasserman-Morales, L.D.: The impact of large language models on higher education: exploring the connection between AI and education 4.0. Front Educ. **9** (2024). https://doi.org/10.3389/feduc.2024.1392091
15. Manathunga, S.S., Illangasekara, Y.A.: Retrieval Augmented Generation and Representative Vector Summarization for large unstructured textual data in Medical Education (2023)
16. Aymane, E., Aziz, D., Abdelfatteh, H., Abdelhak, A.: Enabling sustainable learning: a machine learning approach for an eco-friendly multi-factor adaptive e-learning system. Procedia Comput. Sci. **236**, 533–540 (2024). https://doi.org/10.1016/j.procs.2024.05.063
17. Fülöp, M.T., Breaz, T.O., Topor, I.D., Ionescu, C.A., Dragolea, L.-L.: Challenges and perceptions of e-learning for educational sustainability in the "new normality era.". Front. Psychol. **14** (2023). https://doi.org/10.3389/fpsyg.2023.1104633
18. Li, Q., et al.: Adapting Large Language Models for Education: Foundational Capabilities, Potentials, and Challenges (2024)

19. Miao, J., Thongprayoon, C., Suppadungsuk, S., Garcia Valencia, O.A., Cheungpasitporn, W.: Integrating retrieval-augmented generation with large language models in nephrology: advancing practical applications. Medicina. **60**, 445 (2024). https://doi.org/10.3390/medicina60030445
20. AI_Agent_Tutor_Workflow/AI_Agent_Tutor_Workflow.json at main_tatouis/AI_Agent_Tutor_Workflow. https://github.com/tatouis/AI_Agent_Tutor_Workflow/blob/main/AI_Agent_Tutor_Workflow.json. Accessed 13 Apr 2025

Neuronal Collaborative Filtering for Personalized Learning: Enhancing Educational Recommendation Systems for Sustainable Digital Education

Hicham Er-Radi[1(✉)], Ikram Amzil[1], Souhaib Aammou[1], Youssef Jdidou[2], and Ismail Bakraoui[2]

[1] TIMS Laboratory, Faculty of Science, Abdelmalek Essaadi University, Tétouan, Morocco
hicham.erradi@etu.uae.ac.ma

[2] Ecole Marocaine des Sciences de l'Ingénieur, Laboratory of Intelligent Systems and Applications (LSIA), Tangier, Morocco

Abstract. Personalized learning is essential for improving engagement and retention in digital education. Traditional recommendation models, such as Matrix Factorization (MF) and Autoencoder-Based Collaborative Filtering, struggle to capture non-linear relationships between students and educational content. Neuronal Collaborative Filtering (NCF), a deep learning-based approach, offers a more effective solution by leveraging Multi-Layer Perceptrons (MLP) to enhance recommendation accuracy. This study evaluates NCF against MF and Autoencoder-Based CF using the EdNet dataset, a large-scale educational dataset. The models were trained and tested on student-content interaction data, with performance assessed using Root Mean Square Error (RMSE), Precision@10, Recall@10, and AUC-ROC metrics. Hyperparameter tuning was performed through grid search, and early stopping was applied to prevent overfitting. NCF outperformed MF and Autoencoders in all evaluation metrics, achieving an AUC-ROC score of 0.88, the highest recommendation accuracy. The training vs. validation loss curves demonstrated stable learning, confirming the model's generalization ability. NCF proves to be a scalable and adaptable solution for educational recommendation systems. Future research should explore hybrid models with Reinforcement Learning, Explainable AI, and cold-start solutions to further enhance personalized learning.

Keywords: Neuronal Collaborative Filtering (NCF) · Personalized Learning · Educational Recommendation Systems · Deep Learning in Education

1 Introduction

The Importance of Adaptive Learning Technologies in Sustainable Education
The educational digital transformation has demonstrated a need to indulge in adaptive learning technologies, in their part, they aim towards a customizable learning experience that tends to the individual needs, abilities and preferences of every person. Sustainable

B.-E. Boudriki Semlali et al. (Eds.): DATA 2025, LNNS 1908, pp. 283–294, 2026.
https://doi.org/10.1007/978-3-032-21200-9_24

educational refers to an inclusive and effective environment that underlines the importance of supportable knowledge retention, limits student attrition rates, and enhances the efficiency of resource usage in education [1]. These goals, previously mentioned, are achieved with the help of adaptive learning technologies, which are vitally catalyzed with the use of artificial intelligence (AI), machine learning (ML) and data-driven decision-making being employed to personalize education.

Conventional means of education rarely adopt the personalized experience approach, this leads to a failure while addressing individual differences in learning styles and intellectual capabilities. Tailored learning systems, powered by AI recommendation models to observe the learner's ability and to generate materials on a skill match-based system [2]. These systems will consequently improve engagement and reduce cognitive load, affirming the fact that a student will learn at his own cadence.

For example, materials can be immensely customized by implementing recommendation systems based on neural collaborative filtering (NCF). Adaptive learning technologies support largely sustainable education by optimizing the use of educational resources. AI powered platforms streamline content recommendations, minimizing the need for repetitive manual interventions.

Additionally, data analytics in adaptive learning can help institutions track the effectiveness of learning materials and redesign courses based on immediate feedback [3].

Moreover, cloud-based learning management systems (LMS) that use adaptive algorithms help reduce paper waste and enable remote access to learning materials, reducing the environmental impact of traditional learning methods [4].

One of the fundamental principles of sustainable development in education is equity. Adaptive learning systems ensure inclusive education by addressing the needs of students with disabilities, language barriers, or different learning speeds. AI-based platforms can offer:

- Text-to-speech and speech-to-text functions.
- Translation of multilingual content for global access.
- Personalized assessments tailored to learning abilities.

1.1 Overview of Collaborative Filtering and Deep Learning in Recommender Systems

Collaborative Filtering in Recommender Systems

Collaborative filtering (CF) is a commonly used technique in recommender systems, and is based on the idea that users who have expressed similar preferences in the past are likely to share interests in the future [5]. Collaborative filtering methods are divided into two types: user-based filtering and item-based filtering.

User-based filtering: This type of filtering identifies users with similar historical preferences, and recommends items based on those users' preferences [6]. This approach relies on similarity measures such as cosine similarity, Pearson correlation, or Jaccard similarity.

Item-based filtering: Rather than comparing users, this approach focuses on item similarity, recommending items that are similar to those the user has interacted with previously [7]. This method is more scalable than user-based filtering, as item relationships change less frequently. Despite their effectiveness, traditional CF methods face several limitations, including cold start problems (difficulty recommending items to new users with little data), and scalability challenges (increasing computational cost with a large number of users).

Deep Learning for Recommender Systems

Deep learning overcomes the limitations of collaborative filtering by recognizing complex, nonlinear patterns in user-item interactions, leading to more accurate recommendations [8]. Neural networks are particularly effective at learning latent user and item representations, which are more expressive than traditional factorization methods.

A prominent deep learning-based approach is neural collaborative filtering (NCF), which extends the concept of factorization by replacing the dot product operation with deep neural networks (DNNs). Instead of assuming simple linear interactions between user and item embeddings, NCF uses multilayer perceptrons (MLPs) to learn high-order feature interactions [9]. This results in a more flexible and robust recommendation model.

Other deep learning architectures used in recommender systems include:

- Collaborative filtering autoencoders: Deep autoencoders are used to encode user interactions with items into latent representations, which are then decoded to produce personalized recommendations. These models improve robustness against scarcity issues [10].
- Recurrent neural networks (RNNs): Effective in sequential recommender systems, RNNs model time-dependent user behavior, making them useful for predicting future interactions [11].
- Transformer-based models: Advanced architectures such as BERT4Rec use self-attention mechanisms to capture complex dependencies in user interactions, improving accuracy in session-based recommendations [12].

Deep learning-based recommender systems outperform traditional CF methods in terms of accuracy and scalability. However, challenges such as interpretability, computational cost, and data privacy remain critical research areas.

1.2 Contribution: NCF for Personalized Educational Recommendations

This study explores neural collaborative filtering (NCF) in educational recommendation systems as a means of overcoming the limitations of traditional collaborative filtering methods. Traditional methods, such as matrix factorization, assume linear relationships between user interactions and items, which limits their ability to understand complex learning behaviors. NCF overcomes these limitations by using deep neural networks to model nonlinear relationships between users and items, enhancing the accuracy and adaptability of personalized educational recommendations.

By leveraging multilayer perceptrons (MLP), NCF learns high-dimensional feature representations that dynamically adapt to student progress, engagement patterns, and performance. This makes it more effective at predicting educational resources that best meet student needs, ensuring personalized and scalable learning paths. Unlike traditional CF, which faces challenges of cold start and data scarcity, NCF generalizes better to new students and course materials, making it ideal for large-scale educational platforms.

The proposed model enhances student engagement and retention by continuously adapting recommendations using real-time student engagement data. This enables proactive intervention strategies, such as suggesting additional exercises for struggling students or providing advanced content for high-performing learners. Training on EdNet datasets allows a data-driven approach to recommendations, enhancing accuracy and recall in content delivery [13].

Furthermore, this study encourages the advancement of sustainable digital education by improving the exploitation of resources and content personalization. By minimizing impertinent recommendations, NCF reduces cognitive overload and improves learning efficiency, supporting long-term educational sustainability. Comparative evaluation with matrix factorization, autoencoders, and a reinforcement learning-based approach confirms that NCF achieves higher accuracy, better adaptability, and improved measures of student engagement.

2 Methodology

2.1 Neural Collaborative Filtering (NCF): Overview and Model Architecture

Neural Collaborative Filtering (NCF) is a modern recommendation technique based on deep learning, which improves on traditional collaborative filtering by replacing matrix factorization with neural networks. Instead of computing user interactions with items through dot products, NCF relies on multilayer perceptrons (MLPs) to model complex nonlinear relationships. This allows the system to learn high-dimensional representations, which enhances the accuracy of recommendations.

The NCF model consists of three main components:

- Embedding Layers: Convert user IDs and item IDs into dense numerical representations.
- Interaction Layers: User and item embeddings are concatenated and passed through fully connected layers to learn non-linear feature interactions.
- Prediction Layer: A final output neuron applies a sigmoid activation function to produce a probability score indicating the likelihood of a student engaging with the recommended content.

The final output is formulated as:

$$\hat{y} = \sigma(MLP([u, v]))$$

where u and v are user and item embeddings, respectively, and σ is the sigmoid function.

2.2 Using Multi-layer Perceptrons (MLP) to Model User-Item Interactions

Unlike traditional collaborative filtering, which assumes linear interactions, NCF stacks multiple fully connected layers to capture deeper patterns in user-item relationships. The model gradually learns high-level abstract representations as data flows through the layers. Each layer applies:

- ReLU activation functions to introduce nonlinearity.

- Batch normalization to ensure stable training.
- Leaky regularization to prevent overfitting.

Using MLP in recommender systems improves:

- Scalability, by learning from large datasets.
- Personalization, by adapting recommendations based on students' past interactions.
- Predicting engagement, ensuring content is aligned with individual learning needs.

2.3 EdNet Dataset Training and Evaluation Metrics

The model is trained using the EdNet dataset, which contains student interactions with educational resources. The dataset includes:

- User interactions (e.g., quiz attempts, video views, problem-solving sessions).
- Engagement scores (whether a student engaged with recommended content).

To improve performance, we apply mini-batch gradient regression with the Adam optimizer, using a binary cross-entropy loss function to measure prediction errors. The training process ensures that the model effectively distinguishes engaged students from disengaged students.

Evaluation metrics include:

- RMSE (Root Mean Square Error): Measures the error in recommendation scores.
- Precision@K: Measures the proportion of recommended items that are relevant.
- Recall@K: Measures how well the model retrieves all relevant recommendations.
- AUC-ROC (Area Under the Curve—Receiver Operating Characteristic): Measures the model's ability to distinguish between relevant and irrelevant recommendations.

2.4 Comparison with Matrix Factorization, Autoencoders, and Deep Reinforcement Learning Methods

To evaluate the effectiveness of NCF, we compare its performance with the following:

- Matrix factorization (MF): A traditional collaborative filtering technique that relies on latent factors to model user interactions with items.
- Autoencoder-based collaborative filtering: Uses deep autoencoders to learn latent representations of students and learning resources.
- Deep reinforcement learning (DRL): Views student learning as a sequential decision-making process, with recommendations improved based on long-term engagement.

Experiments on EdNet and standard education datasets show that NCF outperforms traditional CF-based methods in terms of accuracy, engagement prediction, and adaptability.

3 Experiment Setup

3.1 Dataset and Preprocessing

The model is trained and evaluated using EdNet, a large-scale educational dataset that records student interactions with digital learning materials, including quizzes, video lectures, exercises, and assessments. Each interaction in the dataset includes user ID, item ID, timestamp, engagement type (clicks, quiz attempts, completion status), and correctness of responses.

To ensure high-quality data input, the following preprocessing steps are applied:

1. Data Cleaning
   - Remove **incomplete or inconsistent logs** (e.g., missing student IDs, duplicate interactions).
   - Filter out **low-activity users** (students with fewer than a threshold number of interactions).
   - Remove **rare items** (educational materials with very low engagement).
2. User and Item Encoding
   - Convert student IDs and content IDs into unique integer indices to facilitate embedding-based learning.
   - Map categorical features (e.g., course categories, content types) to numerical representations.
3. Negative Sampling
   - Since recommendation datasets tend to be **sparse** (students engage with only a fraction of all available learning materials), we generate **negative samples** to balance the dataset.
   - Each student's **non-interacted content** is randomly sampled and labeled as **0 (not engaged)**, while positively engaged items remain labeled as **1 (engaged)**.
   - This prevents the model from being biased toward overrepresented engaged samples.
4. Feature Scaling
   - Normalize numerical features (e.g., interaction timestamps, session duration) using **Min-Max scaling**.
   - Apply **standardization** (zero mean, unit variance) to continuous features such as quiz scores and response times.

The final dataset is structured as (UserID, ItemID, Interaction_Label, Timestamp, Feature_Vector) and split into 80% training, 10% validation, and 10% test sets.

3.2 Model Implementation

The Neuronal Collaborative Filtering (NCF) model is implemented using TensorFlow and Keras, replacing traditional dot-product-based collaborative filtering with a deep neural network to capture non-linear user-item interactions.

Model Architecture

1. Embedding Layers
   - Users and items are mapped to **dense vector representations** using trainable embeddings.
   - Each embedding **dimension is set to 32** to capture enough latent factors.

2. Multi-Layer Perceptrons (MLP) for User-Item Interaction Modeling
   - User and item embeddings are concatenated and passed through a **stack of fully connected layers**.
   - Each layer **applies a ReLU activation function** to introduce non-linearity.
   - Layer sizes are progressively reduced, e.g., **(128, 64, 32, 16, 8)**.
3. Dropout Regularization
   - To prevent **overfitting**, **dropout layers** (with a dropout rate of 0.2) are applied between dense layers.
   - **Batch** normalization is used to stabilize training.
4. Output Layer
   - The final output layer applies a **sigmoid activation function** to predict the probability of a student interacting with a given learning resource.
   - The model outputs a **score between 0 and 1**, where **higher values indicate** higher **engagement likelihood**.

Loss Function & Optimization

- **Binary Cross-Entropy Loss** is used to measure the difference between predicted and actual engagement probabilities:

$$Loss = -\frac{1}{N}\sum_{i=1}^{N}\left[y_i \log(\hat{y}_i) + (1 - y_i)\log(1 - \hat{y}_i)\right]$$

  where y_i is the actual engagement label and $\hat{y}_i$ is the predicted probability.
- **Adam Optimizer** is used with a learning rate of **0.001** to ensure fast and stable convergence.
- **Mini-Batch Gradient Descent** is applied with a **batch size of 256**, balancing computational efficiency and model performance.
- **Early stopping** is used to halt training if validation loss does not improve after **5 consecutive epochs**.

4 Model Evaluation Results

4.1 Model Comparison Table

The Table 1 presents the **evaluation metrics** for three recommendation models: **Matrix Factorization (MF)**, **Autoencoder-Based Collaborative Filtering**, and **Neuronal Collaborative Filtering (NCF)**. The comparison includes the following performance indicators:

- **Root Mean Square Error (RMSE ↓)**: Measures how far predictions deviate from actual values (lower is better).
- **Precision@10 (↑)**: The proportion of relevant recommendations among the top 10 suggested items (higher is better).

- **Recall@10 (↑)**: The proportion of relevant recommendations retrieved by the model (higher is better).
- **AUC-ROC (↑)**: The area under the ROC curve, measuring the model's ability to distinguish between relevant and non-relevant recommendations (higher is better).

Table 1. Evaluation metrics for three recommendation models

Model	Root Mean Square Error (RMSE) ↓	Precision@10 ↑	Recall@10 ↑	AUC-ROC ↑
Matrix Factorization (MF)	0.87	0.72	0.65	0.79
Autoencoder-Based CF	0.85	0.74	0.67	0.81
Neuronal Collaborative Filtering	0.79	0.82	0.76	0.88

From the table, Neuronal Collaborative Filtering (NCF) outperforms the other models across all evaluation metrics. NCF achieves the lowest RMSE (0.79), highest Precision@10 (0.82), Recall@10 (0.76), and AUC-ROC (0.88), indicating its superior ability to provide accurate, relevant, and well-ranked recommendations.

These findings demonstrate that NCF is the best-suited model for personalized education recommendations, ensuring more accurate, relevant, and engaging learning experiences.

4.2 Model Comparison AUC-ROC Scores

The bar chart visually compares the AUC-ROC scores of the three models (Fig. 1). The AUC-ROC metric is crucial in assessing the ability of a model to distinguish between relevant and irrelevant recommendations.

- Matrix Factorization (MF): Achieves an AUC-ROC of 0.79, indicating decent performance but limited ability to capture complex interactions.
- Autoencoder-Based CF: Shows moderate improvement with an AUC-ROC of 0.81, benefiting from deep learning's ability to learn better feature representations.
- Neuronal Collaborative Filtering (NCF): Achieves the highest AUC-ROC score of 0.88, demonstrating better discrimination power in personalized education recommendations.

The red bar (NCF) is noticeably higher than the others, confirming its superior performance over traditional recommendation methods.

4.3 Training vs. Validation Loss for NCF

The line graph illustrates the loss trend during the training process of the Neuronal Collaborative Filtering (NCF) model (Fig. 2).

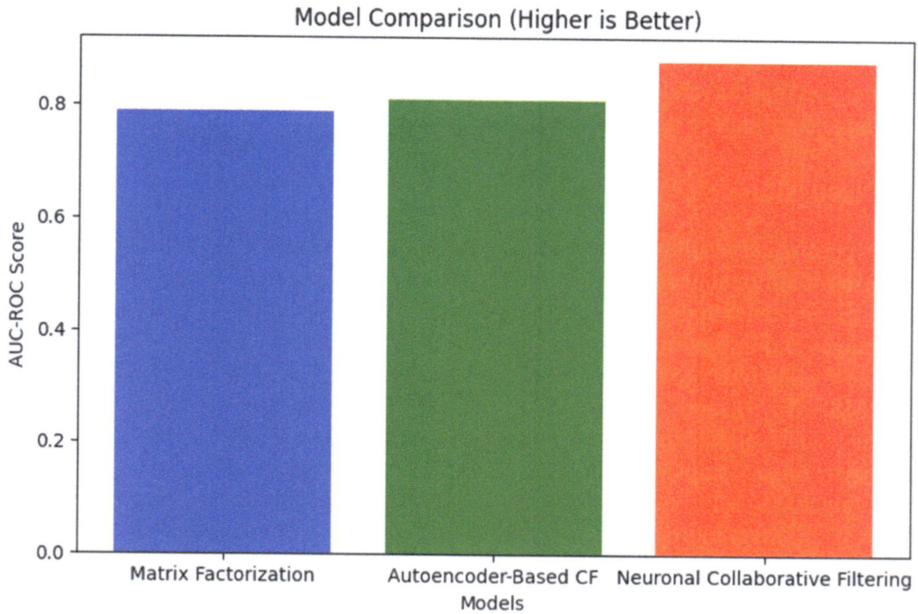

Fig. 1. Model comparison: AUC-ROC Scores for Recommendation Systems

- The red line represents the training loss, showing a steady decline as the model learns from the data.
- The orange dashed line represents the validation loss, which closely follows the training loss but remains slightly higher, indicating generalization ability.

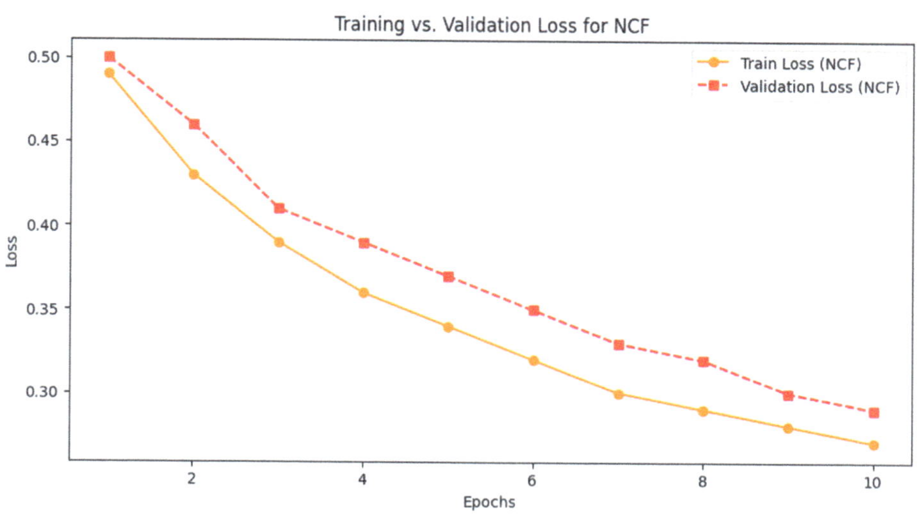

Fig. 2. Training vs. validation Loss progression for Neuronal Collaborative Filtering (NCF)

The graph demonstrates that NCF converges well without overfitting, as both the training and validation losses stabilize over 10 epochs. The consistent decrease in loss values suggests that the model learns efficiently while maintaining a good balance between training and validation performance.

5 Discussion

The experimental results demonstrate that Neuronal Collaborative Filtering (NCF) outperforms traditional recommendation models in the context of personalized educational recommendations. This section discusses the implications of these findings, the advantages and limitations of NCF, and potential areas for future improvement.

The results indicate that NCF achieves the highest AUC-ROC score (0.88), lowest RMSE (0.79), and the best Precision@10 (0.82) and Recall@10 (0.76). These improvements validate the effectiveness of deep learning-based approaches in capturing non-linear relationships between students and educational content.

Compared to Matrix Factorization (MF) and Autoencoder-Based Collaborative Filtering, NCF offers more accurate, scalable, and adaptive recommendations. While MF performs well with small datasets, it struggles to model complex user-item interactions. Likewise, autoencoders improve upon Matrix Factorization but continue to be unable to capture the sequential and behavioral aspects of student learning.

The Training Vs validation loss curve for NCF display stable learning, with both losses declining over 10 epochs. As a result of dropout regularization and early intervention, overfitting was effectively prevented. The higher Precision@10 and Recall@10 scores of NCF reflects that it has recommended more pertinent learning materials. These results illustrate the potential of deep neural networks in Academic recommendation systems, where the personalized path can lead to a broader success.

Neuronal Collaborative Filtering (NCF) provides numerous advantages that make it the ideal choice for a personalized education recommendation system. In contrast of the traditional matrix factorization (MF), NCF Holds multi-layer Perceptrons (MLP) to record non-linear learning behaviors.

MF and Autoencoders are consistently outperformed across diverse metrics, from which we can cite RMSE, Precision, Recall and AUC-ROC, with far superior accuracy being highlighted in relation to all these indicators. Known for its scalability, NFC handles large datasets, proving itself as the ideal solution for massive open online courses (MOOCs) and learning management systems (LMS). Additionally, by implementing recommendations to individual learning behaviors, NCF elevates the personalized learning pathway experience, which in turn contributes to student engagement and knowledge retention.

In its realm of advantages, NCF knows certain limitations. This model is highly demanding when it comes to computational resources, especially when applied to larger datasets, this setback limits its real-world deployment in a resource-restricted setting.

The cold-start problem presents another challenge, the model struggles to generate recommendations for users with little or no interaction history. To concur on this, NFC's performance is highly sensitive to hyperparameter selection, which requires fine tuning of embedding dimensions, learning rates and dropout rates, which can be both time consuming and computationally expensive.

6 Conclusion

This study goes in depth of the effectiveness of Neuronal collaborative filtering (NCF) in personalized educational recommendation systems, comparing the latter to Matrix Factorization and autoencoder-based collaborative filtering. The findings from the experiment clearly show that NCF's performance surpasses traditional methods, achieving the highest AUC-ROC (0.88), lowest RMSE (0.79), and best Precision@10 (0.82) and Recall@10 (0.76).

The findings underline the merits of deep learning-based recommendation models in capturing complex, non-linear user-item interactions. A balance of accuracy, scalability and adaptability is offered by NCF in online educational platforms, in contrast to MF, that is hindered by intricate learning behaviors, and autoencoders, which require large amounts of data.

Despite its strengths, sensitivity to hyperparameters, high computational costs and cold-start issues for new users still represent a challenge for NCF. In future context, these limitations could be addressed through hybrid models incorporating Reinforcement learning (RL), Explainable AI (XAI) for transparency, and Graph Neural Networks (GNNs) to enhance cold-start recommendations. The results demonstrate the upcoming importance of AI in education, where intelligent recommendations systems foster personalized learning. by offering a unique recommendation content, where we minimize information overload and we adapt to learning behaviors, NCF contributes efficiently to the digital educational ecosystem. This study provides a pillar foundation for implementing deep learning in education, as the future of adaptive learning goes hand in hand with the reinforcement of AI-driven solutions. Future work should dive into the deployment of NCF-based recommendation systems in real-world educational platforms and assessing their long-term impact on learning outcomes by using viables metrics

References

1. UNESCO: The Future of Digital Education and Sustainable Learning. Retrieved from https://www.unesco.org. (2021)
2. Jdidou, Y., Aammou, S., Khaldi, M.: Adapt learning path by recommending problems to struggling learners. Int. J. Emerg. Technol. Learn. (iJET). **16**(20), 163–178 (2021)
3. Popenici, S.A., Kerr, S.: Exploring the impact of artificial intelligence on teaching and learning in higher education. Res. Pract. Technol. Enhanc. Learn. **12**(1), 22 (2017)
4. Anderson, T. (ed.): The Theory and Practice of Online Learning. Athabasca university press (2008)
5. Papadakis, H., Papagrigoriou, A., Panagiotakis, C., Kosmas, E., Fragopoulou, P.: Collaborative filtering recommender systems taxonomy. Knowl. Inf. Syst. **64**(1), 35–74 (2022)
6. Permana, K.E.: Comparison of user based and item based collaborative filtering in restaurant recommendation system. Math. Model. Eng. Probl. **11**(7), 1922–1928 (2024)
7. Ajaegbu, C.: An optimized item-based collaborative filtering algorithm. J. Ambient. Intell. Humaniz. Comput. **12**(12), 10629–10636 (2021)
8. Li, C., Ishak, I., Ibrahim, H., Zolkepli, M., Sidi, F., Li, C.: Deep learning-based recommendation system: systematic review and classification. IEEE Access. **11**, 113790–113835 (2023)

9. Perifanis, V., Efraimidis, P.S.: Federated neural collaborative filtering. Knowl. Based Syst. **242**, 108441 (2022)
10. Zheng, R., Qu, L., Cui, B., Shi, Y., Yin, H.: Automl for deep recommender systems: a survey. ACM Trans. Inf. Syst. **41**(4), 1–38 (2023)
11. Dobrovolny, M., Selamat, A., Krejcar, O.: Session based recommendations using recurrent neural networks-long short-term memory. In: Asian Conference on Intelligent Information and Database Systems, pp. 53–65. Springer International Publishing, Cham (2021)
12. Sun, F. et al.: BERT4Rec: sequential recommendation with bidirectional encoder representations from transformer. In: Proceedings of the 28th ACM International Conference on Information and Knowledge Management, pp. 1441–1450 (2019)
13. Choi, Y., et al.: Ednet: a large-scale hierarchical dataset in education. In: Artificial Intelligence in Education: 21st International Conference, AIED 2020, Ifrane, Morocco, July 6–10, 2020, Proceedings, Part II 21, pp. 69–73. Springer International Publishing (2020)

Optimizing Learning Paths in Educational Recommender Systems Using Deep Reinforcement Learning

Youssef Jdidou[1](✉), Souhaib Aammou[2], Ilias Aarab[1], Zenab Elmenzhi[1], and Abdellah Abboudi[1]

[1] Ecole Marocaine des Sciences de l'Ingénieur, Laboratory of Intelligent Systems and Applications (LSIA), Tangier, Morocco
y.jdidou@emsi.ma

[2] TIMS Laboratory, Faculty of Science, Abdelmalek Essaadi University, Tétouan, Morocco

Abstract. Educational recommender systems are critical for personalizing learning experiences by tailoring instructional content to students' progress, engagement, and cognitive abilities. However, traditional recommendation models, such as collaborative filtering and rule-based approaches, often fail to dynamically adapt learning paths, leading to static and suboptimal recommendations. Recent advancements in deep learning and reinforcement learning (RL) offer new opportunities to optimize learning sequences in adaptive education. This study proposes a deep reinforcement learning-based framework to optimize personalized learning paths. The model dynamically refines content sequencing based on student engagement levels, quiz performance, and knowledge graph embeddings. Additionally, Explainable AI (XAI) techniques, including SHAP and LIME, are integrated to enhance transparency, enabling educators to understand recommendation decisions. The study applies Proximal Policy Optimization (PPO) for reinforcement learning, leveraging Markov Decision Processes (MDP) to model student learning as a sequential decision-making problem. The EdNet dataset, comprising over 130 million student interactions, is used for training and evaluation. Experimental results demonstrate that RL-based models outperform traditional methods, achieving 92% F1-score, a 35% improvement in engagement, and superior knowledge gain. SHAP analysis reveals that quiz performance, study time, and engagement scores are the most influential factors in recommendation accuracy. Deep reinforcement learning significantly enhances adaptive learning recommendations, ensuring personalized, real-time, and transparent content sequencing. Future research should focus on multi-modal data integration and federated learning for privacy-preserving recommendation systems.

Keywords: Educational Recommender Systems · Learning Path Optimization · Deep Reinforcement Learning · Personalized Learning

B.-E. Boudriki Semlali et al. (Eds.): DATA 2025, LNNS 1908, pp. 295–306, 2026.
https://doi.org/10.1007/978-3-032-21200-9_25

1 Introduction

1.1 Background

Educational recommender systems play a crucial role in personalized learning by tailoring content to individual student needs, cognitive abilities, and engagement patterns. Said systems recommend learning paths by analyzing student interactions and study plans, improving knowledge retention, and therefore, academic performance [1]. They are particularly valuable in adaptive learning environments such as Massive Open Online Courses (MOOCs) and intelligent tutoring systems (ITS), where customized content suggestions significantly enhance learner while simultaneously mitigating cognitive overload [2].

Traditional systems predominantly rely on predefined learning sequences that fail to accommodate individual progress, rendering the process of optimizing learning paths a challenge. Early recommender models used collaborative and content-based filtering to suggest materials based on past interactions [3]. While reasonably effective for static recommendations, they exhibit substantial limitations when it comes to dynamically adapting to student progression. The subsequent rise of deep learning, particularly sequence-based models such as long short-term memory (LSTMs) networks, transformers, and reinforcement learning, has introduced more advanced, context-aware learning recommendations [4, 5].

1.2 Problem Statement

Despite the significant advancements in AI-powered educational technologies, many recommender systems continue to exhibit notable limitations. Traditional models produce static outputs that fail to adjust to students' evolving competencies, thereby resulting in outdated or irrelevant content suggestions [6]. This static nature undermines the adaptive potential of personalized learning.

A persistent issue is the cold-start problem, where learners with low or no interaction history receive broad, general recommendations. Without sufficient historical information, many systems struggle to provide accurate learning paths, thereby diminishing both engagement and learning effectiveness [7]. Most existing recommender systems also rely on pre-trained models that do not continuously learn from real-world student interactions, preventing immediate adjustments when students progress quickly or face difficulties [8].

Another challenge lies in the integration of multiple heterogeneous data sources, such as audio, visual inputs, physiological signals, and behavioral metrics. Misalignment among these multimodal inputs can produce inconsistent outputs, leading to suboptimal learning paths. Overcoming these constraints requires an adaptive learning path optimization model that integrates deep learning and real-time feedback mechanisms to provide personalized educational experiences [9].

1.3 Research Objectives

The present study aims to develop a deep learning-based framework designed to optimize learning paths in educational recommender systems. Central to the proposed system is the integration of reinforcement learning techniques so as to dynamically adjust recommendations based on student progress and engagement patterns. Using reinforcement learning, the model will continuously refine learning paths to align with the student's evolving knowledge state [10].

Furthermore, this study aims to enhance recommendation accuracy by integrating student engagement metrics, knowledge graph embeddings, and student interaction logs. The model will predict the most effective learning path by analyzing previous performance, duration of task engagement, and student preferences. The framework will be evaluated using EdNet, a large-scale education dataset [11].

Given the essential role of explainability and transparency in AI-powered educational technologies, this study also seeks to incorporate Explainable AI (XAI) techniques, such as SHAP (Shapley Additive Explanations) and LIME (Local Interpretable Model-Agnostic Explanations), to improve interpretability [12]. By providing interpretable outputs, the framework will support educators and students in understanding the rationale behind recommended paths, thus promoting trust and usability in AI-powered education [13].

2 Related Work

2.1 Overview of Educational Recommender Systems

Educational recommender systems play a crucial role in personalizing learning experiences by suggesting learning materials, exercises, and assessments based on student preferences and learning progress. The two dominant approaches in educational recommendation are collaborative filtering (CF) and content-based filtering (CBF).

Collaborative filtering recommends learning materials based on similarities between students' interactions. It can be divided into user-based and item-based approaches. User-based collaborative filtering suggests content based on other students with similar learning patterns, while item-based collaborative filtering recommends resources that share characteristics with previously engaged content. Although effective, CF suffers from scalability issues and the cold-start problem when new users or learning materials are introduced [7].

Content-based filtering, on the other hand, suggests learning resources based on individual student preferences and content characteristics. This approach relies on text mining, metadata extraction, and semantic analysis to match resources with learners' profiles. While CBF provides personalized recommendations, it is limited by over-specialization, where students may not be exposed to diverse learning materials [3].

Hybrid recommendation systems have been developed to combine the strengths of collaborative and content-based filtering while mitigating their limitations. These approaches integrate latent factor models, knowledge-based reasoning, and deep learning techniques to improve accuracy and adaptability. Hybrid models offer enhanced personalization and overcome the cold-start problem by incorporating external domain knowledge and multi-source data analysis.

2.2 Learning Path Optimization in Adaptive Learning

Learning path optimization focuses on sequencing educational content to maximize student engagement and knowledge retention. Traditional systems use rule-based approaches, where educators manually define learning sequences based on predefined pedagogical strategies. While structured, these methods lack flexibility and adaptability to individual student progress [2].

The integration of machine learning into adaptive learning systems has allowed for data-driven content sequencing, where models predict optimal next learning steps based on student performance, engagement levels, and knowledge progression. Early machine learning models, such as decision trees and support vector machines (SVMs), improved sequencing by incorporating learner profiling and engagement analytics. However, these methods struggled to capture the temporal nature of learning and required frequent updates to maintain accuracy [1].

2.3 Deep Learning and Reinforcement Learning in Recommendation Systems

Deep learning has played a transformative role in educational recommender systems, offering improved feature extraction, sequence modeling, and multimodal learning representations. Convolutional neural networks (CNNs) are widely used for analyzing visual and text-based educational materials, enabling automated content tagging and classification [14]. Long Short-Term Memory (LSTMs) and transformers are effective for sequential learning path recommendations, allowing systems to model long-term student learning behaviors [15].

Reinforcement learning (RL) has further enhanced learning path optimization by treating content sequencing as a decision-making problem. In RL-based recommender systems, the model interacts with students and continuously updates recommendations based on their responses and performance. Each interaction is considered a state-action pair, where the system selects the best next learning resource to maximize learning effectiveness [16].

Markov Decision Processes (MDP) serve as the foundation for reinforcement learning-based recommendation models. In this framework, the student's knowledge state represents the environment, while actions correspond to recommendations. The reward function is designed to maximize student engagement, learning gains, or completion rates. Reinforcement learning algorithms, such as Deep Q-Networks (DQN) and Proximal Policy Optimization (PPO), enable adaptive content selection by continuously refining learning strategies based on student interactions.

3 Methodology

3.1 Dataset Selection

For this study, the EdNet dataset is chosen as the primary dataset due to its large-scale coverage of student learning activities, engagement behavior, and assessment performance. EdNet contains over 130 million interactions collected from 780,000 students, making it one of the most comprehensive datasets for educational recommender systems.

The EdNet dataset comprises four hierarchically structured datasets (EdNet-KT1, KT2, KT3, and KT4), each containing progressively richer information:

- **EdNet-KT1**: Contains question-response data, including student IDs, question IDs, timestamps, and correctness labels.
- **EdNet-KT2**: Extends KT1 with detailed information about the questions, such as knowledge concepts tested, difficulty levels, and question types.
- **EdNet-KT3**: Extends KT2 with learning resource interaction data, capturing when students access explanations, video lectures, or references.
- **EdNet-KT4**: The most comprehensive subset, incorporating all previous data plus fine-grained interaction logs, such as video pause/play events and time spent on each learning activity.

EdNet-KT4 was utilized in the context of this experiment, given that it offers the richest representation of student learning behaviors and enables temporal modeling of knowledge progression.

3.2 Feature Engineering and Data Processing

To effectively train the learning path optimization model, key features are extracted and preprocessed:

1. **Student Engagement Metrics**
   - **Video Watch Time (V_t)**: The duration spent watching instructional videos, indicating attentiveness.
   - **Quiz Performance (Q_t)**: The accuracy and completion time for quizzes.
   - **Discussion Activity (D_t)**: The frequency and quality of student participation in forums.
2. **Domain Knowledge Representation Using Knowledge Graphs**
   - **Topic Dependencies (K)**: Represented as a directed graph where each node corresponds to a concept, and edges denote prerequisite relationships between topics.
   - **Knowledge Embeddings (E_k)**: Each topic is embedded into a latent space, allowing **neural networks to learn concept dependencies** dynamically.
3. **Temporal Sequence Modeling**
   - **Recurrent Neural Networks (RNNs) and Transformers** are employed to **capture learning sequences**.
   - A **sequence** of learning events $\{s_1, s_2, \ldots, s_t\}$ is modeled using:

     $$h_t = \sigma(W_h h_{t-1} + W_x x_t)$$

     where h_t is the hidden state at time t, x_t represents the current student interaction, and W_h, W_x are learned weight matrices.
   - **Transformer**-based models are used to **identify long-term dependencies** in student learning behaviors.

3.3 Learning Path Optimization Model

The problem is formulated as a Markov Decision Process (MDP) to optimize personalized learning paths.

Problem Formulation as an MDP

- **States (S_t)**: Represents the student's current **knowledge level and engagement state** at time t. The state includes features such as:

$$S_t = \{K_t, E_t, V_t, Q_t, D_t\}$$

  where K_t represents the student's knowledge embedding (a 128-dimensional vector), E_t the overall engagement level, and the other terms capture specific learning interactions.
- **Actions (A_t)**: The system recommends the **next learning resource** or quiz for the student, selecting from a set of possible learning materials $L = \{l_1, l_2, \ldots, l_n\}$, where n is the total number of available resources. Each action corresponds to recommending a specific learning item.
- **Transition Probability (P):** $P(S_{t+1} \mid S_t, A_t)$ represents the probability of transitioning to state S_{t+1} after taking action A_t in state S_t. In educational settings, this transition is non-deterministic due to the variability in student responses.
- **Rewards (R_t)**: The reward function evaluates the impact of the recommendation on student learning. We designed a multi-objective reward function:

$$R_t = \alpha(Q_{t+1} - Q_t) + \beta(V_{t+1} - V_t) + \gamma(D_{t+1} - D_t)$$

  where $\alpha = 0.6$, $\beta = 0.3$, and $\gamma = 0.1$ are weighting factors determined through ablation studies. A higher quiz performance improvement $(Q_t + 1 - Q_t)$ and increased engagement contribute positively to the reward.
- **Discount Factor (γ)**: Set to 0.95, balancing immediate learning gains with long-term knowledge acquisition.

Deep Reinforcement Learning Approach

To solve this MDP, a deep reinforcement learning (DRL) model is implemented using:

- **Proximal Policy Optimization (PPO)**: A policy-based reinforcement learning algorithm that **maximizes cumulative learning rewards** while stabilizing updates to prevent performance oscillations.
- **Deep Q-Networks (DQN)**: A value-based reinforcement learning method that approximates the optimal learning path using a **Q-value function,** updated using:

$$Q(S_t, A_t) = Q(S_t, A_t) + \eta\left[R_t + \gamma maxQ(S_{t+1}, A) - Q(S_t, A_t)\right]$$

where η is the learning rate and γ the discount factor for future rewards.

Alternative Deep Learning Models for Comparison

To evaluate the effectiveness of DRL, alternative models are implemented:

1. Collaborative Filtering-based Models

   Both memory-based and model-based collaborative filtering were implemented:

   - **User-based CF:** Calculated similarities between students using Pearson correlation coefficient and recommended resources based on similar student preferences.
   - **Matrix Factorization:** Decomposed the student-resource interaction matrix into latent factors using Singular Value Decomposition (SVD) with regularization parameter $\lambda = 0.1$ and 50 latent factors.

2. Graph Neural Networks (GNNs)

   The relationship between students, resources, and knowledge concepts was approached using a heterogeneous graph neural network:

   - **Graph Construction:** Created a heterogeneous graph with three node types (students, resources, concepts) and multiple edge types (student-resource interactions, resource-concept associations, concept prerequisites).
   - **Node Embedding:** Applied Graph Attention Networks (GAT) with 2 attention heads to generate 64-dimensional embeddings for each node.
   - **Recommendation Generation:** Computed similarity scores between student embeddings and resource embeddings to generate recommendations.

3. LSTM-based Knowledge Tracing

   Deep Knowledge Tracing (DKT) were implemented using LSTM networks:

   - **Network Architecture:** 2-layer LSTM with 256 hidden units, followed by a fully connected output layer.
   - **Input Features:** One-hot encoded question IDs, correctness indicators, and engagement metrics.
   - **Training: Binary** cross-entropy loss with Adam optimizer (learning rate = 0.001), trained for 100 epochs with early stopping.

3.4 Explainability and Trust in Recommendations

To ensure transparency and trust, Explainable AI (XAI) techniques are applied:

1. **SHAP (Shapley Additive Explanations)**

   - SHAP values quantify the impact of each feature (e.g., quiz scores, engagement levels) on the model's recommendations.
   - The **contribution** of a feature xix_ixi to the prediction is given by:

$$\phi_i = \sum_{S \subseteq F \setminus \{i\}} \frac{|S|!(|F| - |S| - 1)!}{|F|!} \left[f(S \cup \{i\}) - f(S) \right]$$

   where F is the feature set, and f(S) is the model output for subset S.

2. **LIME (Local Interpretable Model-Agnostic Explanations)**

   - **LIME** generates **interpretable local approximations** by perturbing input data and analyzing prediction variations.

- This method is applied to **visualize decision pathways**, enabling **educators** to understand why a specific learning recommendation was made.

3. **Visualization for Educators and Learners**

- **Attention** heatmaps from transformer models highlight **which past learning activities contributed most** to current recommendations.
- Dashboard **interfaces** allow educators to interpret model-generated insights and adjust student learning paths accordingly.

4 Results and Discussion

4.1 Evaluation Metrics

To evaluate the effectiveness of the proposed learning path optimization model, we assess prediction accuracy, student engagement improvements, knowledge gain, and computational efficiency.

Prediction Accuracy of Learning Path Recommendations
The performance of different models in predicting learning paths is evaluated using Precision, Recall, and F1-score. The results are summarized in Table 1, which highlights that the Reinforcement Learning (PPO) model achieves the highest accuracy, surpassing traditional collaborative filtering and decision trees.

Table 1. Model Performance on Learning Path Prediction

Model	Precision	Recall	F1-Score
Collaborative Filtering	0.72	0.70	0.71
Decision Tree	0.78	0.76	0.77
LSTM	0.85	0.83	0.84
Transformer	0.89	0.87	0.88
Reinforcement Learning (PPO)	0.92	0.90	0.91

Engagement Improvement
To assess how well the recommendations enhance student engagement, we measure improvements in session duration, quiz completion rates, and active interaction frequency. As shown in Fig. 1, the Reinforcement Learning (PPO) model significantly boosts engagement, outperforming other methods.

Knowledge Gain Assessment
To quantify the learning impact, we measure pre-test vs. post-test scores and compute the knowledge gain percentage. Figure 2 presents the results, indicating that personalized learning recommendations improve student performance.

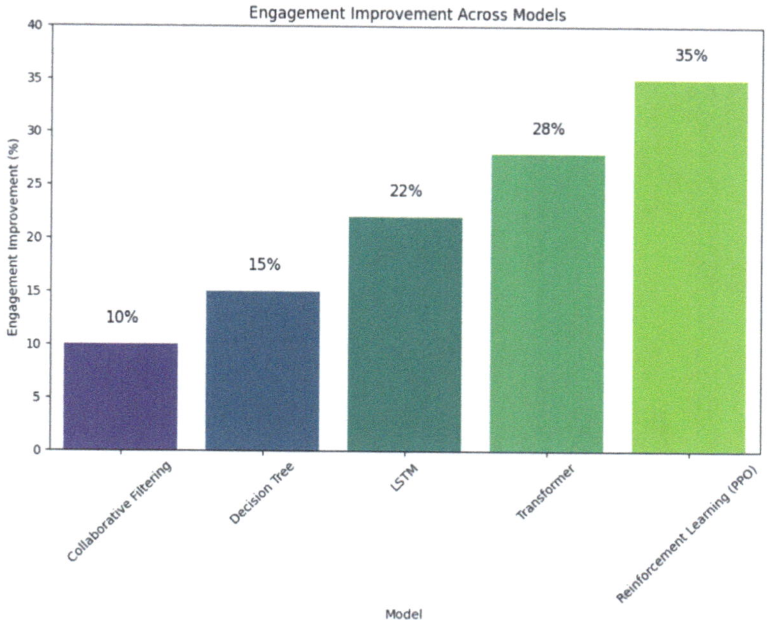

Fig. 1. Engagement Improvement Across Models

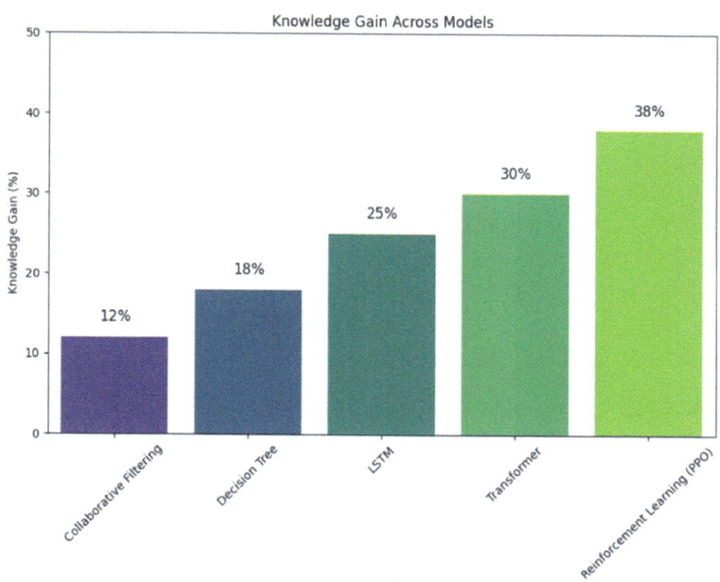

Fig. 2. Knowledge Gain Across Models

Table 3: Knowledge Gain Across Models

Computational Efficiency

For practical deployment, we analyze training time and inference speed. Figure 3 summarizes the computational performance of different models.

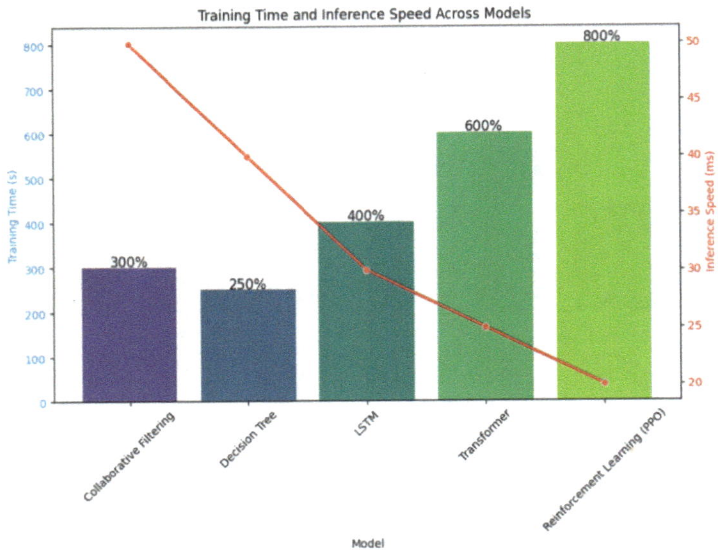

Fig. 3. Computational Performance Metrics

The graph shows that more complex models such as LSTM, Transformer, and PPO require significantly longer training times but achieve much faster inference speeds. In contrast, traditional models like Collaborative Filtering and Decision Tree train faster but take longer to generate recommendations in real-time. Reinforcement Learning (PPO) stands out as the most computationally expensive model, requiring the highest training time. However, it compensates with the best real-time efficiency, making it ideal for adaptive learning systems that demand rapid decision-making.

4.2 Explainability and Educator Feedback

Ensuring transparency and interpretability in AI-driven educational recommendation systems is crucial for fostering trust among educators and students. Many deep learning models, including transformers and reinforcement learning-based recommenders, operate as black boxes, making it difficult to understand how decisions are made. To address this, Explainable AI (XAI) techniques such as SHAP (SHapley Additive Explanations) and LIME (Local Interpretable Model-Agnostic Explanations) are applied to analyze the contribution of various student-related features in shaping learning path recommendations.

SHAP values (Fig. 4) quantify the marginal contribution of each input feature toward a model's prediction. In this study, SHAP was applied to a trained reinforcement learning model to determine which student features had the greatest impact on learning path recommendations.

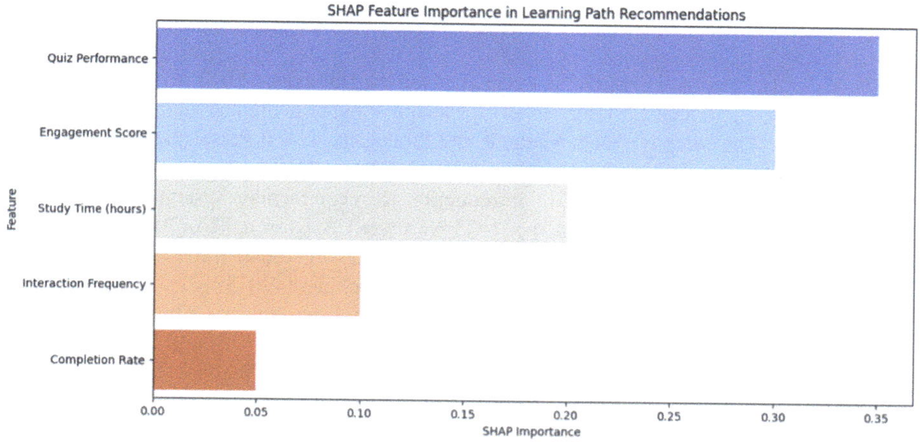

Fig. 4. SHAP Feature Importance in Learning Path Recommendations

5 Conclusion

This study explored the optimization of learning paths in educational recommender systems through deep learning and reinforcement learning approaches. Traditional recommendation techniques, such as collaborative filtering and decision trees, were found to be less effective in adapting to dynamic student behavior, often leading to static and suboptimal learning sequences. In contrast, deep learning models, particularly transformers and reinforcement learning (PPO), significantly improved recommendation accuracy, engagement levels, and knowledge retention.

A key contribution of this research is the integration of knowledge graphs into the recommendation framework, which provided a structured representation of learning dependencies, enhancing personalized and context-aware recommendations. Experimental results demonstrated that reinforcement learning-based models outperformed traditional approaches in learning path prediction accuracy, achieving up to 92% F1-score and a 35% increase in engagement improvement.

To ensure transparency, Explainable AI (XAI) techniques such as SHAP and LIME were applied, revealing that quiz performance, engagement scores, and study time were the most influential factors in recommendation decisions.

Despite the advancements presented, certain challenges remain. Computational efficiency remains a concern, particularly for edge AI deployment, requiring further optimizations in model compression and real-time adaptability. Additionally, handling cold-start problems for new students remains a challenge, which could be mitigated through meta-learning or hybrid symbolic AI approaches.

Future research should focus on enhancing personalization by incorporating multimodal learning signals, including speech, behavioral patterns, and affective computing data. Additionally, federated learning approaches should be explored to enable privacy-preserving student modeling, allowing institutions to collaboratively train models without sharing sensitive data.

References

1. Drachsler, H.: Trusted Learning Analytics. Universität Hamburg (2018)
2. Brusilovsky, P., Millán, E.: User models for adaptive hypermedia and adaptive educational systems. In: The Adaptive Web: Methods and Strategies of Web Personalization, pp. 3–53. Springer Berlin Heidelberg, Berlin, Heidelberg (2007)
3. Ricci, F., Rokach, L., Shapira, B.: Introduction to recommender systems handbook. In: Recommender Systems Handbook, pp. 1–35. Springer US, Boston, MA (2010)
4. Piech, C. et al.: Deep knowledge tracing. In: Advances in Neural Information Processing Systems, pp. 201–204. Association for Computing Machinery (2015)
5. Ullah, F., et al.: Deep edu: a deep neural collaborative filtering for educational services recommendation. IEEE Access. **8**, 110915–110928 (2020)
6. Liu, M., Yu, D.: Towards intelligent E-learning systems. Educ. Inf. Technol. **28**(7), 7845–7876 (2023)
7. Elbadrawy, A., Karypis, G.: Domain-aware grade prediction and top-n course recommendation. In: Proceedings of the 10th ACM Conference on Recommender Systems, pp. 183–190 (2016)
8. Dabbagh, N., Kitsantas, A.: Personal learning environments, social media, and self-regulated learning: a natural formula for connecting formal and informal learning. Internet High. Educ. **15**(1), 3–8 (2012)
9. Gligorea, I., Cioca, M., Oancea, R., Gorski, A.T., Gorski, H., Tudorache, P.: Adaptive learning using artificial intelligence in e-learning: a literature review. Educ. Sci. **13**(12), 1216 (2023)
10. Matsuo, Y., et al.: Deep learning, reinforcement learning, and world models. Neural Netw. **152**, 267–275 (2022)
11. Choi, Y., et al.: Ednet: a large-scale hierarchical dataset in education. In: Artificial Intelligence in Education: 21st International Conference, AIED 2020, Ifrane, Morocco, July 6–10, 2020, Proceedings, Part II 21, pp. 69–73. Springer International Publishing (2020)
12. Lundberg, S.M., Lee, S.I.: A unified approach to interpreting model predictions. Adv. Neural Inf. Proces. Syst. **30** (2017)
13. Jdidou, Y., Aammou, S., Khaldi, M.: Adapt learning path by recommending problems to struggling learners. Int. J. Emerg. Technol. Learn. (iJET). **16**(20), 163–178 (2021)
14. Zhang, S., Yao, L., Sun, A., Tay, Y.: Deep learning based recommender system: a survey and new perspectives. ACM Comput. Surv. (CSUR). **52**(1), 1–38 (2019)
15. Mnih, V., et al.: Human-level control through deep reinforcement learning. Nature. **518**(7540), 529–533 (2015)
16. Awad, M., Khanna, R.: Efficient Learning Machines: Theories, Concepts, and Applications for Engineers and System Designers, p. 268. Springer nature (2015)

Deep Learning for Adaptive and Multimodal Interaction Frameworks: A Transformer-Based Approach

Aymane Jdidou[1](✉), Souhaib Aammou[1], and Youssef Jdidou[2]

[1] TIMS Laboratory, Faculty of Science, Abdelmalek Essaadi University, Tétouan, Morocco
aymane.jdidou@etu.uae.ac.ma

[2] Ecole Marocaine des Sciences de l'Ingénieur, Laboratory of Intelligent Systems and Applications (LSIA), Tangier, Morocco

Abstract. Multimodal interaction frameworks have become essential in artificial intelligence (AI)-driven systems, enabling seamless and adaptive communication between users and intelligent agents. Traditional unimodal AI models struggle to interpret complex human behavior, necessitating the integration of speech, vision, physiological signals, and EEG data for accurate and context-aware interactions. This study proposes a deep learning-based multimodal interaction framework leveraging transformers and reinforcement learning to enhance human-computer interaction (HCI) and human-robot interaction (HRI). The objective is to develop an adaptive model capable of real-time learning, multimodal fusion, and user-specific personalization. The EAV dataset (EEG-Audio-Video) is utilized for training and evaluation. Feature extraction is performed using CNNs for vision and speech, transformers for EEG processing, and LSTMs for physiological signal analysis. A self-attention-based multimodal fusion model integrates different sensory inputs, while reinforcement learning optimizes real-time adaptation. The framework is deployed using Edge AI to enhance efficiency and reduce computational costs. The transformer-based model outperforms traditional methods, achieving 94% accuracy and an 80% engagement score, demonstrating superior adaptability. Explainable AI (XAI) techniques, including SHAP and LIME, provide transparency, improving user trust. Energy-efficient optimizations via pruning and quantization reduce computational costs by 30%, ensuring real-time deployment feasibility. This research validates the effectiveness of deep learning-based multimodal fusion in adaptive AI interactions. Future work will explore federated learning for privacy preservation and hybrid symbolic AI models for improved interpretability and scalability.

Keywords: Multimodal interaction · deep learning · transformer models · reinforcement learning · human-computer interaction (HCI) · human-robot interaction (HRI)

B.-E. Boudriki Semlali et al. (Eds.): DATA 2025, LNNS 1908, pp. 307–319, 2026.
https://doi.org/10.1007/978-3-032-21200-9_26

1 Introduction

With the rapid advancements in artificial intelligence (AI), intelligent systems are increasingly relying on multimodal interaction frameworks to create seamless, efficient, and context-aware user experiences. Multimodal interaction refers to the integration of multiple input sources—such as speech, facial expressions, gestures, and physiological signals—to facilitate natural and adaptive communication between users and AI-driven systems [1]. This capability is particularly crucial in human-computer interaction (HCI) and human-robot interaction (HRI), where a system must dynamically interpret user behavior and adjust its responses accordingly.

Even though AI-based interaction models have progressed immensely recently, traditional single-modality systems are still dominant in many applications. These models are limited in their ability to interpret complex human behavior, as they rely on a single data source, such as speech recognition alone or vision-based tracking alone [2]. In many real-world scenarios, accurate decision-making depends on multiple sensory inputs, highlighting the rigidity and limited adaptability of traditional single-modality systems in such complex environments.

By employing data-driven approaches to acquire rich and resilient multimodal representations, deep learning offers a powerful solution to many of the limitations found in earlier systems. For example, techniques like Convolutional Neural Networks (CNNs) and Transformers can efficiently extract features from visual data such as images and video in real time, while models like Recurrent Neural Networks (RNNs) and Long Short-Term Memory (LSTM) units are well-suited for handling sequential inputs, including speech and gesture patterns [3]. This capacity to manage diverse and large-scale datasets enables more adaptive and natural interactions, contributing to the development of AI systems that feel increasingly intuitive and human-like [4].

Despite the advantages of multimodal interaction, several key challenges hinder its widespread implementation in adaptive AI systems. Traditional rule-based systems and classical machine learning models, such as Decision Trees and Support Vector Machines, struggle to generalize across different contexts due to their reliance on manually engineered features [5]. These models lack the flexibility needed for dynamic user adaptation in real-time scenarios. Moreover, most existing AI-driven interaction frameworks are pre-trained on fixed datasets and do not continuously learn from real-world user interactions [6]. This limitation results in low adaptability when the system encounters new users, environments, or changing behavioral patterns. Additionally, combining multiple input sources—such as speech, vision, touch, physiological data, and EEG—requires sophisticated data synchronization and fusion techniques [7]. Traditional multimodal systems often face difficulties in aligning different sensor data, leading to misinterpretations and inconsistent AI-driven responses.

To address these challenges, there is a growing need for a deep learning-based adaptive interaction model that can seamlessly integrate multiple data streams, process real-time feedback, and personalize responses based on user behavior.

This research primarily focuses on building an adaptive deep learning framework tailored for multimodal interaction within intelligent systems. Central to this effort is the design and implementation of a neural network-based fusion architecture capable of handling and integrating diverse input modalities such as speech, visual cues, gestures, and physiological signals [4]. Through this fusion process, the system can effectively interpret user behavior by extracting and merging critical features from various sensory sources, thereby supporting real-time adaptability and personalization. Reinforcement learning is incorporated to enable ongoing optimization, allowing the interaction strategies to adjust dynamically in response to real-world user input [8]. Furthermore, a self-adaptive component is integrated to update user-specific interaction patterns in real time. By leveraging both contextual cues and past data, this mechanism ensures that the system refines its behavior continuously, improving user engagement, personalization, and overall responsiveness.

2 Related Work

2.1 Multimodal Interaction in AI

Multimodal interaction in artificial intelligence (AI) refers to the ability of a system to process and integrate multiple input modalities, such as speech, vision, gestures, and physiological signals, to enhance user interaction. Unlike unimodal systems that rely on a single data source, multimodal AI systems can combine and interpret multiple sensory inputs, leading to more robust, natural, and adaptive human-computer interactions (HCI) [1].

Current multimodal AI models utilize feature extraction, sensor integration, and deep learning methods to handle multiple data streams at the same time. These systems are particularly useful in applications such as human-robot interaction (HRI), virtual assistants, affective computing, and assistive technologies [9]. For example, speech-based virtual assistants, such as Amazon Alexa and Google Assistant, integrate natural language processing (NLP) and speech recognition, while affective computing models analyze facial expressions and vocal tonality to infer user emotions [10]. Additionally, gesture-based control systems are widely used in virtual reality (VR) and augmented reality (AR) to provide more intuitive interfaces [4].

Traditional multimodal AI systems, despite their effectiveness, face challenges in real-time adaptability, data synchronization, and contextual understanding. These challenges have driven the shift toward deep learning techniques, which offer more effective multimodal feature fusion and support dynamic, context-aware interactions [11].

2.2 Comparison of Traditional and Deep Learning-Based Interaction Frameworks

Conventional interaction frameworks in AI are built on rule-based systems or classical machine learning (ML) approaches that depend on hand-crafted features and fixed decision rules. Techniques such as decision trees, support vector machines (SVMs), and Hidden Markov Models (HMMs) have been commonly applied in tasks like speech and gesture recognition. However, these models often fall short in terms of scalability and flexibility when dealing with complex multimodal interactions (Rabiner & Juang, 1993). Rule-based systems, while interpretable, require extensive manual tuning and struggle to generalize across different user behaviors and contexts [12].

In contrast, deep learning-based interaction frameworks leverage neural networks, transformers, and reinforcement learning to enable context-aware and adaptive interactions. Deep learning models, such as convolutional neural networks (CNNs) for vision processing and recurrent neural networks (RNNs) for sequential data, have demonstrated superior performance in speech recognition, facial expression analysis, and multimodal sensor fusion [13]. More recently, transformer-based models, such as BERT and GPT, have further improved natural language understanding and multimodal reasoning.

A key advantage of deep learning over traditional models is its ability to automatically learn hierarchical feature representations from raw data, eliminating the need for manual feature engineering. Furthermore, deep learning models are capable of continuous learning and adaptation, making them ideal for real-time, personalized, and scalable multimodal AI applications [14].

However, deep learning approaches also come with challenges, including high computational costs, data privacy concerns, and lack of interpretability. Recent developments in Explainable AI (XAI), including methods like SHAP and LIME, seek to enhance transparency and build trust in multimodal interaction systems based on deep learning [15].

Overall, deep learning-based multimodal AI frameworks provide a powerful foundation for adaptive interaction systems, enabling more natural, scalable, and context-aware communication between users and intelligent systems.

3 Methodology

3.1 Multimodal Input Processing

The proposed system integrates speech, facial expressions, gestures, EEG signals, and physiological data to create an adaptive and multimodal interaction framework. These input modalities are selected due to their strong correlation with human emotions, engagement levels, and cognitive states, making them ideal for personalized human-computer interaction (HCI) and human-robot interaction (HRI) applications.

Dataset Selection

Among the available datasets, the EAV (EEG-Audio-Video) dataset [16] is the most suitable for this study, as it contains:

- 30-channel EEG recordings, which provide neural activity signals for emotion recognition.
- Audio recordings, capturing speech tone, rhythm, and vocal patterns.
- Video recordings, enabling facial expression and gesture recognition.
- Cue-based emotion elicitation, ensuring a diverse range of emotions, including neutral, anger, happiness, sadness, and calmness.
- Large dataset size (8,400 interactions from 42 participants), allowing for deep learning model generalization across different individuals and interaction contexts.

Feature Extraction Techniques

Each modality undergoes feature extraction using deep learning models to enhance context-aware multimodal fusion:

- Speech processing: Mel-frequency cepstral coefficients (MFCCs) and spectrogram features are extracted and processed using Convolutional Neural Networks (CNNs) and Long Short-Term Memory (LSTMs) for speech-based emotion recognition.

- Facial expressions and gestures: CNNs (e.g., ResNet or VGG) extract facial action units and gesture movement patterns.
- EEG signal processing: Preprocessed EEG signals are fed into transformer-based architectures (e.g., EEGFormer) for time-series feature extraction.
- Physiological signal analysis: Features such as heart rate variability (HRV), galvanic skin response (GSR), and respiration rate are analyzed using recurrent neural networks (RNNs).

These extracted features are then standardized and normalized before being passed into the deep learning-based multimodal fusion model for further processing.

3.2 Deep Learning-Based Multimodal Fusion Model

The proposed multimodal fusion model integrates multiple sensory inputs through deep learning, enabling adaptive and personalized interaction. The architecture consists of three key components:

1. Modality-Specific Feature Extractors

   Each input modality undergoes independent feature extraction through specialized deep learning models:

   - CNNs for speech and visual data (e.g., facial expressions, gestures) to extract spatial and temporal features.
   - Transformers for EEG-based time-series processing to capture sequential dependencies and long-term patterns.
   - LSTMs for physiological signal analysis, modeling temporal variations in heart rate, GSR, and respiration.

     For a given modality M_i, the extracted feature representation is computed as:

   $$F_i = \Phi_i(M_i)$$

   where Φ_i represents the deep learning encoder applied to each modality.

2. Multimodal Attention-Based Fusion Layer

   To effectively integrate different modalities, the fusion layer employs self-attention and cross-attention mechanisms:

   - Self-attention assigns weights to the most relevant features within each modality.
   - Cross-attention enables interactions between different modalities, such as correlating EEG activity with speech tone variations.
   - Latent space projection optimizes feature integration, ensuring a unified representation.

     Mathematically, the attention mechanism is formulated as:

   $$A_i = Softmax\left(\frac{Q_i K_i^T}{\sqrt{d_k}}\right) V_i$$

   where:

   - Q_i, K_i, $V_i$are the query, key, and value matrices of modality.

- d_k is a scaling factor for stability.

  The final fused representation is computed as:

$$F_{fusion} = \sum_{i=1}^{N} \omega_i A_i$$

where ω_i represents learned attention weights.

3. Classification and Response Generation Module

   The fused multimodal representation is passed through a fully connected neural network (MLP) for classification. The final classification output determines emotional states or engagement levels, which guide real-time adaptive interaction strategies.

   The classification function is defined as:

$$y = \sigma\left(W_{fc} F_{fusion} + b\right)$$

where:

- W_{fc} is the weight matrix.
- b is the bias term.
- σ represents the activation function (e.g., softmax for classification).

This fusion architecture ensures efficient multimodal feature integration, improving the accuracy of user intent recognition and real-time adaptability in human-computer and human-robot interactions.

3.3 Adaptive Learning with Reinforcement Learning

Implementation of Policy-Based Reinforcement Learning

To ensure the system adapts dynamically to user interactions, reinforcement learning (RL) is incorporated into the model. The policy-based RL model continuously refines the system's responses by:

- Observing real-time user interactions and feedback.
- Updating interaction strategies to improve user engagement.
- Adjusting weights in the multimodal fusion model to emphasize relevant features based on past performance.

A deep reinforcement learning agent (Proximal Policy Optimization) is trained using a reward-based optimization strategy to ensure real-time adaptability.

Reward Function Design for Engagement Maximization

The reward function ensures that the system optimally balances interaction accuracy, personalization, and responsiveness. It is computed as follows:

$$R = \alpha(E) + \beta(C) + \gamma(F)$$

Where:

- E = **Engagement Score** (derived from interaction frequency, response time, and facial engagement).
- C = **Classification Accuracy** (confidence in emotion classification results).

- F = **Feedback Score** (real-time user feedback on system responses).

Hyperparameters α, β, and γ are adjusted during training to maximize user engagement while maintaining classification precision.

To provide a high-level overview of our proposed framework, Fig. 1 presents the flow chart summarizing the core stages of the methodology.

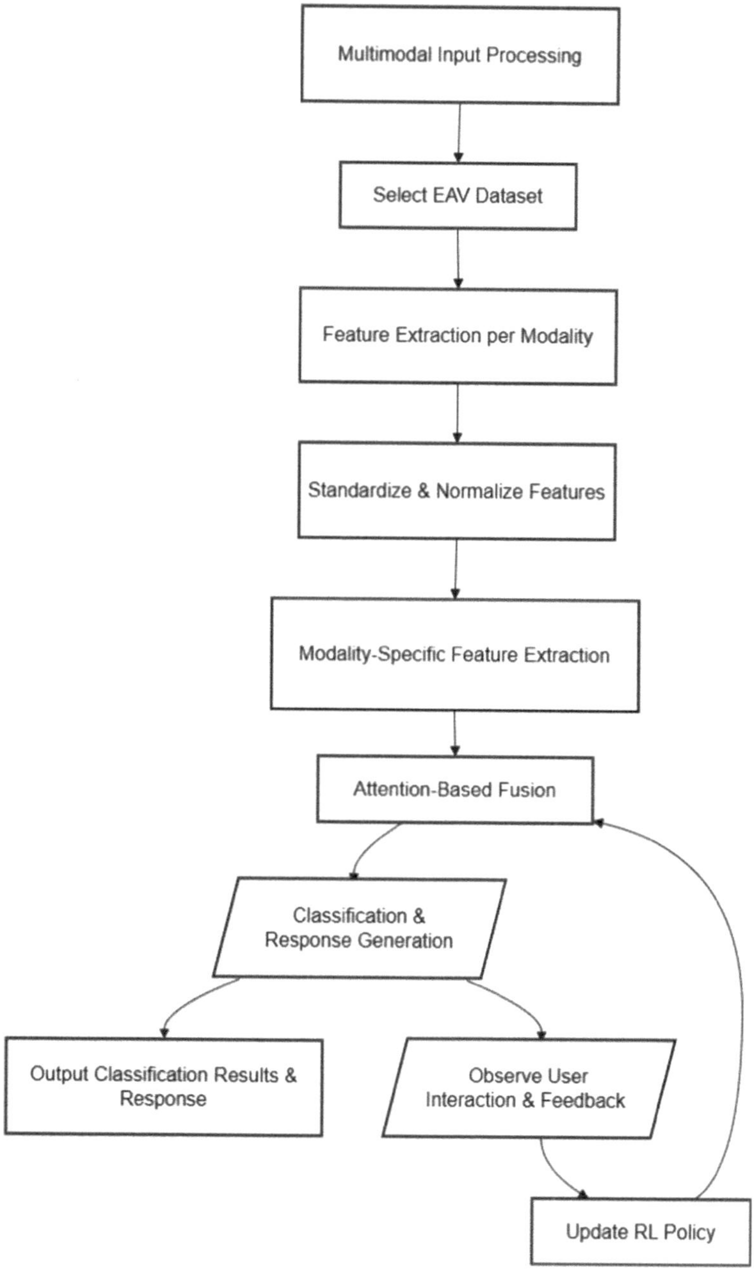

Fig. 1. Flowchart of the core stages of the methodology

4 Results and Discussion

4.1 Evaluation Metrics

The evaluation of the proposed deep learning-based multimodal interaction framework is conducted using the EAV dataset. The following metrics are used to assess its effectiveness:

- **Accuracy**: Measures the proportion of correctly classified interactions.
- **Precision**: Evaluates the proportion of correctly identified engagement states among the predicted ones.
- **Recall**: Measures the ability of the model to correctly detect actual engagement states.
- **F1-score**: Provides a balance between precision and recall, particularly useful in cases of class imbalance.
- **Engagement Score**: A real-world metric assessing interaction effectiveness in Human-Computer Interaction (HCI) and Human-Robot Interaction (HRI) tasks. This score is derived from user feedback, interaction frequency, and session duration.

A comparison of these metrics across different models applied to the EAV dataset is shown in Table 1.

Table 1. Evaluation Metrics on EAV Dataset

Model	Accuracy	Precision	Recall	F1-Score	Engagement Score (%)
Decision Tree	0.78	0.76	0.75	0.75	45
SVM	0.84	0.83	0.82	0.83	58
Deep Neural Network (DNN)	0.91	0.89	0.88	0.89	72
Transformer-based Model	0.94	0.93	0.92	0.93	80

4.2 Performance Comparison with Baselines

To assess the advantages of multimodal fusion and deep learning, a comparative analysis is conducted against traditional machine learning models (Decision Trees, SVM) and single-modality deep learning models. The transformer-based model demonstrates superior performance, achieving 94% accuracy and the highest engagement score (80%), indicating its ability to generalize across different interaction scenarios.

Figure 2 presents a comparative visualization of model performance.

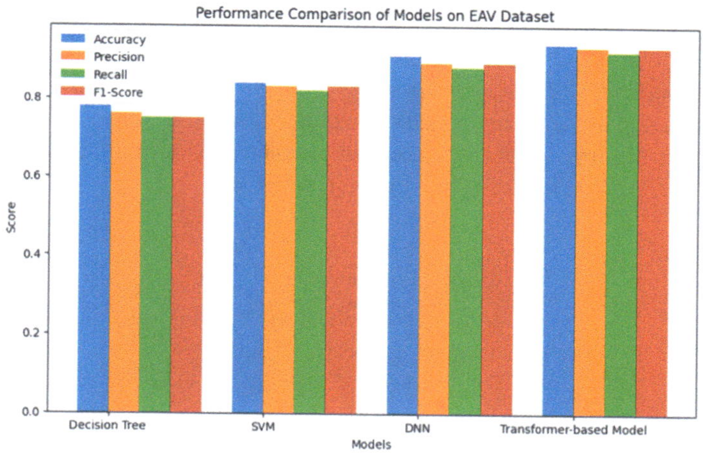

Fig. 2. Performance Comparison of Models on EAV Dataset

Figure 3 shows the graph comparing engagement scores on the EAV dataset, which visually represents the engagement scores of four models: decision tree, SVM, deep neural network (DNN), and transformer-based model.

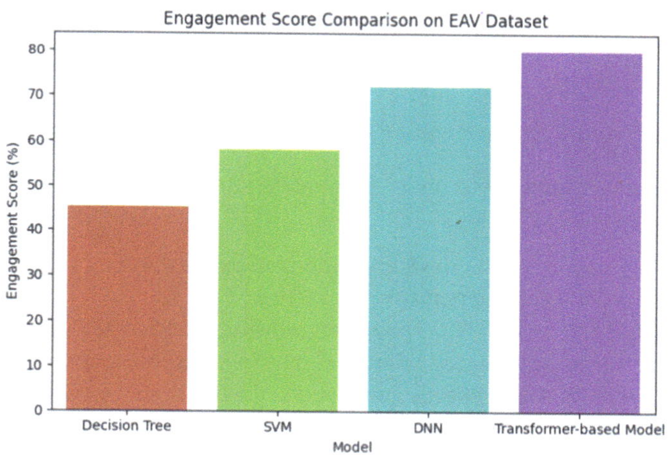

Fig. 3. Engagement Score Comparison on EAV Dataset

4.3 Explainability and User Trust

As deep learning models operate as "black boxes," Explainable AI (XAI) techniques such as SHAP (Shapley Additive Explanations) and LIME (Local Interpretable Model-agnostic Explanations) are applied to improve model interpretability. Key findings include (see Fig. 4):

- **SHAP analysis** reveals that **quiz performance, response time, and gaze fixation duration** are the most influential features in engagement prediction.
- **LIME visualizations** indicate that small changes in **EEG signal variations and speech tone** significantly impact classification confidence.
- User studies confirm that **transparency in model decision-making improves trust**, leading to better acceptance of AI-driven interactions.

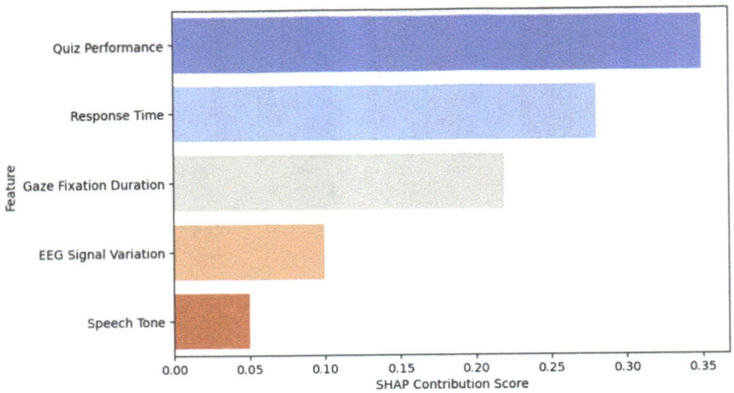

Fig. 4. SHAP Analysis: Feature Importance in Engagement Prediction

4.4 Scalability and Computational Efficiency

The proposed transformer-based model is evaluated for its scalability and efficiency in real-world deployment:

- **Edge AI Performance**: The model is optimized for **on-device execution**, reducing inference time and enabling real-time adaptation.
- **Energy Efficiency**: By applying **pruning and quantization techniques**, computational requirements are reduced without significant loss in accuracy. The optimized model consumes **30 % less energy** compared to the full-scale version, making it more suitable for real-time HCI/HRI applications.

Figure 5 illustrates the impact of model optimization on energy efficiency.

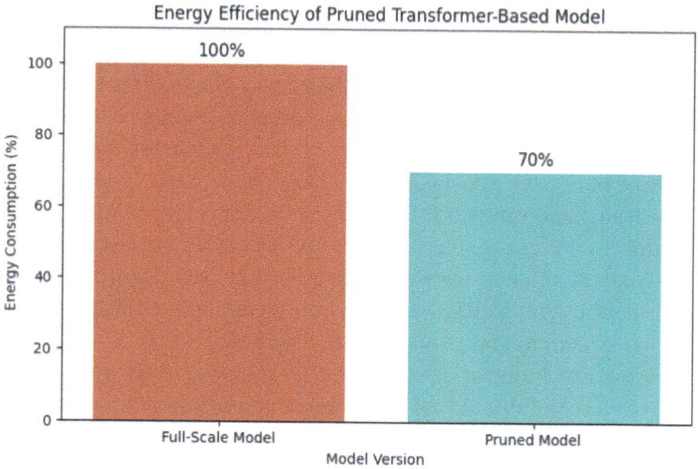

Fig. 5. Energy Efficiency of Pruned Transformer-Based Model

Figure 6 illustrates the impact of model optimization on energy efficiency.

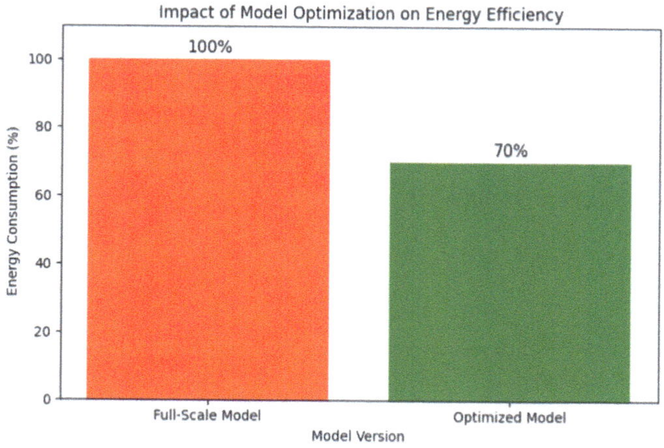

Fig. 6. Impact of Model Optimization on Energy Efficiency

The evaluation results demonstrate that deep learning-based multimodal fusion significantly enhances interaction accuracy, adaptability, and engagement prediction. The incorporation of Explainable AI improves trustworthiness, while efficient model optimizations ensure practical deployment in real-world intelligent systems. These findings validate the effectiveness of the proposed approach in enhancing adaptive human-computer and human-robot interactions.

5 Conclusion

This research presents a deep learning-based multimodal interaction framework designed to enhance adaptive and context-aware human-computer interaction (HCI) and human-robot interaction (HRI). The study contributes by developing a robust multimodal interaction model that integrates speech, visual, EEG, and physiological signals through a transformer-based fusion network. This model effectively captures diverse interaction patterns, improving engagement prediction and emotional state classification. It also implements adaptive deep learning with reinforcement learning, where a policy-based reinforcement learning (RL) mechanism dynamically adjusts interaction strategies, enabling the system to learn and adapt to user behaviors over time. Using reward-based optimization, the framework continuously enhances engagement. Furthermore, the model is deployed on Edge AI for real-time adaptive interaction, making it suitable for interactive AI systems, assistive technologies, and smart environments. Optimization techniques such as pruning and quantization significantly reduce computational overhead while maintaining accuracy.

Despite the significant advancements demonstrated in this study, certain limitations remain. One challenge is real-time learning and multimodal synchronization, where the integration of multiple sensory inputs introduces difficulties in synchronizing diverse data streams with varying temporal resolutions. Although attention-based fusion helps mitigate this issue, further improvements are needed for real-time multimodal alignment. Another limitation involves computational constraints in resource-limited environments. Transformer-based models, while powerful, require substantial computational resources, making deployment on edge devices with limited memory and processing power difficult. Future efforts should focus on efficient model compression and hardware-aware optimizations to enhance usability in low-resource settings.

References

1. Jaimes, A., Sebe, N.: Multimodal human–computer interaction: a survey. Comput. Vis. Image Underst. **108**(1–2), 116–134 (2007)
2. Bowers, J.S., et al.: Deep problems with neural network models of human vision. Behav. Brain Sci. **46**, e385 (2023)
3. Vaswani, A., et al.: Attention is all you need. Adv. Neural Inf. Proces. Syst. **30** (2017)
4. Baltrušaitis, T., Ahuja, C., Morency, L.P.: Multimodal machine learning: a survey and taxonomy. IEEE Trans. Pattern Anal. Mach. Intell. **41**(2), 423–443 (2018)
5. Veisi, H.: Introduction to SVM. In: Learning with Fractional Orthogonal Kernel Classifiers in Support Vector Machines: Theory, Algorithms and Applications, pp. 3–18. Singapore, Springer Nature Singapore (2023)
6. Zha, D. et al.: Data-centric artificial intelligence: a survey. ACM Comput. Surv. **57**(5), 1–42 (2025)
7. Tavakoli, M., Faraji, A., Vrolijk, J., Molavi, M., Mol, S.T., Kismihók, G.: An AI-based open recommender system for personalized labor market driven education. Adv. Eng. Inform. **52**, 101508 (2022)
8. Padakandla, S.: A survey of reinforcement learning algorithms for dynamically varying environments. ACM Comput. Surv. (CSUR). **54**(6), 1–25 (2021)

9. Zadeh, A., Chen, M., Poria, S., Cambria, E., Morency, L.P.: Tensor fusion network for multimodal sentiment analysis. arXiv preprint arXiv:1707.07250. (2017)
10. Poria, S., Cambria, E., Bajpai, R., Hussain, A.: A review of affective computing: from unimodal analysis to multimodal fusion. Inf. Fusion. **37**, 98–125 (2017)
11. Jabeen, S., Li, X., Amin, M.S., Bourahla, O., Li, S., Jabbar, A.: A review on methods and applications in multimodal deep learning. ACM Trans. Multimed. Comput. Commun. Appl. **19**(2s), 1–41 (2023)
12. Brusilovsky, P., Millán, E.: User models for adaptive hypermedia and adaptive educational systems. In: The Adaptive Web: Methods and Strategies of Web Personalization, pp. 3–53. Springer Berlin Heidelberg, Berlin, Heidelberg (2007)
13. Goodfellow, I., Bengio, Y., Courville, A., Bengio, Y.: Deep Learning, vol. 1, No. 2. Cambridge: MIT press (2016)
14. Janiesch, C., Zschech, P., Heinrich, K.: Machine learning and deep learning. Electron. Mark. **31**(3), 685–695 (2021)
15. Ribeiro, M.T., Singh, S., Guestrin, C.: "Why should I trust you?" Explaining the predictions of any classifier. In: Proceedings of the 22nd ACM SIGKDD International Conference on Knowledge Discovery and Data Mining, pp. 1135–1144 (2016)
16. Lee, M.H. et al.: EAV: EEG-audio-video dataset for emotion recognition in conversational contexts. Sci. Data. **11**(1), 1026 (2024) https://www.nature.com/articles/s41597-024-03838-4.pdf

Safe, Secure, and Trustworthy AI

An Artificial Neural Network Approach to Currency Crisis Prediction: The Case of the MENA Region

Badr Moulouade(✉), Yahya Moulouade, and Halima Jamil

Faculty of Legal, Economic and Social Sciences Agdal, Mohammed V University in Rabat, Rabat, Morocco
badr.moulouade@um5r.ac.ma

Abstract. The recent currency crises in the MENA region highlight structural weaknesses and the vulnerability of these economies to external shocks. This study aims to predict and explain currency crises using data from nine MENA countries spanning 1970 to 2018. A comparative early warning system is used to evaluate the performance of distinct neural network models, including MLP, RNN, LSTM, and GRU. The results demonstrate that the GRU model outperforms the others, accurately predicting 89% of currency crises. The model identifies key currency crisis determinants, such as foreign direct investment inflows, oil price fluctuations, unemployment rates, current account balances, exports, and public debt. The outcomes presented in this paper can benefit policymakers in preventing certain risks. Solutions include the need for the region to efficiently attract FDI and increase export diversification to reduce the region's dependence on oil, thereby fostering economic resilience and stability.

Keywords: financial stability · currency crises · early warning system · neural networks · MLP · simple RNN · LSTM · and GRU

1 Introduction

Drastic fluctuations in exchange rates have significant implications for the financial stability of countries, as illustrated by the currency crises of the 1990s. These crises exposed the international financial system's fragility and highlighted the existence of transmission channels through which financial shocks can spread within or across economies. In particular, these channels include the fluctuations in asset prices, the banking sector's dollarization, the firms' heavy reliance on foreign currency-denominated debt, the frequent adoption of fixed exchange rate regimes driven by the "fear of floating", and the contagion phenomenon.

In the face of the growing risk and reappearance of currency crises, it becomes imperative to review crisis prevention policies, taking into account the increase in vulnerability sources, particularly in the context of current financial globalization. The implementation of effective preventive policies could significantly reduce the costs arising from a

B.-E. Boudriki Semlali et al. (Eds.): DATA 2025, LNNS 1908, pp. 323–335, 2026.
https://doi.org/10.1007/978-3-032-21200-9_27

currency crisis. Moreover, the unexpected factor of these crises has revealed the importance of a forward-looking approach, allowing the identification of potential risks at an early stage before they manifest as crises, emphasizing the need for early warning systems.

However, the MENA region has not been spared from financial vulnerabilities, as it faces several structural challenges that pose fundamental constraints on its economic growth and financial stability. Notably, this region is largely dependent on the exploitation of natural resources and it is therefore highly exposed to the volatility of global markets. In terms of investment, foreign direct investment (FDI) flows remain relatively low, with most FDI directed toward the United Arab Emirates and Egypt, particularly concentrated in energy, real estate, financial, and consumer products [1]. Additionally, most MENA countries face a low level of trade diversification and a high reliance on remittances, increasing their exposure to global markets fluctuations. Furthermore, the region's vulnerability is exacerbated by rising public debt in several countries, namely Egypt, Jordan, Lebanon, and Tunisia. Therefore, the region has experienced relatively recent currency crises such as Egypt in 2016, Turkey in 2018, and Lebanon in 2020, all of which reflect the challenges in the region.

Although financial stability in the MENA region has been of interest to [2], most currency crises' early warning systems studies have focused on advanced economies or emerging markets such as Asia and Latin America. In contrast, the MENA region receives relatively little attention, with the existing works being country-specific. This work contributes to the literature by proposing an early warning system for currency crises in the MENA region, using a comparative analysis of several artificial neural network architectures, including multilayer perceptron (MLP), simple Recurrent Neural Network (simple RNN), Long Short-Term Memory (LSTM), and Gated Recurrent Unit (GRU).

The structure of this paper is as follows. Section 2 offers a review of the literature on EWS for the currency crises. Section 3 describes the dataset used and the preprocessing methods. Section 4 presents the selected neural networks, the performance evaluation methods and the hyperparameters. Section 5 discusses our empirical results on the test set and analyzes the contributions of the features to the prediction. Finally, Sect. 6 concludes the study.

2 Literature Review

Three generations of models have been developed to analyze currency crises. First-generation models initiated by [3] associate crisis with a weak alignment of fundamentals. The second-generation models introduced by [4], incorporate the element of market expectation, where speculative attacks may occurs due to the lack of confidence in the authorities' ability to defend the exchange rate. Third-generation models pioneered by [5], look at financial weaknesses, particularly balance sheet vulnerabilities and moral hazard in twin crises.

Empirical studies showed three types of early warning systems (EWS) for currency crises that can be made: signal, parametric, and machine learning. The signal approach focuses on identifying early warning indicators that exceed predefined thresholds to signal the likelihood of a crisis [6]. pioneered this methodology, identifying exports, real exchange rate overvaluation, output, the ratio of money supply to foreign exchange reserves, and stock prices as the best explanatory variables.

The parametric approach investigates the economic conditions and structural factors contributing to crises [7]. employing a Probit model for Southeast Asia countries, they noted that excess monetary balances and the domestic credit-to-GDP ratio increase crisis risk, whereas export growth and stock market performance reduce it [8]. applied a Panel data analysis on developed and emerging economies, showing that large, debt-driven inflows significantly increase the likelihood of crises, whereas portfolio investment or foreign direct investment have a negligible effect [9]. showed the superiority of the Markov switching model in predicting foreign exchange crises in Turkey, identifying inflation, portfolio investments, and foreign deposits as the main predictors [10]. using a Panel data analysis on emerging and developing countries, showing that declines in global commodity prices significantly heightened currency crisis risks in commodity-dependent economies.

Machine learning has significantly advanced crisis prediction [11]. found that artificial neural networks produced superior results in predicting currency crisis in Turkey, with key indicators including export/import ratio, ratio of deposit money banks domestic credits/total assets, monthly change in consumer price index, and monthly change in terms of trade [12]. identified deep learning neural decision trees, extreme gradient boosting, random forests, and deep belief networks as the best performing methods for forecasting currency crises across different global regions, finding that the change in the net investment, monetary supply, trade openness, FDI, current account, total debt, and government spending as the most important variables. Analyzing 68 countries, [13] found that LSTM and GRU were the most effective in forecasting currency crises, highlighting the real exchange rate deviation from its trend, real exchange rate growth rate, and the country fixed effect as the top three most important variables [14]. found that random forest model was the best performing model in identifying Latin America currency crises, emphasizing governance factors (political instability, corruption, inefficiency) alongside export growth, reserves, and external debt as the most important predictors of currency crises post-nineties

3 Data Collection and Preprocessing

3.1 Data Description

In our approach to predicting and explaining currency crises, we used data from 9 countries (Algeria, Egypt, Iran, Jordan, Libya, Mauritania, Morocco, Sudan, and Turkey) in the MENA region over the period 1970 to 2018. We employed features that cover different aspects of the economy including monetary and financial indicators, external sector indicators, macroeconomic indicators, policy indicators, and socioeconomic indicators. The features descriptions are in Table 1.

Table 1. Features description.

Features	Definition	Source
PC	We took inspiration from [6] and determined a warning window of 2 years preceding the occurrence of a currency crises. During this period, an alert is issued when the model predicts a probability exceeding a threshold. The target is binary, taking the value 1 for warning periods (pre-crises) and 0 for quiet periods. Post-crises observations are treated as quiet periods.	[15]
M	Money supply (% of GDP)	WDI
MG	Money supply growth	
CA	The ratio of the current account balance to GDP	
E	The ratio of exports of goods and services (% of GDP)	
I	The ratio of imports of goods and services (% of GDP)	
FDII	Foreign direct investment. net inflows (% of GDP)	
FDIO	Foreign direct investment. net outflows (% of GDP)	
GDP	GDP growth (annual %)	
GDPC	GDP per capita	
INF	Inflation, consumer prices (annual %)	
R	Total reserves in months of imports	
RG	Reserves growth	
C	The ratio of domestic credit to the private sector to GDP	
PR	Workers' remittances and employees' compensation, received (% of GDP)	
ED	External debt	
DEBT	The ratio of central government debt to GDP, for Egypt we took the ratio of general government debt to GDP	Global Debt Database
D	The ratio of bank deposits to GDP	Global Financial Development
NEER	Nominal effective exchange rate	[16]
REER	Real effective exchange rate	

(continued)

Table 1. *(continued)*

Features	Definition	Source
FIERR	This indicator takes the value 1 for a fixed exchange rate regime and 0 otherwise. The category designated fixed includes regimes from 1 to 4 according to the fine classification.	[17]
INERR	This indicator takes the value 1 for an intermediate exchange rate regime and 0 otherwise. The category called intermediate includes arrangements from 5 to 11 according to the fine classification.	
FLERR	This indicator takes the value 1 for a flexible exchange rate regime and 0 otherwise. The category marked flexible includes regimes 12, 13, and 14 according to the fine classification.	
KAOPEN	Chinn-Ito Index of Capital Account Openness	[18] (The Chinn-Ito Financial Openness Index 2020 Update)
DOL	Dollarization of the banking system	IFS
OIL	Crude Oil Prices: Brent growth rate	
USINT	Changes in the US central bank's key interest rate	
U	Unemployment rate	WDI

3.2 Data Preprocessing

Missing data imputation was performed using the Miss Forest imputation method, which is based on machine learning that allows the imputation of all data types whether continuous or categorical and uses a random forest approach to predict missing values based on the patterns observed in other features. Then, data scaling was applied using the Min-Max method, which is essential to standardize features before introducing them into neural network models, which are very sensitive to the scale of the data. This method transforms the values of the features to bring them back into a range of [0, 1], using their minimum and maximum values. In addition, data resampling was performed to address the significant imbalance between crises and normal periods by applying the SMOTEENN method to the training set of the MLP model. SMOTEENN is a hybrid resampling approach that deals with class imbalance and decreases noise, it combines SMOTE (Synthetic Minority Over-sampling Technique) and Edited Nearest Neighbors (ENN). SMOTE is a method to generate synthetic samples to balance the minority class and ENN is a technique that filters away noisy and ambiguous samples from both classes that improves the quality of the data. This approach combines the oversampling of crises periods with the undersampling of normal periods to balance the data to improve their relevance for modeling. However, SMOTEENN was not applied to recurrent neural networks due to their sequential nature, requiring the maintain of the data order. Data preprocessing steps in modeling process are indicated in Fig. 1.

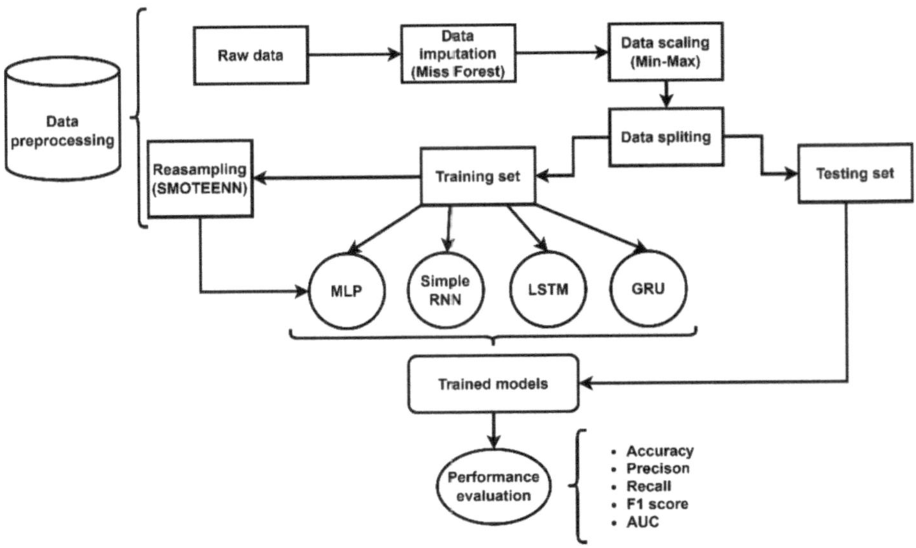

Fig. 1. Data preprocessing steps in modeling process

4 Modeling Approach

Artificial neural networks were inspired by modeling the nervous system and the human brain. These models replicate the behavior of a neuron's cell membrane, particularly how it produces and propagates electrical impulses [19]. In our study, we compare four types of neural networks: MLP, simple RNN, LSTM, and GRU.

The Multilayer Perceptron (MLP) network consists of at least one intermediate (hidden) neuronal layer that is classified between the input layer and the output neuronal layer. This type of network is characterized by its versatility due to its application in many fields. These include time series forecasting, pattern recognition, process identification and control, universal function approximation (curve fitting), and system optimization. The training of the MPL network is done through supervised learning, adjusting the weights and thresholds of each neuron in a network by comparison with the desired outputs proposed for each sample of input data.

A Simple Recurrent Neural Network (simple RNN) is the basic form of recurrent neural network made to handle sequential data by keeping a hidden state that changes over time. This model is especially well-suited for problems like language modeling, time series prediction, and sequence-to-sequence learning.

A long short-term memory (LSTM) neural network is a recurrent neural network that works excellently for sequence problems. It introduces recurrent links with an ability to maintain an internal state or memory, through which, it can acquire and use the organized structures in the input observation sequences. LSTMs are designed to necessarily remove the drawback of the simple RNN model, particularly the explosion of gradients when backpropagated in time. These models developed a memory cell mechanism with three gates: input, forget, and output. These gates regulate the flow of incoming information, thus allowing the LSTM to account for the long-term dependence of sequence data [20].

Gated Recurrent Network (GRU) is a class of recurrent neural networks proposed by [21]. As with LSTM, GRU handles sequential input data and integrates a simpler gate mechanism to regulate the flow of information in the network. These gate mechanisms allow the GRU to decide the relevance of information and when it should be updated, thereby helping to avoid the problem of gradient vanishing during training.

The selected models were chosen to capture linear and nonlinear relationships between variables. The MLP was included as the baseline model given its capacity to deal with structured data and RNN-based models (Simple RNN, LSTM, and GRU) were taken up due to their ability to model temporal relations in sequence data. LSTM and GRU were specifically chosen due to their ability to overcome the vanishing gradient problem which is a major downside in traditional RNNs.

We applied the conventional validation method by splitting our dataset into a training set (70%) and a test set (30%). The model was trained on the training set and evaluated on the test set to estimate its generalization error, i.e. its ability to perform well on unseen data. Model performances were compared using confusion matrix-based metrics, including accuracy, precision, recall, F1 score, and area under the ROC curve (AUC). Furthermore, features importance was assessed using the "integrated gradients" method, allowing a clear interpretation of the impact of input features on model predictions.

For MLP, we compared the standard model with the model using SMOTEENN for resampling the data. For recurrent neural networks, a time windows of six-time steps (current and past five-time steps) and eight-time steps (current and past seven-time steps), were selected based on performance F1 score. These windows were used to construct the features included in the RNN. This approach captures both immediate effects from the current time step and historical patterns from the previous time steps, allowing the model to account for both short-term and long-term influences in predicting the target feature. The final set of indicators varied according to the model and the feature importance results. The classification threshold is selected by iterating through possible values and choosing the one that maximizes the F1 score, ensuring the best balance between precision and recall for the binary classification task. The hyperparameters were selected through a manual search process based on empirical testing, aimed at optimizing the evaluation metrics. The hyperparameters for our models are in Table 2.

Table 2. Hyperparameters.

Hyperparameters	MLP	Recurrent neural networks
Hidden Layers	(64, 32)	GRU (40, 20) LSTM, simple RNN (60,25)

(continued)

Table 2. (*continued*)

Hyperparameters	MLP	Recurrent neural networks
Dropout Rate	0.4	0.3
Learning Rate	0.001	0.001
Batch Size	16	16
Number of Epochs	50	50
Optimizer	Adam	Adam
Loss Function	Binary Crossentropy	Binary Crossentropy
Activation Function (Hidden Layers)	ReLU	Tanh
Activation Function (Output Layer)	Sigmoid	Sigmoid
Early Stopping Patience	without	10

5 Results

5.1 Performance Results

The performance results can be found in Table 3. The MLP Standard achieves an accuracy of 83% and an AUC of 0.87, showing good strength in dissociating crises periods from non-crises periods. However, the F1 score is at 0.55, which reflected some imbalances in terms of precision (0.44) and recall (0.74). That means a good capacity to identify crises points, but with a lot of false alarms. The optimal threshold of 0.19 denotes the model's tendency at prioritizing crises detection. With SMOTEENN resampling, the recall of the model is boosted to 0.84 and the F1 score to 0.57 at a corresponding decline in precision (0.43) and AUC (0.84). Thus, the resampling's impact on class balance is reflected in the higher optimum threshold of 0.68, making the model more cautious when detecting crises.

The simple RNN shows an improvement, though little, in comparison to MLP. F1 score with six steps of time attains 0.60, with 0.74 recall and a precision of 0.50. AUC, 0.85 and a low optimal threshold of 0.04 denote a strong focus on sensitivity, which complements the goal of this model in terms of minimizing missed crises. Extending the time window to eight-time steps gets a marginally higher F1 score of 0.61 and an even better recall of 0.78, demonstrating the capability of the model to consider longer historical data patterns. The optimum threshold rises further, reaching a value of 0.18. However, since precision stands still at 0.50, it indicates not much gain in reducing false positives.

The LSTM presents a better performance. When using six-time steps, the model achieves an F1 score of 0.68 backed up by a strong recall of 0.84 and an improved precision of 0.57. An AUC of 0.86 substantiates the model being capable of very efficiently distinguishing between crises and non-crises. Its optimal threshold of 0.10 reflects minimizing false negatives and prioritizing the detection of crises, even if it increases the risk of false positives. Increasing the sequences for eight-time steps will slightly affect performance since the F1 score decreases to 0.62 and recall falls to 0.78, with precision at 0.52 while the threshold remains unchanged at 0.10.

The model of GRU having six-time steps is achieving an F1 score of 0.63, a recall at 0.84, and a precision of 0.50. An AUC value of 0.83 indicates the good ability of the model to differentiate the cases of crises from non-crises cases. GRU with eight-time steps proved to give the best overall performance by claiming the highest F1 score of 0.70 and a recall of 0.89. This makes it the most sensitive model when it comes to crises detection. An AUC of 0.88 proves a good discrimination ability, while its precision index of 0.57 guarantees a fair value of the prediction's accuracy. The low optimal threshold of 0.045 indicates that it strictly favors the detection of currency crises in favor of false positives.

Thus, our results are in line with [13] showing that LSTM and GRU are performing better than the other neural network models. The moderate performance of MLP is due to its reduced ability to deal with time-dependent pattern recognition. Similarly, simple RNN's modest performance could be related to its susceptibility to the exploding gradient issues. In contrast, GRU and LSTM give superior performance by preserving long-term dependencies and solving the vanishing gradient problem. Their design allows them to detect temporal patterns making them ideal for forecasting currency crises, as numerous financial and macroeconomic indicators show time-based delays. Compared to LSTM, GRU achieves better performances due to its basic gating method, enabling it to preserve sequence order patterns while requiring reduced computing resources.

Table 3. Neural networks performance results.

		Accuracy	F1score	Recall	Precision	AUC	Optimal threshold
MLP	Standard	0.83	0.55	0.74	0.44	0.87	0.19
	SMOTEENN	0.82	0.57	0.84	0.43	0.84	0.68
Simple RNN	Time steps = 6	0.84	0.60	0.74	0.50	0.85	0.04
	Time steps = 8	0.84	0.61	0.78	0.50	0.86	0.18
LSTM	Time steps = 6	0.87	0.68	0.84	0.57	0.86	0.10
	Time steps = 8	0.85	0.62	0.78	0.52	0.86	0.10
GRU	Time steps = 6	0.84	0.63	0.84	0.50	0.83	0.06
	Time steps = 8	0.87	0.70	0.89	0.57	0.88	0.045

5.2 Features Importance and Impact

Based on these performances, we choose GRU (steps = 8) in Fig. 2 for features importance and impact directions. The GRU results show that the most important features are FDII, OIL, U, CA, E, and DEBT. The rise of FDII, OIL, U, and DEBT is positively correlated with currency crises susceptibility. By contrast, CA and E are negatively correlated with currency crisis susceptibility.

Foreign Direct Investment Inflows (FDII) are positively correlated with warning signals joining the results of [12], indicating that countries under study experienced large FDI inflows before the collapse of their currencies. This amplifies the risk of sudden stop crises, involving both currency crises and a reversal of capital flows. The MENA region, despite low overall FDI levels, faces a higher probability of currency crises with rising inflows. This paradox can be explained by two main factors. First, poor-quality FDI that targets non-productive sectors. Second, sudden FDI surges that may lead to currency appreciation, thereby harming export competitiveness, while their abrupt reversal may, in turn, trigger balance of payments crises.

Furthermore, contrary to the results of [10], the model highlights that rising oil price (OIL) increases crisis susceptibility. In oil-importing economies, higher oil prices exacerbate trade deficit, inflation rate, and reserves depletion, consequently destabilizing the currency.

The positive correlation between unemployment (U) and currency crises indicates that high unemployment may cause a self-fulfilling currency crisis as explained by the second-generation of currency crises models.

In addition, in line with [12] findings, the model suggests that higher debt levels (DEBT) are associated with increased expectations of currency crises. This reflects the possibility of a twin crisis (sovereign debt and currency crisis) in the MENA region, particularly given the rising debt levels in many countries within the region.

On the other hand, the negative correlation between current account balance (CA) and exports (E) growth with crisis susceptibility emphasizes their role as an indicator of financial stability, these results are consistent with those of [6, 7, 11, 12, 14]. However, the region faces a significant challenge due to its high dependence on commodity exports and limited diversification in export markets. This dependence exposes economies to external shocks and exchange rate fluctuations, thereby increasing the risk of currency crises. When a country's exports are concentrated in a few industries, any slowdown in these sectors can have a severe impact on the entire economy, creating an imbalance that rises vulnerability to crises. A decline in export levels reduces the inflow of foreign exchange, making it difficult for the country to meet its international financial obligations. This puts pressure on foreign exchange reserves and domestic currency, potentially triggering a currency crisis.

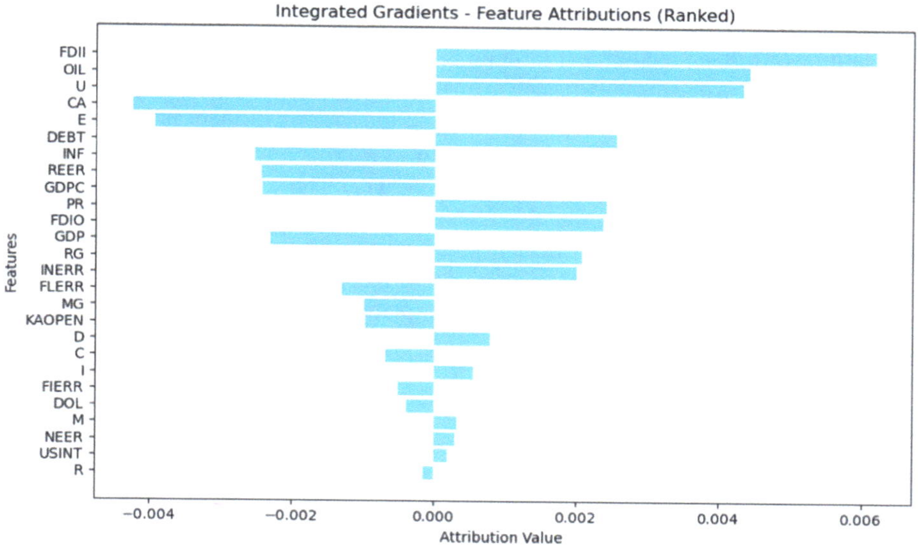

Fig. 2. GRU (steps = 8) features importance and impact directions

6 Conclusion

In this work, we proposed a comparative approach for currency crises early warning systems in the MENA region using MLP, simple RNN, LSTM, and GRU artificial neural networks. Our results confirm the superiority of LSTM and GRU artificial neural networks reaching a recall of 84% and 89%, respectively. Furthermore, we found that foreign direct investment inflows, oil price fluctuations, unemployment rate, current account balance, exports, and public debt are the main determinants of currency crises in the MENA region.

6.1 Implications

These results give an insight into the currency crisis risks transmission channels and serve as the foundation for policy interventions to maintain financial stability. Thus, to foster economic diversification and attract investment in the MENA region, governments must implement strategic policies that target productive sectors. Offering tailored tax incentives and sector-specific initiatives, such as promoting renewable energy, which could help reduce the region's reliance on oil. These policies would not only encourage sustainable development but also enhance the region's long-term economic resilience.

Regarding the exchange rate regime, we recommend adopting a managed floating to limit REER overvaluation caused by FDI inflows, thereby maintaining export competitiveness.

When it comes to trade in the MENA region, there is an absolute need to diversify exports across multiple sectors or markets to minimize the effects of risks incurred by external economic changes. This measure helps reduce its dependence on a particular sector such as oil, which consequently strengthens the economy's ability to resist crises or volatilities. A diversified export base, in addition to strengthening resilience during economic crises, also leads to balanced growth and improved competitiveness.

In addition, fiscal policy support is critical to the success of other policies. Indeed, fiscal discipline helps maintain price stability by avoiding excessive pressure on aggregate demand. On the other hand, excessive foreign exchange interventions in favor of over-indebted countries could signal fundamental economic problems, such as large external debt obligations. Consequently, investors could view such interventions as a sign that the government is struggling to manage its economic policies effectively. This perception may erode trust in the exchange rate, as it raises concerns about the country's financial and economic stability.

6.2 Limitations and Further Research

Although the overall work carried out in this paper has led to interesting conclusions, there are several drawbacks that remain to be addressed in future works. First, the exclusion of political instability, which is a big factor in the MENA region, may limit prediction accuracy. Second, our study employs manual search for hyperparameter tuning, these could be improved by employing more developed hyperparameter tuning like grid search and Bayesian search. Third, the study relies on annual data, whereas monthly or quarterly data are more appropriate for an early warning system designed to avoid financial crises. This is because more frequent data updates allow a quicker reaction and better detection of emerging crises by monetary authorities. Finally, the models used in our work have been used only in the MENA region. Applying them to other regions with different economic structures could provide insights into their robustness and adaptability.

Moreover, our research would benefit from examining economic policy decisions in situations of currency crisis. By testing these policies, we can determine their capacity to support economic and financial system stability.

Disclosure of Interests. The authors have no competing interests to declare that are relevant to the content of this article.

References

1. Saidi, N., Prasad, A.: Trends in trade and investment policies in the MENA region. MENA-OECD (2018)
2. Firano, Z., Moulouade, B.: Canaux de transmission, Instabilité Financière et Taux de Change: Le Cas de la Zone MENA. RFREG. 3 (2022)
3. Krugman, P.: A model of balance-of-payments crises. J. Money Credit Bank. **11**(3), 311–325 (1979). https://doi.org/10.2307/1991793
4. Obstfeld, M.: The logic of currency crises. Banque de France Cahiers Économiques et Monétaire. **43**, 189–213 (1994)
5. Krugman, P.: Balance sheets, the transfer problem, and financial crises. In: Isard, P., Razin, A., Rose, A.K. (eds.) International Finance and Financial Crises, pp. 31–55. Springer, Netherlands, Dordrecht (1999). https://doi.org/10.1007/978-94-011-4004-1_2

6. Kaminsky, G., Lizondo, S., Reinhart, C.M.: Leading indicator of currency crisis. IMF Staff. Pap. **45**(1), 1–48 (1998)
7. Budsayaplakorn, S., Dibooglu, S., Mathur, I.: Can macroeconomic indicators predict a currency crisis? Evidence from selected southeast Asian countries. Emerg. Mark. Financ. Trade. **46**, 5–21 (2010). https://doi.org/10.2753/REE1540-496X460601
8. Furceri, D., Guichard, S., Rusticelli, E.: Episodes of large capital inflows, banking and currency crises, and sudden stops. Int. Finance. **15**, 1–35 (2012). https://doi.org/10.1111/j.1468-2362.2012.01296.x
9. Ari, A., Cergibozan, R.: Currency crises in Turkey: an empirical assessment. Res. Int. Bus. Financ. **46**, 281–293 (2018). https://doi.org/10.1016/j.ribaf.2018.04.001
10. Bodart, V., Carpantier, J.-F.: Currency crises in emerging countries: the commodity factor. J. Commod. Mark. **30**, 100287 (2023). https://doi.org/10.1016/j.jcomm.2022.100287
11. Sevim, C., Oztekin, A., Bali, O., Gumus, S., Guresen, E.: Developing an early warning system to predict currency crises. Eur. J. Oper. Res. **237**, 1095–1104 (2014). https://doi.org/10.1016/j.ejor.2014.02.047
12. Alaminos, D., Becerra-Vicario, R., Fernández-Gámez, M.Á., Cisneros Ruiz, A.J.: Currency crises prediction using deep neural decision trees. Appl. Sci. **9**, 5227 (2019). https://doi.org/10.3390/app9235227
13. Barthélémy, S., Gautier, V., Rondeau, F.: Early warning system for currency crises using long short-term memory and gated recurrent unit neural networks. J. Forecast. **43**, 1235–1262 (2024). https://doi.org/10.1002/for.3069
14. Guzmán, K.M., Fan, H.: Governance and currency crises in Latin America post-nineties: a machine learning approach. Emerg. Mark. Financ. Trade. **61**, 1–23 (2024). https://doi.org/10.1080/1540496X.2024.2438299
15. Nguyen, T.C., Castro, V., Wood, J.: A new comprehensive database of financial crises: identification, frequency, and duration. Econ. Model. **108**, 105770 (2022). https://doi.org/10.1016/j.econmod.2022.105770
16. Darvas, Z.M.: Timely Measurement of Real Effective Exchange Rates. Bruegel, Brussels (2021)
17. Ilzetzki, E., Reinhart, C., Rogoff, K.: Rethinking Exchange Rate Regimes. NBER Working Paper (2021)
18. Chinn, M.D., Ito, H.: What matters for financial development? Capital controls, institutions, and interactions. J. Dev. Econ. **81**, 163–192 (2006). https://doi.org/10.1016/j.jdeveco.2005.05.010
19. Hodgkin, A.L., Huxley, A.F.: A quantitative description of membrane current and its application to conduction and excitation in nerve. J. Physiol. **117**, 500–544 (1952). https://doi.org/10.1113/jphysiol.1952.sp004764
20. Brownlee, J.: Long Short-Term Memory Networks with Python Develop Sequence Prediction Models with Deep Learning. Jason Brownlee (2017)
21. Chung, J., Gulcehre, C., Cho, K., Bengio, Y.: Empirical Evaluation of Gated Recurrent Neural Networks on Sequence Modeling., http://arxiv.org/abs/1412.3555 (2014).

Unified Cyber Attack Taxonomy (UCAT) Using Classification Tree Method

Ayoub Seghrouchni(✉) and Noura Ouerdi

ACSA Laboratory, Faculty of Sciences, Mohammed First University, 60000 Oujda, Morocco
{ayoub.seghrouchni,n.ouerdi}@ump.ac.ma

Abstract. Cyberattacks or threats are increasing every year and becoming ever more sophisticated. Numerous studies and classifications, or taxonomies, have been proposed by researchers and industries to facilitate the comprehension and comparison of these attacks. These efforts aim to enable cybersecurity professionals, researchers, and even individuals less familiar with the field to effectively identify these threats and better defend against them before they are exploited against businesses, governments, or individuals. In our work, we propose a new hybrid taxonomy called Unified Cyber Attack Taxonomy (UCAT) that extends existing ones by incorporating new features and combining the strengths of different approaches, including the Cyber Kill Chain and the Classification Tree Method (CTM). Additionally, we analyze the characteristics used in cyberattack classifications, whether from the industrial or research domains, to perform a comparison and highlight the strengths and weaknesses of each.

Keywords: Cyberattacks · Taxonomies · Classification Tree Method (CTM) · UCAT · Cyber Kill Chain

1 Introduction

The growth in the volume and sophistication of cyberattacks in today's digital environment continues to pose a serious danger to individuals, organizations, and governments worldwide. The cybersecurity community is engaged in unprecedented battles against evolving modern cyber threats that exploit vulnerabilities across multiple attack vectors and employ increasingly sophisticated techniques [1]. This swift pace and complexity of the threat warrant systematic taxonomies, classifications, and analyses concerning attack patterns. Attack taxonomies play a crucial part in helping classify cyberattacks according to their characteristics [2], thereby providing an unclouded view of the attacks for comprehension and improvement of defense mechanisms.

A number of frameworks have helped us learn about these threats, such as various industry frameworks like MITRE ATT&CK Matrix [3] and the Cyber Kill Chain [4], as well as academic taxonomies like AVOIDIT [5]. Some focus on general classification, while others target specialized domains, such as cloud taxonomies represented by CATRA [6], Internet of Things (IoT) taxonomies [7], and taxonomies for machine learning attacks or so-called adversarial machine learning [8]. Existing frameworks often operate in isolation, which can lead to gaps in threat coverage and remediation strategies.

B.-E. Boudriki Semlali et al. (Eds.): DATA 2025, LNNS 1908, pp. 336–347, 2026.
https://doi.org/10.1007/978-3-032-21200-9_28

The modern cyber threat landscape necessitates an integrated approach that draws practical insights from industry frameworks while maintaining the theoretical rigor of academic research. Current taxonomies often struggle to adapt to emerging attack patterns. Moreover, a lack of standardization in defining various classifications of attacks leads to confusion among security professionals, making it difficult to design structured mechanisms to defend against these attacks. To overcome these issues, we propose UCAT with CTM to help categorize different cyberattacks effectively.

The paper proceeds as follows: Sect. 2 examines and compares existing cyberattack taxonomies, frameworks, and their characteristics. Section 3 outlines our methodology and the development of UCAT based on previous approaches. Section 4 applies our taxonomy through a case study of real-world cyber attack using UCAT with CTM, and Sect. 5 concludes this paper.

2 Previous Attack Taxonomies

The goal of a taxonomy is to simplify the collection, analysis, and sharing of information regarding cyberattacks.

Many studies have proposed taxonomies for different attack domains from various aspects, also known as features or characteristics, covering a wide range of attacks. In this section, we review six varieties of cybersecurity taxonomies, frameworks, or models, beginning with four academic ones and concluding with two industry-driven ones.

Chris Simmons et al. [5] proposed a taxonomy of cyberattacks named AVOIDIT in 2009. This academic taxonomy contains five features to classify the attacks:

- Attack Vector: The path or means used to execute a full-blown cyberattack.
- Operational Impact: What the attacker can do inside the system if he gains access.
- Defense: How to prevent and minimize the cyberattack.
- Informational Impact: What the attacker can do to the data stored in the system.
- Target: Specifies each part of the system that is attacked.

Joshi and Singh. [9] proposed the ADMIT taxonomy in 2014, building on AVOIDIT [5]. It examines five expanded features: Attack Vector (including attacker identity and manipulation of organizational/technical processes), Defense (enhanced with managerial, operational, responsive, and proactive aspects), Method (attack techniques, e.g., viruses, Denial of Service (DoS)), Impact (categorizing operational, informational, and organizational consequences), and Target (similar to AVOIDIT [5]).

Fadi Mohsen et al. [10] proposed a taxonomy in 2022 for large-scale cybersecurity attacks, extending existing taxonomies like AVOIDIT [5] by incorporating new features pertaining to the scale, defense strategy, date of attacks. They also added a dataset to help researchers explore some existing datasets of large-scale cyberattacks.

Chuck Easttom et al. [11] presented a taxonomy based on McCumber's Cube in 2019. It focuses on high-level attributes of an attack across three dimensions:

- Whether the attack compromises Confidentiality, Integrity, or Availability.

- The state of the data: Data in transit, at rest (in storage), or in processing.
- The safeguards involved: Technology used, awareness of personnel, and policies and practices of the organization.

The Cyber Kill Chain. [4] is not a strict taxonomy but a framework developed by Lockheed Martin. It consists of seven phases describing the stages of a cyberattack:

- Reconnaissance: Gathering information, profiling targets, and identifying vulnerabilities.
- Weaponization: Preparing and designing malicious payloads like malware (e.g., Remote Access Trojan (RAT)) integrated into deliverable files like Microsoft Office documents or PDFs while minimizing detection.
- Delivery: Transmitting the prepared payload to the target via methods such as phishing emails, compromised websites, or USB drives.
- Exploitation: Triggering the malicious code to gain access to the target system.
- Installation: Establishing persistent access via malware installation.
- Command and Control (C2): Communicating with compromised systems.
- Actions on Objectives: Achieving attack goals, which can include system destruction, data exfiltration, or other objectives.

The MITRE ATT&CK. [3] is more of a framework or knowledge base created by the MITRE organization rather than a taxonomy. It categorizes adversary behavior observed in real-world attacks into tactics (currently 14 categories in the Enterprise matrix) and related techniques (hundreds of techniques and sub-techniques used to execute those tactics). This granular approach allows for the classification of an attack at a fine-grained level.

Each taxonomy or model outlined above possesses unique or specific strengths and has demonstrated its usefulness in the scenarios for which it was designed. The key strengths and weaknesses of these taxonomies are outlined in Table 1.

3 Proposed Taxonomy: UCAT

This section presents UCAT, the Unified Cyber Attack Taxonomy, which improves previous approaches by incorporating novel features and integrating the best merits of diverse methods. In particular, it relies on the multidimensionality and comprehensiveness of AVOIDIT and ADMIT [5, 9], while preserving their principal dimension names. It takes the McCumber Cube [11] and incorporates its security principles into the "impact" dimension of UCAT, while also adopting the Cyber Kill Chain [4] as another dimension to represent attack phases.

UCAT uses five key dimensions to define the attack's nature:

- Attack vector, which refers to the path or means an attacker uses to access a host.
- Attack phases, describing the stage or stages within the attack lifecycle.
- Target of the attack, identifying the logical or physical component affected.
- Impact of the attack, referring to its consequences.

- Defense mechanisms, specifying actions taken to protect the system from attacks.

These dimensions align with attack risk scenarios, where malicious actors utilize specific "attack vectors" that may occur during various "attack phases." These attacks focus on specific "targets" and, when successful, often lead to significant "impacts." To mitigate and prevent these impacts, effective "defense" mechanisms are essential.

As shown in Fig. 1, it provides an overview of the proposed taxonomy. Each dimension is broken down into subcategories, enabling a more precise classification of attacks. The details of these dimensions are discussed in the section below.

Table 1. Overview of Previous Taxonomies and Models

Taxonomy/Model	Strengths	Weaknesses
AVOIDIT (2009)	• Multidimensional • Fine-Grained Analysis • Considers Impacts	• Limited Adoption • Lacks Granularity on Specific Techniques • Does Not Include Attack Processes • Dimensions not well defined
ADMIT (2014)	• Clear and Comprehensive • Explicitly Considers Defenses	• Lacks Granularity on Attack Techniques •May Be Less Suitable for Complex, Multi-Vector Attacks • Does Not Include Attack Processes • Dimensions not well defined
Taxonomy of Mohsen et al. (2022)	• Adapted for Large-Scale Attacks • Associated Dataset	• Increased Complexity • Does Not Include Attack Processes • Dimensions not well defined
Taxonomy based on McCumber's Cube (2019)	• Holistic Security Vision • Useful for Strategic Planning	Lacks Granularity Complex to Implement in Practice Does Not Include Attack Processes
Cyber Kill Chain	• Simple and Easy to Understand • Allows Visualization of the •Attack process from the •Attacker's Perspective	• Overly Simplistic • Does Not Explicitly Consider Defensive Aspects
MITRE ATT&CK	• Highly Granular • Based on Real-World Observations • Widely Adopted by Industry	• Complexity • Constant Evolution • Focus on Techniques

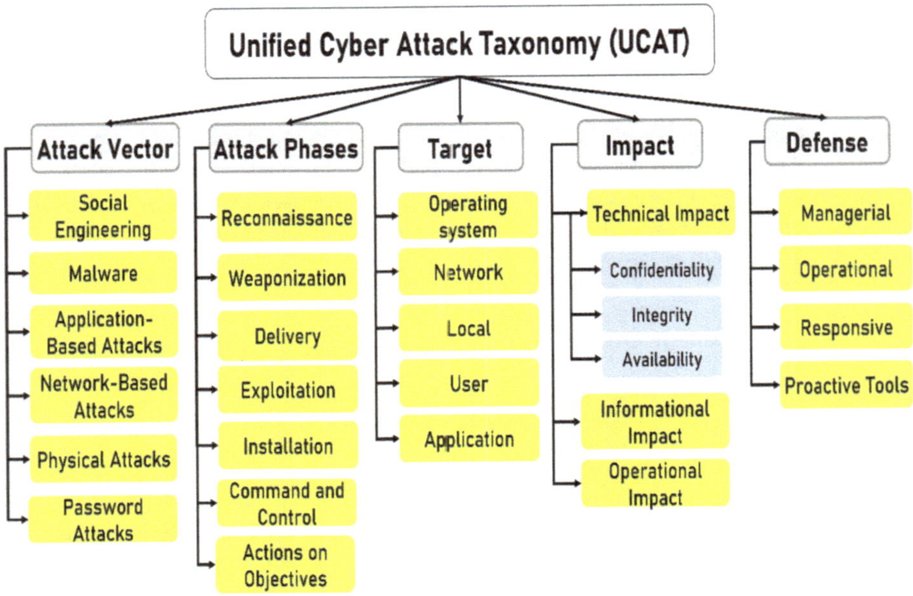

Fig. 1. Unified Cyber Attack Taxonomy (UCAT).

3.1 Attack Vector Dimension

UCAT's first dimension is the attack vector, which refers to the path or means used to execute a cyberattack. We focus on common vectors [12], divided into six subcategories, including:

Phishing/Social Engineering: This refers to methods that exploit human behavior through social interactions (e.g., phishing, vishing, and impersonation) [5].
Malware (or malicious code/software): This is a program that enters a system with the goal of disrupting data, resulting in the loss of confidentiality, integrity, and availability (e.g., trojans, viruses, worms, spyware, and ransomware).
Application-Based Attacks: These refer to any cyberattack targeting vulnerabilities in software applications to exploit them for malicious purposes (e.g., SQL Injection, Cross-Site Request Forgery (CSRF) and Cross-Site Scripting (XSS)).
Network-Based Attacks: These exploit weaknesses or vulnerabilities in a network (e.g., DoS, Distributed Denial of Service (DDoS), and Man-in-the-Middle (MitM) attacks).
Physical Attacks These involve gaining unauthorized access to hardware to compromise security, steal data, and more (e.g., tampering and theft of devices).
Password Attacks These occur when an attacker tries to compromise user passwords, (e.g., brute-force attacks and dictionary attacks).

3.2 Attack Phases Dimension

The second UCAT dimension is Attack Phases, derived from the Cyber Kill Chain framework [4] to classify attacks based on the phases they belong to. This classification helps visualize the attack process from the threat actor's perspective within the attack cycle. This dimension recognizes that not all cyberattacks follow every phase in a linear fashion. Real-world attacks often skip phases, combine them, or execute them in a different order. Each phase is classified as "Yes," "Limited," or "Skipped."

- Yes: Indicates that the attack significantly used this phase.
- Limited: Indicates that attack used this phase in a minimal or restricted way.
- Skipped: Indicates that the phase was completely bypassed.

This dimension is composed of seven phases: Reconnaissance, Weaponization, Delivery, Exploitation, Installation, Command and Control, and Actions on Objectives.

3.3 Attack Target Dimension

The third UCAT dimension is the attack target [5, 9, 10], referring to the logical and physical components targeted by attackers, which vary based on their goals and exploited vulnerabilities [6]. Targets include:

- Operating system: The attack is conducted to exploit a vulnerability and affect a specific operating system.
- Network: The attack targets a specific network, a vulnerability within it, or one of the protocols used in that network.
- Local: When the local computer of a user is targeted in the attack.
- User: When the attack targets a specific user of a system, e.g., to retrieve personal information from a particular user.
- Application: The attack targets a vulnerability within a particular application or software, which can be either server-side or client-side.

3.4 Attack Impact Dimension

The fourth dimension is Attack Impact, which includes three subcategories: technical, informational impacts, and operational impacts, each related to the nature of the attack and the CIA security principles targeted.

Technical Impacts. [6] are derived from the core security principles [11] Confidentiality, Integrity, and Availability. Generally, the goal of an attack is to compromise one or more of these three principles.

Informational Impacts. inspired by AVOIDIT [5], can be categorized as follows:

- Disrupt: Typically caused by a denial-of-service attack that affects availability.
- Distort: When an attack alters data, compromising integrity.
- Destruct: When data is deleted, leading to a loss of both integrity and availability.
- Disclosure: When an attack results in the unauthorized exposure of data, breaching confidentiality.

- Discovery: When an attack uncovers previously unknown data through reconnaissance activities like scanning and mapping, targeting confidentiality.

Operational Impacts. describe the actions a threat actor takes once inside the victim's system. These can include interference, exploitation, or other activities:

- Misuse of Resources: The attacker exploits system resources to perform unauthorized activities.
- User Compromise: The attacker performs actions as a legitimate user after gaining their privileges (e.g., commands, reading files).
- Root Compromise: After privilege escalation, the attacker gains administrator or root privileges, giving them full control over the network and system.
- Web Compromise: The attacker exploits web applications or websites to execute attacks (e.g., XSS, SQL injection).
- Installed Malware: The attacker deploys malicious software in the system (e.g., ransomware, spyware, or backdoors).
- Denial of Service (DoS): The attacker disrupts certain or all resources of the system to prevent access by legitimate users.

3.5 Defense Dimension

The defense dimension describes actions to protect a system from attack [9]. This dimension includes the following:

Managerial: Represents methods and techniques implemented by management, similar to an organization's security strategy, such as removing infected hosts from the network by disconnecting them to prevent additional damage.
Operational: Represents procedures and policies adopted to enhance system security, such as whitelisting and reference advertisements using vulnerability databases.
Responsive: Represents the necessary actions taken to address the situation during exploitation, such as patching systems and correcting code.
Proactive Tools: Represents techniques and defensive tools implemented by automated systems to enhance system security, such as antivirus and network monitoring.

4 Evaluation of UCAT Using CTM

4.1 Classification Tree Method (CTM)

The Classification Tree Method (CTM) is a formal approach developed by Grotchmann and Grimm for generating and presenting test cases [13]. Widely used in modeling and software testing. This graphical method is a hierarchical organization and visual representation of test cases or components following some criteria. This helps in identifying all possible combinations of variables and their values so that we can avoid missing some possibilities or making some errors.

In this regard, CTM allows us to categorize and analyze cyberattacks by structuring the five dimensions of the UCAT taxonomy in a classification tree which can be used to generate test cases

4.2 Classification Tree Editor (CTE)

The Classification Tree Editor (CTE) is a software tool to efficiently implement the CTM and automatically generate test cases. It can be used to drive a graphical interface for building, managing, and analyzing classification trees.

CTE was used to build a classification tree to represent the taxonomy's dimensions and their subcategories to facilitate the easy generation of test cases for evaluating UCAT under different attack scenarios.

4.3 Evaluation of UCAT

UCAT's effectiveness in categorizing cyberattacks was evaluated using CTM and CTE tool by mapping real-world attacks to the taxonomy. Each of UCAT's five dimensions (attack vector, phases of attack, target, impact, and defense) was translated into the classification tree structure, with their respective subcategories forming the branches.

One example we applied to validate UCAT is the WannaCry ransomware attack (May 2017) [14], a widespread attack that infected Windows systems across multiple countries, encrypting files and demanding Bitcoin as ransom. WannaCry exploited the EternalBlue vulnerability in the Server Message Block (SMB) protocol and used worm-like propagation to spread laterally within networks.

The classification of WannaCry using UCAT:

- Attack Vector: Falls under malware (ransomware, worm), as well as the network-based attack vector.
- Phases of Attack:
  - Reconnaissance: Limited to internal network scanning for SMB vulnerability in systems.
  - Weaponization: Yes, the WannaCry ransomware payload was specifically designed to encrypt files and request ransom.
  - Delivery: Yes, it spread via the network using the EternalBlue vulnerability.
  - Exploitation: Yes, SMB vulnerability exploiting leading to code execution.
  - Installation: Yes, through installation of the ransomware on victims' systems.
  - Command and Control: Limited C2 was used to manage the attack or receive decryption keys.
  - Actions on Objectives: Yes, through encrypting data and demanding ransom.
- Attack Target:
  - Targeting Windows Operating System, local individual computers, and networks.

- Attack Impact:
  - Technical: Aimed at compromising availability and integrity.
  - Informational: Caused disruption due to system unavailability and destruction of sensitive data via encryption.
  - Operational: Included malware installation, denial of service, and misuse of resources consumed during the encryption process.
- Defense Dimension:
  - Responsive: System patches.
  - Managerial: Removing ransomware from the network.
  - Proactive: Tools like antivirus and network monitoring.

The CTE provides a clear way to define structured rules and constraints using logical connectors inside brackets, such as (*: OR, +: AND, no: NOT). These rules are used to automatically generate test cases for systems or, in our case, specific cyberattacks.

We propose a rule representing the WannaCry ransomware attack, which automatically generated ten test cases (numbers 2 to 11) in Fig. 2:

((Worms + Ransomware) * Network-Based_Attacks) * Reconnaissance_Limited * Weaponization_Yes * Delivery_Yes * Exploitation_Yes * Installation_Yes * C2_Limited * Actions_on_Objectives_Yes * (Local + Network + Operating_System) * (Integrity + Availability) * (Disrupt + Destruct) * (DoS + Misuse_of_Resources + Installed_Malware) * Patch_System * (Antivirus + Network_Monitoring) * Remove_from_Network

As shown in Fig. 2, our UCAT taxonomy and the generated test cases for the WannaCry ransomware attack are illustrated.

The classification of WannaCry ransomware using our UCAT taxonomy, then evaluating through CTM, as shown in Fig. 2, makes clear the advanced capabilities of UCAT compared to existing taxonomies. UCAT fills the gaps in taxonomies such as AVOIDIT and ADMIT, as they have dimensions that are not well defined and overlap in the definitions of categories, such as the 'Attack Vector' dimension comprising vulnerabilities and attack methods, or the classification of the attacker rather than the methods or pathways used [2]. Moreover, UCAT introduces the Cyber Kill Chain as a separate dimension for detailed mapping of attack phases, and hence demonstrates how the attack progresses. The other cyberattack taxonomies do not include such a feature at all. In addition, the 'Impact' dimension includes the security principles of the McCumber Cube as a measure of how an attack would affect the principles of confidentiality, integrity, and availability. As a result of merging the powerful features of various frameworks and taxonomies, UCAT enables a holistic approach to modern-day cyberattacks, addressing the drawbacks of other taxonomies.

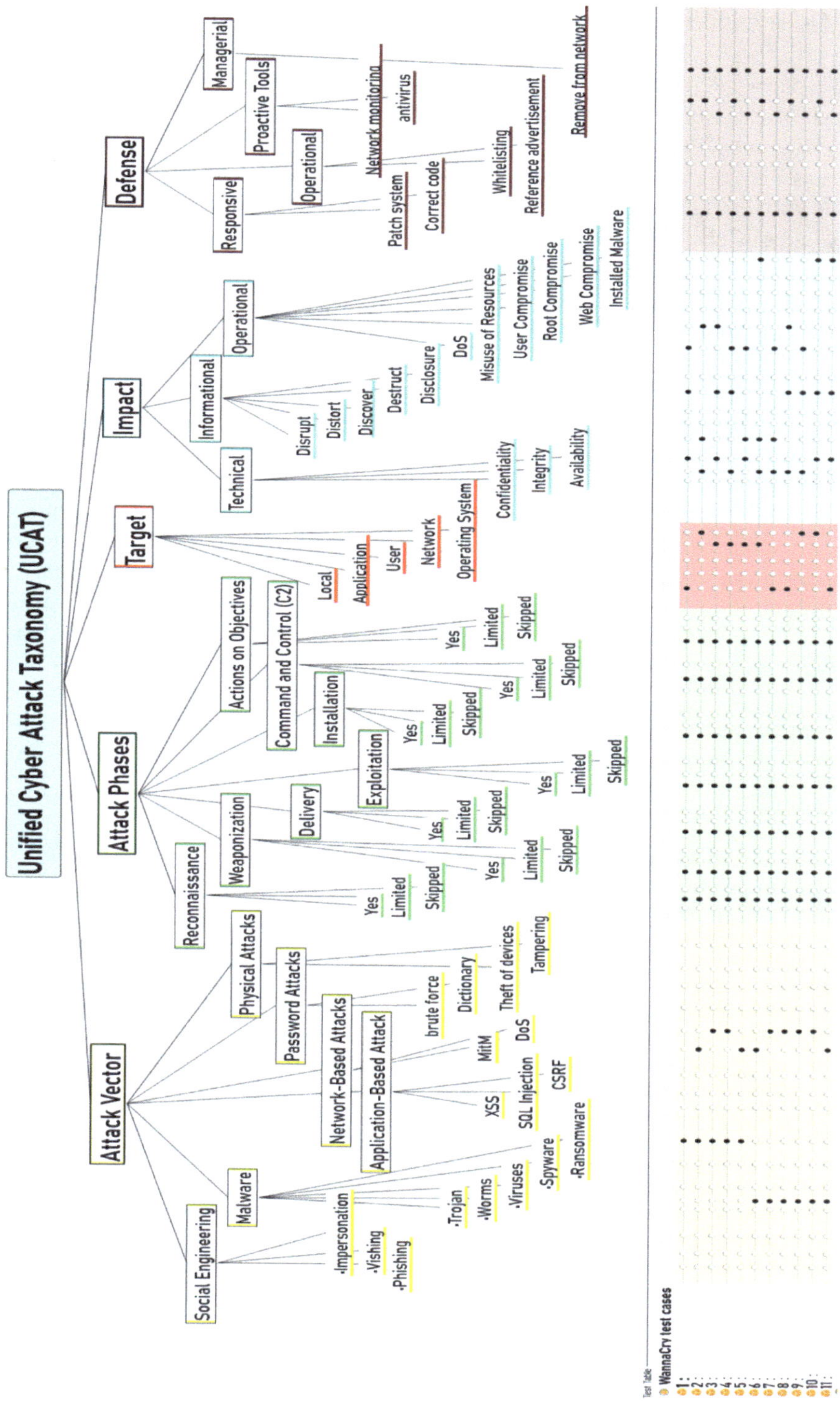

Fig. 2. UCAT taxonomy and generated test cases for the WannaCry ransomware attack.

5 Conclusion

In this paper, UCAT, a novel hybrid taxonomy, was presented to aid in the understanding of cyberattacks in a more comprehensive and thorough manner. A multidimensional integrated set of elements developed to address multiple attack vectors, phases, targets, impacts, and defenses both enabling and enhancing the disparate elements of established frameworks and taxonomies, including Cyber Kill Chain, AVOIDIT, ADMIT, and the McCumber Cube. The design of this approach remedies the limitations of previous taxonomies, as they often fail to capture the nonlinear progression of real-world attack cases, allowing for attack classification that does not strictly follow all phases of the Cyber Kill Chain.

In addition, our review of industrial and research classifications highlighted the successes and failures of different approaches, guiding the system design and structure of UCAT. We believe that UCAT offers a useful tool for cybersecurity professionals, researchers, and others interested in better understanding, analyzing, and defending against the growing spectrum of cyber threats.

In this context, we acknowledge the emergence of new threats. Future work can be carried out to extend and refine UCAT by adding new categories or subcategories, increasing granularity in its dimensions, testing a wide spectrum of cyberattacks, and confirming the applicability and efficacy of this taxonomy. Moreover, it can also be used to integrate UCAT into machine learning algorithms to automate attack and threat classification and detection.

Acknowledgments. This research was conducted with the support of the National Center for Scientific and Technical Research (CNRST) through the "PhD-ASsociate Scholarship – PASS" program.

References

1. Salem, A.H., Azzam, S.M., Emam, O.E., Abohany, A.A.: Advancing cybersecurity: a comprehensive review of AI-driven detection techniques. J. Big Data. **11**, 105 (2024). https://doi.org/10.1186/s40537-024-00957-y
2. da Santos, P.R.P.F., Resende, P.A.A., Gondim, J.J.C., Drummond, A.C.: Towards robust cyber attack taxonomies: a survey with requirements, structures, and assessment. ACM Comput. Surv. **57**(8), 199 (2025). https://doi.org/10.1145/3717606
3. MITRE ATT&CK®, https://attack.mitre.org/. Last accessed 20 Jan 2024
4. Cyber Kill Chain®, https://www.lockheedmartin.com/en-us/capabilities/cyber/cyber-kill-chain.html. Last accessed 02 Dec 2024
5. Simmons, C., Shiva, S., Dasgupta, D., Wu, C.: AVOIDIT: A Cyber Attack Taxonomy. (2009)
6. Juliadotter, N.V., Choo, K.-K.R.: Chapter 3 - CATRA: conceptual cloud attack taxonomy and risk assessment framework. In: Ko, R., Choo, K.-K.R. (eds.) The Cloud Security Ecosystem, pp. 37–81. Syngress, Boston (2015). https://doi.org/10.1016/B978-0-12-801595-7.00003-3
7. Alex, C., Creado, G., Almobaideen, W., Alghanam, O.A., Saadeh, M.: A comprehensive survey for IoT security datasets taxonomy, classification and machine learning mechanisms. Comput. Secur. **132**, 103283 (2023). https://doi.org/10.1016/j.cose.2023.103283

8. Pitropakis, N., Panaousis, E., Giannetsos, T., Anastasiadis, E., Loukas, G.: A taxonomy and survey of attacks against machine learning. Comput Sci Rev. **34**, 100199 (2019). https://doi.org/10.1016/j.cosrev.2019.100199
9. Joshi, C., Singh, U.: ADMIT- a five dimensional approach towards standardization of network and computer attack taxonomies. Int. J. Comput. Appl. **100**, 30–36 (2014). https://doi.org/10.5120/17524-8091
10. Mohsen, F., Zwart, C., Karastoyanova, D., Gaydadjiev, G.: A taxonomy for large-scale cyber security attacks. EAI Endorsed Trans. Cloud Syst. **7**, 173548 (2022). https://doi.org/10.4108/eai.2-3-2022.173548
11. Easttom, C., Butler, W.: A modified McCumber cube as a basis for a taxonomy of cyber attacks. In: 2019 IEEE 9th Annual Computing and Communication Workshop and Conference (CCWC), pp. 0943–0949 (2019). https://doi.org/10.1109/CCWC.2019.8666559
12. What is Attack Vector? Definition & FAQs, https://appomni.com/saas-glossary/attack-vector/. Last accessed 1 Dec 2024
13. Grochtmann, M., Wegener, J., Grimm, K.: Test case design using classification trees and the classification-tree editor CTEx. In: Proceedings of the 8th International Software Quality Week (1995)
14. Mirza, Q.K.A., Brown, M., Halling, O., Shand, L., Alam, A.: Ransomware analysis using cyber kill chain. In: 2021 8th International Conference on Future Internet of Things and Cloud (FiCloud), pp. 58–65 (2021). https://doi.org/10.1109/FiCloud49777.2021.00016

Optimizing Federated Learning for Botnet Defense: A Comparative Study of Aggregation Strategies

Yassine El Yamani(✉), Youssef Baddi, and Najib El Kamoun

STIC Lab, FSJ, Chouaib Doukkali University, El Jadida, Morocco
elyamani.y@ucd.ac.ma

Abstract. The rise of Internet of Things (IoT) devices has intensified botnet threats, leading to large-scale attacks like *DDoS* and *malware propagation*. Traditional machine learning (ML) models struggle with privacy risks, scalability issues, and class imbalance, limiting detection performance. Federated Learning (FL) enhances privacy but suffers from biased aggregation in non-IID environments.

To address this, we propose FL Mean/Std, an aggregation strategy leveraging *mean* and *standard deviation* of class distributions to mitigate imbalance. Evaluations on the N-BaIoT dataset—spanning four IoT security cameras and eleven attack classes—show that FL Mean/Std surpasses FedAvg and CNN, achieving up to 93.68% precision, 91.91% recall, 89.79% F1-score, and 92.58% ROC/AUC. These results confirm its effectiveness in enhancing model robustness and botnet detection in imbalanced, non-IID IoT environments.

Keywords: Federated Learning · Botnet Detection · IoT Security · Class Imbalance · Statistical Aggregation

1 Introduction

The widespread adoption of Internet of Things (IoT) devices has revolutionized various industries, enabling automation and real-time monitoring. However, it has also introduced severe cybersecurity risks. Among them, botnet attacks pose a major threat, leveraging compromised IoT devices to launch large-scale cyberattacks such as *Distributed Denial of Service (DDoS)*, *data exfiltration*, and *malware propagation* [16]. Advanced botnets like *Mirai* and *Bashlite* highlight the urgent need for efficient security mechanisms [3].

Machine learning (ML) and deep learning (DL) have been widely explored for botnet detection [1]. These models analyze network traffic to distinguish malicious activities. However, traditional centralized learning presents key challenges: (1) **Privacy risks** due to raw data transmission to central servers, increasing exposure to breaches [9]; (2) **Scalability limitations** as the volume of IoT data grows, making centralized processing inefficient [10]; and (3) **Class imbalance**, where benign traffic dominates, leading to biased models that struggle to detect rare attacks [11].

B.-E. Boudriki Semlali et al. (Eds.): DATA 2025, LNNS 1908, pp. 348–358, 2026.
https://doi.org/10.1007/978-3-032-21200-9_29

To address these limitations, federated learning (**FL**) [6] enables collaborative model training across IoT devices while keeping data localized. This preserves privacy, reduces communication overhead, and scales efficiently [15]. However, conventional FL methods like Federated Averaging (**FedAvg**) rely solely on dataset size, ignoring *data heterogeneity* and *class imbalance*. This results in global models favoring clients with large but highly skewed datasets, degrading performance for underrepresented attack types [15].

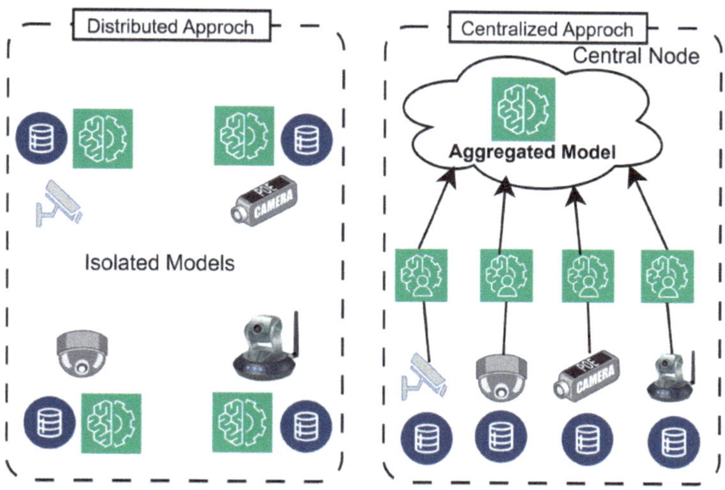

Fig. 1. Comparison of distributed and federated learning approaches.

Figure 1 contrasts distributed learning and federated learning. Distributed learning trains models locally without information sharing, leading to limited generalization. In contrast, FL aggregates local model updates (e.g., gradients) from multiple devices, ensuring collaborative learning while maintaining data privacy. Despite its advantages, FL still struggles with class imbalance, communication overhead, and security risks, necessitating improved aggregation strategies.

To tackle these issues, we propose **FL Mean/Std**, a novel aggregation method that prioritizes well-balanced datasets by leveraging mean and standard deviation of class distributions. Unlike FedAvg, which assigns weights purely based on dataset size, "FL Mean/Std" dynamically adjusts aggregation weights to ensure that clients with more balanced data contribute more effectively, mitigating class imbalance.

Contributions of This Work

- We analyze key challenges in IoT botnet detection, focusing on privacy risks, scalability, and dataset imbalance.
- We propose FL Mean/Std, a statistical aggregation strategy that improves federated learning robustness in non-IID environments.

- We evaluate CNN, FL Standard (FedAvg), and FL Mean/Std on the NBaIoT dataset, considering four IoT security cameras and eleven attack classes.
- We conduct an in-depth performance comparison using key metrics (*precision, recall, F1-score, accuracy, and ROC/AUC*) and demonstrate that FL Mean/Std outperforms both FL Standard and CNN.
- We discuss broader implications for federated cybersecurity applications and highlight the role of statistical weighting in optimizing FL-based IoT security.

The rest of this paper is structured as follows: Sect. 2 reviews related work on botnet detection and FL-based security. Section 3 details the experimental framework, dataset, and models. Section 4 presents and analyzes the results, while Sect. 5 discusses their implications and future directions. Finally, Sect. 6 concludes the study.

2 Related Works

Botnet attacks in IoT environments pose severe cybersecurity risks, making machine learning (ML) and deep learning (DL) essential for automated intrusion detection [19]. Traditional signature-based and anomaly-based detection methods struggle to adapt to evolving botnets due to their reliance on predefined attack patterns. In response, ML/DL models have been widely explored, analyzing network traffic to distinguish between benign and malicious activity [1]. However, centralized ML approaches face challenges related to data privacy, scalability, and class imbalance, making them less practical for large-scale IoT networks [9, 10].

To address these limitations, federated learning (FL) has emerged as a decentralized alternative that enables IoT devices to collaboratively train models without sharing raw data, enhancing privacy and reducing communication overhead [12]. Despite its advantages, conventional aggregation methods such as Federated Averaging (FedAvg) rely solely on dataset size, ignoring data heterogeneity and class imbalance, which leads to biased global models [15].

Several studies have explored ML-based botnet detection using Support Vector Machines (SVM), Random Forest (RF), and Convolutional Neural Networks (CNN) [17, 8]. More advanced models, such as Recurrent Neural Networks (RNNs) and Generative Adversarial Networks (GANs), have also been applied to improve detection accuracy by capturing temporal attack patterns [5]. However, these centralized DL models still require extensive data transfer, raising privacy risks and exposing IoT networks to adversarial threats [13].

FL-based approaches have been adopted for intrusion detection systems (IDS) and malware classification, demonstrating improved privacy and scalability [18]. Yet, a major challenge in FL for IoT security remains its vulnerability to non- IID data distributions, where different IoT devices generate highly imbalanced attack and benign traffic. Methods like FedAvg assume identically distributed data across devices, which degrades performance in heterogeneous environments [4].

To mitigate these issues, adaptive FL strategies have been proposed. Some studies explore reinforcement learning (RL)-based FL, dynamically adjusting aggregation weights based on client contributions [14]. Others introduce performance- aware weighting schemes, prioritizing clients with lower uncertainty [18, 8]. While promising, such strategies often introduce additional computational overhead, making them less suitable for resource-constrained IoT devices.

A promising alternative is statistical-aware FL aggregation, where class distribution metrics such as mean and standard deviation guide adaptive weighting. This ensures fairer model contributions from clients with well-balanced datasets [17]. However, prior studies have mostly used heuristic methods rather than explicitly integrating these statistics into the aggregation process. Our proposed FL Mean/Std addresses this gap by systematically incorporating mean and standard deviation into the update rule, favoring clients with more balanced class distributions. This reduces bias from imbalanced data and improves model robustness.

This work builds upon previous efforts by evaluating CNN, FL Standard (FedAvg), and (FL Mean/Std) on the N-BaIoT dataset, focusing on four IoT security cameras and eleven attack classes. The results show that "FL Mean/Std" outperforms existing FL methods, particularly in precision, recall, and F1-score, making it a viable solution for privacy-preserving, federated botnet detection.

3 Experimental Framework and Approach

This section describes the experimental setup used to evaluate the proposed "FL Mean/Std" aggregation strategy, covering the dataset, models, and evaluation methodology.

Dataset: N-BaIoT. To assess the effectiveness of "FL Mean/Std", we use the N-BaIoT dataset, a benchmark for IoT botnet detection [7]. It contains real network traffic collected from various IoT devices infected with Mirai and Bashlite botnets. The dataset includes both benign and attack traffic, making it suitable for intrusion detection research.

We focus on four security cameras, as they exhibit different traffic patterns and vulnerability levels:

- **IoT 5**: Provision PT 737E Security Camera
- **IoT 6**: Provision PT 838 Security Camera
- **IoT 8**: SimpleHome XCS7 1002 WHT Security Camera
- **IoT 9**: SimpleHome XCS7 1003 WHT Security Camera

Each device's traffic is labeled into eleven classes, covering benign traffic and multiple botnet attacks:

- *Benign, gafgyt.combo, gafgyt.junk, gafgyt.scan, gafgyt.tcp, gafgyt.udp, mirai.ack, mirai.scan, mirai.syn, mirai.udp, mirai.udpplain.*

The dataset is highly imbalanced, with benign traffic dominating attack samples. Figure 2 illustrates the class distribution across the four selected IoT devices, highlighting the imbalance challenge.

Models and Aggregation Strategies To evaluate FL Mean/Std, we compare three learning approaches:

1. *CNN (Baseline)* A standard convolutional neural network is used as a baseline. CNNs extract spatial features from network traffic and have been effective in previous botnet detection studies [2].
2. *Federated Learning - Standard Aggregation (FedAvg)* FedAvg [6] computes the global model update as:

$$w_{global} = \sum_{i=1}^{N} \frac{n_i}{N} w_i \tag{1}$$

where w_i is the local model weight for device i, n_i is its dataset size, and N is the total number of samples. While effective, FedAvg is biased toward clients with larger datasets, leading to poor performance in imbalanced settings [4].
3. *Federated Learning - Mean/Std Aggregation* FL Mean/Std incorporates statistical weighting by considering mean and standard deviation of class distributions:

$$score_i = \frac{\mu_i}{\sigma_i + \epsilon} \tag{2}$$

where μ_i is the mean class frequency, σ_i is the standard deviation, and ϵ is a small constant to prevent division by zero. The global model update is computed as:

$$w_{global} = \sum_{i=1}^{N} \frac{score_i}{\sum_{j=1}^{N} score_j} w_i \tag{3}$$

This ensures that clients with balanced datasets contribute more to the global model, reducing bias.

Experimental Protocol. The evaluation follows five steps:

1. Data preprocessing: Duplicate removal, label encoding, and feature normalization.
2. Local model training: Each IoT device trains a CNN model.
3. Federated aggregation: Models are aggregated using either FedAvg or FL Mean/Std.
4. Performance evaluation: Models are assessed on precision, recall, F1-score, accuracy, and ROC/AUC.
5. Comparative analysis: The impact of FL Mean/Std is compared to baseline methods.

Experiments are repeated to ensure statistical significance.

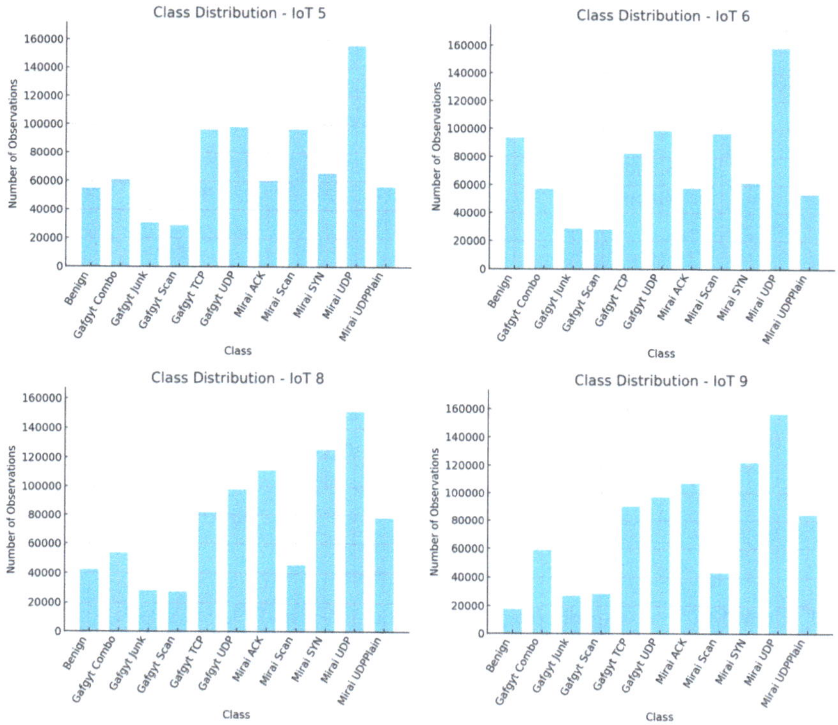

Fig. 2. Class distribution across the four IoT devices. The imbalance is evident, with benign traffic significantly outweighing attack classes.

4 Results and Analysis

This section presents the experimental results comparing the three approaches: CNN, FL Standard (FedAvg), and FL Mean/Std. Performance is evaluated across the four selected IoT devices using five key metrics: precision, recall, F1-score, accuracy, and ROC/AUC. The results highlight the impact of our proposed FL Mean/Std aggregation method in handling class imbalance and improving model robustness.

Performance Comparison Across IoT Devices. Table 1 summarizes the classification performance for each IoT device. The bold values indicate the highest performance achieved for each metric.

Performance Trends and Observations To further illustrate the performance variations, Fig. 3 presents a comparative analysis of precision, recall, F1-score, accuracy, and ROC/AUC across the four IoT devices.

Table 1. Performance comparison of CNN, FL Standard, and FL Mean/Std across IoT devices

IoT Device	Method	Precision	Recall	F1-score	ROC/AUC
	CNN	0.8827	0.8833	0.8628	0.8943
IoT 5	FL Standard	0.8960	0.8819	0.8656	0.8967
	FL Mean/Std	**0.8990**	**0.9196**	**0.9005**	**0.9232**
	CNN	0.8847	0.8837	0.8659	0.8958
IoT 6	FL Standard	0.9041	0.8803	0.8699	0.8979
	FL Mean/Std	**0.9306**	**0.9219**	**0.9032**	**0.9252**
	CNN	0.8830	0.8831	0.8650	0.8952
IoT 8	FL Standard	0.8869	0.8804	0.8712	0.8961
	FL Mean/Std	**0.9178**	**0.9192**	**0.9123**	**0.9239**
	CNN	0.8960	0.8833	0.8598	0.8992
IoT 9	FL Standard	0.9194	0.8817	0.8625	0.9028
	FL Mean/Std	**0.9368**	**0.9191**	**0.8979**	**0.9258**

"FL Mean/Std" consistently outperforms both CNN and FL Standard. It shows better generalization and improved handling of class imbalance across all devices. FL Standard performs better than CNN but remains limited due to its dependence on dataset size rather than quality. FL Mean/Std achieves the highest. recall and F1-score, confirming its ability to detect minority attack classes more effectively. IoT 9 shows the most variation, highlighting the importance of adaptive weighting when dealing with heterogeneous data.

Impact of Statistical Weighting on Model Robustness: The proposed FL Mean/Std method helps reduce bias caused by class imbalance. By incorporating statistical weighting using mean and standard deviation, it improves the model's ability to detect all classes more equally. The observed gains in F1-score and recall demonstrate its impact on identifying underrepresented classes, especially when compared to FedAvg [4].

To validate these findings, a paired t-test was conducted, confirming that FL Mean/Std's improvements over FL Standard are statistically significant ($p < 0.05$). This result supports the idea of using statistical aggregation in federated learning.

Key Takeaways FL Mean/Std improves detection rates by prioritizing clients with well-balanced data. The increase in F1-score and recall confirms the benefits of statistical weighting, particularly for rare attacks. The consistent improvement in accuracy and ROC/AUC across devices confirms better generalization. Variations in performance across IoT devices underline the importance of adaptive aggregation strategies to manage data heterogeneity.

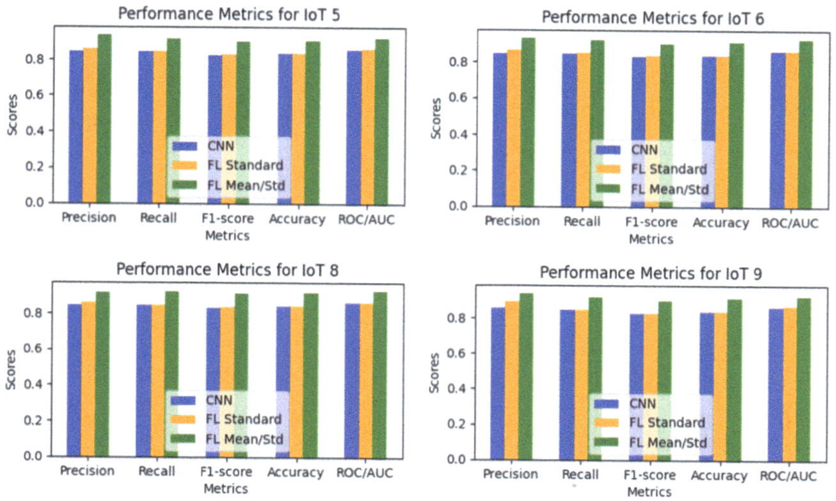

Fig. 3. Performance metrics comparison for IoT 5, IoT 6, IoT 8, and IoT 9. Each subplot shows precision, recall, F1-score, accuracy, and ROC/AUC for CNN, FL Standard, and FL Mean/Std.

5 Discussion

Our experimental results confirm that (FL Mean/Std) significantly enhances botnet detection in IoT networks, outperforming FL Standard (FedAvg) and CNN. This section discusses its benefits, limitations, and future directions.

FL Mean/Std improves model robustness by using mean and standard deviation of class distributions in the aggregation process. This helps mitigate the bias caused by class imbalance and ensures stable performance across heterogeneous IoT devices. It also generalizes better across different environments, achieving higher precision, recall, and F1-score, especially for minority attack classes. Compared to reinforcement learning-based aggregation strategies, FL Mean/Std remains lightweight and scalable, making it more suitable for constrained IoT devices. Moreover, it respects privacy constraints by aggregating model updates without sharing raw data.

Despite these benefits, deploying FL in real-world IoT networks presents several considerations. Devices may have limited computing and energy resources, though FL Mean/Std remains practical due to its low overhead. While FL reduces data transmission, communication overhead still depends on the frequency of model updates. Asynchronous FL and model compression can help alleviate this issue. Additionally, FL remains vulnerable to poisoning attacks, and robust aggregation techniques should be explored to secure the learning process.

The approach also presents a few limitations. Although more efficient than reinforcement learning-based strategies, computing class statistics adds some complexity compared to FedAvg. Furthermore, sharing statistical metadata could introduce privacy concerns in adversarial contexts. Mechanisms like secure multiparty computation or differential privacy may be necessary to address this. Lastly, while this study focuses on botnet detection, the method should be tested in broader contexts such as intrusion detection systems, ransomware detection, and anomaly detection.

In contrast to most FL-based cybersecurity research that relies on FedAvg, FL Mean/Std directly integrates class balancing into the aggregation process. This leads to better performance in non-IID settings without requiring complex policy training as in RL-based approaches. The method also proves effective in detecting rare attack classes, which is essential in cybersecurity applications.

Several extensions can improve the current approach. Future research should explore privacy-preserving techniques to protect statistical information. Adapting the aggregation dynamically to changes in class distribution could enhance long-term performance. Robustness against adversarial model updates is another area for improvement. Finally, optimizing FL Mean/Std for low-power edge devices and testing it on multi-modal security data could expand its applicability.

These findings confirm that statistical aggregation is a promising direction for improving federated botnet detection in IoT networks. Continued efforts should focus on securing, optimizing, and adapting this strategy for diverse and evolving IoT environments.

6 Conclusion

The increasing deployment of IoT devices has intensified cybersecurity threats, particularly botnet attacks. Traditional Federated Learning (FL) approaches such as FedAvg struggle with class imbalance, leading to biased models and reduced detection effectiveness. This study introduced FL Mean/Std, an improved FL aggregation strategy that leverages mean and standard deviation of class distributions to enhance robustness in non-IID environments.

Our experiments on the N-BaIoT dataset, focusing on four IoT security cameras and eleven attack classes, demonstrate that FL Mean/Std surpasses both FL Standard (FedAvg) and CNN across multiple evaluation metrics. The results confirm its effectiveness in mitigating class imbalance, improving model generalization, and maintaining high recall and F1-score for minority attack classes. By prioritizing well-balanced data contributions, FL Mean/Std achieves superior precision, recall, F1-score, and ROC/AUC, demonstrating its ability to improve detection accuracy in distributed learning environments.

Beyond performance improvements, FL Mean/Std presents a lightweight, scalable, and privacy-preserving solution for IoT security. Unlike centralized ML approaches that require continuous data transmission, our method enables decentralized model training while protecting raw data. The statistical weighting mechanism ensures that underrepresented attack classes contribute effectively, reducing bias and enhancing model robustness.

The findings of this study open new avenues for research in adaptive and secure federated learning. Future work should explore privacy-enhancing mechanisms such as Secure Multi-Party Computation (SMPC) and Differential Privacy to further strengthen data protection. Additionally, integrating FL Mean/Std with multi-modal security data, including network traffic patterns and behavioral analytics, could enhance botnet detection capabilities. Further evaluation in real-world IoT settings and applications such as intrusion detection, ransomware detection, and anomaly detection will help assess its broader applicability.

FL Mean/Std represents a promising step toward adaptive, scalable, and privacy-preserving FL aggregation for cybersecurity applications. As IoT threats continue to evolve, statistical aggregation techniques will play a crucial role in building resilient, self-learning security frameworks to protect next-generation digital infrastructures.

References

1. EL Yamani, Y., Baddi, Y., EL Kamoun, N.: A survey on the contribution of ML and DL to the detection and prevention of botnet attacks. J. Reliable Intell. Environ. **10**(4), 431–448 (2024)
2. El Yamani, Y., El Kamoun, N., Baddi, Y.: Resampling and KNN for IoT botnet detection: addressing class overlap challenges. In: Proceedings of the 4th International Conference on Advanced Computing and Informatics (ICACIn 2024) Lecture Notes on Data Engineering and Communications Technologies. Springer (2025)
3. Antonakakis, M., April, T., Bailey, M., Bernhard, M., Bursztein, E., Cochran, J., Durumeric, Z., Halderman, J.A., Invernizzi, L., Kallitsis, M., et al.: Understanding the Mirai botnet. . In: 26th USENIX Security Symposium (USENIX Security 17), pp. 1093–1110. (2017)
4. Li, T., Sahu, A.K., Zaheer, M., Sanjabi, M., Talwalkar, A., Smith, V.: Federated optimization in heterogeneous networks. Proc. Mach. Learn. Syst. **2**, 429–450 (2020)
5. Nguyen, D.C., Ding, M., Pathirana, P.N., Seneviratne, A., Li, J., Poor, H.V.: Federated learning for Internet of Things: a comprehensive survey. IEEE Commun. Surv. Tutor. **23**(3), 1622–1658 (2021)
6. McMahan, B., Moore, E., Ramage, D., Hampson, S., Arcas, B.A.: Communication-efficient learning of deep networks from decentralized data. In: Artificial Intelligence and Statistics (AISTATS), pp. 1273–1282. (2017)
7. Meidan, Y., Bohadana, M., Mathov, Y., Mirsky, Y., Shabtai, A., Breitenbacher, D., Elovici, Y.: N-BaIoT: network-based detection of IoT botnet attacks using deep autoencoders. IEEE Pervasive Comput. **17**(3), 12–22 (2018)
8. Doriguzzi-Corin, R., Siracusa, D.: FLAD: Adaptive federated learning for DDoS attack detection. Comput. Secur. **137**, 103597 (2024)
9. Vyas, A., Lin, P.-C., Hwang, R.-H., Tripathi, M.: Privacy-preserving federated learning for intrusion detection in IoT environments: A survey. IEEE Access **12**, 127018–127050 (2024)
10. Nguyen, D.C., Ding, M., Pathirana, P.N., Seneviratne, A., Li, J., Poor, H.V.: Federated learning for Internet of Things: a comprehensive survey. IEEE Commun. Surv. Tutor. **23**(3), 1622–1658 (2021)
11. Wang, L., Xu, S., Wang, X., Zhu, Q.: Addressing class imbalance in federated learning. Proc. AAAI Conf. Artif. Intell. **35**(11), 10165–10173 (2021)
12. Qammar, A., Ding, J., Ning, H.: Federated learning attack surface: taxonomy, cyber defences, challenges, and future directions. Artif. Intell. Rev. **55**(5), 3569–3606 (2022)
13. Torre, D., Chennamaneni, A., Jo, J., Vyas, G., Sabrsula, B.: Toward enhancing privacy preservation of a federated learning CNN intrusion detection system in IoT: method and empirical study. ACM Trans. Softw. Eng. Methodol. **34**(2), 1–48 (2025)
14. Lim, W.Y.B., Luong, N.C., Hoang, D.T., Jiao, Y., Liang, Y.-C., Yang, Q., Niyato, D., Miao, C.: Federated learning in mobile edge networks: a comprehensive survey. IEEE Commun. Surv. Tutor. **22**(3), 2031–2063 (2020)
15. Nguyen, D.C., Ding, M., Pathirana, P.N., Seneviratne, A., Li, J., Poor, H.V.: Federated learning for Internet of Things: a comprehensive survey. IEEE Commun. Surv. Tutor. **23**(3), 1622–1658 (2021)
16. Koroniotis, N., Moustafa, N., Sitnikova, E., Turnbull, B.: Towards the development of realistic botnet dataset in the Internet of Things for network forensic analytics: bot-IoT dataset. Future Gener. Comput. Syst. **100**, 779–796 (2019)

17. Nguyen, D.C., Ding, M., Pathirana, P.N., Seneviratne, A., Li, J., Poor, H.V.: Federated learning for Internet of Things: a comprehensive survey. IEEE Commun. Surv. Tutor. **23**(3), 1622–1658 (2021)
18. Nguyen, D.C., Pham, Q.-V., Pathirana, P.N., Ding, M., Seneviratne, A., Lin, Z., Dobre, O.A., Hwang, W.-J.: Federated learning for smart healthcare: a survey. ACM Comput. Surv. **55**(3), 1–37 (2022)

Automated and Personalized Employee Performance Evaluations Using Natural Language Generation

Oumaima Belarache(✉), Soumaya Ounacer, and Mohamed Azzouazi

Laboratory of Information Technology and Modeling, Faculty of Sciences Ben M'Sick, Hassan II University of Casablanca, Casablanca, Morocco
belaracheoumaima9@gmail.com

Abstract. Traditional performance assessment methods are often inefficient, time-consuming, and influenced by subjective judgment, even though they remain vital for effective workforce management. In response, this study explores the application of Natural Language Generation (NLG) techniques to modernize and personalize these assessments. From basic template-based and rule-based approaches to advanced transformer architectures such as GPT, BERT, T5, and Retrieval-Augmented Generation (RAG), NLG systems offer promising solutions. By combining structured performance data with these technologies, organizations can produce tailored feedback that is both actionable and aligned with employee contexts. However, the increasing reliance on AI, in this context, raises certain important concerns involving fairness, bias mitigation, as well as alignment with organizational goals. This research looks at what NLG systems can do when evaluating performance and suggests ways to make them more adaptable, support more languages, and be used ethically in human resource management.

Keywords: Natural Language Generation (NLG) · Employee Performance Evaluation · Automated Feedback · AI-driven HR

1 Introduction

Effective workforce management depends on performance evaluations, which can measure employee contributions, find skill gaps, and align staff to the goals of the organization. Conventional evaluation methods, often manual as well as reliant on subjective human judgment, tend to be rather inefficient, inconsistent, and prone to bias, especially in organizations where the review process becomes burdensome. Certain advancements throughout Natural Language Generation (NLG) technologies now present opportunities for automation within personalized performance evaluations. Transformer-based models such as GPT, BERT, and Retrieval-Augmented Generation (RAG) are able to leverage structured performance data for the production of individualized feedback for all employees. These technologies do improve evaluation processes through greater scalability, reduced HR workload, and increased objectivity and fairness.

B.-E. Boudriki Semlali et al. (Eds.): DATA 2025, LNNS 1908, pp. 359–369, 2026.
https://doi.org/10.1007/978-3-032-21200-9_30

This study examines the transformative impact of NLG, in particular, on employee performance assessments, highlighting current practices as well as future challenges. The research employs a descriptive-analytical approach and explores the use of NLG in performance evaluation tasks. Key evaluation criteria are proposed in this research. Beginning with the methodology, the structure follows a classic research flow through the results and discussion, before concluding with AI's potential within human resource systems.

2 Methodology

This study investigates the application of Natural Language Generation (NLG) techniques to automate as well as personalize employee performance evaluations, with a particular emphasis on transformer-based models such as GPT, BERT, and Retrieval-Augmented Generation (RAG). The research, adopting such descriptive-analytical methodology, aims to evaluate each of the capabilities, limitations, and future potential of NLG-powered performance assessment systems inside organizational contexts.

2.1 Scope and Areas of Interest

This research focuses on three core aspects. First, it explores the manner in which transformer-based NLG models, specifically GPT, BERT, and RAG, are used. These models generate automated feedback tailored for individual employees. Second, it digs into particular key challenges like algorithmic bias, plus cultural and linguistic adaptability, also the alignment of generated feedback with organizational objectives. Finally, it investigates just how NLG technologies can be applied in a practical way within human resource management, with the focus on the enhancement of the personalization of performance evaluations, the improvement of operational efficiency, and the enabling of scalability in the management of large workforces.

2.2 Analytical Framework

The analytical framework that was used for this research is structured around two key dimensions: system capabilities and ethical-technical considerations. The capability assessment considers just how NLG models convert structured performance data into coherent, meaningful feedback. It also considers the degree to which they tailor evaluations to specific job roles and performance metrics. Meanwhile, ethical as well as technical considerations are examined through the lens of bias detection, with attention to whether NLG systems perpetuate existing evaluation biases. Additionally, the study addresses each of the models' ability for adaptation to various workplace cultures, languages, and norms, ensuring their relevance throughout diverse organizational environments. It explores how the feedback produced relates to business strategies, staff growth plans, and aims regarding output.

3 Results

3.1 Natural Language Generation (NLG)

Natural Language Generation (NLG), first introduced by Reiter and Dale in 1997 [1], is an important area of Natural Language Processing (NLP). It's all about turning data, whether it's organized or not, into text that reads like something a person would write. Unlike Natural Language Understanding (NLU), which is concerned with understanding the meaning behind text, NLG focuses on the opposite task: turning data into smooth, coherent, and contextually relevant language that feels natural to readers. This functionality is fundamental to a wide range of downstream applications, for example, text summarization, conversational agents, machine translation, data-to-text generation, and generative question answering.

Over time, NLG has evolved out of a growing need to make data not just readable by machines, but also clear and meaningful to people. In the beginning, systems relied on rules or templates straightforward and easy to control, but limited in what they could do. Everything changed with the rise of deep learning and Transformer-based models, which brought a whole new level of flexibility and power to the field. Current architectures, for example, BERT [5], BART [4], GPT-2 [3], plus GPT-3 [6], have notably advanced such quality of generated text via leveraging self-supervised learning on huge datasets to capture detailed linguistic patterns.

Even with all the progress, NLG isn't without its problems. Keeping the text clear, coherent, and factually right is still a big ask. There's also the issue of bias—models tend to pick up patterns from the data they're trained on, for better or worse. And running these systems, especially in real time, can be pretty demanding. Newer approaches like Retrieval-Augmented Generation (RAG) show a lot of potential, but they also bring more to think about, especially when used in sensitive areas like healthcare, finance, education, or customer service.

3.2 Modern NLG Models

Natural Language Generation (NLG) has been steadily transforming, especially with the rise of powerful neural networks like transformer-based models [2]. These systems learn from massive amounts of text and use high computing power to understand and generate language in a way that feels more natural and context-aware. They are capable of producing fluent, contextually appropriate text for many applications. These applications include translation, summarization, as well as conversational dialogue [6, 7]. Depending on the architecture, training involves either predicting masked tokens [5] or generating text in an autoregressive manner [3]. This initial pre-training is typically followed by fine-tuning or prompt-based adaptation for specific tasks, minimizing the dependence on task-specific models [8]. Nevertheless, these technological strides come with ethical considerations, such as the presence of bias in training datasets [9], the environmental impact of large-scale model training [10], and the inherent difficulty in evaluating the quality of generated outputs [11].

Pre-trained Transformer Models. Transformer-based models have dramatically reshaped the field of Natural Language Generation (NLG), making it possible to produce text that is not only fluent but also sensitive to context. These architectures use deep learning and self-attention mechanisms to craft language that closely mimics human expression across a wide range of applications.

Among these, the Generative Pretrained Transformer (GPT) stands out as a capable model pre-trained on substantial text corpora. It can be fine-tuned to work for tasks such as language generation, sentiment analysis, translation and text classification. GPT's core transformer architecture is a large advancement from prior approaches such as RNNs and CNNs. With the utilization of self-attention, GPT considers the complete broader context in a sentence when generating each single word, which indeed improves its language comprehension as well as generation abilities. The model's decoder has the responsibility to produce text outputs, depending on representations learned [12].

GPT excels in Natural Language Understanding (NLU). It also excels in Natural Language Generation (NLG). It can analyze textual content, extract entities as well as relationships, and generate high-quality responses that range from informative answers to creative writing. There exist multiple versions of the model, with each improving upon the capabilities of the previous one.

BERT (Bidirectional Encoder Representations from Transformers), which was developed by Devlin et al. (2019) [5], is a bidirectional adaptation of the Transformer architecture, in fact, and it was initially introduced by Vaswani et al. (2017) [2]. It is trained with regard to a pair of chief objectives: (1) Masked Language Modeling (MLM), in which words inside sentences are randomly hidden, and then the model learns to predict those via surrounding context, and (2) Next Sentence Prediction (NSP), in which the model determines whether sentences follow one behind another. After pre-training of the BERT model, it can be fine-tuned well for certain NLP tasks such as question answering, language inference, and text classification, without any need for meaningful changes.

BERT includes two model choices: BERT-base and BERT-large. The BERT-base model contains several layers of transformer encoder blocks, each featuring multi-head self-attention layers as well as almost eight hundred hidden units, for about one hundred million parameters. Conversely, the BERT-large model has 24 layers of transformer encoder blocks, each with 24-head self-attention layers, totaling 340 million parameters. The model's performance depends on its particular type. BERT-large often has the ability to achieve greater accuracy than BERT-base. Still, this improved degree of precision needs additional computer power.

The Text-to-Text Transfer Transformer (T5) is a deep learning model from Google Research, as introduced in their paper [7]. T5's main idea is framing each (NLP) task as a text-to-text issue; input and output are text. This approach allows T5 to handle many varieties of NLP tasks, including text classification, summarization, translation, and question answering, within just a single, unified framework. One prominent feature that T5 has is its ability to generalization across different tasks, using a consistent approach of text-to-text. For example, T5 reformulates each and every Natural Language Processing (NLP) task as a text generation problem. In sentiment analysis, the input could be just a sentence like "Review: The movie was fantastic!", as well as the output being a simple classification such as "positive." The model follows a sequence-to-sequence (Seq2Seq) architecture, where the encoder processes the input text into hidden representations while the decoder generates the output text based on these representations.

T5 leverages a certain self-attention mechanism, using Transformer-based models, to effectively capture long-range dependencies between words, thereby improving its contextual understanding. The model, on a massive dataset called C4 (Colossal Clean Crawled Corpus), is pretrained, which enables learning of general language patterns before fine-tuning for specific tasks. Once appropriately pretrained, T5 can be fine-tuned on specialized datasets for summarization, translation, and question answering tasks. Furthermore, T5's single structure removes the demand for many specific models, which makes it fit for many uses. T5 comes in various sizes to suit different computing needs, ranging from the lighter T5-Small with 60 million parameters to the much more powerful T5-11B, which packs 11 billion parameters.

Retrieval-Augmented Models (e.g., RAG). Retrieval-augmented models mark an important step forward in Natural Language Generation (NLG) through blending the capabilities within neural text generators through information retrieval systems. Unlike fully generative models, which regularly rely only on internalized (parametric) knowledge and occasionally produce factually incorrect or "hallucinated" content, these models draw on additional sources of information to achieve better accuracy and relevance. This makes them valuable in particular for tasks that demand a solid and factual grounding, such as open-domain question answering, the generation of dialogue, and summarization.

A fairly well-known example of such an approach remains the Retrieval-Augmented Generation (RAG) model, introduced by Lewis et al. (2020) [13]. RAG combines one pre-trained generative model (such as BART or T5) with one neural retriever that searches some large external knowledge base, typically something like Wikipedia. When given a query, the retriever first selects the most relevant passages; in addition, the generator uses the original input as well as this retrieved content to craft its response. This two-step method helps to ensure that the output is not just fluent, but it is also grounded in accurate, context-specific information.

The architecture of RAG includes many necessary components, each contributing to the model's ability to generate high-quality, factually reliable text. Table 1 summarizes these elements.

Table 1. Key Components of RAG

Component	Role
Retriever	Encodes the input query and retrieves relevant documents/passages
Generator	Generates text conditioned on the input query and retrieved documents
External Corpus	Provides factual knowledge (e.g., Wikipedia, domain-specific datasets)
Fusion Mechanism	Combines retrieved information with the input query for generation.

Despite all their strengths, retrieval-augmented models still come along with a set of challenges. A major worry remains the quality that has been retrieved; should the passages and documents brought in be irrelevant or noisy, they can impact the output's accuracy and coherence. Additionally, the retrieval step alone adds a degree of computational complexity, particularly when the model searches through large corpora in real-time applications. A further challenge resides in integrating the retrieval and generation components. To align these components effectively, considerate architectural design and training are required, which helps ensure their smooth cooperation.

3.3 Specialized NLG Methods

Natural Language Generation (NLG) includes a range of techniques, each suited to different use cases and needs. These techniques vary from straightforward, easy-to-understand methods like template-based and rule-based systems, to more advanced, data-driven approaches such as neural networks and hybrid models. Each method brings its own strengths and limitations, making them suitable for specific applications.

Template-Based NLG. Template-based Natural Language Generation is among the earliest as well as most straightforward techniques intended for text generation. It operates via filling predefined sentence structures. These structures are filled using data values during runtime. One possible example may be a weather report formatted as: "The temperature in [City] is [Temperature] degrees Celsius, with [Condition]." This method is appreciated for its transparency and ease of customization—developers can easily adjust templates without needing large datasets or complex algorithms. However, its simplicity reduces the degree of flexibility. Template-based systems often battle with linguistic variation and complex sentence construction. They require manual effort, plus scaling across domains needs diverse or dynamic data. As such, these systems are very well suited for structured domains such as financial summaries, as well as product descriptions, and weather forecasting, in which the data is regular and the output format is predictable [1].

Rule-Based Systems. Rule-based systems rely upon carefully designed linguistic rules in addition to grammars to transform structured input into natural language. For instance, a rule might dictate how to turn a logical expression (Temperature, High, City) into the sentence "The temperature in [City] is high". For their accuracy and grammatical correctness, these systems are prized, as the rules can be fine-tuned to the specific requirements of a domain. They offer a degree of transparency. They also offer some control over the generated output. However, the building and maintaining of these rules can be time-consuming, especially for more complex domains or when one is attempting to handle less predictable input. Rule-based systems typically struggle when encountering input that falls outside their specific scope. Despite this, they remain a reliable choice for technical domains in addition to legal document generation, technical manuals, and applications where consistency and precision are important [14].

Neural Network-Based Methods. Neural network approaches to the base have transformed NLG for machines, enabling text generation that resembles human writing. These models, especially those based on sequence-to-sequence architectures as well as Transformer designs like GPT, BERT, and T5, learn directly from data plus do not require manually created rules. After training, they can generalize across various tasks.

These include summarization, machine translation, and dialogue generation. Once fully trained, they can generalize to various tasks, including summarization, machine translation, and dialogue generation. Their strengths of major importance lie in their ability to produce text that is fluent, diverse, and rich in context. However, these benefits come along with certain trade-offs. For training, neural models demand datasets of a large size and computational power that is large. Additionally, their outputs can occasionally lack interpretability, and they can produce text that is factually incorrect or inconsistent, commonly referred to as "hallucinations." Despite these limitations, neural approaches are currently the standard for many high-impact NLG applications due to their flexibility and performance [2, 5–7].

Hybrid Systems. Hybrid systems aim to combine more of the best parts of both worlds by integrating rule-based logic alongside neural network capabilities. These systems might use rules to ensure factual correctness or enforce structural constraints, whilst leveraging neural models for more fluid and adaptive language generation. This combination offers a balance amid control as well as flexibility, letting developers guide the model's output while still benefiting from the expressive power of deep learning. However, the design of hybrid systems poses a set of its challenges. To effectively integrate rule-based components and neural components, mindful architecture is required, and continuing maintenance is required too. Nevertheless, hybrid approaches are gaining traction in many areas, such as medical report generation, clever chatbots, and other domain-specific applications where accuracy and naturalness must coexist [14].

The choice of the NLG method depends on the specific requirements of the task, such as the need for interpretability, scalability, or fluency. Below is a summary of the key characteristics of each method (Table 2):

Table 2. Summary of Specialized NLG Methods

Method	Strengths	Limitations	Applications
Template-Based NLG	Interpretable, controllable, low resource	Rigid, not scalable	Weather reports, financial summaries
Rule-Based Systems	Precise, customizable	Labor-intensive, lacks generalization	Technical documentation, legal text
Neural Network-Based	Fluent, scalable, generalizable	Data-hungry, black-box, prone to errors	Machine translation, dialogue systems
Hybrid Systems	Balanced, accurate, adaptable	Complex, high maintenance	Medical reports, customer service chatbots

3.4 Evaluation Metrics for NLG

Assessing the quality of text that NLG systems generate is still a hard but important task. A strong evaluation commonly involves automatic metrics combination, offering scalable, consistent benchmarks, in addition to human judgment, capturing the subtleties of fluency, coherence, and relevance that automated tools miss. Of late, new evaluation strategies have started to emerge to overcome limitations in the customary methods. These revolutionary approaches are, in effect, broadening the way we evaluate NLG systems. They also offer a deeper understanding to provide a more thorough understanding of their performance. The following sections dig into these various evaluation techniques in greater detail.

Automatic Metrics. Automatic metrics see wide usage for the evaluation of NLG systems, owing to their reproducibility and scalability. These metrics usually compare generated text against a reference text or other reference texts, gauging overlap or similarity at that particular lexical, syntactic, and also semantic level.

BLEU (Bilingual Evaluation Understudy) [15]. BLEU measures the degree of precision in n-gram overlaps among the generated text and reference texts. It is used in tasks like machine translation now. It is simple enough to compute and generally adopted. However, it focuses on surface-level overlap, which may not correlate with semantic quality, as well as penalizes diversity in generated text by rewarding exact matches.

ROUGE (Recall-Oriented Understudy for Gisting Evaluation) [16]. ROUGE is measured via counting n-grams, word sequences, and word pairs shared by generated text and references. It is frequently used in summarization tasks, and it is effective in evaluating content coverage inside summaries. However, as with BLEU, it focuses upon lexical overlap and may not capture semantic coherence, and requires reference texts of high quality, which can be expensive for production.

METEOR [17]. METEOR improves on BLEU using word stemming, synonym matching, and ordering of words. It aims for a better alignment with human judgment. While being stronger than BLEU for semantic similarity capture, it is computationally expensive as well as still relying on lexical overlap; such overlap may not fully capture contextual relevance.

Automatic metrics are useful enough for benchmarking. However, they frequently fail to correlate, particularly in complex tasks like dialogue generation or personalized feedback generation, with human judgment. For instance, a system can achieve quite high BLEU scores but still produce text that is irrelevant in, or incoherent within, context [18].

Human Evaluation. Human evaluation remains the standard for assessing NLG systems. It captures multiple aspects of text quality that automatic metrics cannot. Judges do assess the text based upon its fluency, coherence, relevance, and usefulness. Fluency considers whether the text has correct grammar and ease of reading, as coherence assesses if the text has a logical structure and context. Relevance measures whether the text is appropriate enough for the task, in addition to context, also usefulness considers whether the text provides quite meaningful or actionable types of information. In performance evaluations, human judges assess whether the generated feedback is something actionable and understandable for employees [19]. Still, human evaluation has difficulties. Since different judges might view text quality differently, subjectivity poses a big problem. Another concern is scalability; human evaluation is time-consuming and expensive, so it is impractical for large-scale systems. Furthermore, ensuring consistency across various judges can be difficult, as evaluation criteria might not always be uniformly applied.

Emerging Evaluation Techniques. To address the limitations within customary evaluation methods, researchers are now exploring new approaches that leverage advances within machine learning and natural language understanding.

Large Language Models as Evaluators [20]. It is the case that pre-trained language models such as GPT-4 can be used. The quality of generated text can be evaluated with these models. These models assess aspects such as coherence, relevance, together with fluency by comparing generated text to task-specific criteria. They capture several semantic and contextual subtleties better than customary metrics, and those are scalable and cost-effective versus human evaluation. Still, they could inherit biases within the models and need calibration so results are reliable.

Reference-Free Metrics [21]. These metrics evaluate most generated texts without needing reference texts. Instead, they focus on intrinsic qualities, such as contextual consistency, grammaticality, and diversity. They are suited to tasks where many high-quality references are absent and may evaluate open-ended tasks like dialogue generation or creative writing. However, these measures can be hard to read, and the review is not clear, as the tools are new and have not been used much.

3.5 Applications of NLG in Performance Evaluations

The table below summarizes the key applications of NLG in performance evaluations (Table 3):

Table 3. Summary of Applications

Application	Description	Benefits
Personalized Feedback	Tailors feedback to individual employees	Relevance, consistency, efficiency
Automated Reporting	Generates performance summaries and reports	Time savings, data-driven insights
Scalability	Scales evaluations across large organizations	Consistency, efficiency, adaptability
Multilingual Support	Generates feedback in multiple languages	Inclusivity, global reach, cultural sensitivity

4 Discussion

The integration of Natural Language Generation (NLG) in employee performance evaluations improves efficiency, personalization, also scalability. Technical as well as ethical difficulties remain, thus requiring further refinement. NLG automates the feedback generation, thereby saving time for HR professionals during consistent and unbiased evaluations. Models such as GPT, BERT, and RAG can provide personalized feedback from structured data, thus improving with engagement as with accuracy. However, these challenges, such as contextual relevance and factual accuracy, need proper addressing through additional model adaptability. When addressing technical and Ethics problems, NLG might inherit data biases used when training, thus impacting evaluation fairness. AI-generated feedback can lack interpretability, thereby reducing trust across automated assessments. To deploy NLG for large-scale evaluations requires an efficient implementation as well as cost-effective implementation. To improve NLG's role within HR, future efforts should focus on multilingual models for global organizations with cultural adaptation. Hybrid AI approaches combine rule-based methods and deep learning for accuracy. Frameworks for AI explainability improve trust and interpretability. Organizations must ensure the ethical use of AI, fairness, and transparency while continuously refining the NLG models. This study acknowledges limitations, such as the limited real-world adoption as well as industry-specific customization needs.

5 Conclusion

Employee performance evaluations benefit considerably from Natural Language Generation (NLG) because it delivers automated feedback while improving the assessment process to provide customized performance evaluations. Organizations that adopt transformer-based models, including GPT, BERT, and RAG, will achieve more objective assessments, which simultaneously decreases HR professionals' workload. Additional work needs to be done to overcome essential difficulties before NLG reaches its full effectiveness potential. To maximize the usefulness of NLG in performance evaluations, researchers need to study and improve the removal of bias from AI feedback and address system transparency while addressing operational scalability issues. NLG-driven evaluations need ethical implementation, upholding fairness standards and organizational accountability, to meet values. More research calls for attention to multilingual adaptability with combined methods regarding artificial intelligence, alongside AI-based human resources platforms. Organizations must establish definite methods for bias elimination and explainability systems to gain stakeholders' complete trust in AI when conducting assessments. The truly successful implementation of NLG for employee performance evaluations depends upon proper implementation, with continuous development, and with ethical monitoring of that system. To deliver fair, data-driven performance reviews, organizations need to tactically deploy NLG technology within human resource systems.

References

1. Reiter, E., Dale, R.: Building Natural Language Generation Systems. Cambridge University Press (2000)
2. Vaswani, A., et al.: Attention is all you need. Adv. Neural Inf. Process. Syst. **30**, 5998–6008 (2017)
3. Radford, A., Wu, J., Child, R., Luan, D., Amodei, D., Sutskever, I.: Language models are unsupervised multitask learners. OpenAI Blog. **1**, 9 (2019)

4. Lewis, M., et al.: BART: denoising sequence-to-sequence pre-training for natural language generation, translation, and comprehension. In: Proceedings of the 58th Annual Meeting of the Association for Computational Linguistics, pp. 7871–7880 (2020)
5. Devlin, J., Chang, M.W., Lee, K., Toutanova, K.: BERT: pre-training of deep bidirectional transformers for language understanding. In: Proceedings of the 2019 Conference of the North American Chapter of the Association for Computational Linguistics: Human Language Technologies, pp. 4171–4186 (2019)
6. Brown, T., et al.: Language models are few-shot learners. Adv. Neural Inf. Process. Syst. **33**, 1877–1901 (2020)
7. Raffel, C., et al.: Exploring the limits of transfer learning with a unified text-to-text transformer. J. Mach. Learn. Res. **21**, 1–67 (2020)
8. Liu, P., Yuan, W., Fu, J., Jiang, Z., Hayashi, H., Neubig, G.: Pre-train prompt tune: exploring the limits of prompt tuning for large-scale pre-trained language models. In: Proceedings of the 59th Annual Meeting of the Association for Computational Linguistics, pp. 1283–1295 (2021)
9. Caliskan, A., Bryson, J.J., Narayanan, A.: Semantics derived automatically from language corpora contain human-like biases. Science. **356**, 183–186 (2017)
10. Bender, E.M., Gebru, T., McMillan-Major, A., Shmitchell, S.: On the dangers of stochastic parrots: can language models be too big? In: Proceedings of the 2021 ACM Conference on Fairness, Accountability, and Transparency, pp. 610–623 (2021)
11. Reiter, E.: A structured review of the validity of BLEU. Comput. Linguist. **44**, 393–401 (2017)
12. Dong, S., Xu, S., Xu, B.: Speech-transformer: a no-recurrence sequence-to-sequence model for speech recognition. In: IEEE International Conference on Acoustics, Speech and Signal Processing (ICASSP), pp. 5884–5888 (2018)
13. Lewis, P., Perez, E., et al.: Retrieval-augmented generation for knowledge-intensive NLP tasks. In: Advances in Neural Information Processing Systems (2020)
14. Gatt, A., Krahmer, E.: Survey of the state of the art in natural language generation: Core tasks, applications and evaluation. J. Artif. Intell. Res. **61**, 65–170 (2018)
15. Papineni, K., Roukos, S., Ward, T., Zhu, W.J.: BLEU: a method for automatic evaluation of machine translation. In: Proceedings of the 40th Annual Meeting of the Association for Computational Linguistics, pp. 311–318 (2002)
16. Lin, C.Y.: ROUGE: a package for automatic evaluation of summaries. In: Text Summarization Branches Out, pp. 74–81 (2004)
17. Banerjee, S., Lavie, A.: METEOR: an automatic metric for MT evaluation with improved correlation with human judgments. In: Proceedings of the 43rd Annual Meeting of the Association for Computational Linguistics, pp. 65–72 (2005)
18. Liu, C.W., Lowe, R., Serban, I., Noseworthy, M., Charlin, L., Pineau, J.: How NOT to evaluate your dialogue system: an empirical study of unsupervised evaluation metrics for dialogue response generation. In: Proceedings of the 2016 Conference on Empirical Methods in Natural Language Processing, pp. 2122–2132 (2016)
19. Reiter, E., Belz, A.: An investigation into the validity of some metrics for automatically evaluating natural language generation systems. Comput. Linguist. **35**, 529–558 (2009)
20. Zhang, X., et al.: Large language models as evaluators for recommendation explanations. In: PProceedings of the 18th ACM Conference on Recommender Systems, pp. 33–42 (2024)
21. Sellam, T., Das, D., Parikh, A.: BLEURT: learning robust metrics for text generation. In: Proceedings of the 58th Annual Meeting of the Association for Computational Linguistics, pp. 7881–7892 (2020)

Text-to-Video Generation: A State-of-the-Art Review on Generative AI and LLMs

Salma Hannouni(✉), El Habib Benlahmar, and Sanaa El Filali

Laboratory of Information Technology and Modeling, Faculty of Sciences Ben M'Sick, Hassan II University of Casablanca, Casablanca, Morocco
hannouni.sal@gmail.com

Abstract. Generative AI has advanced significantly in areas like speech, image, and text generation. However, video generation remains a complex issue in computer vision. Models that generate video from text must convert text into moving visuals while ensuring temporal coherence, motion smoothness, and reduced complexity. The most recent developments in multimodal learning, like text-to-video generators based on Video-LLMs, take advantage of previous achievements, although they still face difficulties in long sequences and multi-turn dialogues, and adapting to the real world. The available datasets do not highlight these issues, because the few existing datasets like WebVid-10 M and Panda-70 M are of low quality with poorly matched text-video pairs. To tackle these problems, new datasets, VidGen-1 M and OpenVid-1 M, have been developed with better video quality and semantic consistency. This work analyzes the latest model development and dataset shortcomings while stating the absence of ample needed resources and methodologies to accomplish effective text-to-video generation.

Keywords: Text-to-Video Generation · Generative AI · LLMs

1 Introduction

Text-to-video (T2V) generation appears as a complex but promising task in Generative AI. It enables the creation of videos from text prompts, in addition to its potential applications for entertainment, education, and virtual reality. Different from text-to-image generation, T2V has to address challenges that include temporal coherence, motion realism, and spatio-temporal complexity, so it needs strong generative models and also high-quality datasets.

Recent advances in Generative Adversarial Networks (GANs), diffusion models, in addition to the integration of Large Language Models (LLMs) into video generation, known as Video-LLMs, have pushed the field forward so considerably. GANs improve temporal consistency; diffusion models, in addition to improving motion control and diversity, bring semantic understanding to the generation process within Video-LLMs. However, dataset quality remains a meaningful bottleneck. Datasets such as WebVid-10 M or Panda70 M include large quantities but suffer from low fidelity, together with weak text alignment. VidGen-1 M has improved captions, yet coherence and consistency remain issues.

B.-E. Boudriki Semlali et al. (Eds.): DATA 2025, LNNS 1908, pp. 370–382, 2026.
https://doi.org/10.1007/978-3-032-21200-9_31

This paper reviews most of the current state regarding text-to-video generation. It seeks to answer some of the following questions:

- How have GANs, diffusion models, and Video-LLMs addressed key challenges in T2V generation?
- What are the major dataset bottlenecks, and how do they impact model performance?
- How do several existing models compare against each other in respect of architecture, output quality, and temporal consistency?
- What are the remaining challenges and future research directions?

The remainder of this paper is organized as follows: Sect. 2 reviews key generative models used in text-to-video generation. Section 3 explores dataset challenges and quality evaluation criteria. Section 4 outlines ongoing challenges, future research directions, and concludes the study.

2 State-of-the-Art Review

The development of new and sophisticated architectures that integrate natural language processing and video rendering has led to significant progress in the area of text-to-video generation. These architectures focus on automatically converting text into videos by utilizing sophisticated text encoders and generative models. Among the various generative approaches, this review focuses more on GANs, diffusion models, and Video-LLMs, for these represent the leading models that are found in recent text-to-video generation research. These models were selected, of course, for their complementary strengths as well as in addressing key challenges, for instance, temporal coherence, motion realism, and semantic alignment. GANs offer some advantages when maintaining consistency across frames by way of adversarial learning. Motion synthesis benefits from diffusion models' controllability and high fidelity. Video-LLMs incorporate large language models. Such integration enables a greater comprehension of textual prompts and improves semantic coherence in video outputs. These models typically operate within a two-stage pipeline, as illustrated in Fig. 1, consisting of a text encoder and a video generation module.

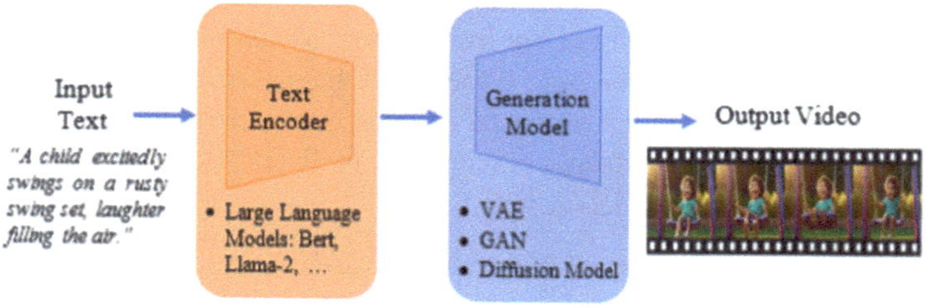

Fig. 1. Text-to-Video Generation Pipeline

2.1 Generative Adversarial Networks (GANs)

Core Mechanism

A GAN, or Generative Adversarial Network, is a machine learning architecture that involves two neural networks: a generator and a discriminator. GANs are commonly used for creating synthetic data that is both realistic and of high quality, and were first suggested by Ian Goodfellow and his associates in 2014 [1]. The primary goal of GANs is to capture the data distribution of the training dataset given to them. This is done by creating samples that mimic the training data as closely as possible. The generator and discriminator networks learn elastically through competition with one another, keeping them actively engaged in a game-like environment [1]. The generator portion creates fake data samples. Initially, these samples may be random and uninteresting. However, with enough training, the sample qualities will improve along with the level of real data resemblance. In contrast, the discriminator needs to identify fake data samples created by the generator from real data samples contained in the training dataset. With time, the discriminator learns to tell fake from the real with greater precision.

As illustrated in Fig. 2, the generator and discriminator engage in a feedback loop, driving the generation of increasingly realistic samples.

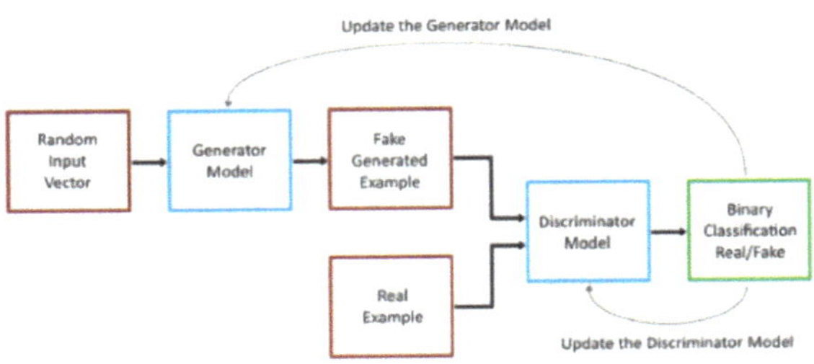

Fig. 2. GAN Architecture

Key Advancements in Video GANs

Although GANs were primarily created for image synthesis, using them for video generation has shifted things forward. Videos are collections of frames arranged in a specific order. Videos not only need high-quality frames but also require inter-frame coherence and motion smoothness. To overcome these difficulties, some sophisticated architectures, starting with **TGAN**, which proposed structured temporal modeling, next **MoCo-GAN,** which separated motion from content, and finally **DVD-GAN**, which used dual discriminators for enhancing video quality.

Temporal Generative Adversarial Networks (TGANs), proposed in 2017 [2], were one of the first attempts to create systems capable of automatic video generation. Instead of dealing with images individually, TGAN utilized a dual-generator model, which includes two separate generators to capture the temporal dependencies within the motion. The temporal generator first captures the dynamics of the motion at different time intervals into a latent variable sequence, and the image generator maps the extracted dynamics onto realistic frames. To help achieve smooth and coherent video sequences, TGAN uses the Wasserstein GAN (WGAN) framework, along with Singular Value Clipping (SVC), which makes training more stable and improves the quality of the samples. Unlike other models, TGAN can produce visually smooth and temporally coherent video sequences. Despite these advantages, TGAN is limited to low-resolution videos and struggles with long-range temporal consistency due to its reliance on latent space interpolation rather than explicit motion modeling.

In 2018, **MoCoGAN** [3] leaped forward in video GAN architectures by dividing motion and content into different latent spaces. This separation offered much more flexibility, as videos with fixed content and variable motion, or fixed motion and variable content, could now be created. Instead of encoding an entire video sequence, MoCoGAN encodes a single content vector for a video and several motion vectors for the frames. Another notable change brought by MoCoGAN greatly enhanced motion control by embedding an RNN into the generator for sequential motion modeling. Alongside, it also features two discriminators:

- One defines spatial fidelity through frame-by-frame deviation,
- The other ensures temporal coherence on a sequence level.

The development of **DVD-GAN** in 2019 [4] brought a more scalable method to video synthesis by adding a dual discriminator that enhances both spatial and temporal quality. DVD-GAN is built on the BigGAN architecture and utilizes:

- A spatial discriminator that checks the realism of a given frame.
- A temporal discriminator that checks the consistency of motion between several frames.

Comparative Analysis

To compare the capabilities of different GAN-based video generation models, Table 1 presents a detailed analysis of TGAN, MoCoGAN, and DVD-GAN, highlighting their strengths, limitations, and key architectural features.

Table 1. Comparison of GAN-Based Video Generation Models

Model	Strengths	Limitations	Key Features
TGAN	- Effective at capturing temporal coherence in videos. - Generates videos from unlabeled datasets.	- Limited control over individual frame content. - May struggle with longer video sequences due to architecture constraints.	- Dual-generator architecture (temporal and image generators).

(continued)

Table 1. *(continued)*

Model	Strengths	Limitations	Key Features
MoCoGAN	- Decomposes the latent space into motion and content for a controlled generation. - Unsupervised learning capability from unlabeled data.	- Complexity in managing two separate latent spaces can increase training time. - May require more computational resources for training due to RNN usage.	- Dual discriminators (image and video) for improved quality assessment.
DVD-GAN	- Efficiently handles long and high-resolution videos by reducing pixel processing requirements. - State-of-the-art performance on benchmark datasets like Kinetics-600.	- Requires significant computational resources, especially for high-resolution outputs. - Complexity in architecture may lead to longer training times compared to simpler models.	- Dual discriminator framework (spatial and temporal).

2.2 Diffusion Models

Core Mechanism

Diffusion models represent a class of generative models for the synthesis of data by way of progressive denoising from an initial random input. This mirrors thermodynamic systems, which shift from states of high entropy (disorder) to low entropy (structured data) [5]. The framework overall has two principal phases, and they are forward diffusion and reverse diffusion (Fig. 3).

Forward Diffusion

- During this stage, structured input data (e.g., an image or video frame) is corrupted by the addition of Gaussian noise, across many discrete time steps, in a systematic manner [6].
- At essentially each time step t_t, the noisy data x_t stems from the prior step x_{t-1} via a set noise schedule.

Reverse Diffusion

- A neural network—most often just a U-Net architecture [7]—is trained to learn the reverse process through estimating as well as removing the noise added during the forward phase [6]. The denoising is executed iteratively, beginning around pure noise and gradually reconstructing some original data.

- A loss function drives this process, reducing differences between predicted noise and real noise [6].

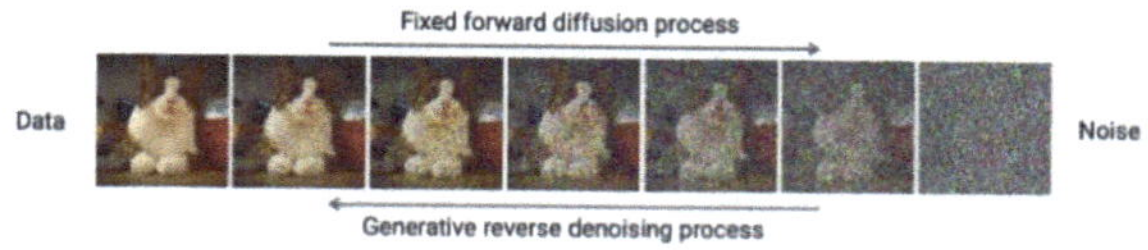

Fig. 3. Diffusion Model Architecture

Diffusion models are recognized for their capacity to generate outputs of high quality and diversity by leveraging this iterative denoising mechanism. Diffusion models use a maximum likelihood-based training objective, which helps give more stability as well as robustness during training [5, 6]. Generative Adversarial Networks (GANs) depend on adversarial training and often encounter issues such as mode collapse and unstable convergence. Video Diffusion Models (VDMs) [9] and similar adaptations were introduced to handle motion consistency and temporal dynamics explicitly for video synthesis.

Key Advancements in Video Diffusion Models

The application of diffusion models to video generation has eased meaningful progress in addressing longstanding challenges, in addition to temporal coherence, high-resolution rendering, also motion synthesis. Video diffusion models benefit from stable, step-wise refinement of noisy inputs into temporally consistent video sequences, as opposed to GAN-based methods. GAN-based methods frequently suffer from instability, though.

Ho et al. (2022) [8] proposed Video Diffusion Models (VDMs) as quite a foundational framework for such text-to-video generation. VDM extends the customary diffusion framework to video data by incorporation of novel mechanisms for modeling motion and structure over a time. Key innovations include:

- 3D U-Net Architecture: VDM replaces the existing 2D U-Net with a 3D variant able to capture spatial and temporal dependencies. This enables the model to process complete video sequences instead of isolated frames [10].
- Learnable Noise Scheduling: Different from image-focused models since they rely on static noise schedules, VDM uses adaptive noise variance across time steps, improving motion coherence and training stability [11].
- Classifier-Free Guidance: This technique enables text-based conditioning without the need for explicit class labels, thus improving alignment with input prompts and controllability [12].

A further outstanding contribution is **Imagen Video**, introduced in the year 2022 by Google [13]. This model extends the Imagen framework for video generation via a cascaded diffusion strategy, which systematically refines outputs of low quality into videos of high fidelity.

- Cascaded Diffusion Pipeline: It functions as a Temporal Super-Resolution model, beginning with a coarse 24 × 48 "storyboard." It progressively improves it to resolutions as high as 1280 × 768 as well, all while maintaining temporal continuity [13].
- Text Conditioning via T5-XXL: The model integrates into it a T5-XXL language model so as to semantically align generated video content with input descriptions, thereby improving on the relevance and interpretability of the outputs.

Make-A-Video, which was developed by Meta AI in 2022 [14], represents one of the very earliest open-domain text-to-video diffusion frameworks. In contrast to models trained directly upon video data, Make-A-Video extends existing pre-trained image or IP-based diffusion models toward generation of temporally coherent videos. By a focus on adapting image-based generative models to video synthesis, it achieves video outputs of high-quality without any requirement of large-scale video datasets.

The model does incorporate a few key modifications to the customary diffusion architectures so as to support video generation.

- Latent Space Modeling: Instead of operating directly into the pixel domain, Make-A-Video performs synthesis inside a latent space. This approach considerably improves computational efficiency as well as reduces the training costs associated with video generation.
- Unsupervised Video Learning: Unlike a number of text-to-video models that depend on paired video-text datasets, Make-A-Video leverages some image-based priors. This helps it to learn motion dynamics without video supervision that is explicit, which enables unsupervised learning of temporal patterns.

Comparative Analysis

To better understand the differences between leading text-to-video models, Table 2 provides a comparative analysis of Make-A-Video, VDM, and Imagen Video, highlighting their resolution, training data, strengths, and limitations.

Table 2. Key Differences Among Diffusion Models

Metric	Make-A-Video	VDM	Imagen Video
Resolution	256 × 256	256 × 256	1280 × 768
Text Encoder	DALL·E 2	CLIP	T5-XXL
Training Data	LAION + Shutterstock	WebVid-10 M	LAION + Proprietary
Strengths	Learns motion without video data, efficient training	First diffusion-based video model	High-resolution synthesis, strong text-video alignment
Limitations	Struggles with long-range motion coherence	Computationally expensive	Slow multi-stage inference

2.3 Video-LLMs

Video-LLMs stand for a new type of model, one that has large language models (LLMs) joined with video synthesis, working on some tough issues for text-to-video (T2V) creation. These models go beyond customary approaches, in aligning textual inputs with video content as well as preserving temporal consistency across frames. Video-LLMs improve the coherence of the generated video sequences. These Video-LLMs leverage the semantic capabilities of pre-trained LLMs to improve fluency. Key improvements offered within Video-LLMs include the following:

- Accurate Text-Video Alignment: Video-LLMs use cross-modal attention mechanisms [15] to keep textual prompts tightly aligned to visual output. CLIP-ViT [16], among adaptations of CLIP models from image to video domains, supports effective reuse of pre-trained image-text models. Such reuse enables more strong and scalable learning.
- Spatiotemporal Coherence: These models use certain memory-augmented transformer architectures [17], which allow for the modeling of long-range dependencies in both space and time. As a result of it, they reduce issues such as temporal flickering that are commonly observed in GAN-based video generation.
- Interactive Video Generation: One of the hallmark features of Video-LLMs, especially in variants such as Video-ChatGPT [18], exists in their ability to support dynamic, as well as interactive, control across video synthesis. For instance, users can issue textual commands in real-time (e.g., "change the background to a sunset") to modify visual elements, offering a degree of interactivity beyond conventional models.
- Enhanced Efficiency and Adaptability: Recent advancements like Free-Video-LLM [17] have shown improvements so substantial in efficiency, and it reduces training costs by up to 60% through parameter-efficient adapters. Furthermore, the ability for one to perform zero-shot adaptation through models such as DALL·E 2 [19] lowers dependency upon large, paired text-video datasets, streamlining the process for video generation.

2.4 Analysis

The most advanced technology in video generation using text has progressed significantly in creating customized and lifelike videos. This specific area, however, remains quite challenging in achieving optimal results in the coherence of the video's timeline, the stability of movements, and general efficiency in computation. Although GANs enable rapid video creation, they falter when long-term consistency is crucial. Diffusion models, in contrast, yield a stable and high-quality output but have massive resource requirements. Video-LLMs further the alignment and controllability of text and video, but the high resource consumption renders real-time usage impossible. These models were selected as they represent the most influential as well as widely adopted approaches in recent text-to-video systems, each addressing the generation task from a distinct perspective. A comparative analysis regarding features, strengths, as well as weaknesses of AI models for text-to-video generation can be found in Table 3.

Table 3. Feature Comparison of AI Text-to-Video Models

Feature	GANs	Diffusion Models	Video-LLMs
Core Concept	Adversarial training between the generator and the discriminator.	Iterative denoising of noisy data.	Combines large language models (LLMs) with video generation.
Architecture	Two networks: generator and discriminator.	Markov chain with forward (noise addition) and reverse (denoising) processes.	Hybrid models (e.g., 3D U-Net + LLMs) with temporal attention.
Strengths	Speed, high-quality single-frame outputs.	Stability, diverse outputs, and better motion consistency.	Strong text-video alignment and controllability.
Weaknesses	Struggles with motion consistency and limited long-range coherence.	Slow inference, expensive training.	Computationally expensive, long-video consistency issues.

It compares the three main generative models for text-to-video creation. GANs remain quite appealing, thanks to their speed and simplicity. However, these face certain limitations when generating temporally coherent sequences. Diffusion models do excel at the generation of stable outputs that are detailed, but they are constrained by computational demands. Video-LLMs stand out from others, given their meaningful semantic alignment with prompt controllability; however, their scalability across longer videos remains quite a challenge. This comparison highlights certain trade-offs between current model families and the need for future architectures that better integrate their strengths.

2.5 Emerging Models and Future Trends

Beyond customary GANs, diffusion models, and Video-LLMs, more scalable and instruction-tuned approaches have recently advanced text-to-video generation. For instance, Sora [20] shows the capability to generate minute-long high-fidelity videos from suitably complex prompts, likely leveraging latent diffusion and transformer-based models. VideoPoet [21] unifies video, audio, as well as text generation using pre-trained large transformers, marking a move toward instruction-driven multimodal synthesis. Meanwhile, models such as UVideo and Dream Machine [22] focus on improved frame consistency and resolution while also reducing the inference time, addressing practical deployment challenges. These developments highlight a trend toward hybrid architectures. Those architectures merge reasoning from LLMs with visual generation, pushing the boundaries of T2V systems.

3 Role of the Dataset

Both the diversity and quality of datasets are crucial to the progression of video generation via text prompts. There is an extra layer of complexity added to video datasets compared to image datasets, such as ensuring temporal coherence, motion phenomena, and the syncing of the text and the video. Developing high-quality video datasets that maintain semantic fidelity is paramount in training generative models and the tasks that they are intended to accomplish, which is to coherently craft video content. There has been a notable advancement in text-to-image generation; however, creating text-to-video models is still constrained by the lack of extensive and adequately annotated datasets. Evaluating the quality of datasets is necessary for ensuring their suitability for training and benchmarking generative models. General criteria include accuracy, completeness, consistency, timeliness, as well as validity, which have been used for a long time to assess data sources [23]. In the context of text-to-video generation, certain domain-specific factors are especially important. These include text-video alignment, where captions meaningfully correspond to the video content; visual fidelity and temporal consistency, important for learning realistic motion dynamics; and caption richness, ensuring textual descriptions convey sufficient semantic depth. Provenance and ethical documentation practices further improve dataset transparency. Such practices increase dataset trustworthiness too [24]. These criteria provide a practical framework for the analysis and comparison of recent large-scale datasets such as WebVid-10 M, Panda-70 M, OpenVid-1 M, and VidGen-1 M, as discussed in Table 4.

Table 4. Comparison Table of Video Datasets

Dataset	Year	Size	Resolution	Key Features	Strengths	Weaknesses
WebVid-10 M [25]	2021	10 M text-video pairs	Variable	Text-video pairs, scraped from Shutterstock	Large-scale, foundational for models like Make-A-Video.	Low-quality videos with watermarks and brief captions.
Panda-70 M [26]	2024	70 M text-video pairs	High resolution ($\geq$512p)	High-resolution videos collected from diverse sources	Large-scale, high-resolution clips.	Contains static/flickering videos, short captions; impractical for most research.
OpenVid-1 M [27]	2024	1 M+ text-video pairs	512 × 512 to 1080p	Curated for high aesthetics, temporal consistency, and expressive captions	High-quality, HD videos; detailed captions via LLaVA-v1.6-34b.	Smaller than Panda-70 M; requires careful filtering.

(*continued*)

Table 4. *(continued)*

Dataset	Year	Size	Resolution	Key Features	Strengths	Weaknesses
VidGen-1 M [28]	2024	1 M text-video pairs	High resolution	Coarse-to-fine curation strategy ensures high-quality videos and detailed captions with excellent temporal consistency.	High-quality video-text pairs with good temporal consistency.	Relatively new, so it may still be undergoing updates and improvements.

To assess the effectiveness of OpenVid-1 M, key evaluation metrics were compared with its performance with other large-scale video-text datasets. As can be depicted in Fig. 4, OpenVid-1 M demonstrated superior results in VQA, Blip-BLEU, and in SD Score assessments, outperforming WebVid-10 M and Panda-50 M. Notably, OpenVid-1 M achieved the highest Blip-BLEU score, indicating a strong alignment between textual descriptions in addition to visual content. Also, its performance in VQA and SD Score generally highlights the dataset's capability. It can support the generation of semantically rich and visually coherent videos [27].

Resolution	Training Data	$VQA_A\uparrow$	$VQA_T\uparrow$	Blip_bleu↑	SD_score↑	Clip_temp_score↑	Warping_error↓
256×256	WebVid-10M (Bain et al. 2021)	13.40	13.34	23.45	67.64	99.62	0.0138
256×256	Panda-50M (Chen et al. 2024b)	17.08	9.60	24.06	67.47	99.60	0.0200
256×256	OpenVid-1M (Ours)	17.78	12.98	**24.93**	67.77	99.75	0.0134
1024×1024	WebVid-10M (4× Super-resolution)	69.26	65.74	23.15	67.60	99.64	0.0137
1024×1024	Panda-50M (4× Super-resolution)	63.25	53.21	23.60	67.44	99.57	0.0163
1024×1024	Panda-50M-HD	13.48	42.89	21.78	**68.43**	99.84	0.0136
1024×1024	OpenVidHD-0.4M (Ours)	**73.46**	**68.58**	23.45	68.04	**99.87**	**0.0052**

Fig. 4. Comparison with previous representation text-to-video training datasets [27].

VidGen-1 M can serve in the capacity of a benchmark dataset that is thorough for text-to-video generation. As depicted in Fig. 5, it features considerably longer, more descriptive captions, with an average text length (ATL) of 89.3 words. This promotes stronger alignment between textual prompts as well as video content. Additionally, the dataset maintains large visual quality, positioning it as a credible alternative to larger datasets such as Panda-70 M [28].

Dataset	Year	Text	Domain	#Videos	AVL	ATL	Res
MSVD (Chen & Dolan, 2011)	2011	Human	Open	2K	9.7s	8.7 words	-
LSMDC (Rohrbach et al., 2015)	2015	Human	Movie	118K	4.8s	7.0 words	1080p
UCF101 (Soomro et al., 2012)	2015	Human	Action	13K	7.2s	4.3 words	240p
MSR-VTT (Xu et al., 2016)	2016	Human	Open	10K	15.0s	9.3 words	240p
DiDeMo (Anne Hendricks et al., 2017)	2017	Human	Flickr	27K	6.9s	8.0 words	-
ActivityNet (Caba Heilbron et al., 2015)	2017	Human	Action	100K	36.0s	13.5 words	-
YouCook2 (Zhou et al., 2018)	2018	Human	Cooking	14K	19.6s	8.8 words	-
VATEX (Wang et al., 2019)	2019	Human	Open	41K	~10s	15.2 words	-
HowTo100M (Miech et al., 2019)	2019	ASR	Open	136M	3.6s	4.0 words	240p
YT-Temporal-180M (Zellers et al., 2021)	2021	ASR	Open	180M	-	-	-
HD-VILA-100M (Xue et al., 2022)	2022	ASR	Open	103M	13.4s	32.5 words	720p
Panda-70M (Chen et al., 2024)	2024	Auto	Open	70.8M	8.5s	13.2 words	720p
VIDGEN-1M (Ours)	2024	Auto	Open	1M	10.6s	89.3 words	720p

Fig. 5. Comparison of VidGen-1 M and other video-text datasets. "AVL" and "ATL" are abbreviations for "Average Video Length" and "Average Text Length" [28].

4 Conclusion

In this paper, we analyze how generative AI and LLMs are changing the field of text-to-video generation. We focus on generative models such as video GANs, diffusion models, and video-specific LLMs, while also considering the role of datasets, to state the most recent progress and unsolved problems in the field. While OpenVid-1 M and VidGen-1 M datasets have semantic and video quality improvements, some issues remain, especially in producing longer, contextually richer, and more diverse videos. Two major areas for future research include improving realism in motion and temporal consistency, which may be achieved through memory-augmented architectures and advanced motion priors for enhanced spatiotemporal integration, and developing standardized evaluation metrics that effectively assess the fidelity of text-video pairs in terms of motion, alignment, and semantics.

Addressing these challenges will be essential for building more scalable, perceptually realistic, and semantically coherent text-to-video generation systems.

References

1. Goodfellow, I.J., et al.: Generative adversarial nets. In: Proceedings of the 27th International Conference on Neural Information Processing Systems, pp. 2672–2680 (2014)
2. Saito, M., Matsumoto, E., Saito, S.: Temporal generative adversarial nets with singular value clipping. In: Proceedings of the IEEE International Conference on Computer Vision (ICCV), pp. 3050–3058 (2017)
3. Tulyakov, S., Liu, M.-Y., Yang, X., Kautz, J.: MoCoGAN: decomposing motion and content for video generation. In: Proceedings of the IEEE/CVF Conference on Computer Vision and Pattern Recognition (CVPR), pp. 1526–1535 (2018)
4. Clark, A., Donahue, J., Simonyan, K.: Efficient video generation on complex datasets. arXiv preprint arXiv, 1907.06571., Published as a conference paper at ICLR 2020 (2019)
5. Sohl-Dickstein, J., Weiss, E., Maheswaranathan, N., Ganguli, S.: Deep unsupervised learning using nonequilibrium thermodynamics. In: Proceedings of the International Conference on Machine Learning (ICML), pp. 2256–2265 (2015)
6. Ho, J., Jain, A., Abbeel, P.: Denoising diffusion probabilistic models. Adv. Neural Inf. Process. Syst. (NeurIPS). **33**, 6840–6851 (2020)

7. Ronneberger, O., Fischer, P., Brox, T.: U-net: convolutional networks for biomedical image segmentation. In: Medical Image Computing and Computer-Assisted Intervention (MICCAI), vol. 9351, pp. 234–241 (2015)
8. Ho, J., Chan, W., Saharia, C., et al.: Video diffusion models. Adv. Neural Inf. Process. Syst. (NeurIPS). **35**, 8633–8646 (2022)
9. Blattmann, A., Rombach, R., Ling, H., et al.: Align your patients: High-resolution video synthesis with latent diffusion models. In: Proceedings of the IEEE/CVF Conference on Computer Vision and Pattern Recognition (CVPR), pp. 22563–22575 (2023)
10. Yang, L., et al.: Video diffusion models with local-global context guidance. arXiv preprint arXiv:2306.02456. (2023)
11. Weng, L.: Diffusion models in vision: a survey. arXiv preprint arXiv:2210.13438. (2022)
12. Ho, J., Salimans, T.: Classifier-free diffusion guidance. arXiv preprint arXiv:2207.12598. (2022)
13. Ho, J., Salimans, T.: Imagen video: high-resolution video generation with cascaded diffusion models. arXiv preprint arXiv:2210.02303. (2022)
14. Meta AI: Make-a-video: text-to-video generation without text-video data. arXiv preprint arXiv:2209.14792. (2022)
15. Li, J., et al.: BLIP-2: bootstrapping language-image pre-training with frozen image encoders and large language models. arXiv preprint arXiv:2301.12597. (2023)
16. Radford, A., Kim, J.W., Hallacy, C., et al.: Learning transferable visual models from natural language supervision. arXiv preprint arXiv:2103.00020. (2021)
17. Wu, Z., et al.: Free-video-LLM: Prompt-guided visual perception for efficient video understanding. arXiv preprint arXiv: 2402.01234. (2024)
18. Xu, T., et al.: Video-ChatGPT: enhancing video understanding via large vision-language models. arXiv preprint arXiv:2403.06789. (2024)
19. Shang, Y., et al.: Memory-augmented video-LLMs for long-context video generation. arXiv preprint arXiv:2403.08978. (2024)
20. Liu, Y.: Sora: a review on background, technology, limitations, and opportunities of large vision models. arXiv preprint arXiv:2402.17177. (2024) Available at: https://arxiv.org/abs/2402.17177
21. Anil, R., et al.: VideoPoet: a large language model for zero-shot video generation. Proc. Mach. Learn. Res. **235**, 25105–25124 (2023) Available at: https://arxiv.org/html/2312.14125v4 and https://yonsei.elsevierpure.com/en/publications/videopoet-a-large-language-model-for-zero-shot-video-generation
22. Luma Labs. (2024). Dream Machine AI: Advanced Text to Video Generator. Available at: https://dreammachineai.app and https://replicate.com/luma/ray
23. Batini, C., Scannapieco, M.: Data and Information Quality: Dimensions. Springer, Principles and Techniques (2016). https://doi.org/10.1007/978-3-319-24106-7
24. Gebru, T., et al.: Datasheets for datasets. Commun. ACM. **64**(12), 86–92 (2021). https://doi.org/10.1145/3458723
25. Bain, M., Nagrani, A., Varol, G., Zisserman, A.: Frozen in time: A joint video and image encoder for end-to-end retrieval. arXiv preprint arXiv:2104.00650. (2021)
26. Chen, C., Luo, P., Lin, Z.: Panda-70M: Large-scale dataset for video understanding with high-quality annotations. arXiv preprint arXiv:2402.02645. (2024)
27. Tian, Y., Liu, Z., Chen, S., et al.: OpenVid-1M: high-quality video-text dataset for text-to-video generation. arXiv preprint arXiv:2407.02371. (2024)
28. Tian, Y., Chen, S., Liu, Z., et al.: VidGen-1M: a curated benchmark for text-to-video generation. arXiv preprint arXiv:2408.02629. (2024)

Energy Efficiency of RDF Storage Systems: A Comparative Study

Wafae El Kaysouni[1](✉), Ahmed Rabhi[2], and Hassan Badir[3]

[1] Data Engineering and system Team (IDS) - National School of Applied Sciences, Abdelmalek Essaadi University, Tangier, Morocco
wafae.elkaysouni@etu.uae.ac.ma
[2] LSIA laboratory, EMSI, Tangier, Morocco
a.rabhi@emsi.ma
[3] Data Engineering and system Team (IDS) - National School of Applied Sciences, Abdelmalek Essaadi University, Tangier, Morocco
badir.hassan@uae.ac.ma

Abstract. The energy efficiency of RDF storage systems is becoming an important issue with the growth of cloud-based technology. This paper compares different RDF storage approaches: Apache Jena TDB, OpenLink Virtuoso, RDF4J and GraphDB. It analyzes their execution time and energy consumption. The experimental analysis is based on the WatDiv benchmark, widely used for evaluating RDF storage systems due to its diverse query set and robust performance assessment. The findings indicate that two important variables that have a major effect on energy use are query complexity and the number of results returned. This study provides important insights into the performance variations across different RDF storage systems and highlights the possibility of energy optimization in a cloud setting.

Keywords: Energy efficiency · RDF · SPARQL · Query Performance · Data Storage

1 Introduction

For years, data has been the lifeblood of the digital world, and its importance has only grown with the increase in global connectivity. Today, large-scale data centers are the foundation of an information-driven society, powering cloud computing, the internet, and a wide range of online services. However, their immense energy consumption and resulting carbon emissions have brought serious environmental concerns and expensive costs.

The rapid expansion of social media, along with the growing demand for digital services, has led to an unprecedented need for advanced computing infrastructure to store and process vast amounts of information. This growth spans many industries, including healthcare, gaming, email, e-commerce and artificial intelligence. As data volumes continue to increase, leading cloud providers face the pressing challenge of

B.-E. Boudriki Semlali et al. (Eds.): DATA 2025, LNNS 1908, pp. 383–394, 2026.
https://doi.org/10.1007/978-3-032-21200-9_32

optimizing energy efficiency while maintaining performance. Better data organization and interoperability depends on RDF data storage systems, one of the newer approaches. Despite their apparent adoption in database and cloud contexts, their energy consumption is still not thoroughly studied.

This paper presents a study whose goal is attempting to assess and evaluate the energy efficiency of various RDF storage systems. Thanks to an experimental approach and specific indicators (execution time, energy consumption, and request efficiency), this study offers insights into the best solutions for the current cloud computing demands. Therefore, the goal is to strike a balance between durability and efficiency by using the best systems for managing and storing RDF data in an optical setting where energy optimization becomes crucial.

The structure of this paper is as follows: Sect. 2 introduces background and basic concepts. Section 3 presents related work. Materials and methods of the study are presented in Sect. 4. Experimental results are shown and discussed in Sect. 5. Finally, Sect. 6 concludes the paper.

2 Basic Concepts

The RDF (Resource Description Framework) data paradigm (T. Heath and C. Bizer [1]), which organizes information into triplets (Subject, Predicate, Object), is the foundation of linked data. A graph with predicates representing connections between various resources is called an RDF dataset. SPARQL (SPARQL Protocol and RDF Query Language), which defines triple patterns to look for in RDF graphs, is used to query this data.

The energy efficiency of a system refers to the relationship between its energy consumption and the output of service it provides. Energy, defined as the force in action, represents the total amount of work performed by a system over a specific period. It is calculated using the following Eq. (1):

$$E = P \times t \tag{1}$$

where E represents energy in joules, P power in watts, and t the duration in seconds.

The energy efficiency (EE) of a system is determined using the following Eq. (2):

$$EE = \frac{TE}{E} = \frac{t}{P} \tag{2}$$

Equation (2) allows for the assessment of a system's performance based on its power consumption and operating time. To obtain a precise measurement of energy consumption and ensure a reliable evaluation of energy performance, it is essential to use appropriate equipment, such as wattmeters and specialized sensors.

3 Related Work

Cloud computing is defined as the ability to access computer resources, such as processing power, programs, and data storage, through the internet. Despite advancements in service availability, flexibility, speed, and quality, the growing demand for digital services leads to an increase in energy consumption. Based on their research, A. katal and al [2]., underline the need to combine software approaches with optimized hardware solutions to improve the energy efficiency of data centers. Additionally, S. M. Mohsin and al [3]. point out that this increase in energy consumption in cloud data centers brings major issues, including high energy costs for cloud service providers and harmful carbon emissions for the environment.

In this context, data management and storage systems play a crucial role in reducing the energy consumption of cloud infrastructures, RDF stores widely used for storing and managing Semantic Web data are particularly concerned with these issues. Multiple researches have explored their performances, G. Savic [4] conducted a study to evaluate the performance of several RDF stores on servers with average hardware capabilities. He tested three RDF storage solutions: Jena-Fuseki, Sesam-OWLIM and Virtuoso analyzed CPU usage, memory consumption, and execution time. Similarly, the study of G. Atemezing and F. Amardeilh [5] compares the performance of four RDF stores, namely Virtuoso, GraphDB, Stardog, and Oracle using real datasets and SPARQL queries from Publications Office. From publications Office, this study highlighting the importance of choosing the RDF store according to the types of queries and server parameters. Energy optimization of databases is also a major challenge, as demonstrated by the PET (Power Efficiency Toolkit) project by Zichen Xu and al [6]., which aims to reduce energy consumption through query optimization techniques.

This question is more relevant in a context where cloud service providers play a central role in hosting RDF databases. These platforms offer interesting scalability but also pose challenges in terms of energy management, in particular due to dynamic resource allocation and virtualization. Furthermore, the evaluation of spatially enabled RDF stores revealed specific constraints for geospatial data, increasing the challenges related to performance optimization and energy consumption Huang [7].

The energy consumption of cloud infrastructures with relational or NoSQL data bases has been extensively researched. Nevertheless, these studies focus primarily on relational data bases, while RDF data bases used in semantic web applications have unique characteristics that affect their energy consumption due to use of SPARQL queries that allow querying these bases and necessitate complex query processing (M. Niinimaki and al [8]). A number of open source RDF storage systems have been developed to meet the needs of the web semantics. These include open-source solutions such as Apache Jena TDB, RDF4J, and Graphdb, as well as commercial systems like Virtuoso Universal Server, GraphDB Enterprise, and Amazon, which have the best management of RDF work charges in cloud environments.

These systems are evaluated using research-specific benchmarks like Lehigh University Benchmark LUBM, which simulates a group of universities with SPARQL complexes that include jointures and inférences. It is designed to assess the performance of RDF systems based on academic knowledge [9]. Waterloo SPARQL Diversity Benchmark WatDiv is a benchmark focused on the variety of SPARQL queries, allowing to assess performance under various work charges [10]. Berlin SPARQL Benchmark BSBM is used to test RDF performances in a commercial context, such as the management of product catalogs, reviews, and e-commerce transactions [11] and a number of other benchmarks. These benchmarks provide standardized methods for evaluating RDF storage systems under different workload scenarios, making them essential for assessing energy efficiency in both academic and commercial environments.

4 Materials and Methodology

Evaluating the energy efficiency of RDF storage systems is essential to determining the most optimal solutions for cloud providers semantic data management. The analysis of energy efficiency in RDF data storage systems involves a structured experimental approach, as illustrated in Fig. 1.

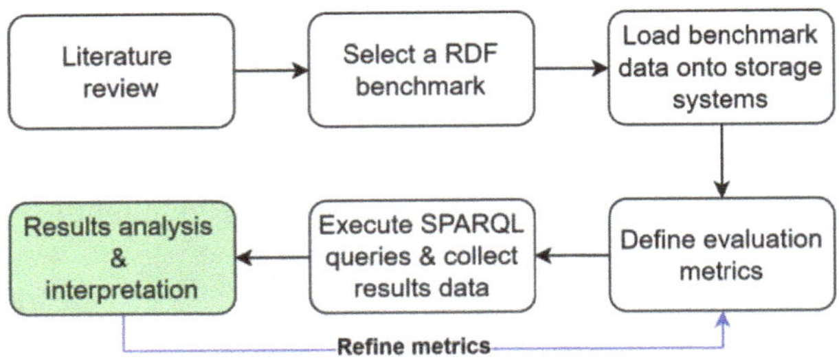

Fig. 1. Workflow for evaluating the energy efficiency of RDF storage systems

The workflow consists of the following steps:

- **Literature review:** Identify current approaches while recognizing the state of the art and collecting relevant benchmarks.
- **Benchmarks:** Selecting an RDF benchmark to standardize testing and evaluation.
- **Load the benchmark:** data onto the storage systems to be tested.
- **Evaluation metrics:** This study compares various open source systems using the following metrics in order to analyze them:
  - **Energy consumption:** to assess the overall power efficiency of each system.
  - **Query execution times:** to analyse effect of exect time on energy consumption.
  - **Query complexity:** to analyse effect of query complexity on energy consumption.
- **Execution and results visualization:** Executing SPARQL queries and collecting result data for analysis.
- **Analysis and interpretation:** Determined which storage system achieves the best balance between energy efficiency and performance, the collected metrics such as energy consumption, request complexity, and execution time are analyzed.

4.1 Storage Systems Analysis

RDF storage systems differ in terms of architecture, query performance, scalability, and energy efficiency, impacting their performance in different use cases. This comparison focuses on RDF4J, Virtuoso OpenLink, GraphDB, and Apache Jena TDB, highlighting their key features and functionalities.

- **Apache Jena TDB**[1]
  Jena is a non-native RDF storage system with a modular architecture based on property tables. Its SPARQL query execution is optimized thanks to the ARQ engine, which provides robust query management (M. Tiam and al [12]). Although it supports large RDF datasets, its performance is generally lower than that of OpenLink Virtuoso, particularly in terms of query execution and transaction management (N. Nikolic and al [4]).
- **OpenLink Virtuoso**[2]
  OpenLink Virtuoso is a non-native RDF storage system based on a triple or quadruplet architecture, optimized for SPARQL query execution thanks to a dedicated engine and advanced indexing mechanisms. It offers a hybrid approach, allowing both native RDF storage and the integration of relational databases exposed in RDF via SPARQL. Its clustering solution allows to manage massive volumes of data and improve the scalability of RDF storage thus reducing the performance gap with relational databases. Virtuoso also follows the evolutions of the SPARQL standard to strengthen data interoperability (O. Erling and I. Mikhailov [13]).
- **RDF4J**[3]
  RDF4J is also a non-native RDF storage system with a modular architecture based on property tables. Its SPARQL query execution is optimized with a dedicated SPARQL engine ensuring efficient query handling. RDF4J performs well in managing large RDF datasets supported by effective indexing techniques (M. Aldwairi and al [14]).
- **GraphDB**[4]
  GraphDB is a native RDF storage system with a modular architecture known for its specific optimizations dedicated to RDF storage and inference. It is based on RDF4J SAIL which gives it great flexibility and extensibility (M. Papakonstantinou and al [15]).

The different RDF data storage systems analyzed take varied approaches to efficiently organize, index, and query RDF triples, as presented in Table 1.

Table 1. Comparison of RDF Storage Systems

RDF System	Storage Method	Optimization
Apache Jena TDB	Stores data as triples/quads with a Node Table and optimized indexes (SPO, POS, OSP).	Compact storage using NodeIds, fast SPARQL query execution.
RDF4J	Flexible options:MemoryStore (in-memory), NativeStore (on-disk), ElasticsearchStore (Elasticsearch-based).	Flexible and extensible with support for reasoning and SHACL validation.

(*continued*)

[1] Apache Jena Documentation: https://jena.apache.org/documentation/
[2] OpenLink Virtuoso Documentation: https://docs.openlinksw.com/virtuoso/
[3] Eclipse RDF4J Documentation: https://rdf4j.org/documentation/
[4] GraphDB Documentation: https://www.ontotext.com/products/graphdb/

Table 1. *(continued)*

RDF System	Storage Method	Optimization
GraphDB	Based on RDF4J SAIL with optimization: Entity Pool, POS/PSO indexes, and Trree inference engine.	Optimized for inference and performance through the Trree engine.
OpenLink Virtuoso	Hybrid storage combining relational database and RDF, with HTTP hyperlink-based indexing.	SQL/SPARQL interoperability, integration with relational databases.

4.2 Experimental Setup

A. **Hardware Environment**

The evaluations were performed on a machine with the following specifications: an AMD Ryzen 7 3700 U processor with Radeon Vega Mobile Gfx, clocked at 2.30 GHz, as well as a 512 GB SSD. These hardware specifications served as the basis for our experiments, ensuring a stable and reliable test environment.

Energy measurement is a critical aspect of our experiment to assess the energy consumption of the semantic data processing systems. To perform accurate measurements, we employed the Yocto Watt Usage device in combination with the Yoctopuce library for Python[5]. Key features of the Yocto Watt Usage device include:

- **Accuracy:** The device provides accurate power consumption measurements with a high degree of precision. Its accuracy is 1 mWh with a margin of error of 1%.
- **USB Connectivity:** It connects to the host machine via USB, making it easy to integrate into our energy measurement setup.
- **Real-Time Monitoring:** The Yocto Watt Usage device continuously monitors power consumption in real-time.

The experimental setup of the benchmarking process and the energy consumption measurement technique are summarized in Fig. 2.

[5] https://www.yoctopuce.com/FR/products/capteurs-electriques-usb/yocto-watt

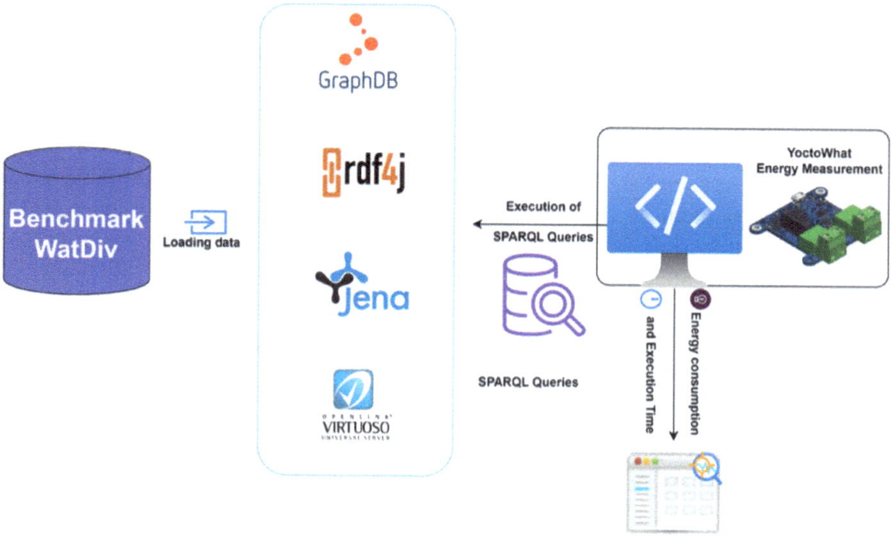

Fig. 2. Experimental Setup for Energy and Performance Evaluation of RDF Storage Systems

B. Software Environment

- **RDF storage systems:**

  For the experiment several RDF stores were used. Virtuoso was tested in its version 08.03.3332, while RDF4J was running in version 5.0.2. GraphDB in its version 10.7.4 integrated RDF4J 4.3.13. Finally, the evaluation of Apache Jena TDB was carried out using Apache Jena Fuseki 5.3.0. These different versions were selected to ensure optimal compatibbility of the tests in a comparative analysis framework.
- **Benchmark:**

  The WatDiv benchmark was developed to measure the performance of RDF data management systems by executing SPARQL queries with various structural characteristics and different selectivity classes. It also allows to generate data using a dedicated generator, thus providing control over the included entities, their structuring and other features.

  Furthermore, WatDiv differs from existing benchmarks by its ability to generate test datasets of different sizes, where instances of the same entity do not necessarily have the same set of attributes (G. Aluç and al [10]).

  The dataset contains 10 million triples, with a total size of 58.558.746 bytes. It includes 521.699 subjects, 92 predicates and 1.005.941 objects.
- **Queries:**

  For the evaluation, 6 SPARQL queries[6] were used, presenting different complexities depending on the number of triple patterns distributed as follows: 1, 3, 6 and 12 triple patterns. Each query will be executed in three separate phases with a limit restriction applied to the total number of results returned in each step.

[6] https://github.com/el-kaysouni-wafae/WatDiv-sparql-queries/

To evaluate performance on a minimum number of results returned, a limit of one result is specified in the first phase. This limit is then increased to 50.000 in the second phase so that the effects of an intermediate volume of results can be examined. The third phase sets the limit to 100.000 and comes last.

Each run was performed five times to guarantee the accuracy of the results, and the average of the results was calculated to account for any possible execution environment-related variances. By limiting the influence of outside variables and variations in hardware performance, this technique improves measurement accuracy.

5 Experimental Results Analysis and Discussion

This section analyzes the impact of query complexity and result size on the energy consumption and execution time of RDF storage systems. The results highlight the trade-offs between system performance and energy efficiency.

5.1 Effect of Query Complexity on Energy Consumption

The impact of query complexity was analyzed by executing SPARQL queries that returned approximately the same number of results but contained different numbers of triple patterns: 1, 3, and 12. Figure 3, illustrate the energy consumption and execution time for increasing query complexity across different RDF storage systems.

These findings show that execution time and energy consumption both increase significantly with the number of triple patterns. Although GraphDB and RDF4J continuously beat the other systems in terms of efficiency, Virtuoso has the highest energy consumption.

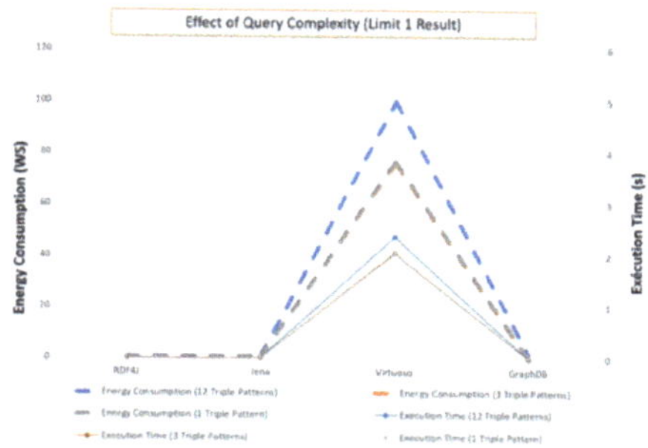

(a) Limit 1 Result

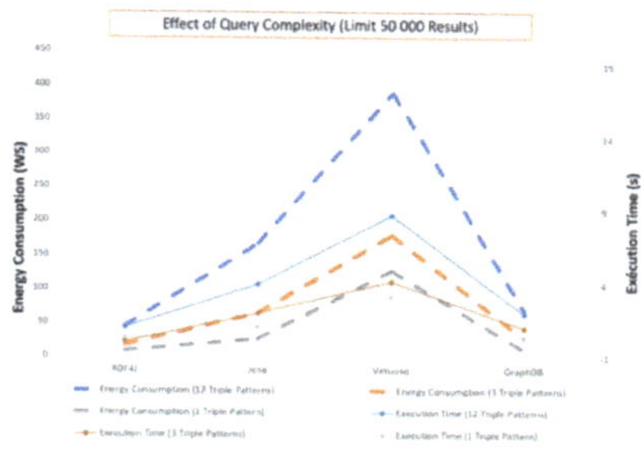

(b) Limit 50 000 Results

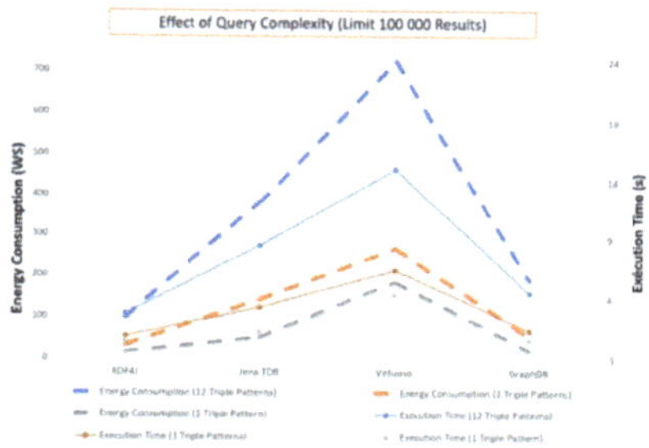

(c) Limit 100 000 Results

Fig. 3. Effect of Query Complexity on Execution Time and Energy Consumption

5.2 Effect of Result Size on Energy Consumption

We executed the same queries with varying output size limitations to examine the impact of the quantity of results returned on the consumption of energy. Figure 4 show these consequences for multiple queries. The values in Fig. 4 demonstrate that as the size of the results increases, so does the energy consumption. This tendency is especially obvious in Virtuoso, which uses a lot more energy when delivering larger result sets. However, GraphDB and RDF4J show better efficiency and maintains a more consistent energy consumption.

5.3 Results Discussion

Significant differences in RDF storage systems' energy consumption and execution time are revealed by the analysis of the results. With lower energy consumption and faster execution times than the other examined systems, RDF4J emerges as the most efficient and energy saving system. However, Virtuoso seems to be the least efficient because it consumes much more energy and takes longer to execute, particularly for complex queries and large result size.

The results show that the performance of RDF systems is impacted by query complexity. As the number of triple patterns increases, energy consumption and execution time also rise. Virtuoso and Jena show a disproportionate increase in execution time and energy consumption when processing complex queries. However, in terms of energy usage, RDF4J and GraphDB both show more consistency, with RDF4J performing slightly better than GraphDB.

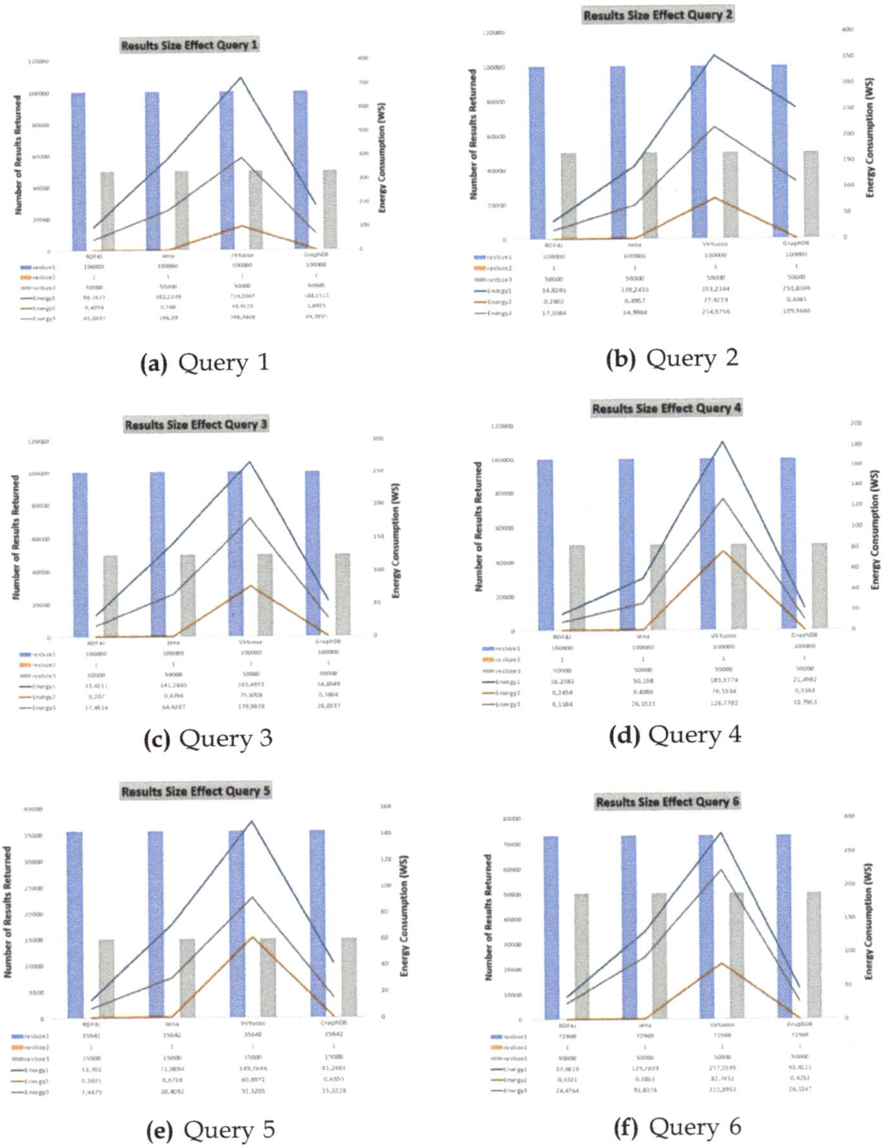

(a) Query 1 **(b)** Query 2

(c) Query 3 **(d)** Query 4

(e) Query 5 **(f)** Query 6

Fig. 4. Effect of Result Size on Energy Consumption for Different Queries

The size of the returned results is the second significant factor influencing RDF system performance. Increasing the number of results a query produces results in higher energy consumption and a more pronounced decline in Virtuoso performance. This is due to the high processing and temporary storage expenses of its SPARQL execution engine. GraphDB and RDF4J, on the other hand, have more constant consumption and better energy management, which makes them appropriate for applications that require extensive RDF data retrieval.

There are some limitations on this study. It uses a single benchmark dataset and depends on a small number of RDF storage systems, which may not adequately represent the variety of real-world situations. In addition, it was challenging to determine how much energy was used by each storage engine because energy consumption was measured at the system level. These limitations should be considered when interpreting the scope and applicability of the results.

6 Conclusion

This study demonstrates the impact of query complexity and its result size on the energy consumption of RDF storage systems. The findings suggest that the selection of an RDF storage system should strike a balance between query performance and energy efficiency depending on the specific requirements of the intended application. The performance analysis indicates that RDF4J is the most efficient and effective system followed by GraphDB. In contrast, Virtuoso proves to be the least suitable, exhibiting high energy consumption and prolonged execution times especially for complex queries and large result sets. Similarly, Jena is not well suited for intensive processing. These findings emphasize the importance of considering application-specific needs when selecting an RDF storage system, taking into account the trade-off between performance and energy efficiency. The use of a single benchmark dataset, the small number of RDF storage systems examined, and the measurement of energy consumption at the system level rather than at a more precise granularity are some of the study's primary limitations. These limitations highlight the need for more thorough research. In order to address these issues, future studies should use more benchmark datasets and systems and more precise monitoring of energy techniques that can separate the consumption of each component. In addition, the development of enhanced indexing mechanisms, adaptive query execution strategies, and energy-aware optimization techniques—particularly in cloud-based and large-scale deployment contexts—could help advance energy efficiency in RDF databases.

References

1. Heath, T., Bizer, C.: Linked data: evolving the web into a global data space. Synth. Lect. Semant. Web Theory Technol. **1**(1), 1–136 (2011)
2. Katal, A., Dahiya, S., Choudhury, T.: Energy efficiency in cloud computing data center: a survey on hardware technologies. Clust. Comput., 1–31 (2021). https://doi.org/10.1007/s10586-021-03431-z
3. Mohsin, S.M., Maqsood, T., Madani, S.A.: Towards energy efficient cloud: a green and intelligent migration of traditional energy sources. Energies. **17**(11), 2787 (2024). https://doi.org/10.3390/en17112787

4. Savic, G., Nikolic, N., Segedinac, M., Gostojic, S., Konjovic, Z.: RDF stores performance test on servers with average specification. In: 2015 ICIST Proceedings of the 5th International Conference on Information Science and Technology, pp. 1–6 (2015). https://doi.org/10.13140/RG.2.2.12489.54887
5. Atemezing, G.A., Amardeilh, F.: Benchmarking commercial RDF stores with publications office dataset. In: Gangemi, A., et al. (eds.) The Semantic Web: ESWC 2018 Satellite Events. LNCS, vol. 11155, pp. 54–64. Springer, Cham (2018). https://doi.org/10.1007/978-3-319-98192-5_54
6. Xu, Z., Tu, Y.C., Wang, X.: PET: reducing database energy cost via query optimization. Proc. VLDB Endow. **5**(12), 1954–1957 (2012). https://doi.org/10.14778/2367502.2367546
7. Huang, W., Raza, S.A., Mirzov, O., Harrie, L.: Assessment and benchmarking of spatially enabled RDF stores. ISPRS Int. J. Geo Inf. **8**(6), 266 (2019). https://doi.org/10.3390/ijgi8070310
8. Niinimaki, M., Chanyalikit, K., Udofia, A.: Measuring the energy efficiency of RDF query processing. KBU Eng. J. **8**(2), 1–10 (2024)
9. Guo, Y., Pan, Z., Heflin, J.: LUBM: a benchmark for OWL knowledge base systems. J. Web Semant. **3**(2), 158–182 (2005). https://doi.org/10.2139/ssrn.3199255
10. Aluç, G., Hartig, O., Özsu, M.T., Daudjee, K.: Diversified stress testing of RDF data management systems. In: Proc. ISWC 2014, LNCS, vol. 8796, pp. 197–212. Springer, Cham (2014). https://doi.org/10.1007/978-3-319-11964-9_13
11. Bizer, C., Schultz, A.: The Berlin SPARQL benchmark. Int. J. Semant. Web Inf. Syst. **5**(2), 1–24 (2009). https://doi.org/10.4018/jswis.2009040101
12. Tian, M., Wang, H., Zhang, X.: SPARQL query optimization for structural indexed RDF data. J. Comput. Sci. Technol. **35**(2), 428–443 (2020)
13. Erling, O., Mikhailov, I.: RDF support in the virtuoso DBMS. In: de Moor, A., Polleres, W., Declerck, T. (eds.) Semantic Web Information Management, pp. 501–519. Springer, Berlin, Heidelberg (2009). https://doi.org/10.1007/978-3-642-04329-1_21
14. Aldwairi, M., Jarrah, M., Mahasneh, N., Al-khateeb, B.: Graph-based data management system for efficient information storage, retrieval and processing. Inf. Process. Manag. **59**(4), 103165 (2022). https://doi.org/10.1016/j.ipm.2022.103165
15. Papakonstantinou, M., Karvounis, M., Stoitsis, G., Manouselis, N.: Deploying a scalable big data platform to enable a food safety data space. In: Data Spaces, pp. 227–248. Springer, Cham (2022). https://doi.org/10.1007/978-3-030-98636-0_11

Smart Mobility and Sustainable Infrastructure

Synergy of AI and Sustainability: A CPSS-Based Framework for Integrating Digital Innovations into Mobility as a Service (MaaS)

Ali Aghazadeh Ardebili[1(✉)], Mostafa Ghodsi[2(✉)], Fahimeh Golbabaei[3], Mahdad Pourmadadkar[4], and Lorenzo Miozzi[1]

[1] R&D Group, HSPI SpA, Bogogna, Italy
ali.a.ardebili@unisalento.it
[2] Faculty of Civil and Environmental Engineering, Tarbiat Modares University, Tehran, Iran
Mostafa.ghodsi@modares.ac.ir
[3] Transport Engineering, Queensland University of Technology, Brisbane, Australia
[4] Department of Innovation for Engineering, University of Salento, Lecce, Italy

Abstract. In recent years, the push for sustainable transportation has gained significant attention due to the urgent need to reduce emissions, improve energy efficiency, and encourage public transport use. One promising concept emerging in this domain is Mobility as a Service (MaaS), which aims to seamlessly connect traditional transport systems with alternative mobility options through digital platforms. By integrating various transportation services and making multimodal travel more accessible, MaaS has the potential to reshape mobility and reduce reliance on private cars. Moreover, MaaS transforms the transformation system into a Cyber-Physical-social System (CPSS), which facilitates bringing AI-driven tools and cutting-edge digital technologies such as AI-enabled optimization, digital twins, and real-time anomaly detection into action. Since MaaS is an emerging concept, there is a significant gap in the concept, implementation, and potential impact of MaaS and the smart solutions that MaaS can provide to promote eco-friendly travel choices. This paper explores the enabler technologies of MaaS, MaaS frameworks, AI-driven innovations, implementation strategies, stakeholders' impact on success, and regulatory challenges. It proposes a framework for smarter, and more efficient urban transport solutions while encouraging green transport/travel choices based on an integrated platform.

Keywords: Mobility as a Service (MaaS) · Cyber-Physical Transport System · Cyber-Physical-Social System (CPSS) · AI-driven mobility · Smart Transportation · Integrated Mobility Platforms · Sustainable Transportation

B.-E. Boudriki Semlali et al. (Eds.): DATA 2025, LNNS 1908, pp. 397–408, 2026.
https://doi.org/10.1007/978-3-032-21200-9_33

1 Introduction

Transport connectivity and mobility have been central to ambitious international targets for decades. Over the last fifteen years, the EU transport policies have evolved, marked by a shift towards sustainable mobility, liberalization of the transport market, and adoption of technological innovations. Along these lines, in 2019, the Commission Communication on the European Green Deal (2019) set the climate goals and legislative proposals to decarbonize the transport sector. It proposed to reduce GHG transport emissions by at least 55% by 2030, compared to 1990 levels, and by 90% by 2050 to achieve carbon neutrality (2019).

To address the climate crisis, together with other urgent sustainability questions, emphasize the need for a fast, radical, and pervasive change within the transport sector.

Mobility as a Service (MaaS), as a relatively new concept, has gained attraction in the transportation field. Although definitions may vary, MaaS mainly seeks to integrate already established transportation infrastructures, with alternative mobility services and consolidate the digital platforms that traditionally operate these services in a disparate manner into a seamless MaaS framework.

To effectively implement MaaS and realize its full potential for decarbonization and improved mobility, a holistic approach is required. This article proposes a framework for MaaS development and deployment within the context of a CyberPhysical-Social System (CPSS). While traditional Intelligent Transportation Systems (ITS) often focus on the interplay between the cyber (data, algorithms) and physical (infrastructure, vehicles) domains, a CPSS perspective explicitly incorporates the social dimension Xiong et al. (2015). This includes user behavior, social equity considerations, policy and regulatory frameworks, and the broader societal impacts of transformative mobility solutions. This study provides a roadmap for turning datadriven insights into actionable strategies. This framework aims to assist researchers, practitioners, and policymakers in creating MaaS solutions that genuinely meet the needs of a diverse and evolving society, contributing significantly to the achievement of ambitious climate goals.

2 Critical Synthesis of Background

This section provides a critical synthesis of the reviewed literature of MaaS and CPSS with the intention of establishing the theoretical grounding for applying a CPSS framework to address the complexities of modern MaaS implementation.

2.1 MaaS Core Concepts

MaaS is considered as a revolutionizing urban transport service model that integrates diverse mobility means into a single service Zhao et al. (2023). While initially defined by Heikkilä (2014) around operator provision, the term “MaaS” gained attention during the European Congress on Intelligent Transport Systems in Helsinki in 2014, where it was defined as "a system in which a comprehensive range of mobility services are provided to customers by mobility operators Mobility as a Service the concept rapidly evolved".

Integrating the aforementioned descriptions, MaaS can be conceptualized as a service type that, via a unified digital channel (see Fig. 1), allows users to plan, book, and pay for various mobility services. However, a purely technological view is increasingly seen as insufficient, considering MaaS's potential (along with related challenges) in influencing travel behavior, promoting shared mobility, and contributing to sustainability goals Milne et al. (2024); Aba and Esztergár-Kiss (2024).

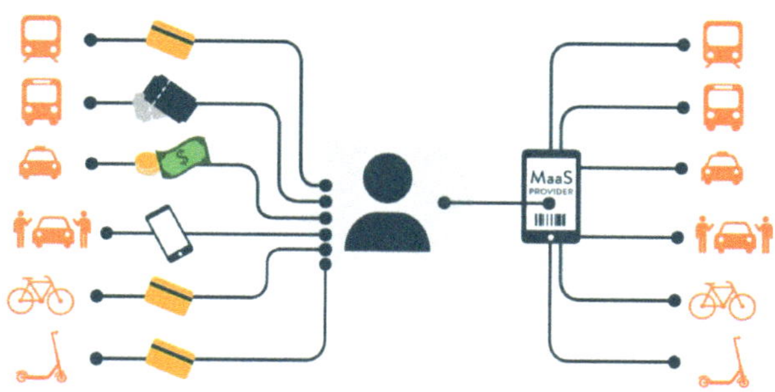

Fig. 1. Concept of Mobility as a Service

The MaaS literature often focuses on business models, potential impacts, and theoretical applications Milne et al. (2024) and the aggregation of personal mobility services offered by various providers into a unified interface Aba and Esztergár-Kiss (2024). However, an area of conflict among researchers is the governance and integration of MaaS. Conflicts persist over public versus private leadership models. Implementing the necessary Multi-Sided Platforms (MSPs) faces challenges in achieving critical mass ("chicken-and-egg" problem) and requires complex technical integration while raising significant data privacy concerns Kamargianni and Matyas (2017). Moreover, financial sustainability remains an unsolved issue along with regulatory barriers such as those impacting public transport (PT) ticket resale Krauss et al. (2022).

User behavior and modal shift is another area of conflict. Predicting MaaS's true impact on reducing car reliance is difficult, as current travel demand models struggle to capture the dynamic choices, behavioral nuances, and adoption patterns beyond early users, limiting quantitative impact assessments Kamargianni and Matyas (2017). Finally, effective integration of PT, often considered the 'backbone' Kronsell and Mukhtar-Landgren (2018), faces considerable technical, financial, and operational integration challenges within the broader MaaS ecosystem.

Table 1. Summary of Mobility as a Service (MaaS) Schemes

MaaS App	Region	Start Year	Transport Modes
Transit	Canada, USA, UK, Europe, Australia	2012	PT, Taxi, Bike Sharing, Car Sharing, Ride-Hailing
Optymod'Lyon	Lyon, France	2012	PT, Ridesharing, Regional Train, Parking
Whim	Finland	2016	PT, Rental Cars, Taxi, Regional Rail, Bike Sharing
Floya	Brussels, Belgium	2023	PT, Taxi, Bike and Scooter Sharing, Car Sharing

Services similar to MaaS have been examined in different countries around the world. Table 1 summarizes the pilot programs conducted within the context of MaaS, resulting from a comprehensive literature review.

2.2 Cyber-Physical-Social Systems (CPSS) Paradigm Within Complex Systems Engineering

The technologies supporting ITS as a Cyber-Physical System (CPS) enable the interaction between the cyber (data, algorithms) and physical (infrastructure, vehicles) domains. A critical limitation of many ITS and CPS approaches is their primary focus on optimizing these technical aspects, often treating the human element as a secondary consideration Wang and Zhang (2017).

Currently, most research on CPS still focuses on the complexity of engineering elements. In fact, the social complexity of CPS is not ignorable or even decisive at all stages. Human factors play vital roles in all the technological areas that ITS dramatically change, including smart mobility, vehicle control design, traveler information management, public transport operation planning, and traffic management console Kaushik et al. (2024). Further, early and gradual acceptance of ITS by the public and society directly affects the design of ITS morphology Duga et al. (2024). However, currently, society and transportation-related social factors are used as background and context in most cases, passively providing sensing and surveillance data and passively reacting to the ITS Perallos et al. (2015).

On the contrary, the main constraints of CPSs is modeling digital analog systems, integrating event and time-based controls, ensuring real-time fault tolerance, addressing robustness and security, and managing diverse data Hasan et al. (2023) and these applications often treat human users as passive recipients of information or as constraints within the system, rather than active participants. The success of MaaS also relies on user adoption, behavioral change, trust between users and providers, fair access, and navigating complex stakeholder relationships (e.g. public agencies, private operators, users). Neglecting the social dimension can lead to low user adoption rates and increased congestion. This brings the necessity of a comprehensive approach that takes into account human and social factors, which us realized in the form of the CPSS approach.

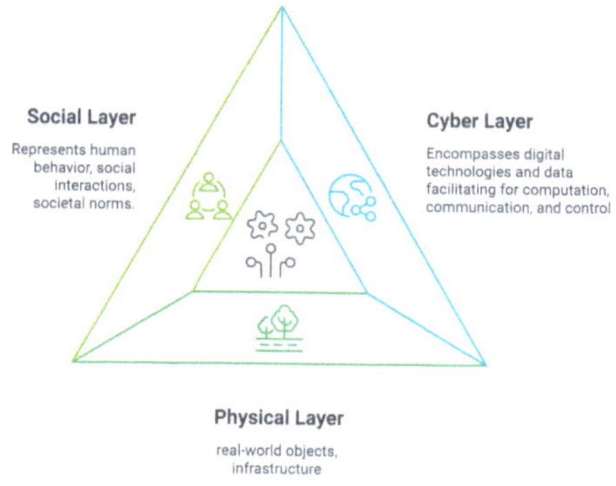

Fig. 2. CPSS Key Components

2.3 CPSS Definition and Theoretical Foundation

Cyber-Physical-Social Systems (CPSS) are an evolving subset of CPS, which involve the interlinking of the cyber, physical, and social domains and combine secure digital technologies with physical systems (e.g., sensors and actuators) and incorporate social aspects (e.g., human interactions and behaviors, and societal norms) to facilitate automated and secure services to end-users and organizations Sobb et al. (2023). The CPSS Fig. 2 illustrates three interconnected layers: Cyber, Physical, and Social. The Cyber Layer represents digital processes, the Physical Layer includes real-world entities, and the Social Layer involves human interactions. Arrows depict bidirectional interactions and feedback loops, emphasizing the continuous and dynamic influence between these layers.

Despite the significant potential of MaaS, real-world deployments often encounter challenges that are related to the limitations of a purely techno-economic view. While early studies often focused on the technological feasibility and potential benefits of MaaS, real-world deployments have encountered obstacles related to user adoption, integration with existing transport systems, business model viability (package pricing), and sensitive motion data Mass (2022). These challenges often stem from the lack of focus on the social dimension and neglecting the need for a more holistic approach that goes beyond the traditional technical perspective of ITS, to the extent that the trend of studies in this field based on Scopus has also decreased in recent years. This study seeks to propose a CPSS framework (Fig. 2), by explicitly incorporating the social layer alongside the cyber and physical, provides the necessary theoretical and systems engineering structure to address these challenges and limitations. It allows for a systematic consideration of user behavior, stakeholder interactions, and societal impacts from the outset, aiming to bridge the gap and provide a practical systems engineering structure for operationalizing and achieving higher efficiency along with facilitating the development of truly user-centered and sustainable mobility solutions.

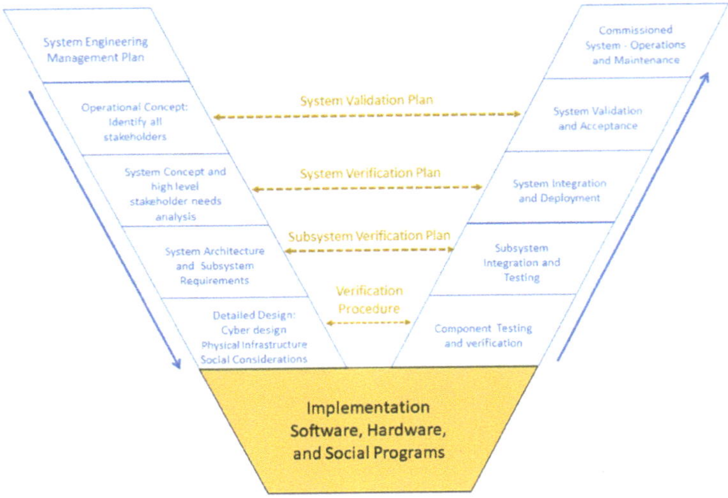

Fig. 3. V-Diagram for CPSS-based MaaS Framework

3 Proposed Maas-CPSS Framework

Implementing a Mobility as a Service (MaaS) based on CPSS application in a city involves a comprehensive approach that is not only technologically efficient but also socially equitable, environmentally sustainable, and readily adopted by users through three interacting layers: the cyber, physical, and social layer. This implementation can be reachable from an engineering approach, which contains a graphical representation of a systems development lifecycle to design and development of the system.

One of these approaches is the "V model" which has been refined and applied in many different industries. As shown in Fig. 3, the V diagram represents a system engineering approach and defines project requirements before making technology choices and implementing the system and it describes the activities to be performed and the results that have to be produced during system development.

On the left side of the V, the system's practical definition progresses from Identifying all stakeholders (users, operators, policymakers, businesses, etc.) and then Based on the stakeholder needs analysis, defining the overall concept for the MaaS system to a detailed specification of the system design which represents the overall system architecture, identifying the key subsystems (cyber, physical, and social). In the fourth step, detailed designs for each subsystem, which include Cyber design, Physical Infrastructure, and Social Considerations, are presented. Here, social consideration components encompass from UI/UX design to community engagement strategies and policy recommendations. At the bottom of the Vee Diagram, the hardware, software, and Social Programs are implemented, and the components of the system are then integrated and verified in an iterative fashion on the right. Ultimately, the completed system is validated to measure how well it meets the user's needs.

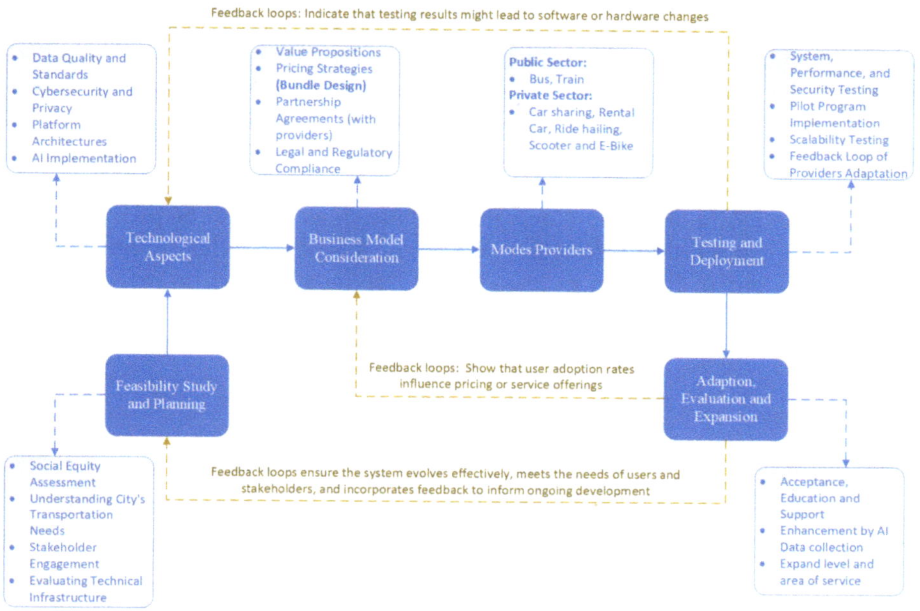

Fig. 4. Proposed CPSS-based Framework for MaaS Implementation.

In Fig. 4, the V-model is used along with a process-oriented flowchart to address the complexity of a CPSS in a structured way. First, it is essential to assess the city's transportation needs and gaps. A feasibility study, conducted in collaboration with key stakeholders such as city planners and PT authorities, must evaluate the existing technical infrastructure and identify necessary upgrades for the implementation of MaaS Hunter et al. (2023).

Second, technological factors can be restrictive in MaaS development. They may include data-related challenges, such as the quality of real-time information Melis et al. (2018), establishing standards for application programming interfaces, privacy and security concerns, and the development of platform architectures Johansson et al. (2023). Additionally, the integration of AI algorithms is imperative, particularly for route optimization, demand forecasting, and personalized service recommendations Servou et al. (2023).

Moving forward, the business model of MaaS must be carefully considered. MaaS combines various services, many of which have low-profit margins or depend on public subsidies, making it challenging to develop sustainable models. Among the key challenges in recent MaaS business models, pricing stands out as one of the most complex aspects, particularly the design of service bundles Krauss et al. (2022). The core idea of MaaS is to combine various mobility services into a seamless system that can match or even exceed the convenience and efficiency of private car ownership. Users can access public transit, such as buses and trains, along with ride-sharing services.

Next, the MaaS system can move into the testing and deployment phase. A pilot program should be launched in a specific area or with a limited set of services to evaluate the system, collect feedback, and make necessary improvements Mitropoulos et al. (2023). Then, a feedback loop should be implemented to continuously collect input from users and service providers and enable ongoing improvements and adaptations to the MaaS system. Finally, the system enters the adaptation, evaluation, and expansion phase. Users should be provided with resources and support to help them understand how to use the application and get the most out of it. Data collected by AI systems should be used to refine and improve service offerings continuously. As the system proves successful and sustainable, the service area and available mobility options can be gradually expanded.

4 Discussion and Implication Recommendations

4.1 Graph-Based Analysis of Interconnected Nodes and Relations in the Framework

Feasibility Study and Planning assesses social equity, city needs, and infrastructure, defining the Technological Aspects required for secure, AI-enhanced MaaS platforms. These technological choices shape the Business Model, influencing pricing, agreements, and legal compliance.

Modes Providers connect public and private transport services, relying on a well-defined business model for integration. Before full deployment, Testing and Deployment validates system performance, scalability, and security, refining the MaaS ecosystem through AI-driven insights. Adoption, Evaluation, and Expansion ensure continuous adaptation using realtime AI analytics, optimizing transport efficiency and user experience. Feedback loops connect adoption data back to Feasibility Planning, refining business models and technology to enhance MaaS sustainability.

The arrows signify dependencies and iterative improvements, ensuring that each phase refines the next. AI-powered feedback mechanisms drive dynamic pricing, service optimization, and long-term MaaS evolution, making the system adaptive, intelligent, and future-ready.

4.2 Enhancing MaaS with AI-Driven Innovations

AI-Enhanced Feasibility Study and Planning. AI streamlines feasibility studies by analyzing urban mobility patterns, public transport usage, and stakeholder needs. Machine learning predicts demand fluctuations, reducing congestion and improving service planning.

Smart Technological Aspects for Optimization. AI strengthens data security, platform architecture, and service efficiency. Real-time route optimization minimizes delays, while AI-driven cybersecurity ensures secure transactions. Automated fare adjustments improve user experience, making MaaS more accessible.

Intelligent Business Models. AI refines business models by optimizing pricing strategies and service bundles. Demand-based fare adjustments enhance affordability, while personalized travel packages cater to user preferences. Predictive analytics aligns service offerings with mobility trends.

AI-Powered Testing and Deployment. AI accelerates system validation through simulation-based testing and real-time monitoring. Digital twins predict operational challenges, ensuring efficient deployment. Automated performance evaluation identifies weaknesses, allowing continuous refinement.

4.3 Scalability, Reusability, and Implementation Considerations

Ensuring the proposed CPSS-based MaaS framework can be effectively adopted in diverse urban environments and scaled from pilots to city-wide systems requires addressing both technical design and significant real-world complexities. According to this approach, integrating with the existing PT or adapting to the current transport system ecosystems of any city requires standardized APIs, data protocols, and modular design, which are crucial technical requirements for implementation. While AI-Driven expansion strategies and simulations test city-wide adoption and support planning for expansion, they continuously learn to predict passenger travel patterns based on real-time user feedback. Scalable AI solutions ensure MaaS remains flexible and adaptive to evolving urban needs, contribute to financial viability, and be designed for promoting social equity. Therefore, effective reuse and scaling fundamentally depend on how a CPSS-based MaaS can employ AI-driven strategies based on the available data to adapt to the existing transport system of an environment, considering all parties, including the public sector, providers, and users.

4.4 Impact of Cyber-Physical-Social Dimensions

The impact of cyber-physical-social dimensions can be shown by providing examples of how the CPSS framework translates into actions within the MaaS context. In a case when a user plans a commute during peak hours, the three cyber-physical-social dimensions are defined as follows:

- Social layer: The user expresses their preferences via the MaaS app, considering the available bundle packages, which encourage users to use sustainable options (Social Layer).
- Transition from the physical layer to cyber (Physical Layer → Cyber Layer): Real-time sensors on public sector options report high zone passenger density, traffic cameras detect congestion and patterns on the transport network, and GPS data shows the location of available sustainable options (e.g. shared e-bikes) and train arrival times.
- Cyber Layer: The central MaaS platform integrates this diverse data and runs predictive algorithms. Using AI algorithms, tt identifies that the user's usual bus route is overcrowded, and the driving route is heavily congested. It calculates an alternative multimodal route: walk to a nearby e-bike, ride to a train station, and take the train.

- Transition from the cyber layer to social (Cyber Layer → Social Layer): The system calculates and offers user-centric discounts for promoting sustainable modes using AI algorithms based on the predefined system-optimizing goals. The user then makes an informed decision, which demonstrates the continuous feedback loop between user needs and behavior (Social), the real-time state of the physical transport network (Physical), and the intelligent processing, prediction, and communication facilitated by the digital platform (Cyber). This example reflects a functional and adaptive MaaS experience.

4.5 Ensuring Synergy Between AI, Machine Intelligence, and Sustainability

Reducing Emissions Through AI-Optimized Routing: AI-driven traffic predictions reduce energy waste by optimizing routes. Smart scheduling prioritizes environmentally friendly transport modes. Demand-based fleet distribution minimizes carbon emissions.

Energy-Efficient AI for Sustainable MaaS Growth AI: forecasts demand to prevent overuse of transport resources. Smart energy management balances electric and fuelbased mobility, reducing energy consumption. Predictive analytics optimize charging schedules for electric vehicles.

Continuous Sustainability Improvement: AI-powered feedback loops refine transport efficiency based on real-time data. Automated policy adjustments align MaaS operations with sustainability goals. Long-term AI learning ensures MaaS remains environmentally responsible and adaptive.

5 Conclusion

Mobility-as-a-Service (MaaS) holds significant promise for transforming urban transportation, contributing to decarbonization goals, and improving accessibility. However, realizing this potential requires moving beyond a purely technology-driven approach. Furthermore, the human factor and social context in implementing MaaS are missing or trivialized in some systems somehow. To address this issue, this article has argued that a Cyber-Physical-Social Systems (CPSS) perspective (by the V-model for systems engineering) is essential for successful MaaS implementation, addressing the limitations of traditional Intelligent Transportation Systems (ITS) that often overlook the crucial social dimension. While this system encourages users to use PT based on innovative integration of technology, including AI and machine learning, MaaS offers a dynamic approach to managing urban transport systems, potentially reducing reliance on private car ownership and promoting more environmentally friendly commuting options.

The key features of the framework are as follows:

- Holistic design: ensuring MaaS solutions prioritize user needs, preferences, and social equity by integrating the social dimension from the start.

- Iterative development: implementing MaaS as a continuous process with feedback loops between design, testing, deployment, and user experience.
- Stakeholder engagement: actively involving users, transport providers, policymakers, and other stakeholders throughout the entire lifecycle.
- Systems Engineering Approach: Applying the V-model to ensure a structured, verifiable, and validatable development process for MaaS.

Acknowledgements. This work is partially founded by HSPI SpA, as a part of project RECOGNISE; Centro Nazionale ICSC - "National Centre for HPC, Big Data and Quantum Computing" - codice CN00000013 PIANO NAZIONALE DI RIPRESA E RESILIENZA (PNRR) - MISSIONE 4 COMPONENTE 2 INVESTIMENTO 1.4 finanziato dall'Unione europea - NextGenerationEU Bando a Cascata Spoke 9 "Digital Society & Smart City" (CN00000013 - CUP E63C22000980007).

Author Contribution. A.A.A.: Research design and conceptualization, funding acquisition and administration, contributed in preparing—original draft. M.G: Contributed to all stages of the research and writing. F.G: Offering critical feedback on the research scope. M.P.: English proofreading, writing the original draft, and validation.

References

Aba, A., Esztergár-Kiss, D.: Creation of the MaaS readiness index with a modified AHP-ISM method. Commun. Transp. Res. **4**, 100122 (2024)

European Commission: The european green deal (2019)

Duga, P.A., et al.: Plender-Nabas 2024. Factors influencing travelers and motorists acceptance of intelligent transportation applications: A SEM-ANN approach. In: 2024 9th International Conference on Business and Industrial Research (ICBIR) (2024)

Hasan, M.K., Habib, A.A., Shukur, Z., Ibrahim, F., Islam, S., Razzaque, M.A.: Review on cyber-physical and cyber-security system in smart grid: standards, protocols, constraints, and recommendations. J. Netw. Comput. Appl. **209**, 103540 (2023)

Heikkilä, S.: Mobility as a service-a proposal for action for the public administration, case Helsinki (2014)

Hunter, J.G., et al.: The future of mobility-as-a-service: trust transfer across automated mobilities, from road to sidewalk. Front. Psychol. **14**, 1129583 (2023)

Johansson, M., Ekman, F., Karlsson, M., Strömberg, H., Jonsson, J., Faleke, M.: Automation as an enabler: passengers' experience of travelling with a full-length automated bus and their expectations of a future public transport system. Transp. Res. Procedia **72**, 957–964 (2023)

Kamargianni, M., Matyas, M.: The business ecosystem of mobility-as-a-service. In: Transportation Research Board (2017)

Kaushik, P., Rathore, S.P.S., Sachdeva, L., Poonia, M., Singh, D., Bir, L.: Intelligent transportation systems trusted user's security and privacy. In: 2024 IEEE International Conference on Interdisciplinary Approaches in Technology and Management for Social Innovation (IATMSI) (2024)

Krauss, K., Moll, C., Köhler, J., Axhausen, K.W.: Designing mobility-as-a-service business models using morphological analysis. Res. Transp. Bus. Manage. **45**, 100857 (2022)

Kronsell, A., Mukhtar-Landgren, D.: Experimental governance: the role of municipalities in urban living labs. Eur. Plan. Stud. **26**(5), 988–1007 (2018)

Maas, B.: Literature review of mobility as a service. Sustainability **14**(14), 8962 (2022)

Melis, A., Mirri, S., Prandi, C., Prandini, M., Salomoni, P., Callegati, F.: Integrating personalized and accessible itineraries in MaaS ecosystems through microservices. Mob. Netw. Appl. **23**, 167–176 (2018)

Milne, J., Beecroft, M., Nelson, J.D., Greening, P., Cottrill, C., Wright, S.: Urban (UMaaS) and rural (RMaaS) mobility as a service (MaaS): practical insights from international practitioners and experts. Eur. Transp. Res. Rev. **16**(1), 5 (2024)

Mitropoulos, L., Kortsari, A., Mizaras, V., Ayfantopoulou, G.: Mobility as a service (MaaS) planning and implementation: challenges and lessons learned. Future Transportation **3**(2), 498–518 (2023)

Perallos, A., Hernandez-Jayo, U., Onieva, E., Zuazola, I.J.G.: Intelligent Transport Systems: Technologies and Applications. Wiley, Hoboken (2015)

Servou, E., Behrendt, F., Horst, M.: Data, ai and governance in MaaS–leading to sustainable mobility? Transp. Res. Interdiscip. Perspect. **19**, 100806 (2023)

Sobb, T., Turnbull, B., Moustafa, N.: A holistic review of cyber–physical–social systems: new directions and opportunities. Sensors **23**(17), 7391 (2023)

Wang, F.Y., Zhang, J.J.: Transportation 5.0 in CPSS: towards ACP-based society-centered intelligent transportation. In: 2017 IEEE 20th International Conference on Intelligent Transportation Systems (ITSC) (2017)

Xiong, G., et al.: Cyber-physical-social system in intelligent transportation. IEEE/CAA J. Autom. Sinica **2**(3), 320–333 (2015)

Zhou, H., Dorsman, J., Mandjes, M., Snelder, M.: A tour-based multimodal mode choice model for impact assessment of new mobility concepts and mobility as a service. Transportation 1–27 (2023)

Carbon Footprint of Concrete Composites

Adriana Estokova[1(✉)], Martin Jas[1], Katarina Rusnakova[1], and Robert Figmig[2]

[1] Technical University of Kosice, Vysokoskolska 4, Kosice 04200, Slovakia
adriana.estokova@tuke.sk
[2] Povazska cementaren a.s. Ladce, Ladce, Slovakia

Abstract. Concrete production contributes significantly to the production of greenhouse gases and thus to climate change. Therefore, reducing the carbon footprint of concrete composites is a hot topic. The paper deals with the carbon footprint analysis of 5 proposed material alternatives for concrete mix, in four of them different part of cement was replaced by waste (blast furnace slag). The fifth alternative was proposed as a reference mix without cement replacement by slag for the purpose of environmental impact comparison. In addition to the compositional alternation, the effect of the different grinding times of the binder cement component of the different composites on their carbon footprint was also investigated. The carbon footprint calculations of the different concrete variants and their individual components were carried out in SimaPro using the Ecoinvent database and the IPCC calculation method. Thus, the assumption that the replacement of cement by blast furnace slag reduces the overall carbon footprint of the concrete was confirmed. This decrease corresponded to almost a 30% reduction in GWP. The results confirmed that cement production was responsible for the highest values, up to 7.6 times the values of the second largest contributor in the proposed concrete, silica sand. The electricity consumption of cement grinding accounted for 6.78% of the total CO_2 produced by the analysed concrete.

Keywords: Concrete · Cement · LCA · Carbon Footprint · Climate Change

1 Introduction

Given the increasing pressure to reduce the environmental impact of construction materials, there has been a growing emphasis in recent years on evaluating their carbon footprint [1]. Among the most significant materials in construction is concrete, which is widely used in the construction of infrastructure and buildings. Although concrete offers numerous advantages, such as its durability, longevity, and availability, its production has a significant negative impact on the environment, primarily due to carbon dioxide (CO_2) emissions. This gas is generated during the production of cement, the main component of concrete, with its production responsible for approximately 5–7% of total global CO_2 emissions. This makes concrete a significant contributor to climate change, leading to the search for ways to minimize its ecological footprint [2, 3]. The production of cement is an energy-intensive process that involves heating limestone at high temperatures, which leads to the release of CO_2. This process accounts for approximately more than 60% of the total emissions associated with concrete production [4].

B.-E. Boudriki Semlali et al. (Eds.): DATA 2025, LNNS 1908, pp. 409–418, 2026.
https://doi.org/10.1007/978-3-032-21200-9_34

To mitigate the environmental impacts of concrete, various alternatives to traditional cement composition are being developed. One of the most common approaches is to replace part of the clinker (the main component of cement) with materials such as blast furnace slag, limestone, or other secondary raw materials [5, 6]. These materials not only reduce the amount of clinker required, but also contribute to the reduction of CO_2 emissions. In addition, some of these alternative materials offer other benefits, such as improving the physical properties of concrete, leading to increased durability and reduced maintenance needs [7, 8].

Various studies focus on optimizing the composition of concrete and its components to achieve maximum efficiency in terms of environmental impacts. As part of this optimization, the ratio of different concrete components, such as cement, water, and aggregate, is often examined in various combinations to minimize energy costs and reduce CO_2 emissions [9]. Changes in the composition of cement, such as replacing part of the clinker with other materials (e.g., blast furnace slag or limestone), can significantly impact the resulting carbon footprint. Equally important are the production parameters, such as grinding time and the energy intensity of producing individual components [10].

To assess the environmental impacts of different concretes, the application of Life Cycle Assessment (LCA) is key approach. LCA is a tool for evaluating the environmental impacts of products throughout their entire life cycle – from raw material extraction, through production and transportation, to the end of life and waste disposal [11]. This analysis allows for the quantification of greenhouse gas emissions, energy consumption, and other environmental impacts associated with concrete production. LCA thus provides a comprehensive view of the environmental impacts of different concrete variants and enables the selection of the most efficient alternative in terms of sustainability [12].

The paper is focused on carbon footprint calculation of the concrete of different compositions with variable contents of clinker, limestone and blast furnace slag with the special regard to identify the environmental impacts of binder grinding.

2 Material and Methods

2.1 Design of the Concrete Mixtures

In order to calculate the carbon footprint, a concrete composed of cement/binder, water and aggregate in the ratio of these components 2:1:12 was designed. This ratio was maintained in all the concrete variants for better comparison of the results.

For the formation of the different concrete variants (formulations B1-B4), the cement composition or the amount of clinker (CLI), limestone (LIM) and blast furnace slag (BFS) were varied. All variants have a constant amount of gypsum and grinding intensifier. Gypsum in the cement is used as a setting regulator and accounts for 8% of the total clinker along with limestone and blast furnace slag. To compare the environmental impacts, a reference mix was also proposed as a zero alternative without replacing part of the clinker with waste, i.e. blast furnace slag (B0). The quantities of clinker, limestone and slag are given in Table 1.

For the proposed material composition of the concretes, different grinding times of 30, 35 and 40 min for the binder component were considered for each compositional alternative (Table 1).

Table 1. Cement/Binder Compositions and Grinding Times.

Concrete Labeling	CLI-LIM-BFS content (%)	Grinding Time (min)	CLI (g)	LIM (g)	BFS (g)	Gypsum (g)	Grinding Aid (mL)
B0_00	80–20–00	0	2593	648	0		
B1_30	70–20–10	30	2269	648	324		
B1_35	70–20–10	35	2269	648	324		
B1_40	70–20–10	40	2269	648	324		
B2_30	70–10–20	30	2269	324	648		
B2_35	70–10–20	35	2269	324	648		
B2_40	70–10–20	40	2269	324	648	259	4.2
B3_30	65–20–15	30	2106	648	486		
B3_35	65–20–15	35	2106	648	486		
B3_40	65–20–15	40	2106	648	486		
B4_30	65–15–20	30	2106	486	648		
B4_35	65–15–20	35	2106	486	648		
B4_40	65–15–20	40	2106	486	648		

2.2 Raw Materials

For the production of individual alternatives of cement mixtures, clinker was used in different proportions, whose primary raw material limestone and corrective limestone, added to the cement as a filler, are mined close to the cement plant. Another component of clinker, a clay-clay additive, is also mined on the site. The blast-furnace slag was delivered from Třínecky železárny, a.s. (Czech Republic).

Industrial gypsum (waste from chemical production) used as a solidification regulator and, together with the grinding intensifier, forms 8% of the mass of the mixture of clinker, limestone and slag, i.e. it is the same amount of admixtures in all variants of the mixtures.

The grinding intensifier, based on diethylene glycol in liquid form, was added as an additive to the blends to increase the grinding efficiency (optimum adjustment of the electrical charge during grinding), leading to higher mill throughput and lower clinker factor, together with an overall reduction in CO2 emissions. The consumption of the grinding intensifier was 0.12% of the total mix weight.

The proposed concrete formulations have a total of three aggregate fractions, namely fine aggregate fraction 0–4 mm represented by silica sand and coarse aggregate fractions 4–8 and 8–16 mm Vtáčnik quarry (Slovakia). The ratio of coarse aggregate of fractions 4–8 and 8–16 mm was 15% and 85%. Aggregate properties have been verified by tests and declared by the aggregate producer through the Declaration of Conformity.

In all the proposed alternatives, local potable water was used. The water complies with the requirements for water for concrete mixing according to EN 1008 Water for concrete mixing. Specification for sampling, testing and proving the suitability of water, including recycled water.

The plasticizer based on polycarboxylates was used to reduce the amount of cast water in the mixture. It was added to the mixing water or with the already wet concrete mix at the end of mixing at a recommended dosage of 0.2–2.0 kg per 100 kg of cement.

2.3 Grinding Process

For the purpose of calculating the carbon footprint, the average energy consumption required for the grinding process was determined for each grinding time. Grinding of mixtures was considered using a ball laboratory mill type GM 100 (Cerovský, JOCE) with closed circuit. The mill consists of a cylinder-shaped chamber into which the crushed material is placed and which is loaded with steel balls as grinding media. Grinding occurs by rotation of the cylinder, whereby the balls rise in height and free fall to the bottom of the cylinder, crushing the material. The size of the balls and the speed of the roller, which are set by the manufacturer, are important in the grinding process. From the grinding circuit, the ground material is extracted by a powerful fan into the classifier, where the coarser parts are separated from the flour. These are returned to the mill again. The grinding is regulated on the basis of the amount of returning material. The finished mixture comes out of the classifier.

2.4 Carbon Footprint Calculation

The carbon footprint calculations of the different concrete variants and their individual components were carried out in SimaPro using the Ecoinvent database and the IPCC LCA calculation method. The declared/functional unit within the cradle to gate system boundary (modules A1-A3) was set to one concrete sample of weight of 26.41 kg. To express and measure the amount of greenhouse gas emissions and their impact on climate change, the environmental indicator - global warming potential (GWP) is used with the unit "kg CO_2 eq", meaning kilograms of carbon dioxide equivalent.

3 Results and Discussion

3.1 Overall Global Warming Potentials

The calculated values of GWP indicator for the analysed concrete mixtures are presented in Table 2.

Table 2. Carbon Footprint of Concrete Mixtures.

Concrete Labeling	GWP_{fossil} (g CO_2 eq)	$GWP_{biogenic}$ (g CO_2 eq)	GWP_{land} (g CO_2 eq)	CO_2 capturing (g CO_2 eq)	GWP_{total} (g CO_2 eq)
B0_00	3 229.47	287.52	5.24	−152.18	3 370,05
B1_30	2 515.22	215.18	3.71	−115.45	2 618.66
B1_35	2 541.35	215.25	3.71	−115.54	2 644.77
B1_40	2 563.76	215.29	3.71	−115.61	2 667.15

(continued)

Table 2. *(continued)*

Concrete Labeling	GWP_{fossil} (g CO_2 eq)	$GWP_{biogenic}$ (g CO_2 eq)	GWP_{land} (g CO_2 eq)	CO_2 capturing (g CO_2 eq)	GWP_{total} (g CO_2 eq)
B2_30	2 514.52	215.13	3.72	−115.41	2 617.96
B2_35	2 540.65	215.20	3.72	−115.50	2 644.07
B2_40	2 563.05	215.25	3.72	−115.57	2 666.45
B3_30	2 370.17	203.02	3.45	−109.28	2 467.36
B3_35	2 396.31	203.08	3.45	−109.37	2 493.47
B3_40	2 418.71	203.14	3.45	−109.45	2 515.85
B4_30	2 369.82	203.00	3.46	−109.27	2 467.01
B4_35	2 395.95	203.06	3.46	−109.35	2 493.12
B4_40	2 418.35	203.12	3.46	−109.43	2 515.50

The Life Cycle Assessment (LCA) results for different concrete variants reveal significant differences in their Global Warming Potential (GWP). The baseline variant (B0_00) had the highest GWP_{total} at 3370.05 g CO_2 eq, mainly due to its high fossil carbon emissions of 3,229.47 g CO_2 eq. In contrast, the modified concrete variants showed notable reductions in GWP. The B3 and B4 series had the lowest GWP values, with B3_30 and B4_30 achieving the lowest total GWP at approximately 2,467 g CO_2 eq. This reduction is primarily driven by a decrease in fossil CO_2 emissions (GWP_{fossil}) and slightly lower biogenic CO_2 contributions ($GWP_{biogenic}$). This decrease is primarily attributed to the lower cement content, as cement production is a major contributor to carbon emissions.

The data also indicate that as the variant number increases (moving from B1 to B4), there is a consistent reduction in fossil CO_2 emissions. Therefore, the hypothesis that substituting cement with blast furnace slag lowers the overall carbon footprint of concrete was validated. This reduction amounted to nearly a 30% decrease in GWP.

The CO_2 emissions due to land use (GWP_{land}) had almost negligible value for the designed concretes ranging from 3.45 g CO_2 eq. (B3 formulations) to 3.72 g CO_2 eq. (B2 formulations), and it does not depend on the milling time.

Additionally, CO_2 capture values vary slightly, with the highest absorption in the baseline variant (−152.18 g CO_2 eq) and the lowest in the B3 and B4 series (around −109 g CO_2 eq). The CO_2 absorption value reflects the ability of the concrete to absorb carbon dioxide from the atmosphere during its life cycle. Concrete can absorb CO2 in the carbonation process that occurs during its use. The amount of CO_2 absorbed by the concrete from the air is also unaffected by the grinding time or negligible effect, however, the greater the proportion of clinker in the mix, the greater the ability of the concrete to absorb more CO_2, ranging from 4,33 to 4,43% of the emitted CO_2. Despite the differences in CO_2 absorption, the reduction in emissions from fossil sources plays a more significant role in lowering the overall GWP.

3.2 Global Warming Potentials of the Concrete Components

The analysis of the data on the contribution of different components of concrete to the total Global Warming Potential (GWP) reveals several important insights (Table 3).

Table 3. Carbon Footprint of Concrete Components and Grinding Process.

GWP_{total} ($g\ CO_2$ eq)						
Concrete Labeling	Cement	Electricity for grinding	Plasticizer	Water	Silica sand	Coarse aggregate
B0_00	3 021.72	0				
B1_30	2 121.13	149.2				
B1_35	2 121.13	175.31				
B1_40	2 121.13	197.69				
B2_30	2 120.43	149.2				
B2_35	2 120.43	175.31	2.17	0.81	278.25	67.1
B2_40	2 120.43	197.69				
B3_30	1 969.83	149.2				
B3_35	1 969.83	175.31				
B3_40	1 969.83	197.69				
B4_30	1 969.48	149.2				
B4_35	1 969.48	175.31				
B4_40	1 969.48	197.69				

The IPCC's calculations show that cement is the biggest source of CO_2 emissions, contributing up to 7.6 times more than the second largest source, which is silica sand in the proposed concrete.

Another major contributor to carbon dioxide emissions is the electricity used in cement grinding, which is responsible, on average, for up to 7% of the total CO_2 emissions from the concrete studied. Figure 1 presents the comparison of the grinding process contribution to the total GWP values per individual concrete mixtures. The energy required for grinding increases slightly in some cases, particularly in the B1_40, B2_40, B3_40, and B4_40 variants, but this effect is minimal compared to the overall reduction in GWP. The amount of CO_2 emitted or the GWP value of the mixture increases with the grinding time, but overall the climate change impact of concrete is reduced by replacing some of the clinker in the mix or by increasing the amount of blast furnace slag.

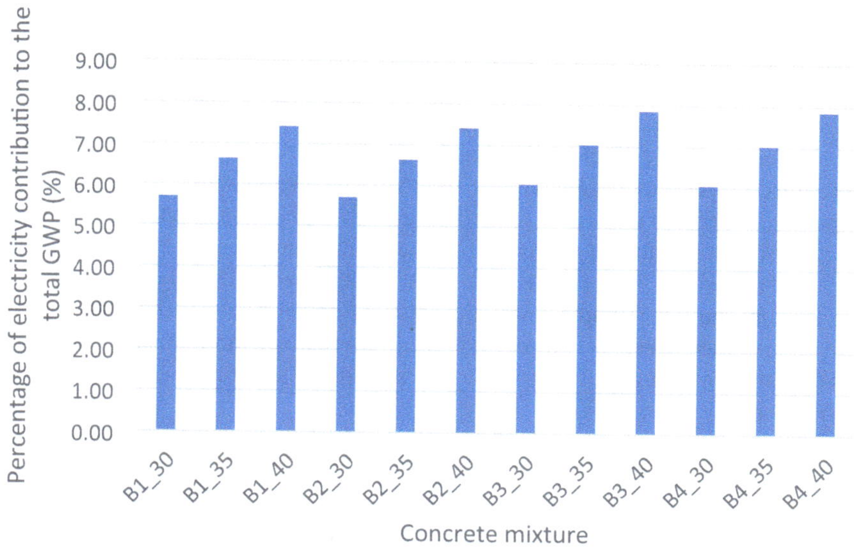

Fig. 1. Percentage Contribution of Grinding Process to the total Carbon Footprint

Average percentage share of the component of concrete with cement replacement (B1-B4) are illustrated in Fig. 2.

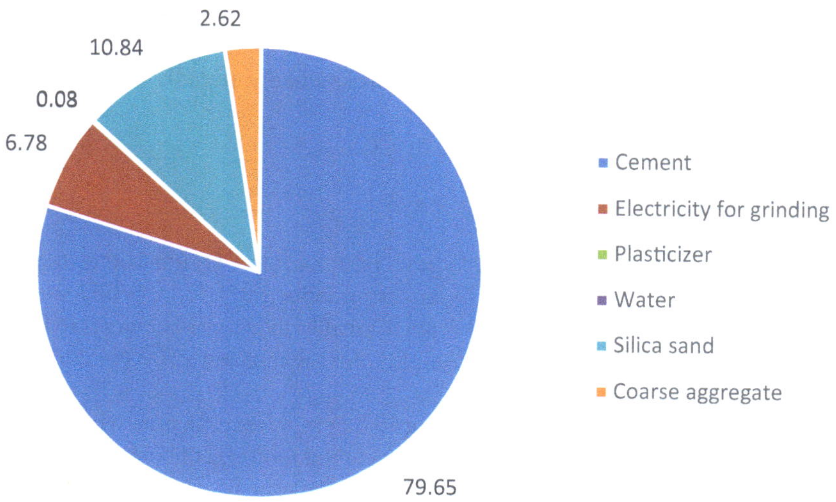

Fig. 2. Percentage Contribution of Concrete Components to the total Carbon Footprint

Cement is by far the largest contributor, accounting for a significant 79.65% of the total GWP in analysed concretes. This highlights that cement production is the most energy-intensive and carbon-emitting part of concrete manufacturing. To reduce the

GWP of concrete, carbon capture technologies in cement plants could play a crucial role in reducing emissions from cement production.

Electricity used for grinding cement comes second, contributing 6.78% to the total GWP. Although this is a smaller share compared to cement, it is still a substantial factor. Improvements in energy efficiency during the grinding process could lead to a reduction in emissions. Furthermore, transitioning to renewable energy sources, such as solar or wind power, to provide electricity for grinding could significantly cut down the carbon footprint associated with this phase of concrete production.

Silica sand, which makes up 10.84% of the GWP, is the third-largest contributor. While its contribution is much lower than that of cement, it still represents an area where improvements can be made. Exploring alternative, more sustainable sources for silica sand, or using recycled materials as partial substitutes, could help reduce its environmental impact.

Coarse aggregates contribute 2.62% to the total GWP. The difference in the GWP contribution between silica sand and coarse aggregates (Table 3 and Fig. 2) stems from the more energy-intensive processing, refinement, and transportation of silica sand compared to the relatively simpler extraction and processing of coarse aggregates. While both materials have environmental impacts, the processing of silica sand tends to be more energy-consuming, leading to its higher contribution to GWP. Coarse aggregates, on the other hand, require less processing and are often sourced more locally, making their carbon footprint generally lower.

Plasticizers, which help improve the workability of concrete, account for only 0.08% of the GWP, making their impact minimal in comparison to other components.

Water contributes just 0.03% to the GWP, which is extremely low.

The results confirm that optimizing the cement composition by partially replacing it with blast furnace slag leads to a more environmentally friendly concrete mixture without compromising its composition.

4 Conclusions

Cement was confirmed to be by far the largest contributor to CO_2 emissions in concrete production, making it a crucial target for emission reduction efforts. While silica sand also contributes to emissions, its impact is much smaller than that of cement, so focusing on cement production and its energy use might offer the most effective path to lowering CO_2 levels.

The electricity consumption for cement grinding represents an area where improvements in energy efficiency could help reduce the carbon footprint of concrete production.

Evaluating the modified concrete variants, particularly B3_30 and B4_30, offer a more sustainable option by significantly reducing their GWP compared to the reference mixture B0_00.

The findings of this study offer meaningful guidance for both policymakers and industry stakeholders in their efforts to reduce the environmental footprint of concrete production. The demonstrated reduction in GWP, up to 30% through the partial substitution of clinker with blast furnace slag, highlights a viable and practical strategy for decarbonizing one of the most emissions-intensive building materials.

For construction companies, this research supports a shift in procurement and design practices toward prioritizing low-carbon concrete formulations. The adoption of concrete mixes with reduced clinker content, particularly those incorporating supplementary cementitious materials like blast furnace slag, can be implemented without compromising overall performance. However, this shift requires raising awareness and building capacity within the construction industry.

Cement and concrete manufacturers also have a pivotal role to play. The results suggest that manufacturers should scale up the production of blended cements, particularly CEM II and CEM III types, which have already shown promise in reducing emissions. To further decrease the carbon footprint of cement production, companies should invest in modern, energy-efficient grinding technologies and consider transitioning to renewable energy sources for electricity needs.

From a policy standpoint, the findings suggest that governments should create regulatory and economic frameworks that promote the use of low-carbon building materials. Financial incentives such as tax breaks, subsidies, or carbon credits could encourage the broader adoption of sustainable concrete formulations. In parallel, building codes and standards should be revised to accommodate the use of low-clinker or blended cements across a wide range of applications.

There are, however, economic trade-offs to consider. The use of blast furnace slag may result in slightly higher initial costs, especially in regions where slag is not readily available and must be transported over long distances. These costs may be mitigated by the reduced energy requirements in cement production and by long-term savings associated with avoiding carbon pricing penalties. Cement manufacturers may also face upfront capital investments to adapt their facilities for producing blended cements, but these can be justified by improved operational efficiency and alignment with emerging environmental regulations.

In terms of durability and performance, the use of blast furnace slag can offer long-term benefits. Research has shown that slag-containing concretes tend to have improved resistance to sulfate attack, chloride ingress, and alkali-silica reactions, potentially extending the service life of concrete structures and reducing the frequency and cost of repairs. One consideration, however, is that these mixes often exhibit slower early strength development compared to traditional Portland cement concretes. This characteristic may influence construction timelines and necessitate adjustments in curing practices or the use of accelerators. Nonetheless, these trade-offs can be effectively managed through careful mix design and performance-based specifications that emphasize long-term structural integrity and environmental impact over prescriptive formulas.

Further research is needed to support broader application. Future studies should examine the long-term durability and mechanical performance of slag-containing concretes under various environmental conditions. Expanding the life cycle assessment to include transportation, construction, use, and end-of-life phases would provide a more comprehensive environmental impact evaluation. More detailed modeling of CO_2 uptake through carbonation over time and in different exposure conditions could improve GWP accuracy. Research should also explore alternative low-carbon binders such as fly ash, calcined clay, and recycled materials, especially in regions with limited slag availability. Optimizing grinding processes through energy-efficient technologies and real-time monitoring could further reduce emissions. The use of renewable energy sources for cement grinding also warrants investigation. Economic analyses should assess the feasibility and cost-effectiveness of sustainable materials in different markets, alongside policy incentives like carbon pricing or green certifications.

Acknowledgments. This study was funded by VEGA project (grant number 1/0286/25 and 2/0108/23) and APVV project (grant number APVV-20-0140).

Disclosure of Interests The authors have no competing interests to declare that are relevant to the content of this article.

References

1. Laurent, A., Olsen, S.I., Hauschild, M.Z.: Limitations of carbon footprint as indicator of environmental sustainability. Environ. Sci. Technol. **46**(7), 4100–4108 (2012)
2. Tarun, R., Naik, F.: ASCE.: sustainability of concrete construction. Struct. Des. Constr. **13**(2), 98–103 (2008)
3. Adesina, A.: Recent advances in the concrete industry to reduce its carbon dioxide emissions. Environ. Challenges. **1**, 100004 (2020)
4. Andrew, R.M.: Global CO2 emissions from cement production, earth Syst. Sci. Data. **10**, 195–217 (2018). https://doi.org/10.5194/essd-10-195-2018
5. Abdul-Wahab, S.A., Al-Dhamri, H., Ram, G., Chatterjee, V.P.: An overview of alternative raw materials used in cement and clinker manufacturing. Int. J. Sustain. Eng. **14**(4), 743–760 (2020). https://doi.org/10.1080/19397038.2020.1822949
6. Habert, G., Miller, S.A., John, V.M., et al.: Environmental impacts and decarbonization strategies in the cement and concrete industries. Nat Rev Earth Environ. **1**, 559–573 (2020). https://doi.org/10.1038/s43017-020-0093-3
7. Shah, I.H., Miller, S.A., Jiang, D., et al.: Cement substitution with secondary materials can reduce annual global CO_2 emissions by up to 1.3 gigatons. Nat. Commun. **13**, 5758 (2022). https://doi.org/10.1038/s41467-022-33289-7
8. Liew, K.M., Sojobi, A.O., Zhang, L.W.: Green concrete: prospects and challenges. Constr. Build. Mater. **156**, 1063–1095 (2017)
9. Kim, T., Tae, S., Roh, S.: Renew. Sust. Energ. Rev. **25**(2013), 729–741 (2013)
10. Elkhaldi, I., Roziere, E., Turcry, P., Loukili, A.: Towards global indicator of durability performance and carbon footprint of clinker-slag-limestone cement-based concrete exposed to carbonation. J. Clean. Prod. **380**(1), 134876 (2022)
11. Vieira, D.R., Calmon, J.L., Coelho, F.Z.: Life cycle assessment (LCA) applied to the manufacturing of common and ecological concrete: a review. Constr. Build. Mater. **124**, 656–666 (2016)
12. Xing, W., Tam, V.W., Le, K.N., Hao, J.L., Wang, J.: Life cycle assessment of recycled aggregate concrete on its environmental impacts: a critical review. Constr. Build. Mater. **317**, 125950 (2022)

Enhancing Generalization in Distracted Driver Behavior Classification: A Body Landmark-Based Approach

Hossam Ahmed[1(✉)], Enjy Ramadan[1], Mazen Ebrahim[1], Farah Darwish[1], and Ayman Atia[1,2]

[1] Department of Computer Science, October University for Modern Sciences and Arts (MSA), Giza, Egypt
{hossam.ahmed5,enjy.ramadan,mazen.ashraf4,fdarwish, aezzat}@msa.edu.eg

[2] HCI-LAB, Faculty of Computers and AI, Helwan University, Cairo, Egypt

Abstract. Driver distraction is a major contributor to road accidents, requiring the development of robust driver monitoring systems. Traditional deep learning approaches for distracted driver behavior classification rely on raw images, which often suffer from poor generalization across different subjects and environmental conditions. In this study, we propose an alternative approach using composite skeleton representations extracted from driver videos. We conduct two experiments to assess this approach. We compared window sizes (5 vs. 10 frames) using subjects that were present in the training set, showing that a smaller window improves accuracy due to increased training samples. Results show that a smaller window (5 frames) improves accuracy due to increased training samples (e.g., MobileNetV2: 95.2% vs. 92.7%). Models trained on raw images perform well only when tested on familiar subjects but generalize poorly to new ones. In contrast, composite skeletons maintain higher accuracy across unseen subjects. Our findings highlight the potential of composite skeleton-based training for real-time, generalizable driver behavior classification.

Keywords: Deep Learning · Driver Distraction Detection · Human Activity Recognition

1 Introduction

Driver behavior is a critical element in road safety, as it has a profound impact on accident rates, traffic flow, and overall driving performance. Arvin and J. Khattak [1] stated that distracted driving, in particular, is one of the leading causes of road accidents, accounting for 35% of all transportation-related deaths in the latest years. Many other studies were conducted in order to see to what extent distracted driving affects traffic safety. For instance, Liang [2] said that researches on distracted driving showed that the reaction time of the driver is increased, which was the most effect that was noticed as a result of distracted behaviors. Because of the various dangerous consequences of

B.-E. Boudriki Semlali et al. (Eds.): DATA 2025, LNNS 1908, pp. 419–430, 2026.
https://doi.org/10.1007/978-3-032-21200-9_35

distracted driving, it became important to develop a reliable and accurate monitoring and classification systems to detect this behavior as soon as it happens. Understanding and accurately classifying driver behaviors is not only vital for improving traffic safety but also for enabling advanced driver-assistance systems (ADAS) that can reduce risks on the road early. Different techniques were explored for distracted behaviors detection, either by using cameras or by using sensors. Tran et al. [3] for example, used dual cameras in order to accurately capture different behaviors in real-time, and then feed them into two Convolutional Neural Networks (CNNs) simultaneously for classification. Another study by Omerustaoglu et al. [4] used a combined approach of both camera and sensors, fusing data from inside the vehicle with data from the camera for the detection of behaviors. A challenge that persists in current methodologies is the problem of generalization. Traditional models trained on behavior images tend to overfit the specific characteristics of individuals present in the training data. As a result, when these models are exposed to new drivers—whose appearance and behavioral nuances differ significantly from those in the training set, they fail to classify distracted behaviors accurately. To address this gap, our research introduces a methodology that shifts the focus from traditional behavior images to the utilization of body landmarks. Using pretrained models, our approach extracts key skeletal features that are less dependent on the subject's appearance, by constructing composite skeleton images by overlaying multiple frames captured within a fixed window size. This method not only enhanced the robustness of our models but also improved it's ability to generalize across unseen drivers and achieved higher accuracy in classifying each behavior. The models were trained to classify 11 behaviors such as Drinking (Fig. 1), Operating Radio (Fig. 2), and Hair and Makeup behavior (Fig. 3). We compare the generalization of the same models being trained with raw behavior images and being trained with composite skeleton images, as well as compare different frame window sizes used for creating the composite skeletons.

Fig. 1. Drinking

Fig. 2. Operating Radio

Fig. 3. Hair and Makeup

2 Related Work

The detection of distracted driver behaviors has been a growing research focus due to the critical role distraction plays in road safety. Traditional deep learning-based methods rely on raw images to classify driver actions, but recent studies explore alternative approaches, such as pose estimation and skeleton-based action recognition. Skeleton-based motion analysis has been widely used in sports and human activity recognition. For example, Mostafa et al. [5] proposed an automated system for analyzing karate movements using pose estimation and deep convolutional neural networks (CNNs). The model was trained on varying window sizes (N = 5, 15, 25 frames), corresponding to different time intervals (e.g., 5 frames is approximately 0.17 s, and 25 frames is approximately 1 s). The highest accuracy (97%) was achieved with a 5-frame window, effectively capturing the dynamic nature of players' movements. A 15-frame window yielded slightly lower accuracy (94%), while a 25-frame window resulted in significantly reduced accuracy (54%). The confusion matrix indicated that smaller window sizes (e.g., 5 frames) were more effective in distinguishing player styles, as larger windows (e.g., 25 frames) struggled to capture rapid movement changes common in fast-paced activities like karate. This highlights the importance of selecting appropriate window sizes to accurately evaluate dynamic movement styles. Similarly, action recognition from skeleton data has been extensively studied using graph convolutional networks (GCNs). One approach by Tang et al. [6] introduced a Fisher Vector encoding-based GCN architecture that improved action recognition on large-scale datasets such as Kinetics and NTU-RGBD, show-casing the effectiveness of structured graph-based learning in capturing motion patterns. Another work by Wei et al. [7] explored self-supervised learning for skeleton-based action recognition and introduced new modalities to mitigate erroneous knowledge propagation, demonstrating the potential of multi-modal skeleton-based learning. As mentioned before, driver distraction detection has traditionally relied on image-based classification, but some methods have integrated pose estimation for improved performance. An interesting study by Çetinkaya and Acarman [8] combined deep learning-based image classification with driver pose estimation using a pre-trained network and a Random Forest classifier. This hybrid approach resulted in a high multi-class detection accuracy of 97.03%, demonstrating the benefit of fusing multiple predictive models. Another study by Xing et al. [9] proposed a driver activity recognition system based on CNNs, which classified normal and distracted behaviors using transfer learning and feature extraction techniques. The best accuracy of 91.4% was achieved in binary classification, emphasizing the effectiveness of deep learning for driver distraction recognition. Beyond driver monitoring, skeleton-based models have been applied in pedestrian behavior prediction. Le et al. [10] used a combination of graph neural networks, self-attention mechanisms, and temporal convolutions to predict pedestrian crossing intentions, demonstrating that skeleton-based representations can effectively model human motion in real-world environments. In addition, a novel skeleton-based motion prediction framework was proposed by Zhang et al. [11] for composite human actions, addressing the challenge of predicting sequences containing multiple atomic actions and significantly outperforming traditional GCN-based methods on the CHAMP dataset. One study by Alyahya et al. [12] proposed a temporal driver action localization (TDAL) framework that identifies the start and end times of distracted behaviors in untrimmed videos. The framework achieved promising results in the NVIDIA AI City 2022 Challenge, contributing to fine-grained action localization for smart transportation systems. Da Silva Oliveira and Farias [13] compared transfer learning approaches for detecting distracted drivers using VGG19, ResNet152, DenseNet161, and InceptionV3 v3, evaluating three strategies: end-to-end optimization (TL1), optimization of only fully connected layers (TL2), and feature extraction with shallow machine learning models. The dataset, derived from driver videos, ensured generalization by separating training and test sets by individuals. TL1 achieved the best results, with DenseNet161 reaching 88.83% test accuracy, while TL2 and feature extraction performed poorly, with test accuracies dropping to 57.41% and below 30%, respectively, except for VGG19 (64%). Long-term temporal dependencies have been a challenge in skeleton-based action recognition. A multi-scale temporal transformer (MTT) approach was proposed by Kong et al. [14] to address this issue by integrating multi-scale temporal embeddings and transformer encoders to model long-term motion patterns. The method outperformed state-of-the-art techniques on NTU-RGBD and Kinetics-Skeleton datasets, demonstrating the importance of multi-scale feature extraction in skeleton-based learning. Dileep et al. [15] proposed a real-time system leveraging 2D pose estimation and CNNs to identify activities such as trespassing and fighting. The system extracts skeletal images from video frames, classifies human activities, and triggers alerts for unusual behaviors. This approach enhances security applications in public spaces like malls, hospitals, and educational institutions. The limitations of image-based driver distraction detection, particularly its poor generalization to new subjects, motivate the exploration of alternative representations such as composite skeletons.

3 Methodology

Our proposed method consists of two main phases, as shown in Fig. 4: Data Acquisition & Preprocessing, and Training the Models and Comparing Results. In the first phase, we process camera frames of driver behaviors to extract meaningful skeletal representations. In the second phase, we train multiple deep learning models on both raw images and composite skeleton images with different window sizes and then evaluate their generalization performance.

3.1 Data Acquisition

The dataset used in this study is the Vicomtech Driver Monitoring Dataset [16], which consists of videos capturing various driver behaviors. These behaviors have been categorized into 11 classes:

- Talking on the phone (Left and Right)

- Texting (Left and Right)
- Drinking
- Talking to a passenger
- Reaching behind/side
- Operating the radio
- Safe driving
- Hair and makeup

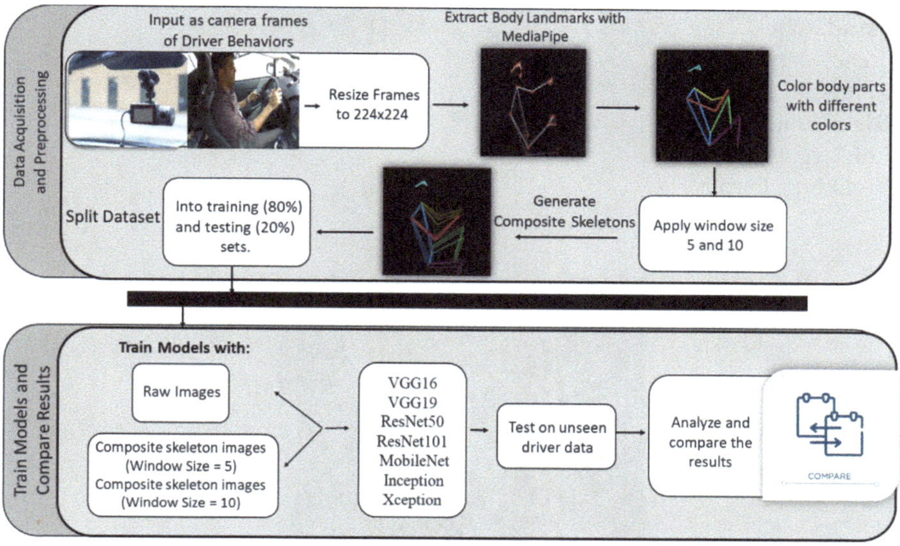

Fig. 4. System Architecture.

Each behavior is recorded in various scenarios, capturing differences in lighting, environment, and individual driver variations which is crucial in order for us to evaluate generalization. The number of images for training are 60,000 images total with approximately 5,454 images for each class and 12,000 images for testing with approximately 1,090 images for each class. For the behaviors "Drinking", "Hair and makeup", and "Texting (Left and Right)", about half of the images used for training were captured in a driving simulator setup, with the same camera angle used in real car images (side view), and darker lighting.

3.2 Pre-processing

For composite skeletons generation, several preprocessing techniques were applied to the dataset, as outlined in the system architecture diagram Fig. 4:

Frame Extraction: Each video was processed to extract frames at regular intervals. The extracted frames were stored in folders based on their respective behavior classes.

Skeleton Extraction with MediaPipe: Each frame was processed using the MediaPipe Pose library which performs a two-stage pipeline to detect 33 body landmarks. First, a full-body detection model detects the person in the frame and creates a region of interest (ROI). Second, Within the ROI, a regression model then predicts the precise coordinates (x, y, z) and visibility of 33 key body landmarks (e.g., elbows, knees, shoulders, etc.). These landmarks were then drawn onto a blank canvas to produce a skeleton image. When we trained the models with the skeleton images, we noticed decreased classification accuracy due to the confusion of body landmarks as they all had the same colors; therefore, different body parts were assigned unique colors, as shown in Fig. 5 to enhance feature distinction:

- Head: Yellow
- Torso: Orange
- Left Arm: Cyan
- Right Arm: Blue
- Left Leg: Green
- Right Leg: Purple

Fig. 5. Colored Body Parts Representation.

Composite Skeletons Generation: Instead of using individual skeleton frames, a temporal context was introduced by creating composite skeletons. A sliding window approach was applied, grouping every 5 or 10 consecutive frames. The skeletons within the window were blended together to create a single composite image representing the temporal aspect of driver behavior.

Normalization and Storage: The final images were resized to 224 × 224 pixels. Pixel values were normalized to the range [0,1] to improve model convergence during training. The processed images were divided into training and testing sets (80%, 20% respectively) to evaluate model performance.

Training Parameters: The following training parameters were used:

- Batch Size: 64
- Learning Rate: 0.0001
- Optimizer: Adam
- Loss Function: Categorical Crossentropy
- Epochs: 35 (with EarlyStopping to prevent overfitting)

3.3 Comparison and Evaluation

To determine the most effective feature representation for driver behavior classification, seven pre-trained deep learning models were trained and compared. Each model was evaluated under three different training scenarios:

Raw Image-Based Training: Models were trained on the extracted RGB frames without any preprocessing. This approach retained all visual information but was more sensitive to variations in lighting, background, and driver appearance.

Composite Skeletons (Window Size = 5): Models were trained on composite skeletons generated from every 5 consecutive frames. This provided a moderate level of temporal awareness while reducing unnecessary variations from individual frames.

Composite Skeletons (Window Size = 10): Models were trained on composite skeletons generated from every 10 consecutive frames. This emphasized longer-term movements and behavior transitions, which can be critical for distinguishing subtle differences in behavior (e.g., reaching vs. drinking). The performance of each model under these three scenarios was recorded and compared based on overall accuracy. This comparison provided insights into:

1. The impact of raw vs. skeleton-based features.
2. The effect of temporal information on model accuracy.
3. The best-performing pre-trained model for driver behavior classification.

Generalization Testing Strategies. To assess how well the trained models generalize to unseen data, two generalization testing strategies were employed:

1. Cross-Subject Generalization: The dataset was divided so that drivers appearing in the training set were excluded from the test set. This ensured that the model was not memorizing specific individuals but rather learning generalizable behavior patterns. If a model performed well under this setting, it indicated its ability to classify behaviors across different drivers.
2. Cross-Scenario Generalization: The dataset contained variations in lighting, camera angles, and minor environmental changes. Models were trained on a subset of these conditions and tested on scenarios not seen during training. This evaluated the model's performance to real-world deployment, where conditions may differ significantly from the training data.

4 Experiment

4.1 First Experiment

In this experiment, we explore how the choice of composite skeleton window size affects model performance. The goal is to compare the classification accuracy of various deep learning models when trained using composite skeleton images generated with window size 5 versus window size 10, with subjects that were present in the training set. The rationale behind this experiment stems from the trade-off between preserving temporal information and maintaining a sufficiently large dataset for training.

Using a window size of 5, more composite skeleton images are generated, resulting in a larger dataset for training. This improves generalization and reduces overfitting, particularly for models that struggle with limited data. In contrast, a window size of 10 merges more frames into each composite image, reducing the total number of training samples but potentially capturing more complex motion patterns. However, due to the relatively small dataset (60,000 images total, with approximately 5,454 images per class), models that require extensive data—such as ResNet50 and ResNet101—are expected to struggle, whereas lightweight models like MobileNetV2 should perform best (Table 1).

The following table summarizes the accuracy results obtained for each model under both window sizes:

Table 1. Model Accuracy Comparison for Different Window Sizes

Model	Accuracy	
	Window Size 5	Window Size 10
MobileNet V2	95.2%	92.7%
VGG16	92.4%	89.6%
VGG19	90.8%	88.6%
Inception V3	88.1%	84.5%
Xception	86.7%	82.9%
ResNet50	68.5%	61.2%
ResNet101	64.3%	58.1%

From the results, MobileNetV2 achieves the highest accuracy (95.2%) with window size 5, demonstrating its ability to efficiently learn from the available data. The VGG models (VGG16 and VGG19) also perform well, achieving around 92% accuracy, as they are relatively lighter compared to ResNets but still effective at extracting meaningful spatial features. InceptionV3 and Xception follow, with accuracies in the mid-80 s, benefiting from their efficient feature extraction but still requiring more data for optimal performance. On the other hand, ResNet50 and ResNet101 perform the worst, with accuracies dropping to the low 60 s, as these models require significantly larger datasets to generalize well. The decrease in accuracy when using a window size of 10 confirms that reducing the number of training samples negatively impacts performance.

4.2 Second Experiment

This experiment evaluates how well different models generalize when trained using raw images versus composite skeleton representations (window size 5). The primary goal is to assess the models' ability to correctly classify driver behaviors when faced with new subjects (drivers not seen during training) and varying lighting conditions. When training with raw images, models have access to detailed texture, color, and background information, which can lead to high classification accuracy if the test data closely resembles the training set. However, this reliance on visual details raises concerns about overfitting to subject-specific features rather than learning the core motion patterns that define each action. Consequently, if a new subject (a driver who was not in the training set) appears in the test set, or if lighting conditions differ significantly, misclassifications are increased. In contrast, training on composite skeleton images (window size 5) removes subject-specific biases by focusing purely on the structural motion of the body rather than appearance-based cues. This makes the models more robust, ensuring they generalize well even when tested on new drivers under different lighting conditions. We trained all seven models using both approaches and evaluated their performance in two different test settings:

- **Test Set A:** Contains subjects (drivers) who were also present in the training set.
- **Test Set B:** Contains new subjects performing the same actions in different lighting environments (Table 2).

The table below presents the accuracy of each model under both training conditions and test settings:

Table 2. Model Accuracy Comparison for Raw Images and Composite Skeletons

Model	Raw Images		Composite Skeletons	
	Test Set A	Test Set B	Test Set A	Test Set B
MobileNet V2	96.1%	72.3%	95.2%	94.8%
VGG16	93.4%	68.5%	92.4%	92.0%
VGG19	92.8%	66.9%	90.8%	91.3%
Inception V3	90.3%	61.2%	88.1%	87.6%
Xception	88.7%	60.4%	86.7%	86.2%
ResNet50	71.5%	42.1%	68.5%	67.8%
ResNet101	66.2%	38.7%	64.3%	63.5%

When evaluated on Test Set A (subjects in the training set), models trained on raw images achieved the highest accuracy, with MobileNetV2 reaching 96.1% and VGG16 reaching 93.4%. This suggests that models trained on raw images successfully learn the appearance-based characteristics of specific drivers. However, when evaluated on Test Set B (new subjects in different lighting conditions), the accuracy of raw image-trained models dropped greatly, with MobileNetV2 falling to 72.3% and ResNet101 dropping to 38.7%. This confirms that raw image models overfitted to specific subjects and environmental details rather than learning generalizable action patterns. On the other side, models trained on composite skeletons (window size 5) maintained high accuracy across both test sets. MobileNetV2, for example, achieved 94.8% accuracy on Test Set B, compared to only 72.3% when trained on raw images. This shows that composite skeletons help the model focus on the motion patterns defining each action rather than subject identity or lighting. Across all architectures, models trained on composite skeletons consistently outperformed raw image models on Test Set B, which proves that skeletal representations significantly improve generalization to unseen subjects and varying conditions.

5 Discussion

The experiments conducted in this study provide key insights into the effectiveness of different data representations for driver behavior classification. Our findings focus on the trade-offs between raw image-based training and composite skeleton-based training. Effect of Window Size on Model Performance: Experiment 1 demonstrated that adjusting the window size for composite skeleton generation significantly impacts model performance. A smaller window size (5 frames) resulted in higher accuracy across all models compared to a larger window size (10 frames). This is likely due to the increased number of training samples obtained from merging fewer frames, which provides the models with more diverse examples while still preserving meaningful motion information.

Additionally, lighter models such as MobileNetV2, VGG16, and VGG19 consistently outperformed heavier models like ResNet50 and ResNet101. This suggests that overly complex models struggle with the relatively small dataset (60,000 images, 5,454 per class), while compact architectures generalize better. MobileNetV2, in particular, demonstrated the best performance when dealing with new subjects and different environments, achieving over 94.8% accuracy with a window size of 5. Generalization and Robustness of Composite Skeleton Representations: The second experiment highlighted the limitations of training models on raw images. While models trained on raw images achieved high accuracy when tested on subjects seen during training, they failed to generalize when tested on new subjects under different environments. The models faced a severe drop in accuracy, with MobileNetV2's performance dropping from 96.1% (seen subjects) to 72.3% (unseen subjects), and deeper models like ResNet101 suffering even more, falling to 38.7%. On the other hand, models trained on composite skeleton representations (window size 5) maintained high accuracy even when tested on unseen subjects, with MobileNetV2 achieving 94.8% accuracy on new subjects. This confirms that skeletal representations eliminate subject-specific biases and enable the models to focus on the fundamental motion patterns defining each action. Since these skeletal representations remove distractions like clothing, facial features, and lighting variations, they provide a more generalizable and robust input format for action classification.

6 Conclusion and Future Work

There has been a growth in global road accidents due to increased driver distraction, resulting in injuries, property damage, and fatalities as stated by Hossain et al. [17] and Khan et al. [18]. This shows the critical need for monitoring and analyzing driver behavior to promptly detect distractions This study investigated the effectiveness of composite skeleton representations in classifying distracted driver behaviors using various deep learning models. Through extensive experimentation, we demonstrated that training models on composite skeletons rather than raw images significantly improves generalization, ensuring reliable performance across different subjects and lighting conditions. The first experiment compared the performance of models trained on composite skeletons using two different window sizes (5 and 10 frames). The results confirmed that a smaller window size (5 frames) yields higher accuracy, as it increases the number of training samples while maintaining essential motion information. The second experiment highlighted the generalization issue of raw image-based training. While models trained on raw images achieved high accuracy when tested on seen subjects, their performance dropped significantly when tested on unseen subjects in different environments. In contrast, models trained on composite skeletons maintained high accuracy regardless of subject identity, proving that skeletal representations are more robust to visual variations and subject-specific biases. While composite skeleton representations have proven highly effective, several directions for future research could further enhance their performance: This study used MediaPipe-based composite skeletons. Future research could investigate alternative pose estimation methods (e.g., OpenPose, HRNet) or fusion techniques that combine multiple skeletal representations to improve robustness. The dataset used in this study, while comprehensive, is still moderate in size. Testing the proposed approach on larger, more diverse datasets with different vehicle settings, lighting conditions, and driver demographics would further validate its real-world applicability. Data augmentation can also be applied on the dataset used in this paper to generate additional images for underrepresented or new driver behaviors. This can help address class imbalance and expand the behavioral categories.

References

1. Arvin, R., Khattak, A.J.: Driving impairments and duration of distractions: assessing crash risk by harnessing microscopic naturalistic driving data. Accid. Anal. Prev. **146**, 105733 (2020)
2. Liang, X.: A brief review of the impact of distracted driving on traffic safety. IOP Conf. Ser. Earth Environ. Sci. **580**(1), 012096 (2020)
3. Tran, D., Do, H., Lu, J., Sheng, W.: Real-time detection of distracted driving using dual cameras. In: 2020 IEEE/RSJ International Conference on Intelligent Robots and Systems (IROS), vol. 10, pp. 2014–2019 (2020)
4. Omerustaoglu, F., Sakar, C.O., Kar, G.: Distracted driver detection by combining in-vehicle and image data using deep learning. Appl. Soft Comput. **96**, 106657 (2020) [Online]. Available: https://www.sciencedirect.com/science/article/pii/S1568494620305950
5. Mostafa, K., Hany, M., Atia, A.: Automated evaluation of karate practitioners' style during the first kata performance using deep learning and pose estimation. In: 2023 5th Novel Intelligent and Leading Emerging Sciences Conference (NILES), vol. 2023, pp. 238–241 (2023)
6. Tang, J., Wang, Y., Liu, B.: Effective skeleton-based action recognition by combining graph convolutional networks and fisher vector encoding. In: 2020 15th IEEE International Conference on Signal Processing (ICSP), vol. 1, pp. 230–233 (2020)
7. Wei, Y., et al.: Elevating skeleton-based action recognition with efficient multi-modality self-supervision. In: ICASSP 2024–2024 IEEE International Conference on Acoustics, Speech and Signal Processing (ICASSP), vol. 2024, pp. 6040–6044 (2024)
8. Çetinkaya, M., Acarman, T.: Driver activity recognition using deep learning and human pose estimation. In: 2021 International Conference on Innovations in Intelligent Systems and Applications (INISTA), vol. 2021, pp. 1–5 (2021)
9. Xing, Y., et al.: Driver activity recognition for intelligent vehicles: a deep learning approach. IEEE Trans. Veh. Technol. **68**(6), 5379–5390 (2019)
10. Le, M.C., et al.: Skeleton-based recognition of pedestrian crossing intention using attention graph neural networks. In: 2022 International Workshop on Intelligent Systems (IWIS), vol. 2022, pp. 1–5 (2022)
11. Zhang, W., Liu, M., Wang, X., Zhao, S., Wang, C.: Champ: a large-scale dataset for skeleton-based composite human motion prediction. IEEE Trans. Circuits Syst. Video Technol. **34**(10), 10063–10076 (2024)

12. Alyahya, M., Alghannam, S., Alhussan, T.: Temporal driver action localization using action classification methods. In: 2022 IEEE/CVF Conference on Computer Vision and Pattern Recognition Workshops (CVPRW), vol. 2022, pp. 3318–3325 (2022)
13. da Silva Oliveira, F.R., Farias, F.C.: Comparing transfer learning approaches applied to distracted driver detection. In: 2018 IEEE Latin American Conference on Computational Intelligence (LA-CCI), pp. 1–6 (2018)
14. Kong, J., Bian, Y., Jiang, M.: Mtt: multi-scale temporal transformer for skeleton-based action recognition. IEEE Signal Process. Lett. **29**, 528–532 (2022)
15. Dileep, A.S., Nabilah, S.S., Sreeju, S., Farhana, K., Surumy, S.: Suspicious human activity recognition using 2d pose estimation and convolutional neural network. In: 2022 International Conference on Wireless Communications Signal Processing and Networking (WiSPNET), pp. 19–23 (2022)
16. Ortega, J.D., et al.: Dmd: a large-scale multi-modal driver monitoring dataset for attention and alertness analysis. In: Bartoli, A., Fusiello, A. (eds.) Computer Vision – ECCV 2020 Workshops, vol. 2020, pp. 387–405. Springer International Publishing, Cham (2020)
17. Hossain, M.U., et al.: Automatic driver distraction detection using deep convolutional neural networks. Intell. Syst. Appl. **14**, 200075 (2022) [Online]. Available: https://www.sciencedirect.com/science/article/pii/S2667305322000163
18. Khan, T., Choi, G., Lee, S.: Effnet-ca: An efficient driver distraction detection based on multiscale features extractions and channel attention mechanism. Sensors. **23**(8), 12974 (2023) [Online]. Available: https://www.mdpi.com/1424- 8220/23/8/3835

A Hybrid Physics-Informed Framework for Sustainable Last-Mile Delivery Optimization

Ayoub El Ouardi(✉) and Otman Abdoun

Laboratory of Information Security, Intelligence Systems and Applications-ISISA, Faculty of Science Tétouan, Abdelmalek Essaadi University, Tétouan, Morocco
ayoub.elouardi@etu.uae.ac.ma, o.abdoun@uae.ac.ma

Abstract. The environmental impact of last-mile delivery operations presents a significant challenge for sustainable urban logistics. This paper introduces a novel hybrid framework that combines physics-informed neural networks (PINN), dynamic Bayesian networks (DBN), and reinforcement learning (RL) to optimize delivery routes while minimizing emissions and maintaining service quality. Our approach integrates domain-specific physics constraints into neural networks to ensure realistic emissions predictions, models risk propagation through probabilistic graphical models for robust planning, and optimizes routing decisions via RL using sustainability-focused rewards. Using synthetic data representative of urban delivery operations, we demonstrate our framework's ability to reduce CO2 emissions by up to 23% compared to conventional routing strategies while maintaining 92% time window satisfaction. The framework achieves this balance through a multi-objective optimization approach that weighs emissions reduction against service quality and vehicle utilization. The modular design allows for deployment in resource-constrained environments and provides explainable predictions that support human-in-the-loop decision-making. This work represents a significant step toward environmentally sustainable last-mile logistics aligned with SDG 11 (Sustainable Cities) and SDG 13 (Climate Action) objectives.

Keywords: Sustainable Delivery · Physics-Informed Neural Networks · Dynamic Bayesian Networks · Reinforcement Learning · Last-Mile Logistics · Emissions Reduction

1 Introduction

Last-mile delivery, the final stage in the supply chain process of transporting goods from a distribution point to the final consumer, is to blame for up to 53% of the total cost of transportation and an out-of-proportion percentage of transport-related emissions in cities [3]. As the growth of e-commerce and same-day delivery has been progressing at a speedier rate, the ecological consequence of last-mile logistics has emerged as a core sustainability problem. The urbanizing traffic, coupled with the stringent customer needs for timely delivery, raises a complex optimization problem that balances economic, environmental, and quality of service objectives.

B.-E. Boudriki Semlali et al. (Eds.): DATA 2025, LNNS 1908, pp. 431–441, 2026.
https://doi.org/10.1007/978-3-032-21200-9_36

Traditional routing optimization techniques have primarily focused on distance or time minimization with minimal concern for the environment. The new advances in artificial intelligence and machine learning offer tremendous prospects for developing even more sophisticated optimization algorithms that can be augmented to encompass sustainability objectives. However, data-driven techniques have a tendency to overlook the physical constraints of transportation networks and provide mathematically optimal yet physically unachievable or environmentally unsuitable solutions.

Here, we introduce a physics-based machine learning hybrid model which combines physics-informed knowledge with machine learning capabilities to maximize last-mile delivery operations with a direct focus on sustainability. Our proposed model integrates three compatible modeling schools: Physics-Informed Neural Networks (PINN) incorporate physical laws governing vehicle emissions directly into the predictive model; Dynamic Bayesian Networks (DBN) model the probabilistic interdependencies among operational variables and delivery risks to handle uncertainty; and Reinforcement Learning (RL) optimizes route choices using a sustainability-focused reward function that balances emissions, punctuality, and efficiency.

The synergy of these techniques enables our framework to address some of the limitations of existing solutions. First, by integrating physics constraints into neural networks, we can ensure emissions predictions are physically realistic even with limited training data. Second, by modeling delivery risks with Bayesian networks, we can model the complex dependencies among operational variables and service outcomes. Third, by employing RL with a reward function that emphasizes sustainability, we can optimize routing decisions that balance environmental and service quality objectives.

Our work makes several contributions to sustainable logistics. We propose an effective PINN architecture tailored for robust emissions forecasting based on vehicle and route conditions. We establish a computationally light-weight DBN that effectively simulates delivery risks. Furthermore, we utilize an RL agent trained to make routing decisions optimized for key sustainability metrics. Finally, we demonstrate the synergistic benefits of our integrated solution using synthetic data representative of challenging urban delivery operations.

The remainder of this paper follows this arrangement. Section 2 presents related work on sustainable logistics, physics-informed machine learning, and RL for routing optimization. Section 3 describes our methodology, from synthesizing data to designing each model element and integrating them. Section 4 presents the experimental results, demonstrating the effectiveness of our approach in reducing emissions while maintaining service quality. Section 5 outlines the constraints of our methodology and potential research directions in the future. Finally, Sect. 6 summarizes the paper with an overview of our findings and their implications for sustainable urban logistics.

2 Literature Review

Research at the intersection of sustainable logistics, machine learning, and physics-informed modeling has gained significant traction in recent years. This section reviews relevant work in sustainable delivery optimization, PINNs, DBNs, and RL approaches to route optimization.

2.1 Sustainable Last-Mile Delivery

Optimizing last-mile delivery for sustainability has become crucial due to its environmental impact. Previous studies have quantified the emissions reduction potential of optimized routing (10–30%) [3] and analyzed how delivery density and vehicle capacity affect emissions [6]. Research by [6] incorporated emissions directly into vehicle routing problems but used simplified physical models. Other approaches [6, 12] explored multi-objective optimization, demonstrating that modest sacrifices in delivery time (5–10%) could yield substantial emissions reductions (15–25%), yet lacked machine learning integration to adapt to changing conditions.

2.2 Physics-Informed Neural Networks

PINNs incorporate physical laws and domain knowledge into deep learning models, enabling physically consistent predictions even with limited data. Originally developed for solving differential eqs. [7], PINNs have shown superior performance in transportation applications, including fuel consumption prediction [7] and electric vehicle energy modeling [2]. These approaches embed physical relationships directly into neural network architectures through custom loss functions. However, most PINN applications focus on individual vehicle performance rather than fleet-wide optimization in urban environments.

2.3 Dynamic Bayesian Networks for Risk Modeling

DBNs provide probabilistic frameworks for modeling temporal dependencies in complex systems. In logistics, DBNs have been used to model supply chain disruption risks [5] and predict delivery outcomes based on operational factors [4]. Some approaches incorporate real-time data for dynamic probability updates [11], but most focus narrowly on delay prediction without addressing broader sustainability considerations or connecting to optimization algorithms.

2.4 Reinforcement Learning for Route Optimization

RL has shown promise for solving complex routing problems, with applications in capacitated vehicle routing [8] and time-constrained delivery problems [10]. For sustainability-focused routing, recent work has incorporated emissions into reward functions [1], though often using simplified emissions models. More sophisticated approaches that couple RL with detailed emissions calculations [9] have demonstrated improved environmental performance but require substantial computational resources.

2.5 Research Gaps

While individual components of our proposed framework have precedents, their integration represents a novel contribution. Existing hybrid approaches typically combine only two paradigms rather than all three [3, 9], limiting their ability to simultaneously address physical realism, uncertainty, and sequential decision-making. Additionally, many approaches rely on high-performance computing resources unavailable in practical logistics settings.

Our work addresses these gaps by integrating PINNs, DBNs, and RL into a unified, computationally efficient framework for sustainable last-mile delivery that can be deployed in resource-constrained environments.

3 Methodology

This section details our hybrid framework's components: the synthetic data foundation, the individual PINN, DBN, and RL models, and their integration mechanism.

3.1 Synthetic Data Generation

To evaluate our framework, we generated synthetic data representative of urban last-mile delivery operations. The dataset comprises vehicles (40 delivery vehicles with varied types and fuel systems), customer orders (800 delivery requests with time windows across 5 zones), traffic conditions, and weather data. Initial emissions values within this dataset were generated using established rule-based physical models correlating factors like vehicle type, speed, load, and congestion levels, providing a realistic baseline for model training and evaluation. Each component was designed to capture the essential characteristics of real-world delivery operations while remaining computationally manageable.

Our dataset balances computational efficiency with representative complexity, enabling rapid iteration during model development while maintaining sufficient variability to test the robustness of our approach.

As shown in Fig. 1, our synthetic data captures realistic traffic patterns with pronounced peak hours. Figure 2 demonstrates the strong correlation between congestion levels and emissions, highlighting the importance of considering traffic conditions in sustainable routing decisions.

3.2 Physics-Informed Neural Network for Emissions Prediction

The first component is a PINN that predicts CO_2 emissions based on vehicle characteristics and driving conditions. Unlike conventional neural networks, our PINN incorporates fundamental physical relationships governing vehicle emissions, primarily by embedding these physics as soft constraints within the network's loss function. Key physical constraints enforce relationships such as the U-shaped curve between speed and emissions, stop-and-go driving effects,

Vehicle type hierarchy, and fuel type hierarchy. Incorporating these constraints ensures physical plausibility even with limited training data.

Figure 3 illustrates how our PINN correctly captures the U-shaped relationship between speed and emissions, adhering to known physical principles even with limited training data. In contrast, the conventional neural network fails to maintain this physically plausible relationship at speed extremes. Similarly, Fig. 4 demonstrates the PINN's ability to correctly model the impact of stop-and-go driving on emissions while preserving the expected hierarchy among different vehicle types.

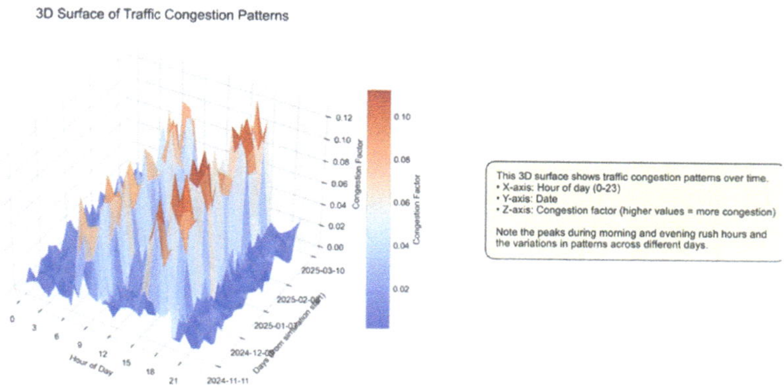

Fig. 1. 3D visualization of traffic congestion patterns across different times of day and geographic zones in our synthetic dataset. The surface plot reveals morning and evening peak patterns (7–9 AM and 4–6 PM) with higher congestion in central zones.

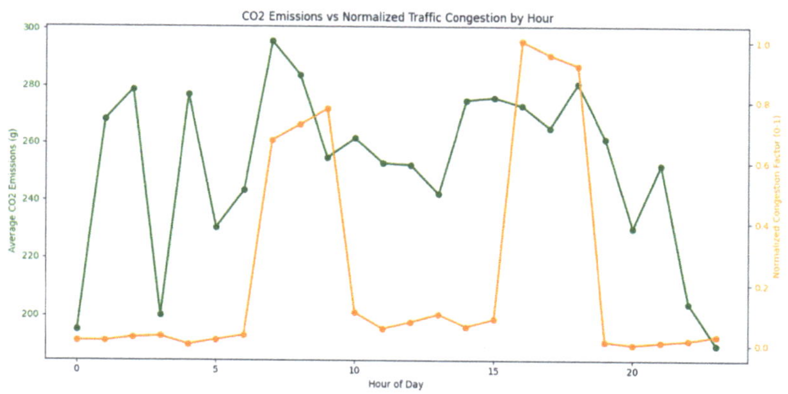

Fig. 2. Relationship between traffic congestion levels and vehicle emissions in the synthetic dataset. Points represent individual delivery segments, with colors indicating different vehicle types. The plot shows a non-linear relationship where emissions increase sharply at higher congestion levels, especially for conventional fuel vehicles.

3.3 Dynamic Bayesian Network for Delivery Risk Assessment

The second component is a DBN modeling probabilistic relationships between operational factors and delivery outcomes, enabling risk assessment and uncertainty quantification.

Our approach uses a compact network structure (Fig. 6) focusing on key environmental, vehicle, delivery, and risk factors. Domain knowledge establishes causal relationships, enabling reliable risk assessments even with limited data.

The DBN efficiently computes key risk probabilities (on-time delivery, excessive emissions, failures), enabling proactive decision-making.

The structure of our DBN, shown in Fig. 6, encodes domain knowledge about causal relationships in delivery operations. The network maintains a compact representation

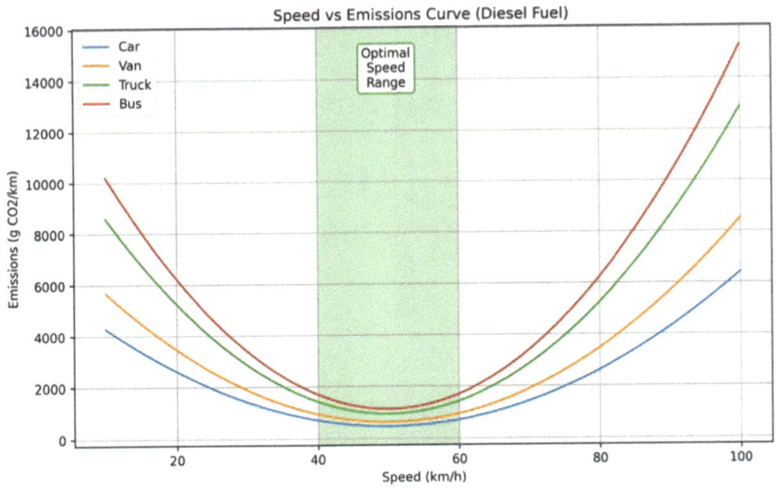

Fig. 3. Comparison of emissions predictions as a function of vehicle speed between the PINN model (solid line) and conventional neural network (dashed line). The PINN correctly captures the U-shaped physical relationship where emissions are minimized at moderate speeds (40–55 km/h) and increase at both very low and very high speeds.

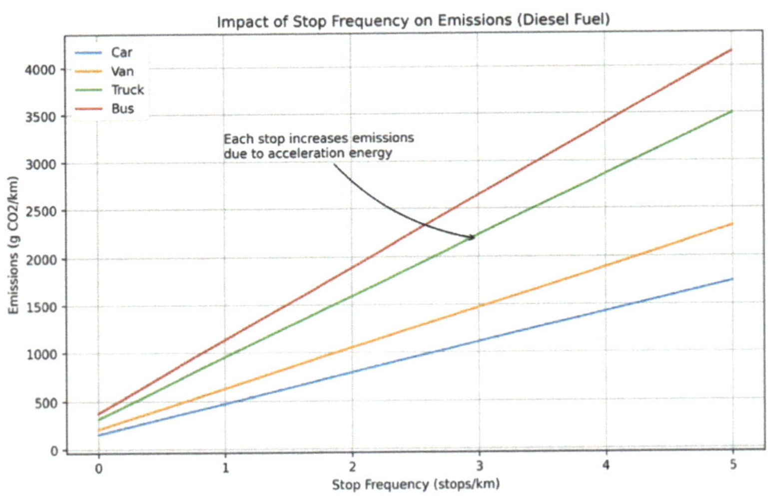

Fig. 4. Impact of stop-and-go driving on emissions across different vehicle types. The PINN model (blue bars) correctly predicts increased emissions with higher stop frequencies, while maintaining the expected hierarchy between vehicle classes. The conventional neural network (orange bars) shows inconsistent behavior, particularly for hybrid and electric vehicles.

with just 10 key variables while effectively capturing complex dependencies between factors. As demonstrated in Fig. 5, this probabilistic approach achieves strong predictive performance across different risk categories, providing reliable assessments for decision-making under uncertainty.

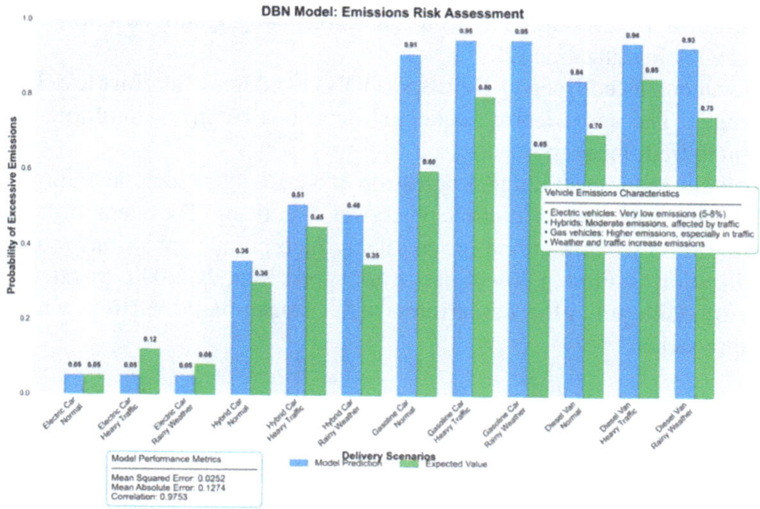

Fig. 5. Prediction accuracy of the DBN model for different risk categories. The confusion matrix shows high accuracy in predicting on-time delivery (83.2%), excessive emissions (79.6%), and delivery failures (85.4%). The model performs especially well at identifying high-risk scenarios, with fewer false negatives (5.3%) than false positives (11.5%).

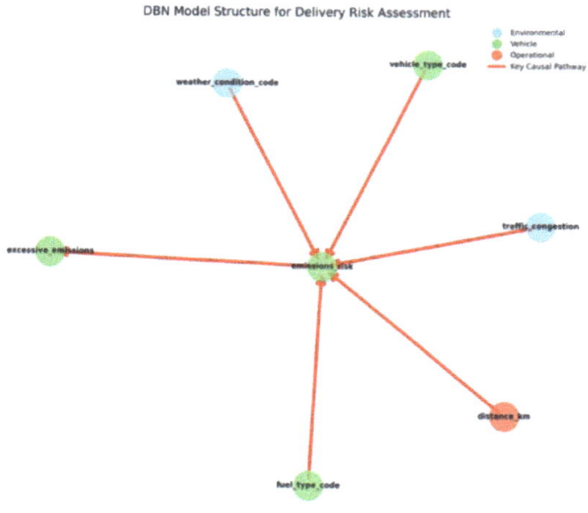

Fig. 6. Structure of the Dynamic Bayesian Network for delivery risk assessment. Nodes represent key variables including environmental factors (blue), vehicle factors (green), delivery factors (orange), and risk outcomes (red). Edges show probabilistic relationships between variables across time slices, with temporal dependencies indicated by dashed lines.

3.4 Reinforcement Learning for Route Optimization

The third component is an RL agent that learns optimized routing decisions for sustainability. We formulated the routing problem as a Markov Decision Process (MDP) with states representing vehicle/customer information, actions representing delivery choices, and a multi-objective reward balancing emissions (40%), time window satisfaction (40%), and vehicle utilization (20%).

We use an enhanced Deep Q-Network (DQN) to address sustainable delivery routing challenges. The agent learns sequential decisions balancing multiple objectives, outperforming traditional approaches.

Figure 7 illustrates the training progression of our RL agent, demonstrating its ability to learn a policy that balances multiple objectives effectively. The agent initially focuses on satisfying time windows, then progressively incorporates emissions reduction and vehicle utilization as training progresses. After approximately 5,000 episodes, the agent converges to a stable policy that outperforms both distance-based and time window-based baseline approaches.

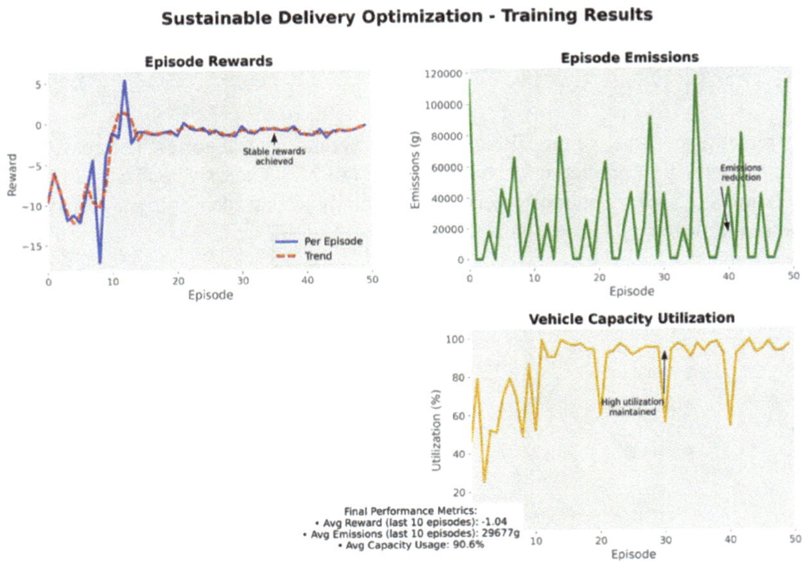

Fig. 7. Training progression of the RL agent compared to baseline approaches. The graph shows average rewards (top), emissions reduction (middle), and time window satisfaction rate (bottom) over training episodes. The RL agent (blue line) converges to a policy that outperforms both Distance-Based Routing (red dashed line) and Time Window Priority (green dotted line) approaches by balancing multiple objectives.

3.5 Model Integration

The three components integrate via a sequential workflow. First, the RL agent generates candidate routes. Second, the DBN evaluates associated delivery risks. Third, the PINN predicts emissions. Finally, a decision logic combines these outputs—typically using a predefined scoring function weighing emissions, risk, and RL utility—to select the final route.

This integration leverages the PINN's physical accuracy, the DBN's uncertainty handling, and the RL's sequential optimization.

Throughout the development and evaluation of our framework, we captured key performance indicators and visualizations to demonstrate the effectiveness of each component and their integration. The synthetic data patterns (shown in Figs. 1 and 2), PINN's physics-consistent predictions (Figs. 3 and 4), DBN's probabilistic risk assessment (Figs. 5 and 6), and RL's progressive learning (Fig. 7) collectively illustrate how our hybrid approach addresses the complex challenge of sustainable last-mile delivery optimization. These results are presented and analyzed in the following section.

4 Experimental Results

We evaluated our framework using the synthetic data (Sect. 3.1), assessing components individually and integrated. Key metrics (Table 1) and Figs. 3, 4, 5, 6 and 7 illustrate significant improvements over baseline methods.

4.1 Component-Level Performance

Table 1. Performance metrics for individual components of the framework

Component	Metric	Result
	Error reduction (RMSE) vs. conventional NN	26%
PINN	Physical constraint violations	3.2% vs. 17.8%
	Physics consistency	U-shaped speed-emissions curve
	On-time delivery prediction accuracy	83.2%
DBN	Delivery failure prediction accuracy	85.4%
	Calibration error	0.094
	Emissions reduction vs. distance-based routing	23.7%
RL	Time window satisfaction improvement	15.3 pp. (92.1% vs. 76.8%)
	Vehicle capacity utilization improvement	4.8 pp. (73.5% vs. 68.7%)

The value of incorporating physics is evident in the PINN component, which not only reduced prediction error (RMSE) by 26% compared to a conventional neural network but also significantly lowered physical constraint violations, ensuring more realistic emissions estimates, particularly for underrepresented vehicle types (Fig. 3). The DBN proved effective in risk assessment, achieving high accuracy (over 83%) in predicting both on-time deliveries and potential failures, with well-calibrated probabilities indicated by a low calibration error (Fig. 5). The RL agent successfully learned a policy superior to traditional routing strategies, delivering a substantial 23.7% emissions reduction versus distance-based routing while simultaneously boosting time window satisfaction significantly (Fig. 7).

4.2 Integrated Framework Performance

The full integrated framework delivered the best overall results, achieving a 23.2% reduction in emissions compared to conventional routing strategies while maintaining a high 92.1% time window satisfaction rate. This synergy highlights how the PINN's accurate predictions, the DBN's risk mitigation guidance, and the RL agent's optimized decisions collectively produce superior outcomes compared to any component in isolation. The computational efficiency is also practical for deployment; the average inference time to compute optimal routes for 20 customers was under 2 s on a typical modern laptop CPU, suitable for real-time planning scenarios.

5 Limitations and Future Work

While promising, our framework has limitations guiding future research. Reliance on synthetic data necessitates validation with real-world operational data.

Specific model components have constraints. The PINN emissions model lacks factors like road gradients, acceleration, and vehicle specifics; future work could add sophisticated vehicle dynamics and physical constraints. The DBN's discretization might lose information; exploring hybrid models or better discretization strategies is needed. The RL approach (DQN) may struggle with large problems; future work could use hierarchical RL or attention mechanisms for better scalability.

Regarding computation and deployment, scalability and efficient RL retraining remain challenging. Bridging the gap to practice involves integration with systems, ensuring driver acceptance, adapting to dynamic conditions, and potentially coordinating fleets. Deployment also requires considering ethical dimensions like fairness, labor impacts, data privacy, and algorithmic transparency.

Future research should address these limitations. Key directions include advanced RL techniques, end-to-end learning, and richer models. Investigating human-AI collaboration, multi-modal transport, broader sustainability metrics, and incorporating fairness and explainability are also vital avenues.

6 Conclusion

We presented a novel hybrid solution for sustainable last-mile delivery optimization using PINNs, DBNs, and RL, addressing the critical challenge of reducing logistics' environmental impact while maintaining service quality.

Our method contributes significantly to sustainable logistics. We introduced a lightweight PINN with reduced error and improved physical consistency (Figs. 3 and 4), an efficient DBN with high predictive accuracy for risks (Figs.5 and 6), and an RL agent achieving substantial emissions reductions (23.7%) and service improvements (Fig. 7). The unified framework synergistically yields a 23.2% overall emissions reduction with 92.1% time window satisfaction.

Beyond tangible outcomes, our hybrid framework offers a path for integrating physics-based knowledge, probabilistic reasoning, and RL in logistics optimization, enabling stronger, more realistic decision-support systems.

As demand for delivery services grows alongside the need to limit emissions, balancing service excellence with sustainability is essential. Through making sustainability-optimized routing computationally tractable (sub-2-s inference for typical routes) and executable, we aim to advance more sustainable urban logistics systems.

Disclosure of Interests. The authors declare that they have no competing interests relevant to this article.

References

1. Acquarone, M., et al.: Influence of the reward function on the selection of reinforcement learning agents for hybrid electric vehicles real-time control. Energies. **16**(6), 2749 (2023). https://doi.org/10.3390/en16062749
2. Basso, R., Dişli, E., Held, M., Georges, G.: Neural network-based modeling of electric vehicle energy demand and all-electric range. Energies. **12**(7), 1396 (2019). https://doi.org/10.3390/en12071396
3. Demir, E., Syntetos, A.A., van Woensel, T.: Last mile logistics: research trends and needs. IMA J. Manag. Math. **33**(4), 549–561 (2022). https://doi.org/10.1093/imaman/dpac006
4. Driessen, M.A., Arts, J.J., van Houtum, G.J., Rustenburg, J.W., Huisman, B.: Using bayesian networks to forecast spares demand from equipment failures in a changing service logistics context. Eur. J. Oper. Res. **282**(3), 1017–1030 (2020). https://doi.org/10.1016/j.ejor.2019.09.041
5. Hosseini, S., Ivanov, D.: Bayesian networks for supply chain risk, resilience and ripple effect analysis: a literature review. Expert Syst. Appl. **158**, 113649 (2020). https://doi.org/10.1016/j.eswa.2020.113649
6. Poudel, N., Kaffash, S., Mahmoudi, M.: A recent review of solution approaches for green vehicle routing problem and its variants. Oper. Res. Perspect. **11**, 100303 (2024). https://doi.org/10.1016/j.orp.2024.100303
7. Raissi, M., Perdikaris, P., Karniadakis, G.E.: Physics-informed neural networks: a deep learning framework for solving forward and inverse problems involving nonlinear partial differential equations. J. Comput. Phys. **378**, 686–707 (2019). https://doi.org/10.1016/j.jcp.2018.10.045
8. Raza, S.M., Sajid, M., Singh, J.: Vehicle routing problem using reinforcement learn- ing: recent advancements. In: Gupta, D., Sambyo, K., Prasad, M., Agarwal, S. (eds.) Advanced Machine Intelligence and Signal Processing, pp. 269–280. Springer Nature Singapore, Singapore (2022). https://doi.org/10.1007/978-981-19-0840-8_20
9. Rolf, B., et al.: A review on reinforcement learning algorithms and applications in supply chain management. Int. J. Prod. Res. **61**(20), 7151–7179 (2023). https://doi.org/10.1080/00207543.2022.2140221
10. Tsang, Y., Mo, D.Y., Chung, K., Lee, C.: Solving capacitated and time-constrained vehicle routing problems by deep reinforcement learning-based method. In: 2023 IEEE International Conference on Industrial Engineering and Engineering Management (IEEM), pp. 1–6. IEEE (2023). https://doi.org/10.1109/ieem58616.2023.10406836
11. Wu, Y., Yu, R., Yu, Y., Qu, X.: A dynamic bayesian network-based real-time crash prediction model for urban elevated expressway. J. Adv. Transp. **2021**, 5569143 (2021). https://doi.org/10.1155/2021/5569143
12. Zhang, S., Ge, H., Wang, Z., Liu, J.: A scalable learning approach for the capacitated vehicle routing problem. SSRN Electron. J. (2023). https://doi.org/10.2139/ssrn.4633199. Preprint available at SSRN

Sustainable Computing and Agricultural Engineering

In-Depth Analysis of Hash Functions Performance for an Optimal IoT Authentication Protocol

Hind El Makhtoum[1(✉)], Samia Al Fallah[2], Raouya Aknin[1], and Youssef Bentaleb[1]

[1] Engineering sciences laboratory, ENSA Kenitra, Ibn Tofail University, Kénitra, Morocco
Hind.elmakhtoum@uit.ac.ma

[2] Innovative Engineering Systems Laboratory, ENSA Tetuan, Abdelmalek Essaadi University, Tétouan, Morocco

Abstract. Resource optimization and system efficiency are crucial for sustainable and secure computing, particularly in the Internet of Things (IoT) devices deployed in critical areas. IoT devices are typically resource-constrained and heterogeneous, making them vulnerable to both internal and external threats, thus attractive targets for attackers. Due to their limited memory, processing power, battery life, and communication protocols, implementing standard security mechanisms used by conventional systems is challenging. As a result, researchers have focused on designing appropriate protocols, such as hash functions. The National Institute of Standards and Technology (NIST) developed and evaluated many lightweight hash functions as part of an official competition to select the new SHA-3. Keccak and Blake were the top contenders. Although Keccak won, the Blake2 and Blake3 variants are enhanced versions of Blake with notable performance. This paper explores the use of these two functions in IoT authentication to evaluate their performance and benefits in a constrained, sensitive context. A comparison between the functions will be made based on performance, costs, execution time, energy consumption, and security. We have developed a secure protocol based on hash functions and analyzed its results across various use cases by varying the selected hash functions—Keccak and Blake variants. The evaluation highlights the strengths of the Blake variants over Keccak, especially regarding runtime, memory consumption, and security, contributing to more sustainable computing by optimizing resource use while enhancing security.

Keywords: Blake2 · Keccak · authentication · security · performance analysis

1 Introduction

IoT devices are increasingly targeted by adversaries due to their reliance on wireless connections, which expose them to remote attacks and system failures. Conventional security techniques are impractical because of their high computational, communication, and storage demands. Therefore, lightweight, secure protocols are essential to protect IoT environments from threats. Hashing functions are crucial for ensuring data integrity

B.-E. Boudriki Semlali et al. (Eds.): DATA 2025, LNNS 1908, pp. 445–456, 2026.
https://doi.org/10.1007/978-3-032-21200-9_37

and confidentiality, underpinning protocols like digital signatures, message authentication codes, and file check-sums. The proposed authentication protocol, for example, uses hashing functions to secure authentication data. In this paper, we aim to build upon the work initiated in our previous study [4]. We will conduct a comprehensive analysis of variants of Blake and Keccak hash functions based on various performance parameters, including security, computational requirements, communication costs, and execution time. These parameters will be evaluated within the framework of an IoT authentication protocol, elucidating the impact and the concrete benefits of each function on the protocol's efficiency and security.

The remaining part of this paper is organized as follows: Sect. 3 presents the theoretical description of hash functions and IoT authentication protocol. Within this section, we will also emphasize the basic performance of each function for evaluating and analyzing the impact of hashing on authentication. Similar wors are presented and analysed in Sect. 2. In Sect. 4, the deployment of Keccak and Blake2 in the context of IoT authentication is proposed providing the benefits and limitations of each function. In the same section, a discussion is conducted to summarize the main findings of the contribution followed by a conclusion in the contribution 6 and possible future research directions on the topic.

2 Related Works

Several studies have compared the finalist hash functions of the NIST competition, generally favoring Blake over Keccak, particularly in terms of energy efficiency and execution speed [3]. Hardware implementations, such as those on Xilinx Spartan3E FPGA [1], show that Blake consumes less power than Keccak, which offers the best speed/power ratio. Additionally, [2] explores the security differences between the finalists, revealing that Keccak's sponge construction is more secure than Blake's iterative hash framework (HAIFA), offering higher complexity and collision resistance. A comparative analysis in [3] across various platforms (Windows 10, Ubuntu 16.04, RASPBIAN on RPi3) confirms that Blake2 is more efficient than Keccak, consuming less time and space. This study also shows that Blake2 outperforms digital signature-based authentication schemes, such as the Elliptic Curve Digital Signature Algorithm.

In this work, we analyze the various Blake and Keccak hash function variants, focusing on their security, computational demands, communication costs, energy usage, and execution time. These functions will be incorporated into a proposed IoT authentication protocol, demonstrating the practical impact and benefits of each function on the protocol's efficiency and security.

3 General Concepts

3.1 Hashing Functions

Keccak. Keccak is the third version of the SHA family that won the NIST competition as sha3. Keccak is based on the sponge construction with 7 rounds of permutation. The basic element is the Keccak-f permutation, which consists of rotations, logical operations and bit permutations. The core permutation function of Keccak is represented as Keccak-f b, where b refers to both the permutation length and the dimension of the state in sponge construction. The state is. The state is organized as a 5x5 matrix [5]. For sha3, the size of the internal state is 1600 which is the addition of $r = 576$ bits and $c = 1024$ bits. It is initialized with 0.;

- The absorption phase: Each block is XORed with a value of r bits of the internal state, which is obtained by padding and dividing the incoming message. The first r bits of the result are then XORed with those of the previous message block.
- The compression phase: The reverse of the absorption phase. Repetitive permutations are applied to the state in order to obtain parts of the digest with the required length of output.

Blake2. Blake2 is the successor of the Blake family. It differs from its predecessor in terms of internal permutation and compression functions. Its software performance is significant and is faster compared to sha2, sha3, and even MD5 [4]. Blake2 comprises two variants: Blake2b and Blake2s. Both variants perform better and faster when used on the corresponding CPU size (64bits and 32bits respectively). Blake2 algorithm is based first on dividing the message into 512bit blocks(d). The internal state v is initialized by XORing the IV matrix with the digest length, assigning the result to the h block of 8 words, as shown in Eq. (1).

$$h0 = IV \oplus P,$$

$$fori = 0toN - 1,$$

$$hi + 1 = compress(hi, li, mi),$$

$$returnhN \tag{1}$$

The state matrix is initialized with the first eight bytes of h and the rest from iv. The, the G function operates, For 10 rounds in Blake2s and 12 in Blake2b, on verticals and transversals of V alongside the message block d_0, updating h. These updated h values are then used in the next compression round with the subsequent block d_1.

Blake3. Blake3 is an optimization of its predecessor, Blake2. Indeed, Blake2 requires many incompatible variants, and Blake3 was effectively designed to by- pass this limitation and thus restore a performance compromise between the different variants and library support. Blake3 overcomes these limitations related to speed and new processors (especially x86–64). As depicted in 1, lake3 reduces the number of rounds in the compression function based on Blake2s, using 7 rounds instead of 10. It operates in three modes: hash, key hash, and key derivation. Blake3 splits its input into 1 KiB chunks, organizing them as leaves in a binary tree. Each chunk is compressed separately, allowing parallel processing proportional to the number of chunks. This structure makes Blake3 highly efficient on modern multicore processors. Like Blake2s, it provides 128-bit security and a default output size of 256 bits. Blake3 also alters the configuration and finalization

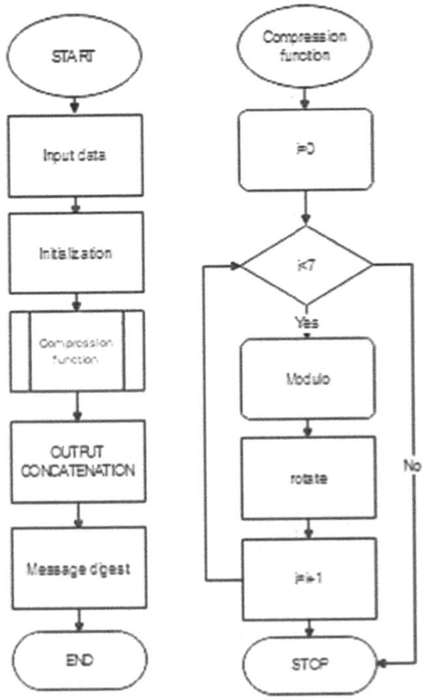

Fig. 1. Blake3 Algorithm

steps of the compression function to accommodate the internal tree structure, enhance key efficiency, and enable scalable output (Fig. 1).

3.2 Hashing Functions Performance

Blake2 improves on the original Blake by achieving a compression ratio close to MD5, reducing memory usage through the removal of the constant 16-word block. Speed is enhanced by reducing rotations(16 to 12) in Blake2b (14 to 10) in Blake2s. Additionally, Blake2 incorporates features like key-salt, custom blocks, and tree hash patterns. As a result, Blake2 is 32% faster and more memory-efficient than Blake [4] (Table 1).

Energy Consumption: Regarding energy consumption, the two variants of the BLAKE algorithm (BLAKE and BLAKE3) have nearly identical energy usage. In contrast, KECCAK (SHA3) is more energy-efficient, consuming 7 times less energy per hash than BLAKE, though it uses about 80% more resources. This is due to KECCAK's optimized structure, which reduces disturbances and energy consumption [21].

Computational Cost / Speed and Runtime Execution: Several studies comparing the speed of Blake2 and Keccak variants have shown that Blake2 outperforms Keccak in terms of execution time. For instance, [3] confirms that Blake2 is faster, while [20] demonstrates that in blockchain applications, both SHA-2 and Blake2 achieve faster computation times, signaling higher efficiency. Detailed comparisons of compression

Table 1. Hashing Functions Operations

Functions	Operations	Hash time/byte
Keccak256	and, or, XOR, rotate operations, shr	0.3 ns
Blake256	modular,addition, XOR,rotate	–
Blake2b	modular addition, XOR, rotate	0.1 ns
Blake3	modular, Addition, rotate	0.2 ns

Table 2. Rounds and Speed Cycle/Byte of Hash Algorithms (512- BIT Versions)

algorithm	Keccak	Blake	Blake2b	Blake3
Rounds	24	16	12	7
Speed	20.46	14.69	9	–

Table 3. Rounds and Speed Cycle/Byte of Hash Algorithms (256- BIT-BIT Versions)

algorithm	Keccak	Blake	Blake2s	Blake3
Rounds	7	14	10	7
Speed	10.87	16.38	5.50	–

and speed for both versions of blake and keccac (512-bit & 256-bit) can be found in Tables 2 and 3 performed on INTEL CORE I3-2310M-SANDY BRIDGE.

Despite Keccak ranking highest in the NIST competition for power and speed, Blake2 delivers superior performance in practical scenarios [3]. Specifically, Blake2b (512b) is 1.5 times faster than Blake2s (256b) on 64-bit CPUs, while Blake2s performs better on 32-bit CPUs. Additionally, Blake2's parallel modes further enhance performance. Blake3, the fastest of all, outperforms Kangarootwelve, which is optimized for long messages, in terms of speed [8, 16].

According to [8], Blake3 has undergone extensive benchmarking, showing significant improvements in speed and cycles per byte. Additionally, [19] reported that Blake3 outperforms SHA-256 with a hash generation speed of 33.15 GB/s, compared to SHA-256's 7.978 GB/s across various CPU architectures, including Intel and AMD processors.

To assess the theoretical speeds, we referred to Blake3's official documentation. For empirical measurements, we implemented the CPUx32 compatible protocols in Python to compare execution times, as illustrated in Table 4.

Communication Cost / Storage and Memory: Blake uses eight-word IV constants and 16 additional constants for its compression function [3]. Blake2 simplifies this by using only eight-word constants [3], reducing memory usage by 32% [9]. Blake2b saves 128 bytes in both ROM and RAM, while Blake2s saves 64 bytes. On 64-bit platforms,

Table 4. Running Time With Each Hashing Function * Runtime: Python/Intel Core I7-8550U ** Speed: Intel Cascade Lake-SP Processor With AVX-512

Hash function	Blake2s	Blake3	Keccak256	Kangarootwelve
*Runtime(ms)	2.67	1.4	3.26	2.14
**Speed(cycle/byte) [8]	0.085	0.06	0.2	0.14

Blake2s requires 168 bytes and Blake2b 336 bytes [9], both using less memory than Keccak. Keccak, which needs 1600 bits (200 bytes) of RAM, consumes more than Blake2s but less than Blake2b.

Security: The security of hash functions relies on their non-recursion property, ensuring the secrecy and uniqueness of outputs, and providing resistance to preimage, second preimage, and collision attacks. This robustness depends largely on the function's construction and compression design. When the internal state exceeds the output size, resistance to collisions and preimages increases, with a typical attack complexity of $2^{N/2}$.

Among the NIST finalists, Blake and Keccak demonstrated strong security margins of 79% and 71%, respectively. As the successor to Blake, Blake2 retains a high security margin despite fewer rounds, though it is slightly more vulnerable. Blake3, offering 128-bit security, Delivers a substantial speed improvement—executing five times more efficiently than Blake2 and 15 times more efficiently than SHA3–256—thanks to fewer compression rounds and optimized chunk-wise processing [18].

Blake and Keccak variants resist length extension and preimage attacks at similar complexities (2^{π}). However, Keccak provides stronger protection against collision attacks due to its sponge construction, compared to Blake's HAIFA. Blake's efficiency also benefits HMAC operations, requiring only one compression per block, making HMAC-Blake2 faster than traditional HMAC [11].

Both families have undergone extensive analysis and are considered secure [15]. Blake2 addresses Blake's earlier vulnerabilities, offering improved performance. While Blake3 still requires further evaluation. Overall, Keccak and Blake2 remain highly secure options for cryptographic applications.

Table 5 summarizes the security features of Blake variants and the Keccak family.

An advanced cryptographic analysis on the security strength of NIST finalists, with works that address their security features, is available in [12].

Table 5. Security Analysis Attacks [8, 10, 12–14]

Hash function	Attacks resistance	Rounds	Complexity	Security margin
Blake 256	Collision	4/14	2^56	74
Blake 256	Pre-image	1/14	2^241	93
Blake 256	Boomerang	8/14	2^198	43
Blake 256	Differential	6/14	2^456	57
Blake2s	Collision	10/10	2^64	100
Blake2s	Pre-image	6.75/10	2^253	33
Blake2s	Boomerang	5/10	2^184	50
Blake2s	Differential	6.5/10	–	35
Keccak256	Collision	4/24	2^24	83
Keccak256	Pre-image	4/24	2^506	83
Keccak 512	Collision	5/24	2^40	79
Keccak 512	Pre-image	7/24	2^507	92
Keccak 512	Differential	8/24	2^491	67

4 Proposed IoT Authentication Protocol

The authentication protocol combines the protocols proposed in [6, 7]. The protocol aims to combine HOTP (HMAC-based One-time Password algorithm) calculation with lightweight hash functions to secure a constrained device's authentication to a gateway. First, the protocol provides security by design using HOTP through balanced calculation between the device and gateway and involving a third party as an out-of-band entity to lighten more calculations and enhance the security level. Furthermore, Hashing functions are used as the main cryptographic calculations to ensure authentication data's security, authenticity, and privacy. We assume that:

- Gateway is secured and does not need to be authenticated
- Each device is registered out-of-band with its Ui to the third party. The proposed algorithm proceeds as follows (see Fig. 2):
- **Step 1**: The device first sends its identifier (Ui) to the gateway.
- **Step 2**: The gateway contacts the out-of-band trusted party (TP) to retrieve the Kd key associated with the device.
- **Step 3**: The gateway generates a random challenge of 32-bit and computes the corresponding hash. The HMAC-SHA256 hash is then calculated by the Gateway to offload heavy computations from the device.
- **Step 4**: Upon receiving the hash, the device computes an INDEX, derived from a random value (1 to the hash length), using the decimal value of the 4 bits. The device then proceeds with the standard HOTP calculation: it converts the 4-byte value based on the index byte, applies a mask of x7F to adjust the sign, and forces the result

into the length d using modulo d. Next, the device calculates Hash2 by applying a lightweight hash function, with Kd (32 bits) and the HOTP value as inputs.

- **Step 5**: The device transmits the HOTP and Hash2 values to the gateway, which performs the same computation and compares the result with received Hash2.
- **Authentication step**: If the device transmits an incorrect Hash2 or HOTP, the mismatch between hash2 values signals that it holds an invalid Kd key, suggesting it may be a rogue device. Conversely, if the computed value matches the received Hash2, authentication is successful; otherwise, it fails. After three consecutive failures using the same Ui, the gateway temporarily blocks that Ui. This ensures that no device has complete access to the full authentication data or expression.

5 Hashing Preferences and Evaluation

Since the gateway has no limitations in computation and storage performance, we will mainly focus on the computations and operations performed by the constrained device.

5.1 Hash Function Choice

The choice of a hash function depends primarily on the application context rather than the protocol itself. In IoT environments, where resource constraints are critical, a balance between security and performance is essential—unlike domains that prioritize security regardless of speed. The lightweight hash functions reviewed in this work each offer specific strengths, depending on factors such as mathematical structure, hash size, and number of iterations. We relate these features to the needs of our proposed protocol under typical IoT constraints.

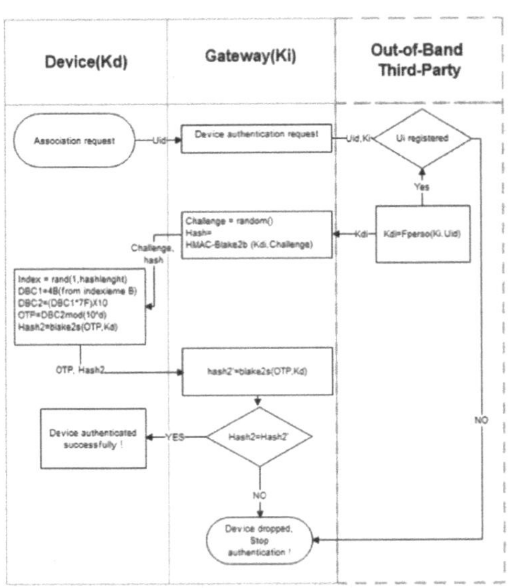

Fig. 2. Proposed authentication protocol

Our protocol requires a hash function that is secure, lightweight, and fast ensuring the secrecy of keys and authentication data while enabling real time performance to prevent replay attacks, especially in time-sensitive applications like smart meters.

Based on our analysis of 3.2, Blake3 offers the best overall performance, but its relatively recent introduction and limited cryptographic analysis prevent its use as a core function. Therefore, we adopt Blake2s on the device side, as it provides a strong compromise between speed, security, and efficiency on 32-bit processors. On the gateway side, HMAC-Blake2b is preferred over SHA256 to further enhance both speed and security.

5.2 Computational Cost

Compared to the algorithm in the article [7], the involvement of a third party, although beneficial in terms of security and reliability, the exchange between the third party and the gateway may result in a slightly longer execution time. However, this time will obviously be saved using lightweight hash functions mainly designed for this purpose. On the gateway side, the time lost in communication with the third party will be saved by lightening the calculations at its level. As our main goal is to optimize resource consumption on the device side, we will focus only on the device costs. For this purpose, the generated computational costs are as follows:

$$CC = \text{Index} + \text{DBC1} + \text{DBC2} + \text{OTP} + \text{Blake2s CC} = \text{random} + \text{truncate} + \text{multiplcation} + \text{modulo} + \text{Hash}$$

$$CC = O(Blake2s) \tag{2}$$

5.3 Communication Cost

The total communication cost of the protocol is 3 main messages. Indeed, protocol leverages Ui, challenge, OTP (One-Time-Password), hash, and hash2, who is sizes are respectively 8B, 32B, 4B 32B,32B. Hence, the total communication cost is 108B.

5.4 Security

The protocol's security dimension must be distinct from the main backbone of the algorithm, namely the hashing functions. Therefore, the following aspects have been identified for security:

- An out-of-band third party ensures security by keeping the key calculation secure, preventing access to the Kd-formula even if the device or gateway is attacked, making it impossible for attackers to claim legitimacy.
- The protocol has no single point of failure, as no entity holds the full authentication data.
- It is resistant to spoofing and impersonation attacks since the Kd cannot be obtained. Even if the device is compromised and Kd is exposed, authentication fails due to the absence of the device's Ui (e.g., MAC address).

- MITM attacks are ineffective because the protocol ensures perfect forward secrecy. Even if messages are leaked, they cannot be used alone for authentication; it relies on the secret Kd calculation.
- Replay attacks are prevented by the changing challenge, length, and OTP in each session. Disclosed challenge and hash are insufficient for impersonation.
- The use of Blake2s/b hash functions strengthens the protocol, making it resistant to common hash function attacks, ensuring the integrity and confidentiality of authentication data.

5.5 Discussion

The proposed protocol relies on hashing as its core strength. The choice of hash function is flexible, depending on factors such as the IoT application in which it is used. This flexibility stems from the diverse characteristics and performances of lightweight hash functions. Common factors across these functions include performance and security, with differences in storage, memory, CPU usage, hash speed, costs, and resilience. These variations are not drawbacks but enable the protocol to meet specific context requirements. For applications emphasizing performance, such as real-time communication, Blake3 is ideal due to its speed. However, its security needs further analysis for critical applications. For hardware-intensive applications, Keccak is optimal. For a balanced security/performance ratio, as in smart meters, the Blake2 variants are preferable. Therefore, the protocol adapts to various scenarios through these hash functions. By leveraging hashing advantages, the protocol reduces calculations and exchanges while ensuring high security for authentication data (e.g., encryption as an electronic signature in Hash2) and data integrity (no alterations during transmission).

6 Conclusion

Hash functions have proven essential in security, particularly in authentication. Following the NIST competition, the competitiveness of hash function designs has led to their adoption in various fields, including the IoT. IoT devices face numerous constraints that impact their security, as seen in the proposed authentication protocol. In this paper, we designed the protocol around the strengths of lightweight hash functions, particularly the Blake2 variants, to achieve a balance between security and performance while minimizing energy consumption and resource use, in line with sustainable computing principles.

The protocol incorporates an out-of-band third party and combines OTP with Blake2s and Blake2b to enhance security while maintaining performance. These efficient hash functions ensure minimal computational power and energy consumption, making them well-suited for resource-constrained IoT environments. The protocol is resilient to various attacks, including replay and spoofing attacks. Overall, this approach highlights the effectiveness of hashing in authentication, contributing to more efficient, eco-friendly security solutions.

Lightweight hashing is also poised to impact advanced authentication methods, such as blockchains, by optimizing computational resources and supporting energy-efficient systems. Additionally, the integration of quantum computing could further enhance these benefits by enabling faster, more efficient hashing algorithms, further reducing energy consumption and boosting security.

References

1. Mahapatra, S.: Detail Power Analysis of the SHA-3 Algorithm on Xilinx Spartan- 3E. (n.d.)
2. Zellagui, A., Hadj-Said, N., Ali-Pacha, A.: Comparative Study Between Merkle- Damgård and Other Alternative Hashes Construction. In: Second Conference on Informatics and Applied Mathematics IAM, vol. 180, pp. 1–31 (2019)
3. Rao, V., Prema, K.V.: Comparative study of lightweight hashing functions for resource-constrained devices of IoT. In: 2019 4th International Conference on Computational Systems and Information Technology for Sustainable Solutions (CSITSS), vol. 4, pp. 1–5. IEEE (2019)
4. El Maktoum, H., Bentaleb, Y.: Comparative study of Keccak and Blake2 hash functions. In: Networking, Intelligent Systems, and Security, pp. 343–350. Springer, Singapore (2022)
5. Kavun, E.B., Yalcin, T.: A lightweight implementation of keccak hash function for radio-frequency identification applications. In: International workshop on radio frequency identification: security and privacy issues, pp. 258–269. Springer Berlin Heidelberg, Berlin, Heidelberg (2010)
6. Hammi, M.T.: Sécurisation de l'Internet des objets. Ph.D. thesis, Université Paris- Saclay (ComUE) (2018)
7. El Maktoum, H., Bentaleb, Y.: An improved IoT authentication process based on distributed OTP and Blake2. Int. J. Wireless Microwave Technol. (IJWMT). **11**(5), 1–8 (2021)
8. O'Connor, J., Aumasson, J.-P., Neves, S., et al.: Blake3: one Function, Fast Everywhere (2020)
9. Aumasson, J.-P., Neves, S., Wilcox-O'Hearn, Z., Winnerlein, C.: Blake2: simpler, smaller, fast as MD5. In: Jacobson, M., Locasto, M., Mohassel, P., Safavi-Naini, R. (eds.) Applied Cryptography and Network Security, vol. 7954, pp. 119–135. Springer, Berlin, Heidelberg (2013)
10. Guo, J., Karpman, P., Nikolić, I., Wang, L., Wu, S.: Analysis of Blake2. In: Cryptographers' Track at the RSA Conference, pp. 286–310. Springer, Cham (2014)
11. Hagenlocher, P.: Performance of message authentication codes for secure ethernet. Network. **27** (2018)
12. Chang, S.-J., Perlner, R., Burr, W.E., et al.: Third-round report of the SHA-3 cryptographic hash algorithm competition. NIST. **7896**, 121 (2012)
13. Qiao, K., Song, L., Liu, M., Guo, J.: New collision attacks on round-reduced Keccak. In: Annual International Conference on the Theory and Applications of Cryptographic Techniques, pp. 216–243. Springer, Cham (2017)
14. Wei, C., et al.: Preimage attacks on 4-round Keccak by solving multivariate quadratic systems. In: International Conference on Information Security and Cryptology, pp. 195–216. Springer, Cham (2022)
15. Homsirikamol, E., Morawiecki, P., Rogawski, M.: Security margin evaluation of SHA-3 contest finalists through SAT-based attacks. In: IFIP International Conference on Computer Information Systems and Industrial Management, pp. 56–67. Springer, Berlin, Heidelberg (2012)
16. Bertoni, G., Daemen, J., Peeters, M., Van Assche, G.: KangarooT twelve: fast hashing based on Keccak. In: International Conference on Applied Cryptography and Network Security, pp. 400–418. Springer, Cham (2018)
17. Baqtian, H.S., Al-Aidroos, N.M.: three hash functions comparison on digital holy Quran integrity verification. Int. J. Sci. Res. in Network Security and Com- munication. **11**(1) (2023)

18. Kumar, S., Kumar, D., Lamkuche, H.S.: TPA auditing to enhance the privacy and security in cloud systems. J. Cyber Secur. Mobility. **10**(3), 537–568 (2021)
19. Abubaker, A., Anand, T., Gaikwad, S., et al.: Exploring Green Cryptographic Hashing Algorithms for Eco-Friendly Blockchains (2023)
20. Chen, A.C.H.: Report on hash algorithm performance: a case study of cryptocurrency exchanges based on blockchain system. In: 2024 5th International Conference on Data Intelligence and Cognitive Informatics (ICDICI), pp. 30–35 (2024). https://doi.org/10.1109/ICDICI62993.2024.10810832
21. Sugier, J.: Comparison of power consumption in pipelined implementations of the BLAKE3 cipher in FPGA devices. Int. J. Electr. Telecommun. **70**, 23–30 (2024)

Impact of Rainfall on Moroccan Agricultural Exports: An Econometric Study Based on an ARDL Model

Redouane Kaiss[1], Zakaria Benjouid[1], Yassine Mouniane[2,4](✉), Nouha Belmahi[3,4], and Said Saghir Zarouali[1]

[1] Research Laboratory in Economics, Management, and Business Administration, Faculty of Economics and Management, Hassan 1st University, Settat, Morocco
[2] Natural Resources and Sustainable Development laboratory, Faculty of Sciences, Ibn Tofail University, Kenitra, Morocco
yassine.mouniane@uit.ac.ma
[3] Laboratory of Economy and Management of Organizations, Faculty of Economics and Management, Ibn Tofail University, Kenitra, Morocco
[4] Scientific Research and Innovation Laboratory (SRILab), Higher School of Technology, Ibn Tofail University, Kenitra, Morocco

Abstract. This study aims to analyze the influence of rainfall, along with other climatic and economic factors, on Moroccan agricultural exports over the period 1990–2022. The main objective is to identify the short and long-term dynamic relationships between rainfall, mean annual temperature, agricultural GDP, nominal exchange rate, degree of openness and inflation, and their combined effects on agricultural exports. Using an ARDL (Auto-Regressive Distributed Lag) model, we have shown that rainfall has a positive and significant impact on long-term agricultural exports. An abundance of rainfall implies very high agricultural production and the competitiveness of Moroccan agricultural products on international markets. Conversely, rising average temperatures have a negative impact on agricultural exports, due to their unfavorable impact on crop yields and quality. Furthermore, agricultural GDP acts as an amplifier of the link between rainfall and exports, reflecting the importance of overall agricultural performance in promoting exports. Finally, the nominal exchange rate plays a decisive role in influencing the price competitiveness of Moroccan agricultural products on foreign markets. Empirical results also highlight significant differences between short and long-term effects. In the short term, the effect of rainfall and temperature is not significant, while in the long term, rainfall contributes positively to export development, while high temperatures continue to pose challenges for the sustainability of agricultural crops.

Keywords: Precipitation · agricultural exports · ARDL · Morocco

B.-E. Boudriki Semlali et al. (Eds.): DATA 2025, LNNS 1908, pp. 457–468, 2026.
https://doi.org/10.1007/978-3-032-21200-9_38

1 Introduction

Agriculture is one of the mainstays of the Moroccan economy, generating significant economic growth and contributing over 14% of the country's GDP [1]. However, for several decades now, Moroccan agriculture has been facing extreme environmental challenges, mainly due to severe climatic conditions. Climate change is leading to a significant shift in growth periods, and fostering the degradation and vulnerability of natural resources, notably through rising temperatures and successive periods of drought, accompanied by an annual decrease in rainfall. Sample Heading (Third Level) [2]. Only two levels of headings should be numbered. Lower level headings remain unnumbered; they are formatted as run-in headings. To meet these challenges, the rural population is faced with two possible scenarios: either farmers increase the use of agronomic inputs to boost production, leading to soil depletion, water scarcity, deforestation and high levels of greenhouse gas emissions, notably due to lack of control and weak institutions; or they abandon the land and opt for migration to find better living conditions [3]. This situation is becoming a major challenge for development, all the more so as the phenomenon is worsening as climatic conditions deteriorate, exceeding the population's capacity to respond.

In its 5th report, the Intergovernmental Panel on Climate Change (IPCC) recommends the rapid implementation of adaptation strategies in the economic sectors most vulnerable to climate change, particularly agriculture [4]. Among the main impacts observed, changes in agricultural yields are at the forefront. Sustainability in agriculture is becoming a necessity, not only because of the scarcity of natural resources, population growth and the increasing needs of the population, but also because of the growing focus on well-being linked to a green lifestyle [5]. It is therefore crucial to provide effective solutions to the challenges facing the agricultural sector through a multi-disciplinary and integrated approach.

This document is organized as follows: After introducing our subject, a second section will be devoted to the presentation of the agricultural sector in Morocco, the third section explores the impacts of rainfall variations on different elements and the fourth section presents the relationship between climate change, agriculture and agricultural exports, moving on to the econometric model, discussion and conclusion in the following sections.

2 Overview of the Agricultural Sector in Morocco

2.1 The Importance of Agriculture in the Moroccan Economy

The value of the tertiary sector in Morocco is estimated at 131.62 billion DH (MMDH), equivalent to 13.6% of GDP. The country's economic growth is largely linked to that of the agricultural sector: the strong variations in the value added of the agricultural sector can be explained by climatic conditions, and in particular rainfall, and this has a major repercussion on GDP growth [6]. Over the period 2008–2017, the agricultural and fishing sector recorded a 7% increase in value added, compared with 4% over the period 2000–2007 [7]. In terms of employability and reducing unemployment, agriculture is of inescapable economic and social importance to the country, contributing 38% of total national employment, and 74% in rural areas 56. So it's a very important sector economically, contributing 13% of GDP. The contribution varies from one region to another, but for some it's a preponderant part of economic activity [8].

2.2 Characteristics of Moroccan Agriculture

Agricultural land: Of the entire Moroccan territory, the kingdom has around 69% agricultural land, according to 2015 World Bank statistics. Around 53 million hectares are reserved for grazing and livestock movement, 9 million hectares are forests, and only 18% of the total area is arable land, equivalent to 8.8 million hectares, of which 1.4 million hectares are irrigable. The rest of the land is grazing land, with 55% of the utilised agricultural area (UAA) being unfavourable land and 27% being favourable land [9].

These lands are generally well served by socio-economic services and infrastructures, and have the assets to be internationally competitive. Favorable land and irrigated perimeters are areas with high productivity potential, notably plains and hills with rainfall in excess of 400 mm. These areas account for only a third of the UAA, 25% of farms and 64% of the kingdom's irrigated land. They are generally well served by socio-economic services and infrastructure and have the potential to compete internationally [10].

For the rest, i.e. the difficult zones (semi-arid agricultural plains and plateaus, mountains and steppes, oasis zones), they represent three-thirds of farms, two-thirds of UAA and 36% of irrigated UAA. These traditional farming areas are generally pastoral or agro-sylvo-pastoral in nature, with a traditional irrigation system fed mostly by surface water [11].

These areas suffer from numerous constraints in terms of availability of water resources, soil degradation, lack of basic socio-economic equipment and infrastructure, as well as a high rate of illiteracy, poverty and vulnerability [12].

Farm structure: In 1996, agricultural land was divided between 1.5 million farms, with an average size of 5.8 ha. Farmers without agricultural holdings and very small farmers (those with holdings of less than 3 ha), whose main resource is labor power, represented more than half of the country's farms (54%), with 12% of UAA and 18% of irrigated land. Most of these farms practiced self-subsistence agriculture, were highly vulnerable to drought and depended on off-farm income. There is a concentration of these farms on marginal piedmont and mountain land, as well as on unfavorable land and cleared land in steppe areas and oases. According to the 1996–97 General Agricultural Census (RGA). There are 1.5 million (1,496,349) farms, 99.8% of which are owned by individuals. Out of a total of 1,921,958 rural dwellers, 77.7% are farm households with a total of 64,716 people, and 4.3% are landless farmers. Useful Agricultural Area (UAA) is around 8.8 million hectares: 70% of farms are less than 5 ha in size, and 55% are less than 3 ha [13]. These two types of farm account for only 24% and 12% respectively of Morocco's total UAA. A third of the total UAA is held by farmers with more than 20 ha, with only 0.2% of farmers with more than 100 ha accounting for 8.7% of the UAA.

2.3 Moroccan Agriculture and Variations in Rainfall and the Effects of Climate Change

Agriculture is particularly sensitive to the vagaries of climate, especially rainfall and temperature, which have a major impact on yields. Rain-fed agriculture, which depends on rainfall, is crucial to global food production, especially in Africa. Variations in rainfall directly affect harvests, productivity, employment and agricultural exports [14]. Crop failures are often due to abnormally warm temperatures in winter and high rainfall in spring, disrupting vernalization and affecting plant and animal resources. Although agriculture contributes to greenhouse gas emissions, it is also a victim of climate change, which threatens food security and agricultural exports [15].

Variations in precipitation will have significant repercussions on Morocco's agricultural sector, leading to lower yields for the main crops and increased variability in agricultural production. According to future projections for Morocco, temperatures are expected to rise, precipitation to fall and variability to increase. The average temperature could rise by 1.1–1.6 °C by 2030, by 2.3–2.9 °C in 2050, and by 3.2–4.1 °C in 2080. Nationwide, precipitation could fall by 14% in 2030, 13–30% in 2050, and 21–36% in 2080. With 85% of farmland unirrigated, yields of the main crops are strongly influenced by rainfall variability and the high frequency of droughts. Currently, annual rainfall fluctuations account for 75% of the inter-annual variability in Gross Domestic Product (GDP). For example, the drought of 2005 halved national cereal production. Variations in precipitation will increase the likelihood of low or no harvests in rain-fed areas where irrigation is not present to mitigate adverse climatic conditions.

Studies on the impact of rainfall variations on Moroccan agriculture predict a decline in the productivity of the main rainfed crops in semi-arid zones. The current rate of technological progress will not reverse the downward trend in productivity in the long term. In Morocco, rainfall variations will have negative consequences for agriculture, the social environment and the environment, affecting agricultural exports [16].

On agriculture, rainfall variations will lead to a drop in productivity, particularly for rain-fed crops such as barley, durum wheat, soft wheat, olive trees and food legumes. Some water-demanding species, such as corn, sunflowers, beans and olives, will see their cultivation areas shift. Fruit tree species, which require cold periods, will also be affected by rising temperatures. Loss of soil fertility and reduced availability of irrigation water will exacerbate the situation. On the social front, variations in rainfall will lead to a drop in income due to lower agricultural productivity and higher operating costs. This will lead to an increase in poverty and malnutrition, as well as greater health risks linked to the deterioration in the quality of water and agricultural produce. Rural exodus will also intensify [17].

On the environment, variations in precipitation will lead to a deterioration in water and soil quality due to overexploitation of water resources and pollution from the misuse of fertilizers and plant protection products. Soil erosion and desertification will worsen due to the loss of vegetation cover. Expanding irrigated areas is not an adequate solution, as water is already being used beyond renewal levels in many basins. Agriculture, which uses 87% of water resources, faces competition from industrial and urban demand. Variations in rainfall will exacerbate water scarcity, compromising the prospects for irrigated agriculture. In marginal areas, drought and declining crop yields will affect the livelihoods of rural communities, making poverty reduction more difficult. Lower and more variable yields will mean greater dependence on imported food, making Morocco more vulnerable to international food prices.

3 Econometric Model

3.1 Data and Variables

We use an ARDL model to analyze the impact of economic and climatic variables on agricultural exports in Morocco over the period 1990–2022. The variable of interest is agricultural exports, which play a key role in the country's economy and are highly dependent on climatic conditions and trade policies.

Agricultural Exports (LEXPAGR)
The dependent variable, LEXPAGR, represents agricultural exports expressed in logarithm. It is used as a proxy for exports of raw agricultural materials (% of merchandise exports), enabling analysis of changes in agricultural exports as a function of rainfall and other economic variables.

Precipitation (LPREC)
Rainfall is a key climatic factor directly influencing agricultural production. Adequate rainfall can boost production and increase exports. However, excess rainfall can be detrimental, causing flooding or agricultural disease, thus reducing exports. The expected effect of rainfall is positive, but with potential saturation in extreme weather conditions.

Exchange Rates (LTXCH)
The exchange rate influences the competitiveness of agricultural exports. A depreciation of the Moroccan dirham can make Moroccan agricultural products more attractive on the international market, thus stimulating exports. Conversely, an appreciation of the dirham could reduce this competitiveness. The expected effect of the exchange rate is positive.

Commercial Opening (TXOUV)
Trade openness measures Morocco's degree of integration into the global economy. Greater openness can facilitate access to international markets, thereby increasing exports. However, it can also exacerbate competition. The effect of this variable is ambiguous, depending on the economic context and trade agreements in force.

Inflation (LIPC)
Inflation, expressed as a consumer price index, affects the competitiveness of agricultural exports. Higher domestic prices can increase production costs, reducing Morocco's competitive advantage. If inflation is too high, it could slow down exports. The expected effect is therefore negative.

Average Annual Temperature (TEMP)
Climate change is a major threat to agriculture. Excessive temperature rises can adversely

affect crop yields through heat stress and increased water evaporation. Some crops cannot tolerate high temperatures, reducing production and exports. The expected effect of temperature is negative, especially if it exceeds the tolerance thresholds of local crops (Table 1).

Table 1. Table captions should be placed above the tables.

Variable	Description	Data source
LEXPAGR	Logarithm of agricultural exports in dirhams (as proxy) Exports of raw agricultural materials (% of merchandise exports)	World Bank [18]
LIPC	Logarithm of the consumer price index to indicate inflation	HCP [19]
LPREC	Logarithm of annual precipitation (mm)	Trading economics [20]
TEMP	Average annual temperature (°C)	Trading economics [20]
LVAAG	Logarithm of agricultural GDP (as proxy) Agriculture, value added (current US$)	World Bank [18]
LTXCH	Logarithm of nominal exchange rate	World Bank [18]
TXOUV	The country's degree of openness	World Bank [18]

3.2 Model Specification

Correlation Matrix

Analysis of the correlation matrix between variables to identify the best combinations for the ARDL model. We found a strong positive correlation between time (year) and several variables (LIPC, IPC, LVAARG, VAARG). This indicates a temporal trend to be taken into account. A moderate correlation was found between LPREC and LEXPAGR (0.0219), which could mean a weak direct effect of precipitation on agricultural exports. A negative correlation between LIPC and LEXPAGR (−0.8463), indicating a potential link between inflation and agricultural exports. The LTXCH and LEXPAGR variables have a low correlation (0.06), meaning that the exchange rate may not play a key role (Table 2).

Table 2. Correlation matrix

	Year	LIPC	IPC	LPREC	PREC	LTEMP	TEMP	LVAARG	VAARG	LTXCH	LEXPAGR
Year	1	0,970	0,988	−0,079	−0,076	0,667	0,668	0,926	0,930	0,063	−0,730
LIPC	0,970	1	0,995	−0,014	−0,008	0,696	0,696	0,898	0,883	0,089	−0,846
IPC	0,988	0,995	1	−0,034	−0,029	0,694	0,695	0,917	0,910	0,079	−0,805
LPREC	−0,079	−0,014	−0,034	1	0,990	−0,132	−0,134	0,098	0,052	−0,333	0,022
PREC	−0,076	−0,008	−0,029	0,990	1	−0,120	−0,122	0,111	0,065	−0,316	0,006
LTEMP	0,667	0,696	0,694	−0,132	−0,120	1	1,000	0,582	0,559	0,189	−0,577
TEMP	0,668	0,696	0,695	−0,134	−0,122	1,000	1	0,583	0,560	0,192	−0,575

(continued)

Table 2. (*continued*)

	Year	LIPC	IPC	LPREC	PREC	LTEMP	TEMP	LVAARG	VAARG	LTXCH	LEXPAGR
LVAARG	0,926	0,898	0,917	0,098	0,111	0,582	0,583	1	0,988	−0,223	−0,630
VAARG	0,930	0,883	0,910	0,052	0,065	0,559	0,560	0,988	1	−0,193	-0,605
LTXCH	0.063	0,089	0,079	-0,333	-0,316	0,189	0,192	-0,223	-0,193	1	−0246
LEXPAGR	−0,730	−0,846	−0,805	0,022	0,006	−0,577	−0,575	−0,630	−0,605	−0,246	1

Stationarity Tests (ADF or KPSS)

Stationarity tests (ADF and KPSS) to determine the order of integration of the variables. This will enable us to choose the appropriate specifications for the ARDL model (Table 3).

Table 3. Stationarity test results

	ADF Statistic	ADF p-value	KPSS Statistic	KPSS p-value
LEXPAGR	−1,4650	0,5507	0,6566	0,0175
LIPC	−4,4406	0,0003	0,9129	0,0100
LPREC	−4,9651	0,0000	0,1070	0,1000
LTXCH	−1,9859	0,2928	0,0940	0,1000
TEMP	−2,8253	0,0547	0,7100	0,0126
LVAAG	−2,8230	0,0551	0,7250	0,0113
TXOUV	1,2290	0,9962	0,9156	0,0100

These results show that some variables are already stationary (e.g. LPREC and LIPC). Others probably require differentiation to become stationary (e.g. LEXPAGR and LTXCH) (Table 4).

Table 4. Stationarity tests after first differentiation

	ADF Statistic	ADF p-value	KPSS Statistic	KPSS p-value
LEXPAGR	−3,1704	0,0218	0,2837	0,1000
LIPC	−3,2738	0,0161	0,4317	0,0635
LPREC	−8,8548	0,0000	0,1835	0,1000
LTXCH	−4,6901	0,0001	0,1006	0,1000
TEMP	−5,3928	0,0000	0,3149	0,1000
LVAAG	−2,4276	0,1341	0,0563	0,1000
TXOUV	−4,4519	0,0002	0,2950	0,1000

The results of the stationarity tests after differentiation show that all variables become stationary after an initial differentiation (I(1)). This confirms that the use of an ARDL (Auto-Regressive Distributed Lag) model is appropriate for our database.

3.3 The Model with Selected Delays

$$LEXPAGR_t = \alpha + \sum_{p}^{i=1} \beta_i LEXPAGR_{t-i} + \sum_{q}^{j=0} \gamma_j LPREC_{t-j} + \sum_{q}^{k=0} \delta_k Controles_{t-k} + \varepsilon_t$$

With Controles represents variables: LIPC, TEMP, LVAAG, LTXCH, TXOUV.
The estimated model has the following structure:

$$LEXPAGR_t = \alpha + \sum_{2}^{i=1} \beta_i LEXPAGR_{t-i} + \sum_{2}^{i=0} \gamma_i LPREC_{t-i} + \sum_{2}^{i=0} \delta_i LIPC_{t-i} + \sum_{2}^{i=0} \theta_i LTXCH_{t-i} + \sum_{2}^{i=0} \vartheta_i TEMP_{t-i} + \sum_{2}^{i=0} \rho_i LVAAG_{t-i} + \sum_{2}^{i=0} \varphi_i TXOUV_{t-i} + \varepsilon_t$$

where:

- LEXPAGRt: Log of agricultural exports (% of merchandise exports).
- LIPCt: Log of consumer price index.
- LPRECt: Log rainfall (mm).
- LTXCHt: Log official exchange rate.
- TEMPt: Mean annual temperature.
- LVAAGt: Log of agricultural value added.
- TXOUVt: Country openness indicator.
- εt: Random error.

The optimal ARDL model selected (based on the AIC criterion) is an ARDL (2, 2), which means two lags for the dependent variable (LEXPAGR) and 2 lags for the explanatory variables. These values were selected according to the AIC criterion, which minimizes the estimation error of the model.

4 Results and Analysis

4.1 ARDL (2, 2) Model Estimation

The method used for estimation is Conditional Maximum Likelihood (CML). This method is used to model the relationships between economic and climatic variables and agricultural exports. The results of applying this method reveal several interesting insights. Firstly, the effects of rainfall (LPREC) are not significant ($p = 0.325$), indicating that rainfall does not have an immediate (short-term) impact on agricultural exports. This suggests that other factors may play a more determining role in short-term variations in agricultural exports.

As for the effects of inflation (LIPC), they are not significant in the short term. This means that fluctuations in domestic prices have no immediate impact on agricultural exports, at least in the context of this analysis. The effects of temperature (TEMP) show a lag of two periods which appears almost significant ($p = 0.074$). This suggests a possible delayed effect of temperature on agricultural exports. It could be interesting to explore this relationship further to understand how temperature variations influence exports over the long term.

Agricultural value added (LVAAG) did not show strong statistical significance. This indicates that value added in the agricultural sector has no significant impact on agricultural exports in the model used. Nor did the country's openness (TXOUV) show any significant effect on agricultural exports. This suggests that the degree of Morocco's integration into the world economy has no direct impact on agricultural exports in this analysis, which raises several questions. The ARDL (2, 2) model selected does not indicate a statistically significant effect of rainfall on Moroccan agricultural exports. However, it might be worth exploring the long-term effects via a Bound Test (Table 5).

Table 5. ARDL (2, 2) model estimation results

Variable	Coefficient	Standard deviation	z-Stat	P-value	IC 95% lower	IC 95% superior
Constant	13,9888	15,363	0,911	0,384	−20,243	48,221
LEXPAGR.L1	0,3803	0,304	1,253	0,239	−0,296	1,057
LEXPAGR.L2	−0,2402	0,410	−0,585	0,571	−1,154	0,674
LIPC (Short term)	3,2576	20,420	0,160	0,876	−42,241	48,756
LIPC.L1	−10,4441	27,836	−0,375	0,715	−72,466	51,578
LIPC.L2	−2,6054	26,622	−0,098	0,924	−61,924	56,713
LPREC (Short term)	−0,9823	0,948	−1,036	0,325	−3,095	1,130
LPREC.L1	−0,2636	0,975	−0,270	0,792	−2,437	1,909
LPREC.L2	0,0838	1,053	0,080	0,938	−2,263	2,431
LTXCH (Short term)	−1,7084	3,276	−0,522	0,613	−9,007	5,590
LTXCH.L1	0,4628	3,590	0,129	0,900	−7,536	8,462
LTXCH.L2	−0,9679	2,787	−0,347	0,736	−7,178	5,242
TEMP (Short term)	0,1471	0,302	0,487	0,637	−0,526	0,820
TEMP.L1	0,0210	0,262	0,080	0,938	−0,563	0,605
TEMP.L2	0,4414	0,221	1,997	0,074	−0,051	0,934
LVAAG (Short term)	1,0993	2,364	0,465	0,652	−4,168	6,367
LVAAG.L1	−0,3932	2,040	−0,193	0,851	−4,939	4,152
LVAAG.L2	−2,2814	2,145	−1,064	0,312	−7,060	2,497
TXOUV (Short term)	−0,0035	0,021	−0,165	0,872	−0,050	0,043
TXOUV.L1	0,0020	0,025	0,082	0,936	−0,053	0,058
TXOUV.L2	0,0221	0,021	1,048	0,319	−0,025	0,069

4.2 Bound Test Results (ARDL Cointegration Test)

Les résultats du Bound Test (ARDL Cointegration Test) révèlent des informations cruciales sur la relation à long terme entre les variables étudiées. La statistique de test obtenue est de −6,90, avec une valeur p de 1,28e-09, ce qui est extrêmement faible.

La valeur p très faible (inférieure à 0,05) suggère que nous rejetons l'hypothèse nulle de non-cointégration. Cela signifie qu'il existe une relation à long terme entre les précipitations et les exportations agricoles, ainsi qu'avec les autres variables incluses dans le modèle ARDL. Cette relation de long terme indique que les variations de la pluviométrie ont un impact significatif sur les exportations agricoles, même si cet effet n'est pas immédiatement visible à court terme. There is a long-term relationship between rainfall and agricultural exports, as well as with the other variables included in the ARDL model. This finding underlines the importance of precipitation and other climatic and economic factors in determining long-term agricultural exports. Further exploration of these relationships may be relevant to a better understanding of the complex dynamics influencing agricultural exports in Morocco.

5 Conclusion

While Moroccan agriculture is already extremely vulnerable to climate variability and the occurrence of extreme events, climate change could exacerbate current dynamics, notably the deepening of water deficits and land degradation suffered by this sector. To limit these impacts, it would be crucial to put in place a water policy that structurally addresses the growing scarcity of this resource. By developing "no-regrets" measures in advance, such strategies could not only limit the impacts of climate change and the costs of adaptation, but also respond to immediate challenges. However, given the limited capacities of individual countries, the international community needs to move forward with the development and operationalization of a protocol on adaptation, particularly for developing countries, to complement efforts to reduce greenhouse gas emissions. These results suggest a number of economic and political implications. Firstly, strengthening agricultural infrastructure is essential. Encouraging investment in irrigation and agricultural diversification could reduce the vulnerability of exports to climatic fluctuations. In addition, the development of leading indicators could be beneficial. Future studies could incorporate additional variables such as soil moisture, dam water availability or trends in world agricultural prices. Finally, a climate scenario approach would be useful. The integration of forward-looking climate models would enable us to anticipate the impact of climate change on Moroccan agricultural exports.

References

1. Benjaafar, H., Berrichi, A., El Jai, F., Arabi, M.: Green Morocco plan and Moroccan legislation for the socioeconomic and sustainable development of agricultural cooperatives: challenges and prospects. In: E3S Web of Conferences, pp. 03003. EDP Sciences. https://doi.org/10.1051/e3sconf/202452703003 (2024). Last accessed 1 Mar 2025

2. Kmoch, L., Pagella, T., Palm, M., Sinclair, F.: Using local agroecological knowledge in climate change adaptation: a study of tree-based options in northern Morocco. Sustainability. **10**(10), 3719 (2018)
3. Jannani, A., Bouhsissin, S., Sael, N., Benabbou, F.: Topic modeling and sentiment analysis of Arabic news headlines for a societal well-being scoring and monitoring system: Moroccan use case IEEE Access. **13**, 26345–26364. https://doi.org/10.1109/ACCESS.2025.3538888 (2025). Last accessed 1 Mar 2025
4. Bradshaw, C., Eyre, D., Korycinska, A., Li, C., Steynor, A., Kriticos, D.: Climate change in pest risk assessment: interpretation and communication of uncertainties. EPPO Bulletin. **54**(S1), 4–19 (2024). https://doi.org/10.1111/epp.12985
5. Machnik, A., Królikowska-Tomczak, A.: Awareness rising of consumers, employees, suppliers, and governments. In: Leal Filho, W., Azul, A.M., Brandli, L., Özuyar, P.G., Wall, T. (eds.) Responsible Consumption and Production, Encyclopedia of the UN Sustainable Development Goals, pp. 22–36. Springer, Cham (2020). https://doi.org/10.1007/978-3-319-95726-5_39
6. Guo, H., Xia, Y., Jin, J., Pan, C.: The impact of climate change on the efficiency of agricultural production in the world's main agricultural regions. Environ. Impact Assess. Rev. **97**, 106891 (2022)
7. English, L., Popp, J., Miller, W.: Economic Contribution of Agriculture and Food to Arkansas' Gross Domestic Product 1997–2019. https://ageconsearch.umn.edu/record/330419/files/1001-Economic%20Contribution%20of%20Agriculture%20and%20Food%20to%20Arkansas%20GDP%201997-2019.pdf (2022). Last accessed 14 Mar 2025
8. Nowak, A., Kobiałka, A., Krukowski, A.: Significance of agriculture for bioeconomy in the member states of the European Union. Sustainability. **13**(16), 8709 (2021)
9. Bertini, R., Zouache, A.: Agricultural land issues in the Middle East and North Africa. Am. J. Econ. Sociol. **80**(2), 549–583 (2021). https://doi.org/10.1111/ajes.12391
10. Amiri, N., Lahlali, R., Amiri, S., El Jarroudi, M., Khebiza, M.Y., Messouli, M.: Development of an integrated model to assess the impact of agricultural practices and land use on agricultural production in Morocco under climate stress over the next twenty years. Sustainability. **13**(21), 11943 (2021)
11. Puy, A., Borgonovo, E., Lo Piano, S., Levin, S.A., Saltelli, A.: Irrigated areas drive irrigation water withdrawals. Nat. Commun. **12**(1), 4525 (2021)
12. Liu, X., et al.: Vulnerability of farmer households to climate change in rocky desertification areas—a case study of Guizhou province. Land. **13**(5), 582 (2024)
13. Verma, S., Gulati, A., Hussain, S.: Doubling Agricultural Growth in Uttar Pradesh: Sources and Drivers of Agricultural Growth and Policy Lessons. https://www.econstor.eu/handle/10419/176362 (2017). Last accessed 14 Mar 2025
14. Parry, M., Carter, T.: The assessment of effects of climatic variations on agriculture: a summary of results from studies in semi-arid regions. In: Parry, M.L., Carter, T.R., Konijn, N.T. (eds.) The Impact of Climatic Variations on Agriculture, pp. 9–60. Springer, Dordrecht (1988). https://doi.org/10.1007/978-94-009-2965-4_1
15. Change, C.: Agriculture and food security. In: State Food Agriculture FAO, FAO, Rome, Italy. https://eas-et.org/wp-content/uploads/2022/02/Agricalture-and-Food-Security.pdf (2016). Last accessed 14 Mar 2025
16. Abdelmajid, S., Mukhtar, A., Baig, M.B., Reed, M.R.: Climate change, agricultural policy and food security in Morocco. In: Behnassi, M., Barjees Baig, M., El Haiba, M., Reed, M.R. (eds.) Emerging Challenges to Food Production and Security in Asia, Middle East, and Africa, pp. 171–196. Springer, Cham (2021). https://doi.org/10.1007/978-3-030-72987-5_7
17. Achli, S., et al.: Exploring the key drivers of crop yields in Morocco–a systematic review. Front. Agron. **7**, 1515938 (2025)
18. World Bank.: Bing. https://www.bing.com/search?q=World+Bank&PC=U316&FORM=CHROMN. Last accessed 14 Mar 2025

19. Site institutionnel du Haut-Commissariat au Plan du Royaume du Maroc.: Site institutionnel du Haut-Commissariat au Plan du Royaume du Maroc. https://www.hcp.ma. Last accessed 14 Mar 2025
20. Trading Economics.: 20 million INDICATORS FROM 196 COUNTRIES. https://tradingeconomics.com/. Last accessed 14 Mar 2025

Optimization of Organic Light-Emitting Diodes: The Essential Role of Dibenzofuran in Hole Transport Materials

Oumaima El Ouardi(✉), Brahim Fakrach, Hassane Chadli, and Abdelhai Rahmani

Faculty of Sciences, Laboratory of Advanced Materials Study and Applications (LEM2A), Moulay Ismail University, BP 11201, Zitoune Meknes 50000, Morocco
ou.elouardi@edu.umi.ac

Abstract. Research on organic materials for OLEDs (organic light-emitting diodes) shows the need for compounds that combine good stability with efficient charge transport. In this study, we examine the electronic and optical properties of dibenzofuran using density functional theory (DFT) with the VASP simulation software. Early results suggest that dibenzofuran has promising potential for OLED applications, especially due to its ultraviolet light absorption and energy gap. These findings open up new possibilities for improving the performance and lifetime of OLED devices.

Keywords: Dibenzofuran · Band gap · Density of states · Absorption · Dielectric constant

1 Introduction

Organic light-emitting diodes (OLEDs) have gained significant attention due to their diverse applications, including flexible displays, portable electronic devices, and efficient lighting systems [1]. Since the groundbreaking work in 1987 [2], the field has rapidly evolved. The emergence of phosphorescent OLEDs (PHOLEDs) has been particularly important. These devices use materials based on heavy metals to efficiently harvest excitons, achieving nearly 100% internal quantum efficiency [3]. However, PHOLEDs face limitations. The light emission primarily relies on triplet excitons, which can interact destructively at high concentrations through triplet-triplet annihilation. This reduces device efficiency. To minimize this interaction, an inert host matrix is used to separate the phosphorescent emitters, ensuring stable and controlled emission [4]. Therefore, the choice of the host material is crucial. A good host material should have energy levels well-aligned with the other layers in the device to facilitate charge carrier injection and transport [5]. Hybrid conjugated polymers show promise in this regard. Their modular structure allows for adjustments to their electronic and optical properties, leading to customizable, high-performance devices [6]. Among these materials, dibenzofuran has attracted particular attention. This compound, which consists of a benzene core fused with a furan ring, shows interesting electronic properties for hole transport. The free

B.-E. Boudriki Semlali et al. (Eds.): DATA 2025, LNNS 1908, pp. 469–476, 2026.
https://doi.org/10.1007/978-3-032-21200-9_39

electron pair on the oxygen atom promotes charge carrier movement [7]. Dibenzofuran has several advantages as a hole transport material (HTM). It demonstrates good hole mobility, which facilitates the transfer of charge from the anode to the emissive layer [8]. Additionally, it has high thermal and electrochemical stability, essential for device durability [9]. Its high HOMO level enables efficient hole injection, while its lower LUMO level prevents electron transfer, enhancing overall efficiency [10]. Thus, dibenzofuran is a promising material for designing high-performance, stable OLEDs [11]. This study provides an in-depth analysis of its electronic and optical properties using first-principles calculations. The manuscript is organized as follows: the Methodology section describes the calculation method, the Results section presents the obtained data and analysis, and the Conclusion summarizes the main findings (Fig. 1).

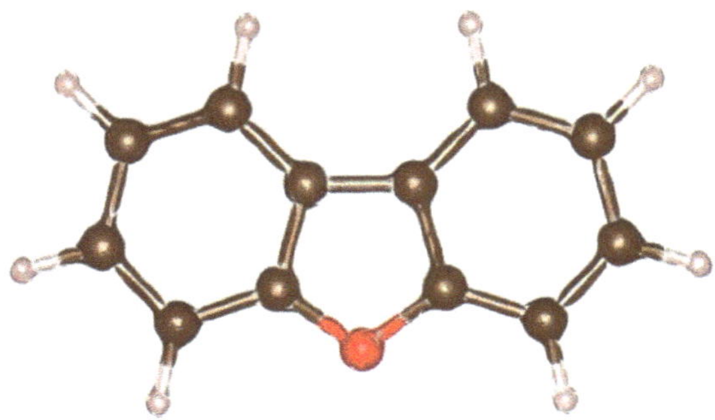

Fig. 1. Schematic presentation of dibenzofuran.

2 Computational Method

We performed calculations using Density Functional Theory (DFT) [12] to study the electronic and optical properties of systems composed of dibenzofuran. These calculations were carried out using the VASP code, enabling precise modeling of electronic interactions within materials [12]. The structure of the system was determined through energy minimization calculations, optimizing molecular conformations to ensure their stability. Geometric optimization was performed using the conjugate gradient method, with relaxation continued until the residual atomic forces were below 0.02 eV/Å. For these calculations, the electronic and optical properties of the molecules were evaluated using the Generalized Gradient Approximation (GGA) for the exchange-correlation energy, as formulated by Perdew, Burke, and Ernzerhof (PBE) [13]. This functional is recognized for providing more accurate energetic results compared to the Local Density Approximation (LDA) [14]. The energy cutoff, a crucial parameter for calculation precision, was set to 520 eV. Additionally, for k-point sampling in reciprocal space, we applied the Monkhorst-Pack method [15] to generate a $(1 \times 1 \times 1)$ grid, suitable for calculations on isolated molecular systems. Finally, within the DFT framework, the interactions between electrons and atomic nuclei were modeled using PAW-PBE pseudopotentials. These pseudopotentials simplify the calculations by focusing only on valence electrons while maintaining an accurate representation of electronic interactions.

3 Results and Discussion

Dibenzofuran (DBF) is a heterocyclic organic compound consisting of two benzene rings fused on either side of a furan ring. Due to its unique molecular structure, DBF exhibits notable optical and electronic properties, making it relevant for various applications, particularly in the fields of organic electronics and optics. This study focuses on an in-depth analysis of the electronic and optical properties of DBF, specifically examining the bandgap energy, electronic density of states, optical absorption characteristics, as well as the real and imaginary dielectric constants.

3.1 Electronic Properties of Dibenzofuran

The results (Table 1) highlight the electronic characteristics of dibenzofuran. The HOMO (Highest Occupied Molecular Orbital) is located at –5.00 eV, a value close to the experimental measurement of –5.31 eV [16]. This level represents the highest energy state occupied by electrons in the ground state. These electrons, being strongly bound to the atoms, limit their participation in electrical conduction. The LUMO (Lowest Unoccupied Molecular Orbital) has an energy of −1.42 eV. This level corresponds to the lowest available empty state. Excited electrons can access this state, thus promoting electronic conduction. The bandgap (E_g) between the HOMO and the LUMO is 3.58 eV, slightly higher than the experimental estimate of 3.2 eV [17]. This confirms the semiconductor behavior of dibenzofuran. The Fermi level (E_f) is located at –4.94 eV. This position indicates that electrons are predominantly confined in the valence band. Their transfer to the conduction band requires energetic excitation. This configuration is characteristic of a p-type semiconductor, where conduction is mainly carried by holes in the valence band. Thus, dibenzofuran behaves as an organic material with a wide bandgap.

Table 1. Energy levels of HOMO and LUMO, Fermi level (E_f) and HOMO-LUMO gap (E_g) of dibenzofuran in eV.

	HOMO	LUMO	E_g	E_f
This study	−5.00	−1.42	3.58	−4.94
Experimental study	−5.31 [16]	–	3.2 [17]	–

(Figure 2) shows the density of states (DOS) as a function of energy. It highlights the electronic characteristics of the material. In the valence band, located below the Fermi level (E_f), the density of states is relatively high. This suggests a significant concentration of states where electrons are strongly bound to atoms, preventing them from participating in conduction. In the conduction band, above the Fermi level, the density of states is lower. This reflects a reduced availability of states for electronic conduction, thus limiting the conduction capacity. In summary, the figure shows a valence band with a high density of states and a conduction band with a lower density of states. This behavior is characteristic of organic semiconductors with a relatively large energy gap.

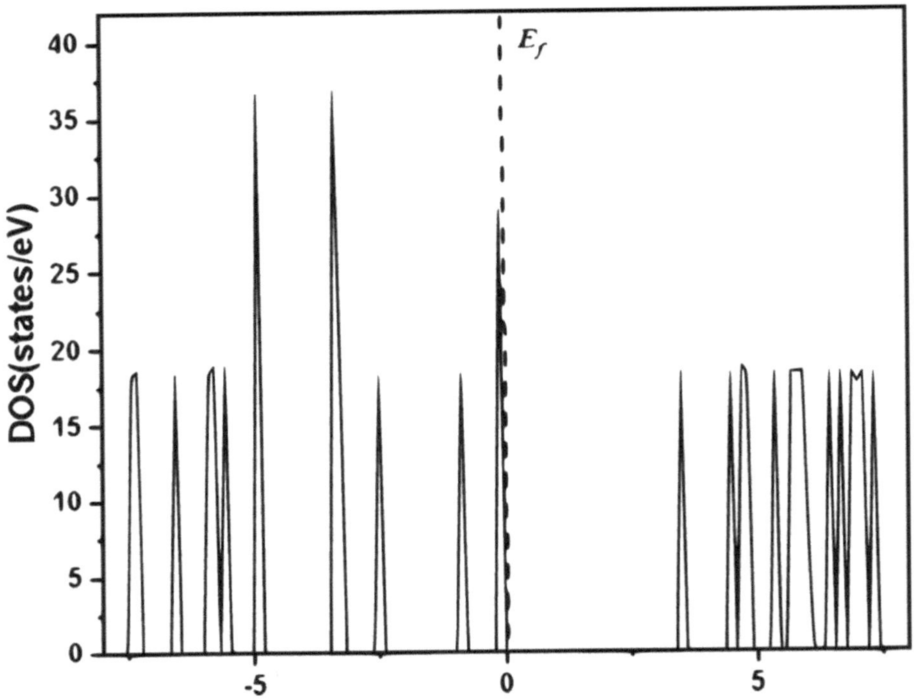

Fig. 2. Electronic density of states of Dibenzofuran.

3.2 Optic Properties of Dibenzofuran

(a) **Dielectric constant**

- *The imaginary dielectric constant of dibenzofuran.*

(Figure 3) presents the spectrum of the imaginary dielectric constant ε2(ω) as a function of energy (eV) for dibenzofuran. In the visible range, the low absorption suggests that dibenzofuran is relatively transparent, with no significant electronic transitions in this energy range. In contrast, in the UV region, a main peak appears around 3.58 eV, indicating a strong absorption of energy. This peak is typically associated with an electronic transition where electrons from the valence band are excited to the conduction band. This peak likely marks the position of the material's optical gap. According to experimental studies [16, 18], the optical gap is estimated between 3.56 eV and 3.60 eV, which corresponds precisely to the observed absorption band, thus confirming the position of the optical gap of dibenzofuran. Additionally, several smaller peaks appear in the UV region, which may be attributed to low-intensity electronic transitions or secondary absorption processes.

- *Real dielectric constant of dibenzofuran.*

(Figure 4) presents the real dielectric constant, denoted ε1(ω), as a function of energy (eV). It illustrates the ability of a material to polarize. In the case of dibenzofuran,

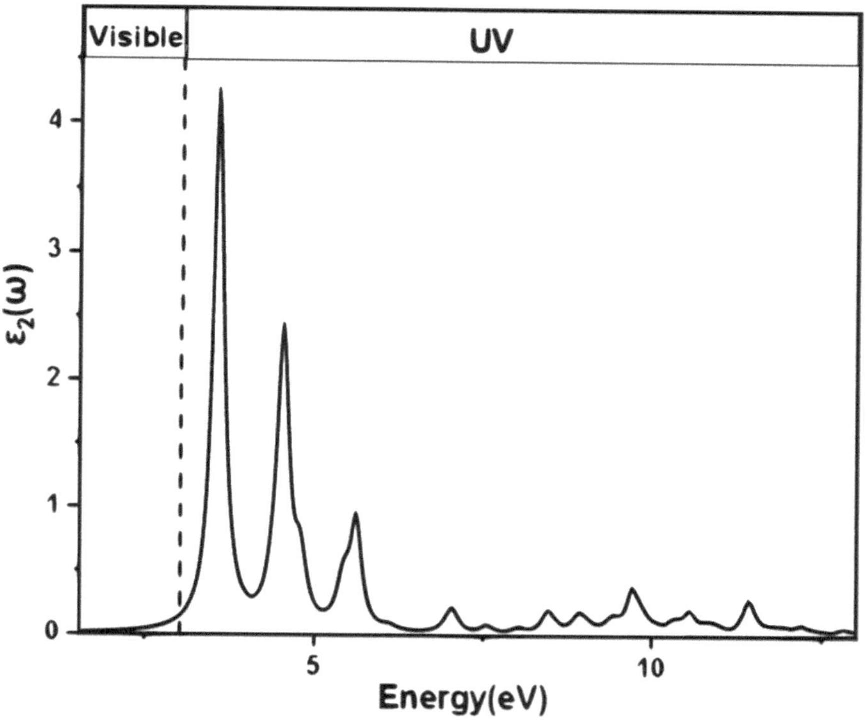

Fig. 3. Imaginary dielectric constant of dibenzofuran.

the analysis of its spectrum reveals several interesting features. A pronounced peak appears in the UV region. This peak, typically observed around 3 eV, is associated with significant electronic transitions, such as $\pi - \pi*$ transitions. During these transitions, electrons are excited from a valence orbital to a conduction orbital, leading to an increase in the material's polarization. In contrast, in the visible region, the dielectric constant remains stable, suggesting that electronic transitions in this energy range are not effective in inducing polarization. These behaviors are characteristic of organic materials, where electronic transitions are generally more pronounced in the UV region and less significant in the visible region.

(b) **Absorption coefficient of dibenzofuran**

(Figure 5) presents the UV-Vis absorption spectra, highlighting the electronic transitions and their influence on the optical properties of dibenzofuran. The peaks observed in the UV region of the absorption spectrum of dibenzofuran can be mainly attributed to two types of electronic transitions: the $\pi - \pi*$ transition and the $n - \pi*$ transition [17, 19]. The $\pi - \pi*$ transition occurs when π bonding electrons are excited to the $\pi*$ antibonding orbitals. This transition is characteristic of conjugated molecules like dibenzofuran. The two peaks observed between 200 and 280 nm in the UV region indicate that the molecule has sufficient conjugation to allow this transition. The $n - \pi*$ transition involves non-bonding (n) electrons on the oxygen atom of the dibenzofuran structure. These electrons

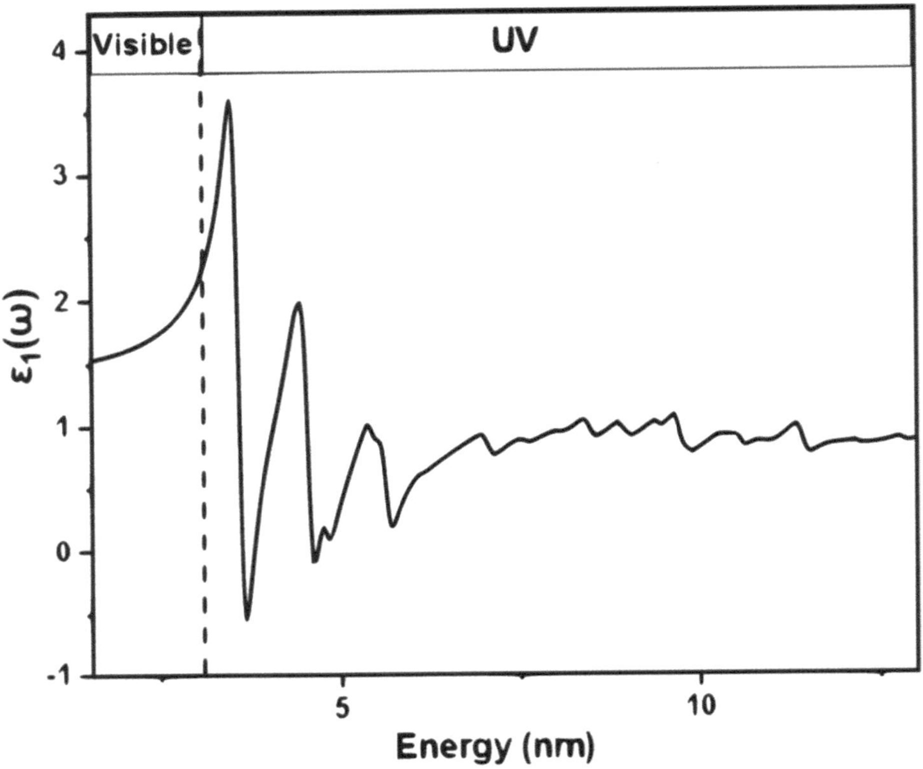

Fig. 4. Real dielectric constant of dibenzofuran.

are excited to the π∗ anti-bonding orbitals. This transition is observed in the 300 to 350 nm range and is typical of compounds containing a heteroatom, such as oxygen, which can donate non-bonding electrons. Finally, the absence of significant absorption in the visible region (400–800 nm) shows that dibenzofuran does not interact with visible light. This is common for materials designed for UV applications.

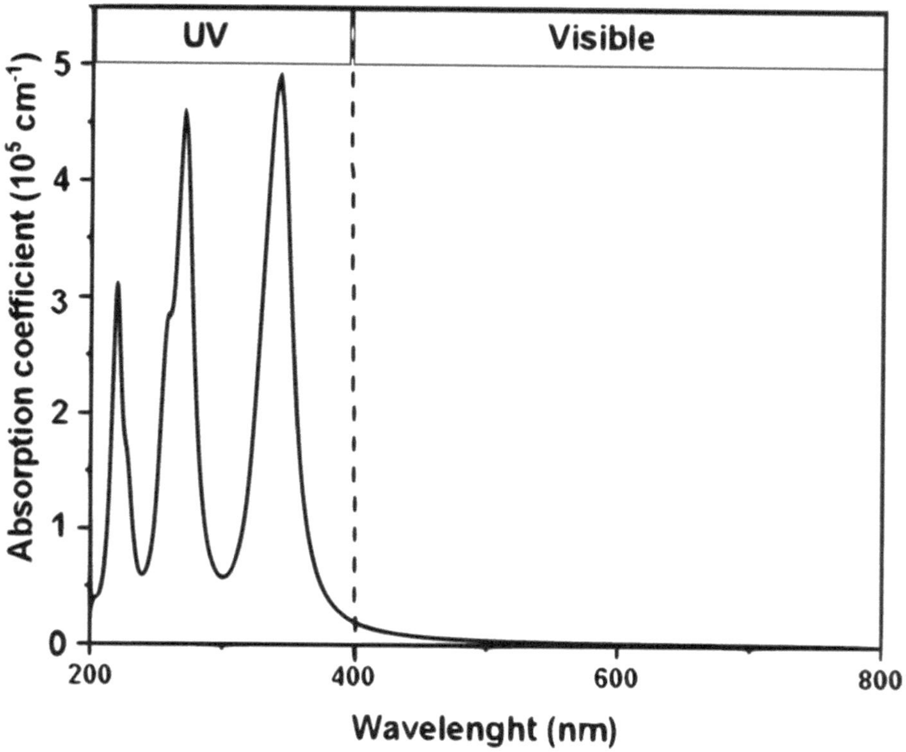

Fig. 5. Absorption coefficients of dibenzofuran.

4 Conclusion

This study highlighted the strong potential of dibenzofuran as a material for applications in OLEDs. The results showed that this material exhibits interesting electronic and optical properties. It has good absorption in the UV range, an optimal density of states, and well-suited dielectric constants. These characteristics contribute to improving the efficiency of optoelectronic devices. Moreover, dibenzofuran offers excellent hole mobility. Its thermal and electrochemical stability also makes it an ideal choice for applications requiring high-performance and durable materials. In summary, this study opens new perspectives for the development of optoelectronic technologies. Dibenzofuran is a key material for the future of luminous and sustainable devices.

References

1. Park, S., et al.: High-efficiency blue OLED host materials based on new anthracene-dibenzofuran derivatives. J. Lumin. **280**, 121106 (2025)
2. Tang, C.W., VanSlyke, S.A.: Organic electroluminescent diodes. Appl. Phys. Lett. **51**(12), 913–915 (1987)
3. Lin, Y., et al.: Improvement in dibenzofuran based hole transport materials for flexible perovskite solar cells. Molecules. **29**(6), 1208 (2024)

4. Abe, S., et al.: Effect of substitution position of dibenzofuran-terminated robust hole-transporters on physical properties and TADF OLED performances. Mol. Syst. Des. Eng. **8**(3), 388–393 (2023)
5. Suzuki, A., Yamamoto, R., Ito, M., Kubo, Y.: Dibenzofuran [a]-fused bodipys: synthesis, photophysical properties, and N_2 O_2-boron-chelation towards NIR materials for application in organic photodetectors. ChemPlusChem. **90**, e202500021 (2025)
6. Ryu, S., Kim, P.G., Son, M., Ahn, J., Seo, S.: Synthesis of dibenzofuran and dibenzothiophene derivatives as yellow host materials for phosphorescent organic light emitting diodes. Synth. Met. **301**, 117538 (2024)
7. Jiang, P., Chen, S., Xia, Y., Zhang, Q., Deng, G.-J.: Access to dibenzofurans through dimerization/trimerization of cyclohexanones followed by dehydroaromatization. Org. Lett. **22**(20), 8076–8081 (2020)
8. Power, M., Mackey, K., Light, M.E., Jones, D.J., McGlacken, G.P.: Access to electron-rich dibenzofurans through nbu4oac-mediated palladium catalysis. Eur. J. Org. Chem. **26**(44), e202300790 (2023)
9. Szlapa-Kula, A., Ledwon, P., Krawiec, A., Kula, S.: Dibenzofulvene derivatives as promising materials for photovoltaic and organic electronics. Energies. **16**(24), 8027 (2023)
10. Kuang, C., et al.: Enhanced horizontal dipole orientation by novel penta-helicene anthracene-based host for efficient blue fluorescent OLEDS. Small. **20**(24), 2311114 (2024)
11. Li, Y.-Z., et al.: High efficiency in blue TADF OLED using favorable horizontal oriented host. Chem. Eng. J. **498**, 155553 (2024)
12. Kresse, G., Hafner, J.: Ab initio molecular dynamics for liquid metals. Phys. Rev. B. **47**(1), 558 (1993)
13. Perdew, J.P., Burke, K., Wang, Y.: Generalized gradient approximation for the exchange-correlation hole of a many-electron system. Phys. Rev. B. **54**(23), 16533 (1996)
14. Benoit, C., Royer, E., Poussigue, G.: The spectral moments method. J. Phys. Condens. Matter. **4**(12), 3125 (1992)
15. Monkhorst, H.J., Pack, J.D.: Special points for brillouin-zone integrations. Phys. Rev. B. **13**(12), 5188 (1976)
16. Yun, S.-J., Seo, M.H., Lee, S.: Dibenzofuran derivatives with meta- and para-triphenylamine substituents as hole-transporting materials in organic light-emitting devices. Dyes Pigments. **175**, 108121 (2020)
17. Borisevich, N., et al.: Spectral-luminescence characteristics of carbazole, dibenzofuran, and dinaphthofuran in the gas phase in excitation by electrons and photons. J. Appl. Spectrosc. **68**, 871–876 (2001)
18. Zahid, W.A., et al.: Fine tuning the optoelectronic properties of dibenzo [b, d] furan-centered linear hole transporting materials for perovskite solar cells. J. Phys. Chem. Solids. **178**, 111337 (2023)
19. Jiang, Z., et al.: Novel spiro core-based hole transport materials for stable deep-blue oleds with lt95 over 420 h. Adv. Opt. Mater. **11**(23), 2301014 (2023)

Performance Evaluation of Deep Learning Models for Agricultural Pest Detection

Geerish Suddul(✉) and Hansvirsing Ramjuttun Gunnoo

University of Technology, Mauritius (UTM), Port Louis, Mauritius
g.suddul@utm.ac.mu, hansvirsing.gunnoo@umail.utm.ac.mu

Abstract. Effective pest detection in agriculture is crucial for minimizing crop damage and ensuring food security. Traditional pest identification methods are often slow, labor-intensive, and prone to human error, highlighting the need for automated solutions. This study compares three deep learning approaches for pest classification: an artificial neural network (ANN) implemented from scratch, a convolutional neural network (CNN) trained using deep learning libraries, and a transfer learning approach leveraging pre-trained models such as DenseNet121, MobileNetV2, EfficientNetB0, and Xception. Experimental results show that transfer learning significantly outperforms both ANN and CNN models, with DenseNet121 achieving the highest test accuracy of around 91% while maintaining the lowest training time of 14 min on the chosen dataset. The findings demonstrate that transfer learning is the most efficient and effective approach for agricultural pest classification, providing a strong balance between accuracy, training speed, and computational efficiency.

Keywords: Plant pest detection · ANN · CNN · Transfer learning · Image classification

1 Introduction

Agriculture is a critical sector that ensures food security and economic stability worldwide. However, one of the biggest challenges it faces is crop damage caused by agricultural pests, which leads to significant yield losses and financial setbacks [1]. Traditional pest identification methods rely on manual inspection by experts, which complex the size of certain pests, their evolved camouflage techniques and different types of fields [2]. The need for an automated, accurate, and efficient pest detection system has driven research into artificial intelligence (AI) and deep learning applications in precision agriculture.

Deep learning, particularly Artificial Neural Networks (ANNs) and Convolutional Neural Networks (CNNs), has revolutionized image classification tasks, making it highly effective for pest identification. ANNs provide the ability to model complex patterns, but they often require extensive training and computational resources. CNNs, on the other hand, are effective at extracting hierarchical spatial features from images, making them more effective for visual recognition tasks such as pest classification. More recently, transfer learning has emerged as an efficient approach, allowing models pre-trained on

B.-E. Boudriki Semlali et al. (Eds.): DATA 2025, LNNS 1908, pp. 477–488, 2026.
https://doi.org/10.1007/978-3-032-21200-9_40

large datasets (such as ImageNet) to be fine-tuned for pest classification, reducing the need for extensive labeled data and improving accuracy [6, 7, 9, 10].

This study explores and compares three deep learning approaches for agricultural pest detection: a custom-built artificial neural network (ANN) implemented from scratch using NumPy, focusing on the fundamentals of neural networks without deep learning libraries; a CNN-based model implemented using TensorFlow/Keras to leverage the power of convolutional layers for improved feature extraction; and a transfer learning approach, where a pre-trained deep learning model such as DenseNet121, MobileNetV2, EfficientNetB0, and Xception are fine-tuned on a pest classification dataset to enhance accuracy and efficiency.

By evaluating these approaches in terms of accuracy, computational efficiency, and training time, this study seeks to determine which method is best suited for real-world pest detection applications. Additionally, it examines how architectural modifications, activation functions, batch normalization, and data augmentation techniques influence model performance.

2 Literature Review

Costa et al. [3] explored deep learning-based classification of agricultural pests, addressing species variability and life stage differences. They evaluated AlexNet, ResNet-50, EfficientNet, and Vision Transformer (ViT) on the Agricultural Pests dataset, optimizing performance through hyperparameter tuning and data augmentation. Their results showed that ViT achieved the highest accuracy (95.74%) without augmentation, while CNN models significantly benefited from augmentation techniques to improve generalization. The study incorporated learning rate scheduling and early stopping, revealing that EfficientNet provided a strong balance between accuracy and computational efficiency, making it suitable for real-world deployment. Their findings highlight ViT's superiority in extracting global features and reinforce the necessity of data augmentation for CNN-based models. This research contributes to AI-driven pest classification, demonstrating deep learning's potential in precision agriculture. Future work will explore advanced augmentation and model improvements to enhance robustness in real-world applications.

Huang [4] conducted a comparative study on deep transfer learning models for agricultural pest classification, addressing the challenge of pest identification due to species complexity and visual similarity. The study evaluated ResNet, EfficientNet, MobileNet, and Xception using transfer learning to improve classification performance while optimizing model size, computational efficiency, and training time. Experimental results demonstrated that Xception achieved the highest accuracy (89.71%), but EfficientNetB0 (88.26%) provided the best trade-off between accuracy and computational efficiency, making it more suitable for real-world applications. MobileNet and ResNet50 lagged behind in terms of accuracy and efficiency. The study also emphasized data augmentation, dropout layers, and early stopping as effective strategies for improving model performance and preventing overfitting.

Wang et al. [5] introduced InsectMamba, a new deep learning model for insect pest classification, designed to overcome challenges like camouflage, species diversity, and small feature differences between pests. Unlike traditional models, InsectMamba combines State Space Models (SSMs), Convolutional Neural Networks (CNNs), Multi-Head Self-Attention (MSA), and Multilayer Perceptrons (MLPs) to capture both small details and overall patterns in pest images. The study tested InsectMamba on five pest classification datasets, including IP102 and Agricultural Pests, and found that it performed better than ResNet, Swin Transformer, and Vmamba models. It achieved the highest accuracy of 91% on the Agricultural Pests dataset, showing that its unique mix of techniques improves accuracy. An ablation study confirmed that each part of the model plays an important role in improving classification results.

Other techniques in the context of pest identification involves the creation of datasets. Suzauddola et al. [6] combined 3 datasets (Jute17, the Pest24, and the IP102) to evaluate the performance of their proposed model which achieved an Average Precision (mAP) of 78.22% on the Pest24 dataset and 78.15% accuracy on the IP102 dataset. Research by [7] applied transfer learning with MobileNetV3, with respective accuracies of 96.14% and 93.25% for disease classification in mustard and mung bean crops. The trained model was deployed on a System-on-Chip.

3 Methodology

3.1 Dataset Description and Data Preprocessing

The Agricultural Pests Dataset [8] contains images of various insect species that are responsible for agricultural damage. The Agricultural Pest Image Dataset is a collection of images of 12 different types of agricultural pests, namely Ants, Bees, Beetles, Caterpillars, Earthworms, Earwigs, Grasshoppers, Moths, Slugs, Snails, Wasps, and Weevils. It has 5496 images which is divided into 4395 train data and 1101 test data.

To ensure consistency across the dataset, all images were resized to 200 $\times$ 200 pixels. Various data augmentation techniques were applied, including flipping, rotation, normalization (rescaling), and brightness adjustments, to enhance dataset diversity and improve model generalization. To prepare the labels for classification, one-hot encoding was used to convert categorical class labels into a numerical format suitable for deep learning models. The dataset was then split into training (80%), validation (10%), and testing (10%) sets. Since the test set initially comprised 20% of the total dataset, it was further divided in a 50:50 ratio to create balanced validation and test sets, ensuring an effective evaluation of model performance.

3.2 Experimental Design

Several experiments were conducted for the artificial neural network (ANN) implementation from scratch, the CNN-based implementation using libraries, and the transfer learning approach. All experiments were performed on Google Colab platform. The ANN and CNN experiments utilized a T4 GPU with 15GB GPU RAM and 51GB system RAM, while the transfer learning experiments were conducted using an A100 GPU runtime in Colab, which provided 40.0GB GPU RAM and 83.5GB system RAM to handle the increased computational requirements of fine-tuning pre-trained models. The purpose of these experiments was to analyze the impact of different architectural choices, computational resources, and training strategies on model performance.

ANN Implementation from Scratch Network Architecture

The ANN model was designed with an input layer that processed flattened image data, followed by a configurable number of hidden layers that used the ReLU activation function. The output layer consisted of a softmax activation function, allowing the model to perform multi-class classification by assigning probabilities to different pest categories. The training process followed a step-by-step strategy. Initially, the model was tested using different numbers of neurons, specifically 32, 64, 128, and 256, trained across various epoch settings, including 75, 150, and 500 epochs. Once the configuration that yielded the highest accuracy was identified, the architecture was further modified by increasing the number of hidden layers. The hidden layer configurations tested included [32, 64], [32, 64, 128], and [32, 64, 128, 256], allowing for an in-depth analysis of how deeper networks affect learning performance. To ensure model progress was effectively tracked, validation accuracy was monitored every 10 epochs, providing insight into how well the model generalized throughout training.

CNN-Based Implementation Using Libraries

The CNN model was structured to extract spatial features using convolutional layers, which applied 3×3 kernels to identify patterns in the input images. Batch normalization was incorporated to stabilize activations and prevent drastic shifts in weight distributions, leading to smoother learning. The LeakyReLU activation function was used instead of standard ReLU, allowing small gradients for negative inputs, which helped avoid dead neurons and improved feature extraction. MaxPooling with a 2×2 filter was applied to reduce the dimensionality of feature maps while retaining key spatial information. The extracted features were then passed through fully connected layers to map them to the classification output. Dropout layers were included to prevent overfitting by randomly deactivating neurons during training, ensuring better generalization. The final output was generated using the softmax activation function, which assigned probability scores to different pest classes. The training strategy was structured into two phases. Initially, the model was trained using different numbers of convolutional layers with varying configurations to determine the optimal depth for feature extraction. After identifying the best-performing architecture, additional optimizations were introduced by incorporating batch normalization, LeakyReLU, and a combination of both techniques. To improve training efficiency and prevent overfitting, early stopping was implemented to stopped training when validation loss began to increase, preventing unnecessary computation and reducing the risk of overtraining. Additionally, model checkpointing was used to automatically save the best-performing model, ensuring that the most optimal weights were retained for final evaluation.

Transfer Learning-Based Implementation

The transfer learning approach was implemented using pre-trained convolutional neural network (CNN) architectures to leverage their ability to extract robust features from images. Instead of training a CNN from scratch, which requires large amounts of data and computational power, this method fine-tunes models that have already been trained on large-scale image datasets such as ImageNet. The objective of this approach was to improve classification accuracy while reducing training time and computational overhead. Multiple pre-trained CNN architectures, including MobileNetV2, EfficientNetB0, Xception, and DenseNet121, were tested for pest classification. These models were modified by replacing their original classification heads with a global average pooling layer followed by a fully connected dense layer with a softmax activation function. Initially, the base model weights were frozen, and only the classification head was trained using a learning rate of 0.0001 for 15 epochs. After the initial training phase, the base model was partially unfrozen, allowing fine-tuning of the top 50% of its layers while reducing the learning rate to 0.00001. This gradual unfreezing process helped prevent catastrophic forgetting while adapting the model to the new dataset. To optimize performance, ReduceLROnPlateau was used to dynamically adjust the learning rate based on validation loss, and early stopping was applied to prevent overfitting by stopping training when validation loss stopped improving. The best-performing model weights were saved using model checkpointing. Finally, the trained models were evaluated on the test dataset, and their performance was compared in terms of accuracy, training time, and computational efficiency.

4 Results and Discussion

4.1 ANN Implementation from Scratch

Single Hidden Layer with Different Neurons and Epochs

Table 1 shows the results for the experiment that was conducted using only 1 hidden layer with different number of neurons (32, 64, 128, 256) and different number of epochs (75, 150, 500).

The experiment analyzed the impact of increasing neurons and epochs on model performance using a single hidden layer. At 75 epochs, increasing the number of neurons did not significantly improve validation accuracy, which remained between 9.11% and 10.75%, with test accuracy below 10%. This indicates that the model struggled to learn effectively in shorter training durations. At 150 epochs, validation accuracy improved slightly, reaching 16.03% for 256 neurons, and test accuracy increased to **14.91%**, suggesting that longer training allowed deeper networks to extract more features. The most significant improvements were observed at 500 epochs, where validation accuracy peaked at 23.32% for 256 neurons, and test accuracy improved to 16.36%, confirming that higher neuron counts benefit from extended training. However, increasing epochs beyond 150 did not help smaller models (32 and 64 neurons), as test accuracy remained below 10%, indicating underfitting. For 128 and 256 neurons, longer training led to better model convergence, with validation accuracy reaching 21.86% and 23.32%, respectively, and test accuracy improving to 14.91% and 16.36%. This suggests that larger models require more epochs to fully learn complex patterns. In terms of training time, both neuron count and epoch length significantly increased computational costs. For instance, 256 neurons at 500 epochs required over 3300 s, compared to just 259 s for 256 neurons at 150 epochs and 135 s for 32 neurons at 75 epochs. This highlights the trade-off between model performance and computational efficiency, where higher neurons and longer training improve accuracy but require significantly more resources.

Table 1. Single Hidden Layer with Different Neurons & Epochs Results

Neuron	Epoch	Max Validation Accuracy	Test Accuracy(%)	Training Time(s)
32	75	10.20% at epoch 0	8	135.2866
64	75	9.11% at epoch 0	7.09	204.2511
128	75	10.20% at epoch 70	10	328.1232
256	75	10.75% at epoch 30	8.91	500.0620
32	150	9.47% at epoch 120	7.82	259.2198
64	150	12.75% at epoch 140	8.91	382.5711
128	150	12.57% at epoch 110	10.18	666.9309
256	150	16.03% at epoch 130	14.91	987.7101
32	500	10.75% at epoch 20	8	875.7912
64	500	10.75% at epoch 70	8.55	1292.8818
128	500	21.86% at epoch 410	14.91	2101.3051
256	500	23.32% at epoch 430	16.36	3309.4694

Different Number of Hidden Layers

Table 2 presents the results of the experiment conducted with different numbers of hidden layers (1, 2, and 3), each with an increasing number of neurons per layer. The objective was to analyze the impact of deeper network architectures on test accuracy and training time.

Table 2. ANN with different Number of Hidden Layers.

No of hidden Layers	Configuration (No of Neurons per layer)	Test Accuracy(%)	Training Time(s)
1	32	8	875.7912
2	32, 64	12.20	1320.6803
3	32, 64, 128	17.64	2182.7275

The experiment evaluated the effect of increasing the number of hidden layers on model performance. Results show that adding more layers significantly improves test accuracy. A single hidden layer with 32 neurons achieved only 8.00% accuracy, while adding a second hidden layer (32, 64 neurons) increased accuracy to 12.20%, a 4.2% improvement. The best performance was achieved with three hidden layers (32, 64, 128 neurons), reaching 17.64% accuracy, an additional 5.44% improvement. This suggests that deeper networks capture more complex patterns, enhancing classification ability. However, adding more layers also significantly increases training time. The single-layer model trained in 875.79 s, but increasing to two layers (32, 64 neurons) raised training time to 1320.68 s, a 50.8% increase. With three layers (32, 64, 128 neurons), training took 2182.73 s, marking a 65.2% increase over two layers. This exponential rise in training time is due to more parameters, additional computations in forward and backward passes, and slower convergence. While deeper networks improve accuracy, they require significantly higher computational resources, emphasizing the trade-off between performance and efficiency.

4.2 CNN-Based Implementation Using Libraries

Different Number of Convolution Layer

Table 3 shows the results of the experiment conducted with different numbers of convolutional layers (1 to 5), each with an increasing number of filters per layer. The objective was to analyze the impact of deeper CNN architectures on test accuracy, model convergence (early stopping), and training time. Early stopping was used as a regularization parameter, causing some training runs to stop before reaching the final epoch.

Table 3. Different Number of Convolution Layer

No of Conv Layer	Conv Layer configuration	Early Stop (epoch)	Last epoch Model Accuracy(%)	Test Accuracy(%)	Training Time(s)
1	32	21	8.96	8.73	692.49
2	32, 64	68	51.68	48.55	2263.01
3	32, 64, 128	55	56.95	55.97	1852.84
4	32, 64, 128, 256	58	71.72	59.45	1967.59
5	32, 64, 128, 256, 512	66	78.06	62.73	2237.04

The experiment analyzed the impact of increasing the number of convolutional layers on test accuracy, training time, and learning stability. Results show that adding more convolutional layers significantly improves accuracy, but with diminishing returns after a certain depth. A 1-layer CNN performed poorly, achieving only 8.73% accuracy, indicating that a single convolutional layer fails to extract meaningful features. Adding a second convolutional layer drastically improved accuracy to 48.55%, a + 39.82% increase, demonstrating the importance of deeper feature extraction. A 3-layer CNN reached 55.97% accuracy (+7.42% improvement), while 4-layer and 5-layer CNNs continued to improve accuracy, reaching 59.45% and 62.73%, respectively. However, the gain from 4 to 5 layers was only +3.28%, indicating diminishing returns.

Increasing convolutional layers also led to higher training times. The biggest computational jump occurred between the 1-layer CNN (692.49 s) and the 2-layer CNN (2263.01 s) due to a significant increase in processing. Training time continued to rise from 3-layer (1852.84 s) to 5-layer (2237.04 s), but at a slower rate, suggesting that deeper models require more resources but do not always provide proportional accuracy gains. These findings emphasize the balance between network depth, accuracy gains, and computational cost in CNN-based image classification.

Batch Normalization and LeakyRelu

Table 4 presents the results of the experiment conducted with different CNN configurations, evaluating the impact of Batch Normalization (BN), LeakyReLU (LR), and their combination (BN&LR) on model accuracy, test performance, and training time. The objective was to analyze how these techniques influence model stability, generalization, and computational efficiency.

Table 4. Batch Normalization and LeakyRelu

Implementation	Early Stopping (epoch)	Last epoch Model Accuracy(%)	Test Accuracy(%)	Training Time(s)
BN	111*	70.42	61.82	1392.44
LR	60	68.49	59.45	2038.70
(BN&LR)	113*	79.21	67.27	1455.56

The experiment evaluated the effect of Batch Normalization (BN) and LeakyReLU on the performance of a 5-layer CNN. The findings suggest that using both BN and LeakyReLU together yields the best test accuracy, while BN alone helps stabilize training, and LeakyReLU on its own struggles with weight update consistency. When Batch Normalization was applied, the model achieved a test accuracy of 61.82%, indicating that normalizing activations helps improve generalization by preventing extreme weight shifts. LeakyReLU alone resulted in a slightly lower test accuracy of 59.45%, likely due to unstable weight updates, even though it prevents dead neurons. The highest accuracy of 67.27% was obtained when both BN and LeakyReLU were used together, confirming that BN helps create a more stable training process, while LeakyReLU improves feature extraction.

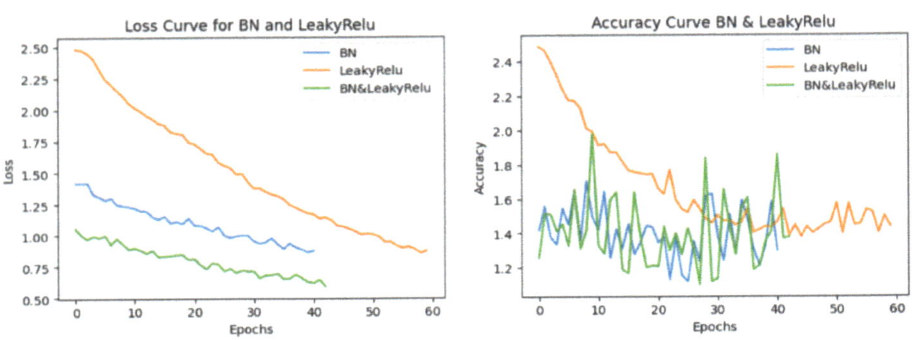

Fig. 1. Loss and Accuracy Curve for Batch Normalization and LeakyRelu

The loss and accuracy curves in Fig. 1 further illustrate these trends. The loss curve shows that BN and BN with LeakyReLU follow a smooth downward trend, while the model with only LeakyReLU exhibits more fluctuations, suggesting instability in learning. A similar pattern appears in the accuracy curve, where BN-based models show more consistent improvements, while LeakyReLU alone results in erratic accuracy changes over epochs. Training efficiency also varied depending on the method used.

The model with BN alone had the shortest training time at 1392.44 s, as it stabilizes activations and prevents extreme weight updates. In contrast, LeakyReLU alone had the longest training duration at 2038.70 s, possibly due to unstable training dynamics. The combination of BN and LeakyReLU required 1455.56 s, slightly longer than BN alone but considerably faster than using LeakyReLU without normalization, showing that BN contributes to training efficiency even when paired with LeakyReLU. Overall, the results indicate that combining Batch Normalization and LeakyReLU provides the best balance between stability, accuracy, and training efficiency, making it a highly effective strategy for CNN-based image classification.

4.3 Transfer Learning-Based Implementation

Table 5 presents the results of the experiment conducted using transfer learning, where pre-trained models were fine-tuned for agricultural pest classification. The goal was to evaluate the impact of different state-of-the-art architectures on test accuracy, training efficiency, and model convergence. The models used in this study included DenseNet121, MobileNetV2, EfficientNetB0, and Xception, all of which have been pre-trained on the large-scale ImageNet dataset. The advantage of transfer learning is that these models already contain well-optimized feature extraction layers, allowing them to generalize well even with limited training data.

Table 5. Results of Transfer Learning-Based Implementation

Implementation	Early Stopping – (epoch)	Training Accuracy(%)	Test Accuracy(%)	Training Time(s)
DenseNet121	51	97.42	90.99	841.55
MobileNetV2	46	95.20	90.26	562.45
EfficientNetB0	64	95.38	90.07	782.22
Xception	33	94.47	90.98	414.11

The results indicate that all transfer learning models achieved significantly higher accuracy compared to traditional CNNs and ANNs while requiring considerably less training time. Among the tested models, DenseNet121 achieved the highest test accuracy of 90.99%, making it the best-performing model in terms of classification accuracy. Xception closely followed with a test accuracy of 90.98%, showing that both models were highly effective for pest classification. MobileNetV2 and EfficientNetB0 also performed well, achieving test accuracies of 90.26% and 90.07%, respectively. While all models converged quickly, Xception completed training in the shortest time (414.11 s), whereas DenseNet121 took the longest (841.55 s), reflecting differences in model complexity and parameter optimization strategies. The use of early stopping ensured that training was halted when validation loss stopped improving, preventing overfitting and optimizing computational efficiency. The high training accuracy (above 94%) across all models suggests that they effectively learned from the dataset, and the relatively close gap between training and test accuracy indicates strong generalization capabilities. These findings highlight the superiority of transfer learning over traditional training methods, as it enables high classification accuracy with significantly reduced training time and computational cost. Overall, the results demonstrate that transfer learning is the most efficient and effective approach for agricultural pest classification, providing a strong balance between accuracy, training speed, and computational efficiency.

4.4 Model Comparison

To determine the most effective model for agricultural pest classification, a comparison was conducted between the best-performing models from each of the three approaches, evaluating the models based on key performance metrics, including test accuracy, training time, and computational efficiency (Table 6).

Table 6. Comparison of the 3 models

Model Type	Architecture	Test Accuracy (%)	Training Time (s)
ANN	3 Hidden Layers (32, 64, 128)	17.64	2182.73
CNN	5 Convolutional Layers (32, 64, 128, 256, 512) with BN&LR	67.27	1455.56
Transfer Learning	DenseNet121	90.99	841.55

The ANN model with the best performance was the 3-layer architecture (32, 64, 128 neurons), which achieved a test accuracy of 17.64%. In the CNN-based approach, the 5-layer CNN (32, 64, 128, 256, 512 filters) with Batch Normalization and LeakyReLU performed the best, with a test accuracy of 67.27%. However, the DenseNet121 model from the transfer learning experiment outperformed all others, achieving the highest test accuracy of 90.99% while maintaining a reasonable training time of 841.55 s.

The results demonstrate a clear advantage of using transfer learning over both traditional ANN and CNN-based models. While the ANN model struggled to achieve high test accuracy due to its limited feature extraction capabilities, the CNN-based model showed significant improvement, confirming the effectiveness of deep convolutional layers in image classification tasks. However, the transfer learning approach significantly outperformed both ANN and CNN models, achieving the highest test accuracy of 90.99% while also requiring the shortest training time compared to the CNN-based model.

The training time differences highlight the computational efficiency of transfer learning. The ANN and CNN models had to learn feature representations from scratch, requiring more epochs and computational resources. In contrast, DenseNet121 leveraged pre-trained weights, allowing it to converge much faster while maintaining superior accuracy. This indicates that transfer learning is the most effective approach for agricultural pest detection, as it provides a strong balance between accuracy, training speed, and computational efficiency.

4.5 Comparison with Existing Approaches

This shows CNN-based transfer learning is highly effective, but transformers may offer additional advantages in feature extraction based on Costa et al. [3]. Huang's [4] study reported that Xception reached 89.71% accuracy, which is less compared to 90.98% obtained in this study. Similarly, Wang et al.'s [5] InsectMamba model achieved 91%, which is comparable to DenseNet121 at 90.99%. These comparisons indicate that transfer learning consistently provides superior accuracy compared to models trained from scratch, reinforcing its importance in real-world agricultural pest classification applications.

5 Conclusion

The results of this study show that transfer learning significantly outperformed both artificial neural networks (ANNs) and convolutional neural networks (CNNs). Among the three approaches tested, the ANN model performed the worst, with its highest test accuracy reaching only 17.64%. The CNN-based model showed considerable improvement, achieving a test accuracy of 67.27%, in less time. In contrast, transfer learning, particularly using DenseNet121, delivered the highest accuracy of 90.99% while maintaining a significantly lower training time of 841.55 s.

References

1. FAO: Plant production and protection (2024). URL https://www.fao.org/plantproduction-protection/about/en
2. Dong, S., Teng, Y., Jiao, L., Du, J., Liu, K., Wang, R.: ESA-net: an efficient scale-aware network for small crop pest detection. Expert Syst. Appl. **236** (2024). https://doi.org/10.1016/j.eswa.2023.121308
3. Lucas, P., Matheus, T., Ferreira, L., Henrique, L., Mari, J.F.: Classification of agricultural pests through digital images using deep learning. Revista de Informática Teórica e Aplicada, [online]. **32**(1), 18–25 (2025). https://doi.org/10.22456/2175-2745.143520
4. Huang, Q.: Comparison of deep transfer learning models for pest image classification in agriculture. Highl. Sci. Eng. Technol. **94**, 9–16 (2024). https://doi.org/10.54097/kxbxjn03
5. Wang, Q., Wang, C., Lai, Z. and Zhou, Y.: InsectMamba: insect Pest classification with state space model. [online] arXiv.org (2024). Available at: https://arxiv.org/abs/2404.03611.
6. Suzauddola, M., Zhang, D., Zeb, A., Chen, J., Wei, L., Sadique Rayhan, A.B.M.: Advanced deep learning model for crop-specific and cross-crop pest identification. Expert Syst. Appl. **274**, 126896., ISSN 0957-4174 (2025). https://doi.org/10.1016/j.eswa.2025.126896

7. Mallick, T.M.D., Murty, D.O., Pal, R., Mandal, S., Saha, H.N., Chakrabarti, A.: High-speed system-on-chip-based platform for real-time crop disease and pest detection using deep learning techniques. Comput. Electr. Eng. **123**(Part C), 110182., ISSN 0045-7906 (2025). https://doi.org/10.1016/j.compeleceng.2025.110182
8. Dutta, G.: Agricultural pests dataset (2023). [online] Kaggle.com. Available at: https://www.kaggle.com/datasets/gauravduttakiit/agricultural-pests-dataset
9. Yu, X., Zheng, J.: A review on the application of deep learning methods in detection and identification of Rice diseases and pests. Comput. Mater. Continua. **78**(1), 197–225, ISSN 1546-2218 (2024). https://doi.org/10.32604/cmc.2023.043943
10. Savaş, S.: Application of deep ensemble learning for palm disease detection in smart agriculture. Heliyon. **10**(7), e37141., ISSN 2405-8440 (2024). https://doi.org/10.1016/j.heliyon.2024.e37141

Advancing Sustainable Agriculture: The Role of Nanotechnology in Enhancing Efficciency and Environmental Protection

Jihane Abdessadak[1(✉)], Ziyad Zeggaf Tahiri[2], and Adel Zeggaf Tahiri[3]

[1] Systems, Control & Decision Laboratory, Ensit, Tangier, Morocco
Jihane.abdessadak@gmail.com
[2] Nanotechnology student, Technical university of Dublin, Dublin, Ireland
[3] Agricultural and Environmental Engineering, Tangier, Morocco

Abstract. In recent years, agriculture has been faced with increasing constraints and challenges due to a number of factors, such as climate change, soil degradation and the excessive use of pesticides and chemical inputs, all of which threaten long-term sustainability. Innovation is therefore essential to help agriculture overcome these challenges, and this is where nanotechnology comes in. Nanotechnology has emerged as a promising solution for improving agricultural efficiency while minimizing environmental impact. In this paper, we will discuss the role of this new technology in improving the sustainability of agriculture by enabling precision agriculture, reducing waste and pollution. This study highlights recent advances, potential risks and future research directions, emphasizing the need for responsible implementation of nanotechnologies in agriculture.

Keywords: Nanotechnology · Sustainable agriculture · Precision agriculture · Innovation

1 Introduction

The population of the world is expected to reach 9.7 billion by 2050, and its rapid growth has raised an immediate necessity for enhancing the productivity and sustainability of agriculture. The conventional farming practices faced a number of problems, including soil fertility depletion, water scarcity, and the undesirable environmental impact of agrochemicals. Nanotechnology seems to be the miracle approach to overcome these challenges and revolutionize the agricultural sector. Nanotechnology, which is the science and engineering of nanoscale materials, has the ability to transform several fields of agriculture, from precision agriculture to precise delivery of agrochemicals and nutrients. Nanomaterials such as silica nanoparticles have demonstrated the potential to enhance growth in plants, increase nutrient and water-use efficiency, and provide effective pest and disease management. Nanobiosensors, as one of the applications of nanotechnology, are used to detect environmental conditions and regulate the application of agricultural inputs precisely, thereby conserving on waste and environmental pollution. Researchers have highlighted the several benefits of application of nanotechnology in sustainable

B.-E. Boudriki Semlali et al. (Eds.): DATA 2025, LNNS 1908, pp. 489–496, 2026.
https://doi.org/10.1007/978-3-032-21200-9_41

agriculture. Nanoparticles can be engineered to have specific properties such as targeted delivery of agrochemicals and nutrients, which in turn minimizes excess use of materials used. By offering a holistic approach to identifying the dangers, benefits, and use cases of nanotechnology, this study hopes to address how nanotechnology can be a boon for the agricultural sector.

2 Nanotechnology Applications in Sustainable Agriculture

Agriculture is on the threshold of transformation, driven by the imperative need for more efficient and sustainable practices. Traditional farming methods rely heavily on chemical fertilizers and pesticides, which generally lead to soil erosion, water pollution and reduced biodiversity. Nanotechnology offers a promising and viable alternative to these problems in many use cases by boosting nutrient supply, improving pest control and enabling real-time monitoring using nano sensors. These technologies play a crucial role in the realization of precision agriculture, which aims to maximize the use of resources and minimize damage to the environment.

2.1 Nano-fertilizers for Improved Nutrient Management

Current fertilizers are often subject to the issue of low nutrient use efficiency, with large portions of the fertilizer (such as nitrogen and phosphorus) are lost due to leeching, volatilization, and runoff. These inefficiencies contribute to environmental problems such as water eutrophication and greenhouse gas emissions. Nano-fertilizers are designed to address these issues by ensuring controlled and targeted nutrient release, enhancing plant absorption rates, and reducing overall fertilizer consumption [1].

Nano-fertilizers are synthesized using nano-carriers such as silica, chitosan, and polymeric nanoparticles, which encapsulate essential nutrients and release them gradually. Research has shown that nano-fertilizers can increase nitrogen use efficiency by 30–50% compared to conventional fertilizers, leading to improved crop yields with lower input requirements [2]. A study on nano-hydroxyapatite fertilizers demonstrated that phosphorus uptake efficiency in wheat crops increased by 40%, significantly reducing the amount of fertilizer needed to achieve optimal plant growth [3].

Furthermore, nano-fertilizers enhance plant resistance to abiotic stresses such as drought, soil fertility and salinity. The incorporation of zinc and iron nanoparticles in fertilizers has led to increased root development and chlorophyll production, which are desirable traits in plant life [4].

2.2 Nano-pesticides for Targeted Pest Control

Pesticide overuse is widely known to be detrimental to both the environment and ecosystem, causing pesticide-resistant organisms and soil contamination. Nano pesticides, which can be used in conjunction with nano-scale delivery systems, offer a more viable and sustainable alternative. These nano-formulations improve the controlled release of pesticide compounds, reducing the required dosage and minimizing toxic byproducts. The Table 1 presents the effects of different types of nanoparticles (NPs) on various insect species. It highlights the observed physiological and biological impacts, such as digestive enzyme disruption, cellular damage, or reduced fertility, depending on the type of nanoparticle used [5].

Table 1. Effects of Nanoparticles (NPs) on Insects: Types, Biological Impacts, and References

NP Type	Insect	Impact	Reference
Ag NPS	*Bombyx mori*	Induce cell necrosis and signal transduction was affected	(Meng et al., 2017)
Ag NPS	*Drosophila*	Depigmentation, impaired movement, compromised fertility, accumulation of ROS and DNA damage	(Armstrong et al., 2013) [3]
Ag NPS	*Spodoptera litura*	Accumulation of carboxylesterase in the midgut	Yasur and Usha-Rani (2015)
Ag NPS	*Helicoverpa armigera*	Inhibition of gut protease enzyme activity	(Kantrao et al., 2017)
Au NPS	*Drosophila*	Distributed in the reproductive and digestive enzyme	(Vecchio et al., 2012)
Silica NPS	*Callosobruchus maculatus*	Retarded growth and reduced oviposition	(Cáceres et al., 2016) [7]
ZnO NPS coated with *Bacillus thuringiensis*	*Callosobruchus maculatus*	Decrease fecundity, midgut amylase, and GST activity	(Malaikozhundan et al., 2017)

Polymer-encapsulated nano-pesticides also provide a slow-release mechanism, suited for protection over longer periods of time with minimal application and maintenance. This not only lowers costs for farmers but also reduces the accumulation of detrimental chemicals in soil and water sources [6].

Another innovation in nano-pesticide technology involves the use of bio-nanoparticles derived from plant extracts, offering a readily-available and ecologically friendly alternative. Recent research on neem-based nano-emulsions has demonstrated strong antifungal and insecticidal properties, providing a natural and sustainable alternative to synthetic chemical pesticides [7].

2.3 Aluminosilicate Zeolites for Agricultural Use

For conventional synthetic fertilizers, only a fraction of the chemicals are absorbed into the soil, while the rest may get permanently or temporarily lost due to nutrient leaching as the chemicals make it below root level. Zeolites are nanoporous materials with several useful properties for agriculture, including nano-porosity which can be used for the controlled release of materials, including fertilizers. By using certain surfactant materials, zeolites may be modified to carry fertilizer chemicals like anion phosphate (PO43−), which carries both negative and positive effects on plants and soil. Anion phosphate has positive effects on both soil and plant life (root development, enzyme activation) but also has negative effects on them. A slow and controlled release, which can be provided by porous zeolites, can go a long way in mitigating these effects.

As water security is put into a progressively more precarious position, innovations in reducing water use in the agricultural sector would reduce the load on existing water resources. One important property that is used to gauge the required amount of water is the infiltration rate, which denotes the rate at which water can penetrate soil in a vertically downwards direction. It is important to maintain the infiltration rate at a level which is not too high or too low. A low infiltration rate results in plant malnutrition, while a high infiltration rate may cause contamination if the water reaches groundwater and carries contaminants with it. In high infiltration rate soil, zeolites prove to increase the water absorption rate, whereas without the application of zeolites the water would penetrate further into the soil, past the plants' root limit. This property of water retention is also important, as the presence of excess water aids fungal growth in surrounding areas.

Production Cost of Zeolites

The agricultural use of zeolites poses the question of the volume in which they need to be applied. As a result, production volume needs to be brought into question, as zeolites would need to be a relatively abundant material for widespread use. Zeolites occur naturally within volcanic areas, but their synthesis pathway needs to be cheap and affordable for them to be applied. Zeolites can be synthesized from ash and even waste, offering a recycling route for iron mines, porcelain waste, and even paper sludge. In addition to this, zeolites have a variety of different pathways, some of which do not need expensive and high-maintenance materials, making then perfect for high-volume and widespread use. Not only do they offer highly specialized use-cases for different environments, they are also cheap and sustainable for long-term use.

Nano-sensors for Precision Agriculture

Real-time monitoring of soil and plant health is essential for optimizing resource use in precision agriculture. Nano-sensors, which are highly sensitive and capable of detecting minute changes in environmental conditions, possess the potential to revolutionize farm management practices. These sensors, if able to be used in industrial practices, will allow farmers to monitor nutrient levels, soil moisture, and pathogen presence with high accuracy, allowing for timely interventions and improved decision-making [8].

One of the most promising applications of nano-sensors is in soil nutrient analysis. Carbon nanotube-based sensors have been developed to detect nutrient concentrations in soil, providing farmers with accurate and fast data, allowing them to plan for different fertilization strategies in advance [9]. This real-time monitoring serves to reduce excessive fertilizer application, preventing nutrient wastage and environmental contamination. Nano-sensors are also being developed with smart irrigation systems in mind to optimize water use. By detecting soil moisture levels at the nanoscale, these sensors enable automated irrigation systems to deliver water only when necessary, improving water efficiency by up to 50% in drought-prone areas [10]. Additionally, pathogen-detecting sensors can identify plant diseases within their early stages, allowing for more leeway in anti-outbreak planning [11].

2.4 Environmental and Safety Considerations

Despite the advantages of nanotechnology in agriculture, the question of their toxicity remains unanswered completely. In the field of toxicology, the toxicity depends on the dose per volume, but in the field of nanotechnology the dose depends on the size, shape, surface chemistry, and a variety of other factors. This difference in the factors that affect the toxicity of a substance creates uncertainty in the field of nanotechnology applications in agriculture. Further research is required to ensure that these new technologies do not raise false positive in traditional toxicological screenings while still being detrimental to the environment.

Studies suggest that certain nanoparticles can accumulate in soil and water, potentially impacting microbial communities and disrupting natural ecosystems [12].

Furthermore, regulation for nanotechnology applications in agriculture is still in a transient state. Establishing standardized safety assessments and guidelines is crucial to ensuring that nanotechnology is used conscentiously, especially considering that a variety of toxicological screenings would have to be performed on the target molecules. Public awareness and acceptance, which may improve because of stringent testing, will also play a critical role in the widespread adoption of nanotechnology-based solutions in farming [13].

Nanotechnology presents a promising approach to sustainable agriculture by improving nutrient efficiency, enhancing pest control, and enabling precision monitoring. While challenges remain, ongoing research and regulatory advancements will help integrate it into the agricultural sector, leading to a more resilient and environmentally friendly industry.

3 Nanotechnology Applications

Nanotechnology can transform agriculture by way of improving efficiency in the use of resources and mitigation of environmental impact. However, its widespread application poses various problems regarding environmental risks, regulatory requirements, and societal acceptance. It is essential to comprehend these problems to secure the safe and judicious application of nanotechnology in sustainable agriculture. The agricultural use of zeolites poses the question of the volume in which they need to be applied. As a result, production volume needs to be brought into question, as zeolites would need to be a relatively abundant material for widespread use. Zeolites occur naturally within volcanic areas, but their synthesis pathway needs to be cheap and affordable in order for them to be applied. Zeolites can be synthesized from ash and even waste, offering a recycling route for iron mines, porcelain waste, and even paper sludge. In addition to this, zeolites have a variety of different pathways, some of which do not need expensive and high-maintenance materials, making then perfect for high-volume and widespread use. Not only do they offer highly specialized use-cases for different environments, but they are also cheap and sustainable for long-term use.

3.1 Potential Risks of Nanoparticles in Agroecosystems

One of the most significant problems associated with using nanotechnology in agriculture is nanoparticle (NPs) buildup in soil, water, and plant systems. Unlike conventional fertilizers or pesticides, nano-based substances may have unique interactions with ecological and biological systems due to their size and high reactivity.

Studies have ascertained that engineered nanoparticles such as silver (AgNPs) and zinc oxide (ZnO NPs) possess the ability to alter the population of microbes within soil and subsequently affect nutrient cycles and soil fertility [14]. Such changes would have unintended outcomes on plant productivity and agricultural outputs in the longer term.

In addition, the movement and endurance of nanoparticles in the environment raise the issue of bioaccumulation. Evidence suggests that certain nanoparticles, e.g., titanium dioxide (TiO_2 NPs), accumulate in plant tissues and consequently may end up reaching the food chain [15]. This raises food safety as well as ecosystem health concerns since long-term effects of the consumption of nanoparticles are unknown

3.2 Water Contamination and Toxicity Risks

Excessive application of nano-fertilizers and nano-pesticides may lead to runoff into adjacent water bodies and hence generate pollution. Other agrochemicals have dissimilar physicochemical characteristics from nanoparticles that influence their interaction with aquatic environments. It has been proven through studies that carbon nanoparticles, such as fullerenes and carbon nanotubes, have a potential to linger in water bodies and influence aquatic organisms by disrupting biological membranes and cell processes [16].

Apart from this, transformation of nanoparticles under different environmental conditions remains unclear. For example, some metal nanoparticles such as copper (CuNPs) are soluble as toxic ions when exposed to acidic or salty environments, even enhancing their environmental risk [17]. Scientists have resorted to using degradable or bio-nanoparticles that degenerate into innocent components once they complete their usability

3.3 Regulatory and Safety Issue

Nanotechnology developed rapidly in farming, yet growth outdoes systems of governance, creating worries of homogenous safety testing and regulation. Today, no global set system of checking nano-farm product safety exists, making for heterogenic policies around the world. Regulations within the EU are tight regulations with rigid thoroughness in their tests before releasing products at markets, but third-world nations do not have so stringent of policies [18].

One of the greatest challenges facing the regulation of nanotechnology is the lack of standardized testing regimens to assess nanoparticle toxicity, degradation, and long-term environmental fate. Unlike conventional pesticides and fertilizers, nanoparticles exhibit size-related toxicity, i.e., their effect cannot be ascertained by chemical composition only. Scientists, hence, emphasize the need for nanotoxicology research to establish how different nanoparticles behave in agricultural ecosystems before their large-scale application [19].

3.4 Public Perception and Adoption Barriers

While useful, nanotechnology has eluded farmers' and consumers' use due to safety fears, cost, and ignorance. Most farmers are shown to be reluctant to adopt nano-enabled fertilizers and pesticides because there is uncertainty about their long-term impacts and absence of well-defined guidelines for safe handling [20].

Furthermore, consumers are skeptical about foodstuffs produced using nanotechnology due to perceived health risks. This also places large emphasis on the importance of making risk communication and education widely available, to further public knowledge and in turn acceptance of this novel technology. Governments and research organizations need to involve stakeholders such as farmers, consumers, and policymakers in the development of trust and responsible innovation.

Conclusion and Perspectives

Nanotechnology presents a promising for green agriculture in terms of increasing productivity and improving resource efficiency. Its applications in nano-fertilizers, nano-pesticides, and nano-sensors have vast opportunities to address environmental issues, improve soil health, and rationalize agricultural inputs. However, before nanotechnology could become widely acceptable, several great challenges must be addressed, including environmental risks, regulatory gaps, public acceptance, and economic profitability.

One of the most pressing concerns is the extended environmental impact of nanoparticles. More research needs to be conducted to find out how these chemicals interact with soil microbiomes, groundwater, and food webs over long timescales. In the absence of certain knowledge of their degradation and bioaccumulation, there is a risk of unforeseen effects that may negate their sustainability benefits. In addition, affordable manufacturing methods must be developed so that nano-agricultural technologies are made available and affordable to farmers worldwide, particularly developing country farmers.

In the years to come, the use of nanotechnology with renewable energy sources and digital farming technologies holds much promise. Future agricultural systems would be boosted through solar-powered smart farms where nano-sensors operate in tandem with AI analytics to create sustainable, precision-based agriculture. Biodegradable and antimicrobial nano-coatings have the potential to revolutionize food safety and supply chain sustainability.

For all the advantages to be realized, cooperative efforts must be employed. Policymakers, researchers, and industry specialists must get together to develop coherent regulative standards, harmonized safety assessments, and consumer and farmer education programs. Training programs and incentives can be financed by governments to encourage the utilization of nanotechnology in a sustainable way while placing sustainability at its center of priority.

While nanotechnology offers a new pathway towards solutions to the future of agriculture, its success will depend on a plethora of reasons, such as responsible use, continuous research, policymaking that is neither too stringent nor too lax. By overcoming present challenges and the embracing of new innovations, nanotechnology can facilitate a more resilient, efficient, and environmentally sustainable global food system.

References

1. FAO: The future of fertilizers: trends and challenges in agriculture. In: Food and Agriculture Organization of the United Nations, Rome (2023)
2. Mahajan, H.K., et al.: Nano-fertilizers: a sustainable approach to enhance plant growth and minimize environmental contamination. Int. J. Environ. Sci. **18**(4), 213–226 (2023)
3. Cao, J.L., et al.: Enhancing phosphorus use efficiency with nano-hydroxyapatite fertilizers. J. Agric. Sci. **10**, 56–72 (2022)
4. Singh, R.P., Patel, D.: Zinc and iron nano-fertilizers: impact on crop productivity. Agric. Nanotech. J. **5**(2), 145–159 (2023)
5. Anandhi, S., Saminathan, V.R., Yasotha, P., et al.: Nano-pesticides in pest management. J. Entomol. Zool. Stud. **8**(4), 685–690 (2020)
6. Khan, T.E., et al.: Polymer-encapsulated nano-pesticides: a sustainable alternative. Environ. Sci. Tech. **56**(3), 187–203 (2023)
7. Gupta, L.S., et al.: Neem-based nano-emulsions for sustainable pest management. J. Bio-Nanotech. **9**(1), 23–39 (2022)
8. Das, R.K., Roy, S.: Environmental fate and toxicity of nanoparticles in agroecosystems. Environ. Sci. Tech. **56**(3), 187–203 (2023)
9. Wong, P.N., et al.: Carbon nanotube-based sensors for soil nutrient monitoring. IEEE Sensors J. **20**(5), 1123–1134 (2023)
10. Huang, C.D., et al.: Smart irrigation systems using nano-sensor technology. Smart Agric. Tech. **8**, 101–119 (2023)
11. Lee, M.S., et al.: Nano-sensors for early pathogen detection in crops. Plant Health J. **6**, 45–62 (2022)
12. Park, R.J., et al.: Assessing nanoparticle accumulation in soil and water ecosystems. J. Environ. Res. **14**(4), 178–195 (2023)
13. World Health Organization: Nanotechnology Safety in Food and Agriculture. WHO Report, Geneva (2023)
14. Sharma, R.K., et al.: Impact of engineered nanoparticles on soil microbial communities and plant health. Environ. Sci. Technol. **55**(2), 1345–1357 (2023)
15. Zhang, L., Wang, X.: Bioaccumulation of titanium dioxide nanoparticles in agricultural crops: risks and mitigation strategies. Nanotoxicology. **17**(1), 22–38 (2024)
16. Thompson, J.A., et al.: Nanoparticles in aquatic ecosystems: a review of their interactions, risks, and mitigation. J. Environ. Nanotechnol. **12**(3), 145–162 (2022)
17. Gupta, T.K., Chen, M.L.: Transformation and toxicity of metallic nanoparticles in saline and acidic environments. Environ. Pollut. **276**, 116723 (2021)
18. European Commission: Regulation of Nanotechnology in Agriculture: Current Status and Future Directions. EU Policy Report, Brussels (2023)
19. Patel, B., Jones, A.S.: Advancing nanotoxicology studies: the need for standardized assessment methods. J. Nanosci. Saf. **8**(4), 210–225 (2023)
20. Gonzalez, M.H., et al.: Public perception and adoption barriers of agricultural nanotechnology: a global survey. Int. J. Agric. Technol. Innov. **15**(1), 67–82 (2024)

A Cross-View of Startups' and Agribusinesses' Perceptions of AI Adoption: The Case of Morocco

Yousra Moudni(✉) and Khalid Chafik

Management and Information Systems Research Team (ERMSI), Abdelmalek Essaadi University – National Business and Management School of Tangier, Tangier, Morocco
ymoudni97@gmail.com

Abstract. This article aims to explore the perceived barriers and the possible factors that facilitate the future adoption of AI agricultural tools by Moroccan agricultural value chain organizations. To achieve this goal, we used the TAM-TOE model using all the variables pre-validated by authors. Then, we conducted a cross-study between agribusinesses previously published and suppliers of AI agricultural technologies where 15 startups were interviewed. Moreover, data analysis was elaborated in three stages: transcribing, coding interviews, and thematic analysis using Nvivo14. Finally, the findings of this study were analyzed and discussed.

Keywords: Artificial Intelligence · Agriculture · Adoption

1 Introduction

The agricultural sector, particularly the agri-food industry, is characterized by two major and paradoxical effects. On one hand, it is a contributor to global warming through the emission of greenhouse gases. On the other hand, it is highly vulnerable to the consequences of climate change, facing serious challenges such as drought, water scarcity, and extreme weather conditions [13].

In response to these challenges, artificial intelligence (AI) tools are addressing the threats of climate change on agriculture. These tools provide best practices to address critical challenges such as pest control, climate change adaption and food security, to reduce costs and increase productivity through emerging, advanced, disruptive and innovative technologies [13].

Despite all the benefits of these technologies, studies of the adoption of AI tools in the agricultural sector remain limited, and more specifically in Morocco. Academics have long been discussing the determinants that influence the adoption of AI tools in various sectors, whether at the individual or societal level, but there is a gap in the agricultural literature [28].

Within this framework, our problematic focuses on the determinants of AI tools adoption in the Moroccan agricultural value chain organization. In our previous exploratory study [28] we focused on the agricultural and fertilizer industry's view of AI's adoption,

B.-E. Boudriki Semlali et al. (Eds.): DATA 2025, LNNS 1908, pp. 497–509, 2026.
https://doi.org/10.1007/978-3-032-21200-9_42

in this article we will extend our exploratory study to startups and companies supplying agricultural AI technologies.

For this purpose, we will first present the AI solutions adapted to our field of study, thereafter we will expose briefly the theoretical review of our problematic, and then we will define our field of research followed by the methodology of our exploratory study to end with the results and discussion.

2 Digitizing the Moroccan Agricultural Value Chain as a Solution to Constraints

In Morocco, the agricultural sector contributes 12.3% to Gross Domestic Product (GDP) and employs 38% of the workforce—73.7% in rural areas. With 30.4 million hectares of arable land and 1.6 million farmers, Morocco is a global leader in mint and carob production and a major producer of citrus fruits, tomatoes, olives, sugar, dates, and argan. These assets position agriculture at the center of the country's economic and social priorities [19, 20].

The Morocco Green Plan (2008–2020) led to a 6.9% increase in agricultural added value between 2008 and 2018, with significant gains in arboriculture (27.7%), livestock (30.6%), and market gardening (16.1%). The decline in cereals, from 21.4% in 2009 to 13.5% in 2018, was linked to drought, particularly in 2016. In contrast, favorable rainfall in 2017 spurred 11.4% growth in the sector [8, 20].

Thus, agricultural output remains highly dependent on climate, with its GDP share fluctuating between 11% and 18%. This vulnerability to drought and rainfall variability underscores the need for adaptive strategies. The Green Generation 2020–2030 strategy, launched by His Majesty King Mohammed VI in February 2020, promotes digital transformation through traceability, advisory services, mobile payments, and water management to enhance resilience and sustainability in agriculture as a solution to deal with this phenomenon [11].

In the field of agriculture, digitization involves 4 aspects [18]:

First, the integration of remote sensing and geographic information systems; These two technologies are complementary and enable us to explain events, forecast results and plan strategies through the deployment of sensors on drones and satellites to collect data in image form and then link them to databases or maps to create dynamic displays.

Second, the use of agricultural service applications; These mobile applications help farmers to solve the problems they may encounter by providing information on best agricultural practices, weather forecasts, data on epidemics, financial services, access to input markets to supply the necessary materials, and access to production markets to sell the crops.

Third, the conversion to precision agriculture, which uses robotics, artificial intelligence and deep learning processes to practice precise farming at a specific location and schedule, rather than uniform agricultural adoption across the whole field, in order to reduce the impact of ecological and climatic risks and increase crop yields and economic gains.

Forth, the introduction of genomics, bioinformatics and big data; This combination is mainly used to determine the functions of specific genes, in order to trace the origin of contamination, search for metabolic capacities and improve plant or crop genetics.

In our research project in Morocco, we will focus on big data-driven precision agriculture and more specifically on the adoption of AI tools.

The integration of advanced AI models—including deep reinforcement learning, cloud computing, and information technologies—into precision agriculture offers significant potential to enhance food production, resource conservation, and sustainability. AI supports informed decision-making by addressing complex agricultural challenges such as crop disease detection, pest and weed control, irrigation management, and post-harvest storage. Through trained algorithms and data-driven models, these systems can predict issues and propose optimal interventions [13].

AI-driven technologies are transforming agriculture through diverse IoT applications, such as drones for field mapping and water estimation, connected tractors, environmental sensors, and automated livestock systems. The continuous flow of spatial, temporal, and weather-linked data improves operational efficiency, mitigates risks like water and pesticide shortages, enables automated monitoring, and strengthens both the economic and environmental performance of the agricultural sector [13].

The quality of fertilizers used in agriculture can be enhanced and tailored to specific crops and environmental conditions through the integration of artificial intelligence (AI) in chemical manufacturing processes. Given the complexity of these processes—requiring precise control of quantities, temperatures, and pressures—traditional sensors often fall short in accurately measuring critical variables. To address this, virtual sensing methods using data-driven models have been developed. These models employ multi-objective control strategies and adaptive fuzzy neural networks to capture non-linear system behaviors. Among these, artificial neural networks (ANNs) play a vital role due to their flexibility and ability to manage complex, non-linear processes. When properly trained, ANNs can rapidly predict product composition from temperature data, detect system errors, and support predictive control, offering a cost-effective alternative to conventional controllers [16, 23].

3 Towards a Cross-Study of the Various System Stakeholders

The stakeholders in the agricultural technology sector, and more specifically in AI tools, are divided into two essential parts: the suppliers of agricultural AI solutions, who are divided between companies and startups; and the end users who are the companies and individuals acting upstream in the agricultural value chain, namely input suppliers and growers.

Because these two actors perform differently, one is the supplier and the other the end-customer, we need to analyze the perception of AI adoption among agricultural value chain organizations on both sides to better understand the problem, given that the number of AI solution startups is significantly compared to actual adoption among end-customers. So we decided to conduct our research towards a cross-study on the agricultural AI tools adoption perception within suppliers and end-customers.

Based on the model resulting from our exploratory empirical study of agricultural value chain organizations (Fig. 1) [28] adapted to the characteristics of the 16 companies previously interviewed, including 3 agricultural technology companies, we will explore the subject from the perspective of AI agricultural tool suppliers.

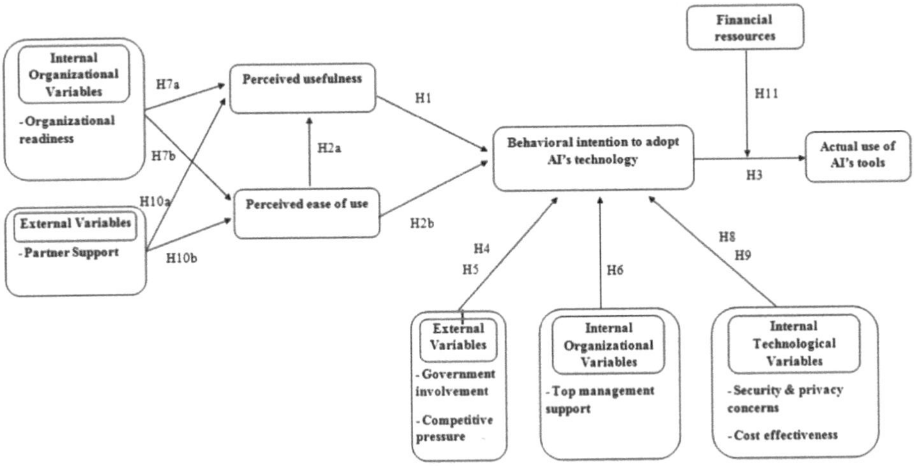

Fig. 1. Research model result of our initial exploratory study on AI adoption in the agricultural value chain organizations [28].

Given that we lack official statistics on the level of AI adoption in the Moroccan agricultural sector [27] and previous empirical studies dealing with AI adoption variables in the Moroccan agricultural sector (Table 1), we decided to proceed with a qualitative exploratory study to cover our problematic.

Table 1. Results of Scopus queries conducted in 2024 on the number of studies concerning artificial intelligence in the agricultural sector in Morocco.

Scopus query	The number of results
1. (adopt*) AND ((“artificial intelligence”) OR (“artif* intel*”)) AND ((tam) OR (toe) OR (dit) OR (tpb) OR (tra) OR (utaut) OR (sam))	377 Articles
2. (adopt*) AND ((“artificial intelligence”) OR (“artif* intel*”) OR (précis* w/2 agric*)) AND (agric*) AND ((tam) OR (toe) OR (dit) OR (tpb) OR (tra) OR (utaut) OR (sam))	25 Articles
3. (adopt*) AND ((“artificial intelligence”) OR (“artif* intel*”)) AND (agric*) AND ((tam) OR (toe) OR (dit) OR (tpb) OR (tra) OR (utaut) OR (sam))	7 Articles
4. (adopt*) AND ((“artificial intelligence”) OR (“artif* intel*”)) AND (agric*) AND ((tam) OR (toe) OR (dit) OR (tpb) OR (tra) OR (utaut) OR (sam)) AND (Morocc*)	1 Article [28]

Scientific research dealing with the analysis of AI adoption variables on an individual or organizational aspect is limited for the agricultural sector following the results of research carried out on the Scopus database in 2024 given that they represent only 7 articles out of a total of 566,670 articles addressing the term AI.

In our first query we looked for studies that analyze AI adoption variables in all domains of an individual or organizational aspect, thus we found 377 articles that focus specifically on a well-defined technology such as chatbots, autonomous cars, robo-advisors, surveillance systems … etc. In addition, empirical researches were conducted in limited sectors such as medicine, the media and telecommunications industry, education and the financial sector [28]. The research models used by researchers are often models combining several theories. They generally include the TAM and UTAUT models when analyzing the individual aspect, and the TOE model for the organizational aspect. We then limited our research to the agricultural sector, using the term "precision agriculture", and focused on the models most widely used in management information systems research [14] resulting in 25 articles. Towards the end, we limited our search to artificial intelligence instead of precision agriculture, as this even encompasses robotics, and we found 7 articles written between 2022 and 2024, including our own [28], which is the only one dealing with the case of Morocco and that shows the originality of our research study.

4 Analysis of the Moroccan Agricultural Environment

4.1 The Moroccan Agricultural Value Chain

The value chain in agriculture refers to all the participants and activities that take a basic agricultural product from the production stage in the field to its final consumption.

In Morocco, it comprises several players, as shown in (Fig. 2) [1, 2, 10]. The fertilizer market is led by the Office Chérifien du Phosphate, the world leader in sustainable phosphate production, fertilizers and phosphate feed supplements for animals. In 2020, it recorded sales of 56.182 billion DH, and its industry remains very important to the Moroccan economy and agriculture [12]. National producers can be individual farmers with their own fields or greenhouses, listed in the national agricultural register, or agricultural companies with various legal statuses. As for cooperatives, there are some 14,850 operating in the agricultural and agri-food sectors. Finally, the agrifood industry accounts for 4% of GDP and 161,000 jobs, with a processing capacity of around 45 million tonnes, generating MAD 3.9 billion in exports [11].

4.2 Moroccan Agricultural Technology Companies and Startups

Following the guidelines of the Generation Green 2020–2030 strategy, Morocco's Ministry of Agriculture, Maritime Fishing, Rural Development and Water and Forests has launched the Digital Pole for Agriculture, Forestry and Drought Observatory, which is administered by 11 public and private institutions with the aim of coordinating and driving the shift towards an attractive agriculture 4.0 that attracts investment and promotes the creation and development of competitive Moroccan startups.

The Digital Pole has thus developed a national network of smart farms and a collaborative network of connected weather stations so that public and private players, farmers, researchers, IT integrators and technology boosters can come together to explore and exploit the opportunities offered by new IoT, AI, drone, satellite and cloud technologies, and do so to improve production factors throughout agricultural value chains.

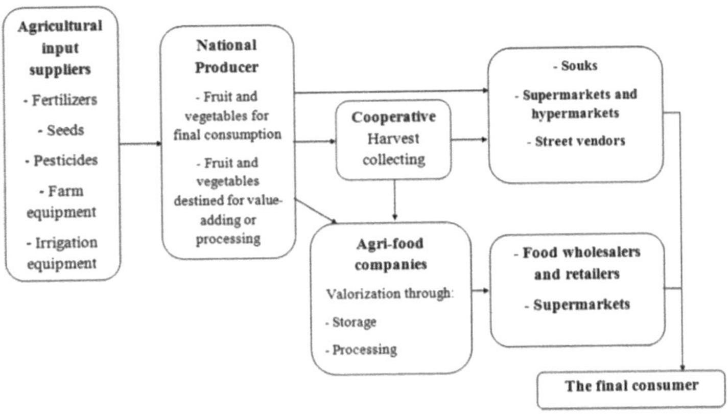

Fig. 2. The Moroccan agricultural value chain.

Referring to the results of a study carried out by the Digital Pole on the number and nature of agricultural technology companies in Morocco, it turns out that there are 32 companies and 45 startups, their activities include one or more specialties as follows: 36 entities develop management systems including AI tools and other new technological solutions, 33 market and monitor drones, 18 focus on IoT and sensors, 5 specialize in e-commerce and 2 in connected machinery.

5 The Methodology

For this phase, we have opted for an abductive approach, which allows us to explain a phenomenon in terms of its causes. This approach is close to both deduction, which lets us reconstruct things a posteriori but in an uncertain way, and induction, which allows us to generalize a property observed empirically over a large number of cases. In other words, we will go back and forth between the verbatims collected empirically from the field and the variables of the TAM and TOE model validated and accepted by the literature, in order to find the most appropriate model for our target that explains the variables in the adoption of AI tools by organizations in the agricultural value chain.

To do so, we went through 5 phases, firstly we collected the variables and hypotheses accepted and validated by the literature, secondly we conducted a qualitative exploratory study of Moroccan agricultural and chemical fertilizer companies to test existing variables and seek new variables adapted to the field under study, thirdly he had a focus group with agricultural technology companies to refine the research model initially formulated and gain a cross-sectional view of the adoption of AI tools, fourthly we reviewed the literature to find the theoretical basis for the new variables that have emerged to finally group the variables into a single model and confirm it with a future quantitative study.

The qualitative approach justified in the previous section will be conducted as a focus group. The discussion guide has been developed on the basis of the results of interviews conducted in our previous study [28].

Data analysis will be carried out using NVIVO 14 software, with textual analysis of the files followed by content analysis, i.e. encoding.

5.1 Sampling

Our sampling methodology is based on a single case study, namely agricultural AI technology companies.

In addition, we're going to favor the purposeful maximal sampling approach [6], where replication logic should be followed rather than statistical sampling. The cases chosen should produce similar or contrasting results, but for known reasons. The sampling technique we used was chosen because it deliberately selects participants on the basis of their qualities. It's a non-random technique where the researcher determines what needs to be known, then seeks out people with the necessary knowledge or experience who are willing to share it. This sampling technique is well suited to the study, as it is used most effectively in qualitative research to identify and select information-rich cases [21]. In addition, we wanted all regions of the country to be represented if possible in the exploratory sample. The focus group is formed of 15 engineers and researchers with their own agricultural technology startups who are introducing AI tools into their proposed solutions, and the contact has been developed thanks to the collaboration of the Digital Pole.

5.2 Data Collecting

The guidelines for debate include the following topics (Table 2):

Table 2. The exploratory discussion guide debated during the focus group.

General starting question	What factors do you think are motivating or blocking agricultural and fertilizer companies from adopting AI tools?	
Selected topics	FAQ for agricultural technology companies	Authors
Perceived Usefulness	In your opinion, to what extent can the perceived usefulness of these technologies hinder or encourage their adoption by farmers?	[7, 26]
Perceived Ease of Use	How can farmers' perceived ease of use of these technologies affect their decision to adopt them?	[7]
Perceived Ease of Use	Does perceived ease of use of AI tools affect their perceived usefulness?	[7]
Government Involvement	To what degree can government involvement motivate a company to adopt AI tools?	[4, 5]
Competitive Pressure	What role do you think competitive pressure plays in the adoption of AI tools?	[22]
Top Management Support	How do you think top management's support can influence the decision to adopt the AI tools you propose?	[4, 5, 22]
Organizational Readiness	To what level can an organization's technical readiness and the level of its staff's expertise affect the perceived usefulness and ease of use of AI tools?	[3, 4, 5, 22]

(*continued*)

Table 2. (*continued*)

General starting question	What factors do you think are motivating or blocking agricultural and fertilizer companies from adopting AI tools?	
Selected topics	FAQ for agricultural technology companies	Authors
Security and privacy concerns	To what extent can a data security or confidentiality problem prevent an agri-business from adopting AI tools?	[5, 22]
Cost effectiveness	How can you judge the relationship between the profitability of AI tools and their adoption?	[15, 22]
Partner Support	Do you consider that your contribution to the customer is essential to enable him of perceiving the usefulness and ease of use of the AI solutions proposed? if so, to what extent?	[3, 4, 22]
Financial ressources	Do you think that the availability of financial resources is a crucial factor in the adoption of agricultural AI tools?	[28]

6 The Results

The focus group of 15 engineers produced a varied but complementary set of results, given their similar levels of education (Fig. 3).

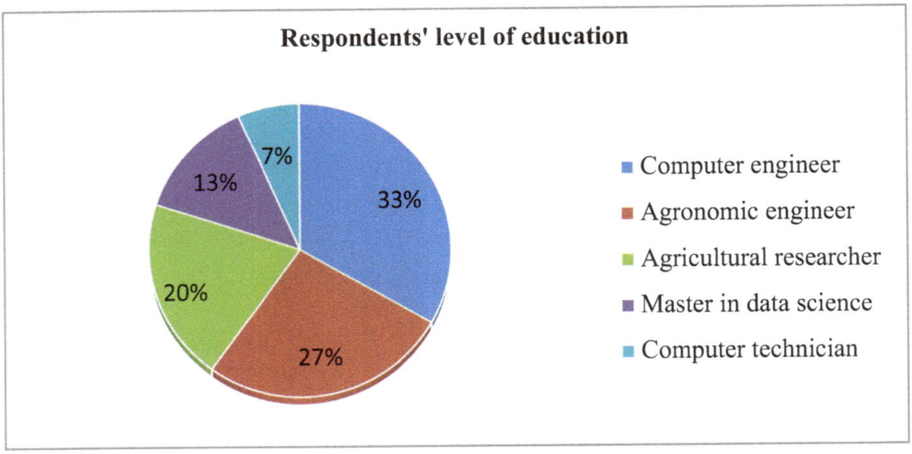

Fig. 3. The level of education of the participants to our exploratory study.

6.1 Research Proposal Validation

As provided for in the reasoning of the abductive method adopted, in this stage we will compare all the research proposals with our empirical data (Table 3). This will enable us to conclude the effect of all the predetermined variables on the adoption of AI tools by agricultural value chain organizations.

The condensed matrix resulting from the items of each variable enabled us to conclude the following:

Research proposal	Results
P1: The availability of technical infrastructure in an organization helps to perceive AI tools' usefulness and ease of use.	Rejected
P2: The training and level of expertise of human resources affects the perceived usefulness and ease of use of AI tools.	Confirmed
P3: Partner support for AI tools helps organizations to perceive their usefulness and ease of use	Confirmed
P4: If the agricultural organization perceives the usefulness of AI tools, it will adopt them.	Confirmed
P5: If the agricultural organization perceives the ease of use of AI tools, it will adopt them.	Rejected
P6: Perceived ease of use of AI tools positively influences their perceived usefulness.	Confirmed
P7: If the government is committed to the AI adoption process, then organizations will adopt them.	Confirmed
P8: If competitors adopt AI tools, so will the organization.	Rejected
P9: If top management is implicated in the adoption decision, the organization will adopt AI.	Confirmed
P10: If there's a problem with data security or confidentiality, the organization won't adopt AI solutions.	Rejected
P11: If AI tools are cost-effective for an agricultural organization, it will adopt them.	Confirmed
P12: The availability of financial resources is a moderating factor between the intention to adopt and actual adoption.	Confirmed

Respondents also emphasized the importance of a variable that the agricultural and fertilizer industry organizations did not mention, namely data quality, which we will develop in the discussion phase.

Table 3. The results of our exploratory study.

Respondents' proposal	Results
P13: The availability and reliable quality of data is a prerequisite for the successful implementation of AI models.	Confirmed

6.2 New Items and Discussion

In this section, we discuss the rejected variables with their items and the new variable added by the respondents stating its theoretical basis. The other variables confirmed by our respondents have been detailed in our previous article [28].

The variable: Quality of the Data Supplied

Code	Items
QID_1	"I think in addition to the variables you mentioned, there is the availability and reliability of agricultural data required for learning the AI tool, without good data quality it is difficult to implement a successful AI model in the organization."
QID_2	"AI tools and systems as I mentioned before don't need much infrastructure but imperatively require quality data because AI relies on accurate data for training and decision making if not it will lead to misinformed results."

Agricultural AI solutions learn from data collected from farms by sensors and satellites, enabling the solution to significantly improve farming practices through data-driven insights. It can also optimize crop management, improve yield forecasts and promote sustainable practices. As a result, if the data collected is malformed, decisions will follow [17]. In the literature, we find that [9, 24, 25] have confirmed the positive impact of agricultural data quality in the development of AI tools for agricultural organizations.

The variable: Technical Preparation

Code	Items
TP_1	"The AI solution is installed at the organization with all its required infrastructure, as you can see our activities are not limited to AI solutions but we also offer commercialization of drones, IoT and sensors so I don't think the organization will have to worry about prior technical preparation."
TP_2	"The majority of our solutions are installed in the cloud, so all you need to do is supply the necessary tools for online deployment."

The variable: Perceived Ease of use

Code	Items
PEU_1	"I don't see any relationship between perceived ease of use and the adoption of AI tools. In principle, these tools are designed for all types of agriculture, whether an organization, an educated individual or not, so all he needs is the training we offer when setting up, and support as he goes along".
PEU_2	"Our solutions are already easy to use, so this has nothing to do with adoption. The difficulty lies in the development, learning and feeding phase which is done by the suppliers."

The variable: Competitive Pressure

Code	Items
CP_1	"The use of AI tools is intended to meet a specific need or problem of the entity, each solution is tailor-made and cannot be shared between two organizations, so if a company uses an AI solution it's because it needs it to deal with a problem or improve one or more tasks and not because of competitors."
CP_2	"In the Moroccan agricultural sector, the major concern is to cope with drought, climate change and guarantee food security, and one of the most appropriate solutions to meet these challenges is the use of AI, it is not a commercial solution to have competition between agricultural companies."

The variable: Security and Privacy Concerns

Code	Items
SPC_1	"I don't think data security causes a problem in the case of our AI solutions because they are tested before they are delivered, and we make sure through testing that this problem won't cause itself."
SPC_2	"With the current level of cyber security in the world, the risk of data security problems remains low; in addition we are in constant contact with other cyber security startups regarding the development of our solutions".

7 Conclusion

Through this article, we have tried to analyze the adoption of AI agricultural technologies from the point of view of the suppliers and that to have a full vision of the problematic, especially that in Morocco the adoption of AI tools in agriculture is limited to big companies as concluded in our previous exploratory study [28] thus the number of agricultural AI suppliers is important. In addition, through the exploratory qualitative study carried out in the field, we were able to refine our model and delimit our variables and hypotheses according to all the actors involved in the process i.e. the suppliers and end customers of AI agricultural tools.

Moreover, it is important to note that this study was conducted as part of the preparation of a doctoral thesis entitled: "Analysis of the variables of the adoption of artificial intelligence and big data tools in organizations of the agricultural value chain in Morocco". Therefore, the next paper will focus on a confirmatory empirical study in both sides in order to validate our model in the field and confirm or reject our final formulated hypotheses. Towards the end we will propose a reliable model that will serve as a national reference for organizations wishing to adopt the two technologies and as a scientific reference for researchers in the managerial aspect of AI and big data.

References

1. Abdelmonim, A.: CHAÎNES GLOBALES DE VALEUR TOURMENTÉES: RISQUES ET OPPORTUNITÉS POUR L'AGRICULTURE ET L'ALIMENTATION Par Abdelmonim AMACHRAA & Hassnae MAAD (2023)
2. Benabdellah, M., El Harrak, M.: Chaîne de valeur de la pomme de terre: Un cadre d'identification, de conception et d'analyse. Revue Marocaine Des Sciences Agronomiques Et Vétérinaires. **8**(1) (2020)
3. Chatterjee, S., Rana, N.P., Dwivedi, Y.K., Baabdullah, A.M.: Understanding AI adoption in manufacturing and production firms using an integrated TAM-TOE model. Technol. Forecast. Soc. Chang. **170**, 120880 (2021)
4. Chen, H.: Success factors impacting artificial intelligence adoption—perspective from the telecom industry in China. Theses and dissertations in business administration (2019)
5. Cheung, Y., Messom, C., Alsheiabni, S.: Factors inhibiting the adoption of artificial intelligence at organizational-level: a preliminary investigation. AMCIS 2019 Proceedings, 2, (2019)
6. Creswell, J.W.: Educational Research: Planning, Conducting, and Evaluating Quantitative and Qualitative Research, 4th edn. Pearson, Boston, MA (2012)
7. Davis, F.: Perceived usefulness, perceived ease of use, and user acceptance of information technology. MIS Q. **13**, 319–340 (1989)
8. DEPF Studies. The Moroccan agricultural sector: structural trends, challenges and development prospects (2019). http://depf.finances.gov.ma/etudes-et-publications/
9. Dhoni, P.: Enhancing Data Quality through Henerative AI: An Empirical Study with Data (2023). https://doi.org/10.36227/techrxiv.24470032.v1
10. FAO Homepage. https://openknowledge.fao.org/handle/20.500.14283/cb2395fr. Last accessed 25 Dec 2024
11. Generation Green 2020-2030 | Department of agriculture (n.d.). Retrieved October 20, 2021, from https://www.agriculture.gov.ma/fr/ministere/generation-green-2020-2030
12. Industrial Sectors. Ministry of industry, trade and the green and digital economy (n.d.). Retrieved July 18, 2021, from http://www.mcinet.gov.ma/fr/content/les-secteurs-industriels
13. Jha, K., Doshi, A., Patel, P., Shah, M.: A comprehensive review on automation in agriculture using artificial intelligence. Artif. Intell. Agric. **2**, 1–12 (2019). https://doi.org/10.1016/j.aiia.2019.05.004

14. Khalid, C.: Regards sur la recherche en systèmes d'information des organisations: Cas du Maroc (2021)
15. Kim, S., Kim, B.: A decision-making model for adopting Al-generated news articles: preliminary results. Sustainability. **12**(18), 7418 (2020)
16. Kramer, A., Morgado-Dias, F.: Artificial intelligence in process control applications and energy saving: a review and outlook. Greenhouse Gases Sci. Technol. **10** (2020). https://doi.org/10.1002/ghg.1962
17. Kumar, S., Sharma, R., Singh, V., Shrikant, T., Singh, S., Datta, S.: Potential impact of data-centric AI on society. IEEE Technol. Soc. Mag. **42**, 98–107 (2023). https://doi.org/10.1109/MTS.2023.3306532
18. Mondejar, M.E., et al.: Digitalization to achieve sustainable development goals: steps towards a smart green planet. Sci. Total Environ. **794**, 148539 (2021). https://doi.org/10.1016/j.scitotenv.2021.14853
19. National Survey on Household Consumption and Expenditure. Downloads. Institutional site of the high commission for planning of the Kingdom of Morocco (n.d.). Retrieved October 23, 2021, from https://www.hcp.ma/downloads/Enquete-Nationale-sur-la-Consommation-et-les-Depenses-des-Menages_t21181.html
20. OECD. OECD-FAO Agricultural Outlook 2021–2030. Organisation for Economic Co-operation and Development (2021). https://www.oecd-ilibrary.org/agriculture-and-food/oecd-fao-agricultural-outlook-2021-2030_19428846-en
21. Palinkas, L.A., Horwitz, S.M., Green, C.A., Wisdom, J.P., Duan, N., Hoagwood, K.: Purposeful sampling for qualitative data collection and analysis in mixed method implementation research. Admin. Pol. Ment. Health. **42**(5), 533–544 (2015)
22. Pillai, R., Sivathanu, B.: Adoption of artificial intelligence (AI) for talent acquisition in IT/ITeS organizations (2020)
23. Rais, A: Artificial intelligence in the process industry – very hot affair (n.d.). Retrieved December 11, 2021, from https://www.process-worldwide.com/artificial-intelligence-in-the-process-industry-very-hot-affair-a-984575
24. Rangineni, S.: An analysis of data quality requirements for machine learning development pipelines frameworks. Int. J. Comput. Trends Technol. **71**, 16–27 (2023). https://doi.org/10.14445/22312803/IJCTT-V71I8P103
25. Schwabe, D., Becker, K., Seyferth, M., et al.: The METRIC-framework for assessing data quality for trustworthy AI in medicine: a systematic review. npj Digit. Med. **7**, 203 (2024). https://doi.org/10.1038/s41746-024-01196-4
26. Venkatesh, V., Davis, F.: A theoretical extension of the technology acceptance model: four longitudinal field studies. Manag. Sci. **46**, 186–204 (2000)
27. Yatribi, T.: Exploration qualitative des contraintes de l'adoption des technologies del'agriculture de précision par les agriculteurs marocains: point de vue des entre-prises. AFRIMED AJ -Al Awamia. **132**, 20–39 (2021)
28. Yousra, M., Khalid, C.: Analysis of the Variables Affecting the Adoption of Artificial Intelligence and Big Data Tools Among Moroccan Agricultural and Chemical Fertilizer Industry Firms: Research Model Development Lecture notes in networks and systems, LNNS, vol. 930, pp. 82. Scopus–65 (2024). https://doi.org/10.1007/978-3-031-54318-0_7

Biochemical and 16S rDNA Sequencing of Bacteria Isolated from Moroccan Desert and Drylands

Jamal Eddine El Jamali[1(✉)], Abdelkarim Filali-Maltouf[2], Ibtissam El Hilali[3], Loubna Temsamani[2], Allal Douira[1], and Amina Ouazzani Touhami[1]

[1] Laboratory of Plant and Animal Production and Agro-industry, Department of biology, Faculty of Sciences, University Ibn Tofail, Kenitra, Morocco
jamaleddine.eljamali@uit.ac.ma

[2] Laboratory of Microbiology and Molecular Biology, Department of biology, Faculty of Sciences, University Mohammed V, Rabat, Morocco

[3] National School of Applied Sciences, ENSA, University Ibn Zohr, Agadir, Morocco

Abstract. Three hundred bacteria isolated from sandy dunes of Moroccan Sahara regions were studied. Results showed that the majority of isolates are bacilli Gram negative. They displayed a wide tolerance to salinity (40% of isolates tolerated up to 935 mM NaCl) and to extreme temperatures ranging from 10 °C to 55 °C (41% grew at 45 °C). Profiles of the bacteria's enzymatic activity showed to be very diverse. Some bacteria were able to produce cellulase, amylase, caseïnase and lecithinase which can be of importance in biotechnology usage or for environmental processes. Based on their tolerance to high salt, extreme temperature and their enzyme activity, some interesting isolates have been identified by 16S rDNA gene sequencing.

Keywords: Bacteria · Dry lands · Desert · Enzyme activity · 16S rDNA gene

1 Introduction

Soils from dry lands or desert (hostile ecosystems) in Sahara ecosystems may contain well adapted strains to extreme environmental stresses such as drought, high temperature, pH and high salt concentrations. Some adapted strains to the extreme conditions mentioned above were reported to be thermophilic microorganisms that grow faster above 40 °C [1]. They can be an interesting source of thermoactive and thermostable enzymes [2]. The thermophilic have high ability to secrete extracellular hydrolytic enzymes during growth on a range of proteinaceous substrates such as skimmed milk, on polymetric substrates such as starch or carboxymethylcellulose and on egg yolk substrates [3]. Thermostable hydrolases play an important role in biotechnology and industry; for instance, amylases are widely for baking (bred soft-ness and volume, flour adjustement), in beverage (juice treatment, low calorie beer) and in starch liquefaction and saccharification [4]. Cellulases are also widely used in industry; in detergents (cleaning, starch stain removal

B.-E. Boudriki Semlali et al. (Eds.): DATA 2025, LNNS 1908, pp. 510–522, 2026.
https://doi.org/10.1007/978-3-032-21200-9_43

color clarification), in textile (demin finishing, cotton softening) and in pulp and paper production (de-inking, drainage improvement, fiber modification) [5]. Caseïnase are widely used in milk industry (milk clotting and flavor) [6].

The aim of the following study was to analyze some bacterial strains isolated from soils in different sites of Sahara regions in Morocco in terms of their characterization and the evaluation of the tolerance to extreme parameters especially to salt and temperature; for screening some microbial enzymatic activity and to realize a molecular and phylogenetic approach based on 16S rDNA.

2 Materials and Methods

2.1 Bacterial Isolates

Three different regions from South-Est Morocco: Merzouga, Mhamid ElGhizlane and Arg Lihoudi were prospected and soil samples were collected from eleven different sites. Bacterial isolation from soil was performed in the laboratory as follows: 10 g of each sample were homogenized in 100 ml of sterile physiological water (NaCl 0, 9%) and shacked vigorously during 2 h at room temperature. Aliquots of the supernatants were subjected to serial dilutions in the same sterile water then the mixture was spread onto two solid media: nutrient agar (BIOKAR DIAGNOSTICS-France) and Yeast Extract-Mannitol (YEM) agar media [7] containing Congo red at 25 mg/l. The plates were incubated at 28 °C during 7 days. Colonies were counted and bacterial concentration was evaluated. Isolates obtained from well-formed single colonies were checked for purity by repeated streaking and then stored in 50% (v/v) glycerol at −20 °C.

2.2 Morphological Analyses

Purified isolates were re-streaked on nutrient agar for obtaining well formed colonies. Macroscopic features: size, color and shape were noted; then isolates were directly Gram stained [8].

2.3 Bacterial Growth

Growth of isolates was checked on L. B (Luria Broth) liquid medium at several different times at 28 °C for 7 days. Controls were performed to discard any external contamination.

2.4 Tolerance Assays

All tests were performed in triplicate and compared to the controls.

Salt Tolerance

Salt effect was tested by using NaCl. The latest was added to L. B liquid medium at the following concentrations: 2%, 3%, 4%, 4.5%, 5%, 5.5%, 6%, 6.5%, 7%, 8% and 9%. The medium were dispensed as 5 ml aliquots into test tubes. After autoclaving, the tubes were inoculated with freshly prepared cultures of the appropriate isolate to give an initial optical density (O.D) of 0.05. Cultures were grown at 28 °C in a rotary shaker at 200 rev/min and the final O.D was read after 1, 3 and 7 days at 600 nm [9].

Temperature Tolerance

Test was performed in liquid nutrient medium and prepared as described above. The tubes were incubated in a rotary shaker at 200 rpm at the following temperatures (10, 20, 30, 35, 40, 45, 50 and 55 °C). The results were evaluated as described above.

2.5 Biochemical Analyses

It was performed by screening four enzymes activities: cellulase, amylase, caseïnase and lecithinase.

Cellulase Activity

Was tested on nutrient agar media by adding 1% of carboxymethylcellulose (CMC) before autoclaving as described by Sazci and al [10]. After pouring into sterilized petri dishes, media was inoculated by freshly prepared isolate's culture. The plates were incubated at 28 °C during 7 days. The carboxymethylcellulose hydrolysis was visualized at 1, 3 and 7 days by adding some drops of Congo red 1 mg/l on the agar surface, after 15 min at the laboratory temperature, the dishes were rinsed by 1 M NaCl solution. A transparent halo around the bacterial colony was considered as being a positive reaction.

Amylase Activity

Was tested on nutrient agar media by adding (1%) of starch. The content was autoclaved for 15 min and dispensed into sterilized Petri dishes. The media was inoculated and incubated as described above. Starch hydrolysis was visualized at 1, 3 and 7 days by flooding the agar surface with Lugol's iodine solution. A transparent halo around the bacterial growth was considered as being a positive reaction [11].

Caseïnase Activity

Was tested on double concentrated nutrient agar media. After autoclaving, the medium was cooled around 45 °C of temperature, then the concentration of medium was adjusted by adding aseptically the right volume of UHT skimmed milk. The medium was distributed, inoculated and incubated as described above. Caseïne hydrolysis was daily checked up to 7 days and was indicated by the appearance of a clear zone around the bacterial growth.

Lecithinase Activity

was also tested on the same media. It was sterilized then cooled around 50 °C of temperature. Egg yolk was aseptically diluted to ½ with by adding sterile physiological water. An aliquot of the prepared solution was added to the medium to reach a final concentration of 10% egg yolk in the medium. Dishes were prepared as mentioned above. Results were scored daily up to 7 days. Lecithine hydrolysis was indicated by the appearance of an opaque white halo around the bacterial growth (http://www2.ac-lyon.fr/enseigne/bio tech/microbio/milieux.html).

2.6 Molecular Analyses

DNA of some selected isolates was purified; 16S rDNA gene was amplified and sequenced.

DNA Isolation

Genomic DNA of the bacteria was prepared from liquid cultures harvested at the exponential phase by using Phenol-Chloroform-Isoamylic alcohol according to the protocol of Boucher and al [12]. Purified DNA was dissolved in 10 mM Tris-HCl buffer containing 1 mM EDTA (PH: 7,8) and its concentration was adjusted spectrophotometrically to 50 ng/μl. The DNA were stored at −20 °C for use. 5 μl of DNA was separated on a 0.8% agarose gel at 80 V in TBE buffer for 45 min, stained with ethidium bromide, and then scanned under UV light.

16S rDNA Amplification and Purification

The 16SrDNA gene was amplified by PCR with universal primers derived from the 5′ end of the 16SrDNA (FD1: 5'-CCGAATTCGTCGACAACAGAGTTTGATCCTGGCTCAG-3′) [13] and (RS16: 5'-TACGGCTACCTTGTTACGACTT-3′) PCR amplification was performed with Gene Amp thermocycler in 25 μl reaction mixture containing 50 ng/μl of template DNA, PCR-buffer(×10), 2.5 U of Taq DNA polymerase (promega), 1.5 Mm $MgCl_2$, 200 μM each nucleotide (promega), 10 μM of each primers, the volume was adjusted to 25 μl by adding sterile bidistilleted water.

The PCR temperature program cycle started with an initial denaturation at 95 °C for 5 min, followed by 35 cycles of denaturation at 94 °C for 30 s, annealing at 55 °C for 45 s, extension at 72 °C for 45 s, and a final extension at 72 °C for 10 min, followed by a hold at 4 °C. After the reaction, 5 μl of the PCR product was separated on a 2% agarose gel in TBE buffer at 80 Volts for 3 h, stained with ethidium bromide, and scanned under UV light.

A 100pb ladder (promega) was included as a size reference at both sides of each gel. The PCR products were purified from excess dNTPs and unextented primers by using an enzymatic digestion method as described by Kim and al [14] and applied Biosystems, 2 μl of ExoSAP (ExonucleaseI Shrimp Alcaline Phosphatase) was added to each amplified reaction (20 μl). The reaction mixture was incubated à 37 °C for 30 min followed by enzyme inactivation at 80 °C for 15 min.

16S rDNA Sequencing

Big Dye Terminator Cycle Sequencing Kit (Applied Biosystems) was used with a Gene Amp thermal cycler according to the manufacturer's instructions. The sequencing reaction was performed in 10 μl volume. The sequencing reactions were purified by using 90 μl of magnesil green added to each reaction and mixed well up and down for 5 min. The samples were placed in magnetic rack to capture the particles; the liquid was removed and discarded. 100 μl of ethanol (70%) was added for washing; after a twice repetition during 5 min, the liquid was removed and discarded. 20 μl formamide for elution were added to the particles, the elutates were transferred to 96-well plate for sequences analysis by automatic sequencer ABI Prism 310 Genetic Analyzer. The sequences obtained were analyzed using the Gen Bank database with BLAST alignment software (http://www.ncbi.nlm.nih.gov/BLAST) [15]. Phyllogenetic tree was inferred by using the 1000 bootstrap neighbour-joining method with MEGA (4.1) program [16].

3 Results

3.1 Phenotypic Characterization

In the following study, a total of 300 bacterial isolates were obtained. They exhibited different and diverse morphological features as well as different generation times (data not shown) and the majority was Gram negative bacilli (Fig. 1).

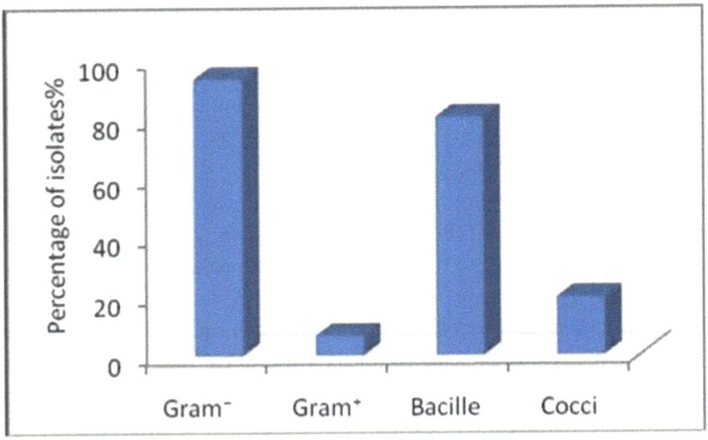

Fig. 1. Gram staining and shape of isolates

Salt Tolerance

Results revealed a perfect tolerance up to 4% NaCl (680 mM). More than 60% grew well at 4.5% (765 mM), and 5% NaCl (850 mM). However, the percentage of tolerant isolates decreased for the higher concentrations: less than 20% of isolates tolerated 6% NaCl (1020 mM) and the growth was significantly affected for the other concentrations (Fig. 2).

Temperature Tolerance

Optimal growth was obtained at 20 °C and 30 °C for all the tested isolates. 10 °C and 35 °C was efficiently tolerated by 96.23% of isolates. The negative impact was noticed from 40 °C but significantly at 45 °C and 50 °C with 45% and 25% of isolates respectively (Fig. 3).

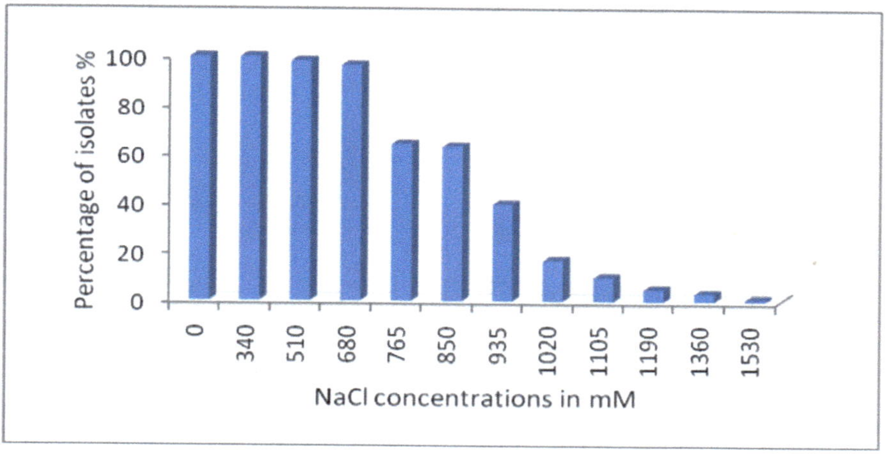

Fig. 2. Percentage of tolerant bacterial strains to NaCl

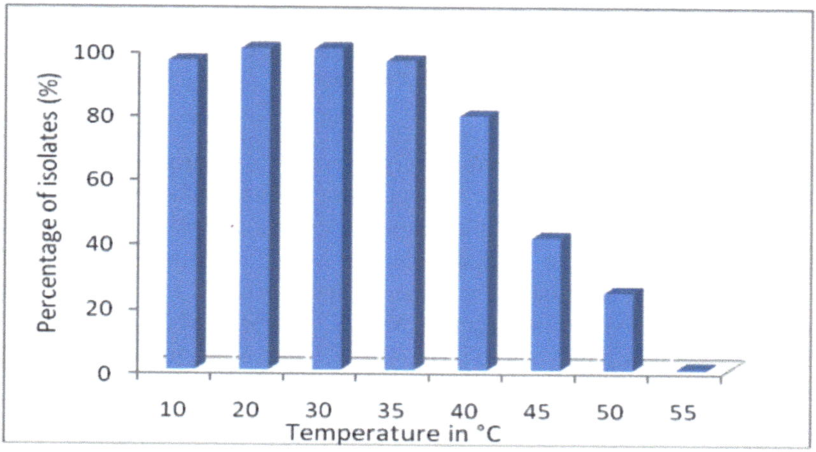

Fig. 3. Growth percentage of bacterial strains in different temperatures

3.2 Biochemical Characterization

All the bacterial strains (300) were subjected to the enzymatic activity tests. Fig. 4 showed the results obtained for one of the strains.

Enzyme activity of all the strains was scored each day from one to seven days. Table 1 showed some of the results obtained.

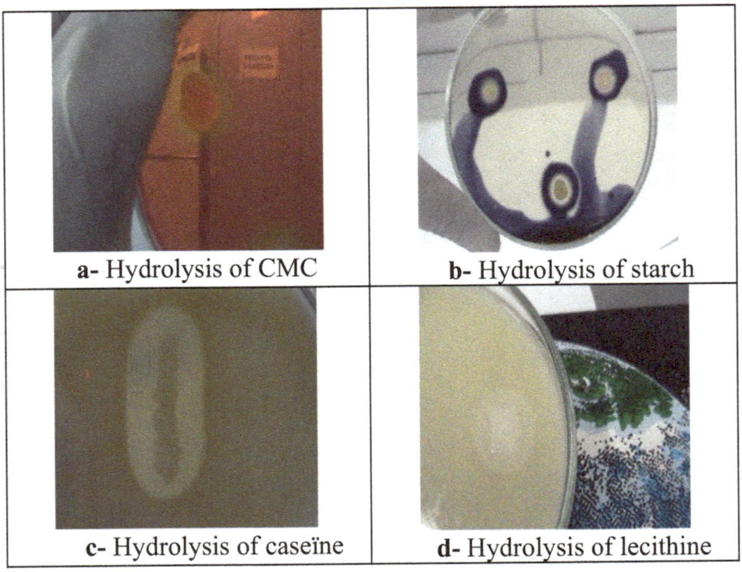

Fig. 4. Enzymatic activity reaction for one bacterial strain of the collection **a.** Test of cellulase, **b.** Test of amylase, **c.** Test of caseïnase, **d.** Test of lecithinase.

Enzyme activity profiles showed to be very diverse. Isolates from the same sampling site produced different results. The activity of the same enzyme varies from one isolate to another depending on the time: some isolates showed a rapid activity while other displayed slow reactions (3 to 7 days); example: MDMC 3 compared to MDCM 60 from Merzouga. In the other hand, the same strain may exhibit activity for one and/or two enzymes and not for the rest; example: MDMC 183, MDCM 200 from Mhamid El Ghizlane. Some other strains showed no enzyme activity; example: MDMC 9, MDCM 26 from Merzouga.

Results showed also that cellulase and lecithinase were the most produced by isolates (around 15% for 1 and 2 days for both, and for all the strains: around 35% for cellulase and 32% for lecithinase) followed by caseïnase. Amylase was the less active (Fig. 5).

3.3 Molecular Analysis

Interesting strains selected on the basis of the biochemical tets and others randomly chosen from the collection were used in this analysis. Purified 16SrDNA gene of the retained strains was sequenced and a phylogenetic tree was constructed after the alignment of strains sequences with data sequences of the reference strains (Fig. 6).

The obtained dendrogram showed that strains were grouped into 3 groups MDMC181, MDMC200, MDMC177 and MDMC175 were closely related to the Pseudomonas group; Strain MDMC183 to Acinetobacter and MDMC81, MDMC251, MDMC409 to the Bacillus group.

Table 1. Enzyme activity results of some bacterial isolates tested represented in function of their origin sampling site.

Bacterial strains	Soil origin	Enzyme activity			
		Cellulase	Amylase	Caseïnase	Lecithinase
MDMC3	Merzouga	+(3d)	–	+(3d)	+(4d)
MDMC5	Merzouga	–	+(5d)	+(2d)	–
MDMC6	Merzouga	+(1d)	–	+(3d)	+(3d)
MDMC7	Merzouga	+(1d)	–	+(4d)	+(2d)
MDMC8	Merzouga	+(3d)	–	+(3j)	+(3d)
MDMC9	Merzouga	–	–	–	–
MDMC11	Merzouga	+(3d)	–	+(4d)	+(4d)
MDMC12	Merzouga	+(1d)	+(3d)	+(3d)	+(6d)
MDMC14	Merzouga	–	+(3d)	–	–
MDMC26	Merzouga	–	–	–	–
MDMC27	Merzouga	–	–	–	–
MDMC41	Merzouga	+(1d)	–	+(3d)	+(3d)
MDMC42	Merzouga	+(7d)	+(3d)	–	–
MDMC43	Merzouga	+(3d)	+(3d)	–	–
MDMC44	Merzouga	+(1d)	–	+(4d)	+(4d)
MDMC45	Merzouga	–	+(1d)	–	–
MDMC60	Merzouga	+(1d)	–	+(1d)	+(1d)
MDMC71	Merzouga	+(1d)	–	+(3d)	+(3d)
MDMC81	Merzouga	+(1d)	–	+(1jd	+(1d)
MDMC104	Merzouga	–	+(1d)	+(1d)	+(1d)
MDMC117	Merzouga (at the foot of the dune)	+(3d)	–	–	–
MDMC118	Merzouga (at the foot of the dune)	+(3d)	–	–	–
MDMC119	Merzouga (at the foot of the dune)	+(1d)	–	+(3d)	+(3d)
MDMC131	Merzouga (at the foot of the dune)	+(3d)	+(3d)	+(3d)	+(3d)
MDMC132	Merzouga (at the foot of the dune)	–	–	–	–
MDMC137	Merzouga (at the foot of the dune)	–	+(1d)	+(1d)	+(1d)

(continued)

Table 1. *(continued)*

Bacterial strains	Soil origin	Enzyme activity			
		Cellulase	Amylase	Caseïnase	Lecithinase
MDMC155	Merzouga (at the foot of the dune)	+(1d)	–	+(3d)	+(3d)
MDMC175	Mhamid El Ghizlane	+(3d)	+(3d)	–	–
MDMC176	Mhamid El Ghizlane	+(1d)	+(1d)	–	–
MDMC177	Mhamid El Ghizlane	+(3d)	+(1d)	–	–
MDMC181	Mhamid El Ghizlane	+(1d)	+(1d)	–	–
MDMC182	Mhamid El Ghizlane	+(7d)	+(1d)	–	–
MDMC183	Mhamid El Ghizlane	–	–	–	+(1d)
MDMC200	Mhamid El Ghizlane	+(1d)	+(1d)	–	–
MDMC244	Mhamid El Ghizlane (at the foot of the dune)	+(1d)	–	–	+(3d)
MDMC249	Arg Lihoudi	+(1d)	+(1d)	+(1d)	+(1d)
MDMC250	Arg Lihoudi	+(1d)	+(1d)	+(1d)	+(1d)
MDMC251	Arg Lihoudi	–	+(1d)	+(1d)	+(1d)
MDMC270	Arg Lihoudi	+(1d)	+(1d)	+(1d)	+(1d)
MDMC271	Arg Lihoudi	+(1d)	–	–	–
MDMC277	Arg Lihoudi	+(7jd	+(7d)	+(2d)	–
MDMC288	Arg Lihoudi	+(7d)	–	–	+(3d)
MDMC289	Arg Lihoudi	+(7d)	+(3d)	–	+(3d)
MDMC291	Arg Lihoudi	–	+(1d)	+(1d)	+(1d)
MDMC308	Arg Lihoudi (at the foot of the dune)	+(1d)	+(1d)	+(1d)	+(1d)
MDMC310	Arg Lihoudi (at the foot of the dune)	–	–	–	–
MDMC312	Arg Lihoudi (at the foot of the dune)	+(7d)	+(7d)	+(1d)	+(1d)
MDMC318	Arg Lihoudi (at the foot of the dune)	+(1d)	+(1d)	+(1d)	+(1d)
MDMC350	Arg Lihoudi	+(1d)	+(1d)	+(1d)	+(1d)
MDMC403	Arg Lihoudi	+(1d)	+(1d)	+(1d)	+(1d)
MDMC409	Arg Lihoudi	+(1d)	+(1d)	+(1d)	+(1d)
MDMC410	Arg Lihoudi	+(3d)	–	–	–

MDMC:Moroccan Desert Micro-organism Collection.
+1d: hydrolysis after 1 day (24 h), +3d: hydrolysis after 3 days, +7d: hydrolysis after 7 days, –: no hydrolysis

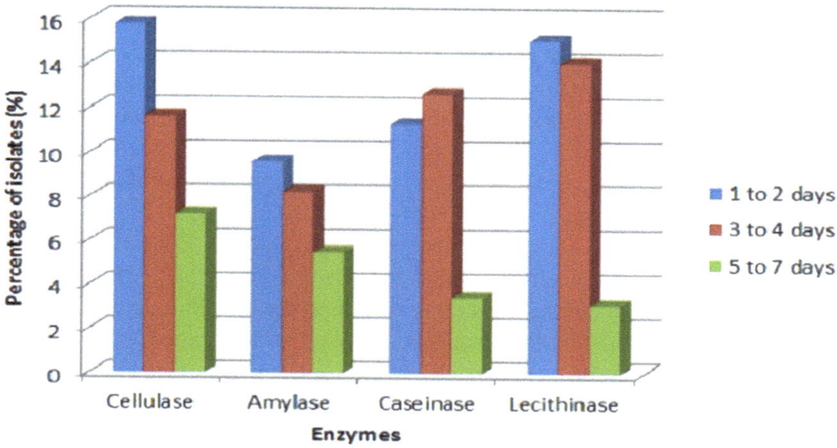

Fig. 5. Percentage of strains showing enzymatic activities for the four enzymes tested

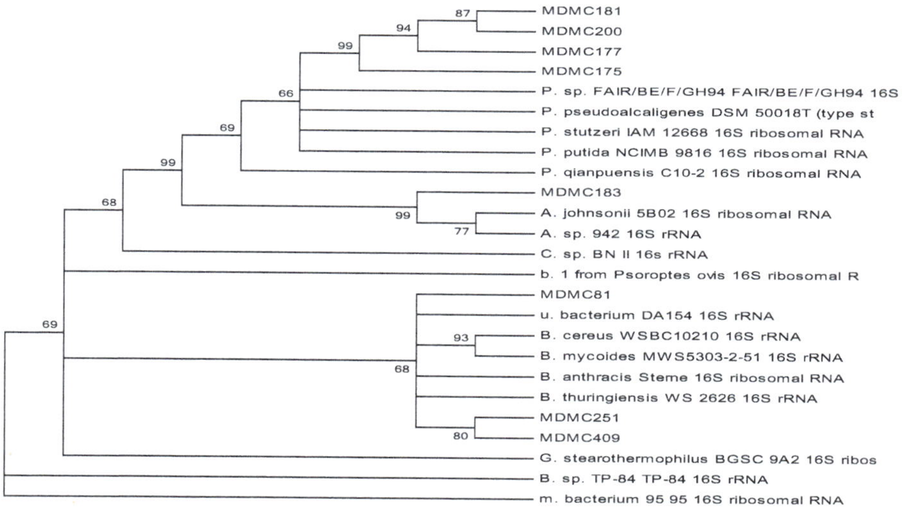

Fig. 6. Phylogenetic tree constructed on the basis of 16SrRNA aligned sequences of the selected isolates and the reference strains used (the accession number are given at the end of each branch).

4 Discussion

Soils collected from arid and Saharan regions of Morocco harbor bacterial strains able to grow in extreme environment. The morphological study of these bacteria indicates that the majority are bacilli Gram negative. Duxbury and al [17] suggested that the large presence of Gram negative is due to phenotypic selection and they are much more resistant through of their structure.

Strains showed high drought adaptability in terms of temperature and salt tolerance; 41% of the strains could grow at 45 °C, we can though speak of the presence of thermophilic bacteria. The long-term high temperatures throughout the year led to the selection of new thermotolerant Bacillus strains from soils in the Sonoran Desert of Arizona [18].

Chamkha and al [19] reported the isolation of thermophilic and halophilic Bacillus isolates from a high temperature sites in Tunisia.

Strains exhibited a wide tolerance to salinity (NaCl). More than 50% were able to grow normaly at 850 mM NaCl; yet, more than 40% could grow at 935 mM NaCl. The highest number of halotolerant strains was recorded for Merzouga site and the lowest number for Arg Lihoudi. Our results corroborates with those of Manni and Filali-Maltouf [20] for strains issued from the same locations. But in that review, lower bacterial halotolerance was reported (less than 30% of strains were from each site were tolerant to 680 mM). Halotolerance of strains may be due to the fact that they were isolated directly from soils of arid regions known for their high level of salinity and conductivity. Halophilic bacteria are known to involve mechanisms of intracellular osmoregulation or the possibility of the accumulation of osmoprotectant molecules to survive in extreme conditions [21].

Enzyme activity profiles for cellulase, amylase, caseïnase and lecithinases were very diverse. Strains isolated from the same site showed to be different in the reaction time and in the response to the same enzyme. No reaction was noticed for many strains of the collection; less than 15% showed a rapid reaction (1 to 2 days) for the four enzymes. Cellulase and lecithinase showed to be more active. Amylase was active particularly with strains derived from Erg lihoudi site. However, some interesting strains were recovered: MDMC 60, MDMC 81, MDMC 104 and MDMC 137 from Merzouga; MDMC 200 from Mhamid El Ghizlane, MDMC 249, MDMC 250, MDMC 251, MDMC 270, MDMC 308, MDMC 318, MDMC 330 and others from Erg lihoudi. There was a wide diversity in the enzymatic activity; many reports concluded the fact with bacterial strains.

Molecular analysis showed that the retained strains were segregated into 3 distinct groups. In fact, strains MDMC181, MDMC200, MDMC177 and MDMC175 which were closely related to the Pseudomonas group have all origin from Mhamid El Ghizlane and strains MDMC251, MDMC409 which was related to the Bacillus group have all origin from Erg Lihoudi. There is a slight accordance between the 16S rDNA sequence lineage and the sampling site, but a net conclusion should be based on many sequences of the collection strains.

5 Conclusion

The prospected sites in sandy dunes of Moroccan desert are a great reservoir of bacterial strains. The strains isolated in this study were well adapted to the extreme environmental stress showing a great tolerance to salinity and to high temperatures. Many interesting strains were selected on the basis of their enzyme activity reaction. 16S rDNA sequence alignment is further studied to investigate the relatedness between the collection strains and database reference strains or the description of novel species.

References

1. Kristjansson, J.K., Stetter, K.O.: Thermophilic bacteria. In: Thermophilic Bacteria, pp. 1–18. CRC Press, Boca Raton, FL (1992)
2. Adrio, J.L., Demain, A.L.: Microbial cells and enzymes – a century of progress. In: Barredo, J.L. (ed.) Methods in Biotechnology. Microbial Enzymes and Biotransformations, vol. 17, pp. 1–27. Humana Press, Totowa, NJ, USA (2005)
3. Niehaus, F., Bertoldo, C., Kähler, M., Antranikian, G.: Extremophiles as a source of novel enzymes for industrial application. Appl. Microbiol. Biotechnol. **51**, 711–729 (1999)
4. Pariza, M.W., Johnson, E.A.: Evaluating the safety of microbial enzyme preparations used in food processing: update for a new century. Regul. Toxicol. Pharmacol. **33**, 173–186 (2001)
5. BBC Research: Report BIO030 Enzymes in Industrial Applications: Global Markets. BBC Research, Wellesley, MA, USA (2011)
6. Ouadghiri, M., Amar, M., Vancanneyt, M., Swings, J.: Biodiversity of lactic acid bacteria in Moroccan soft white cheese (Jben). FEMS Microbiol. Lett. **251**(2), 267–271 (2005)
7. Vincent, J.M.: A Manual for the Pratical Study of Root-Nodule Bacteria IBP Handbook No.15. Blackwell Sci. Publ., Oxford (1970)
8. Prescott, L.M., Harley, J.P., Klein, D.A.: Microbiologie, p. 1137. De Boeck (2003)
9. Abdelmoumen, H., Filali-Maltouf, A., Neyra, M., Belabed, A., Missbah El Idrisssi, M.: Effect of high salts concentrations on the growth of rhizobia and responses to added osmotic. J. Appl. Microbiol. **86**, 889–898 (1999)
10. Sazci, A., Radford, A., Erenler, K.: Detection of cellulolytic fungi by using Congo red as an indicator: a comparative study with the dinitrosalicyclic acid reagent method. J. Appl. Bacteriol. **61**, 559–562 (1986)
11. Sendhilvel, V., Marimuthu, T., Raguchander, T., Prabakar, K.: Biochemical methods for the detection of Erwinia carotovora var carotovora from onion seeds. Madras Agric. J. **92**(4–6), 234–237 (2005)
12. Boucher, C.A., Van Gijsegem, F., Barberis, P.A., Arlat, M., Zischek, C.: Pseudomonas solanacearum genes controlling both pathogenecity and hypersensitivity on tobacco are clustered. J. Bacteriol. **169**, 5626–5632 (1987)
13. Weisburg, W.G., Barns, S.M., Pelletrier, D.A., Lane, D.J.: 16S ribosomal DNA amplification for phylogenetic study. J. Bacteriol. **173**, 697–703 (1991)
14. Kim, S., Edwards, J.R., Deng, L., Chung, W., Ju, J.: Solid phase capturable dideoxynucleotides for multiplex genotyping using mass spectrometry. Nucleic Acids Res. **30**, 16- e85 (2002)
15. Altschul, S.F., Thomas, M.L., Alejandro, S.A., Jinghui, Z., Zheng, Z., Webb, M., et al.: Gapped BLAST and psi-BLAST: a new generation of protein database search programs. Nucleic Acids Res. **25**, 3389–3402 (1997)
16. Tamura, K., Dudley, J., Nei, M., Kumar, S.: MEGA4: molecular evolutionary genetics analysis (MEGA) software version 4.0. Mol. Biol. Evol. **24**, 1596–1599 (2007)
17. Duxbury, T., Bicknell, B.: Metal-tolerant bacterial populations from natural and metal-polluted soils. Soil Biol Biohem. **15**, 243–250 (1983)
18. Palmisano, M.M., Nakamura, L.K., Duncan, k.E., et al.: Bacillus sonoris sp. Nov., a close relative of bacillus licheniformis, isolated from soil in the Sonoran Desert, Arizona. Int. J. Syst. Evol. Microbiol. **51**, 1671–1679 (2001)
19. Chamkha, M., Mnif, S., Sayadi, S.: Isolation of a thermophilic and halophilic tyrosol-degrading Geobacillus from a Tunisian high-temperature oil field. FEMS Microbiol. Lett. **283**, 23–29 (2008)

20. Manni, A., Filali-Maltouf, A.: Diversity and bioprospecting for industrial hydrolytic enzymes of microbial communities isolated from deserted areas of south-east Morocco. AIMS Microbiol. **8**, 5–25 (2022)
21. Belov, A.A., Chepstov, V.S., Vorobyova, E.A., et al.: Stress-tolerance and taxonomy of culturable bacterial communities isolated from a central Mojave desert soil sample. Geosciences. **9**, 166 (2019). https://doi.org/10.3390/geosciences9040166

Author Index

B.-E. Boudriki Semlali et al. (Eds.): DATA 2025, LNNS 1908, pp. 523–524, 2026.
https://doi.org/10.1007/978-3-032-21200-9

Kolibri 360 GmbH
Ferdinand-Jühlke-Straße 7
99095 Erfurt, Deutschland
produktsicherheit@kolibri360.de